AF247538

CHEMISTRY IN WATER REUSE
Volume 1

CHEMISTRY IN WATER REUSE
Volume 1

Edited by
William J. Cooper

Library of Congress Catalog Card Number 80-79321
ISBN 0-250-40377-3

PREFACE

Indirect or unintentional reuse of water is commonplace throughout the world where river systems, and to a lesser extent ground water, are used as water supplies. Intentional or direct reuse of waste water is practiced, and its practice will continue to increase. The major driving forces for water reuse are dwindling local and regional water supplies, and at the other end of the pipe, increasingly strict environmental legislation resulting in costly treatment of wastewater for discharge.

Within an industrial complex, internal recycle/reuse could be translated into increased process efficiency, energy and treatment cost (water and waste water) savings. Agriculture, probably the largest user of reused water, presents opportunities as a customer for reused water. Municipal reuse, planned on a regional basis, can conserve a precious resource for the continued well-being of those affected. The most logical targets for reuse, initially, are the large nonpotable-water users, with potable reuse in some cases the goal for the future.

Chemistry and chemical analysis are integral to successful implementation of water reuse. As reuse increases, it will be necessary to increase our understanding of chemical systems in the natural aqueous environments. Chemical analysis, off-line and on-line, is important in ascertaining the quality of waters intended for reuse. Chemical treatment processes will also be important in reuse, both in supplying water of a desired quality and in wastewater treatment to meet stringent effluent discharge permits or as pretreatment for recycle/reuse. Adequate quality assurance programs will have to be developed to ensure the chemical quality of water intended for resuse.

To provide a forum for discussion of the above issues, the Division of Environmental Chemistry of the American Chemical Society sponsored a symposium entitled "Chemistry and Chemical Analysis of Water/Wastewater Intended for Reuse." This two-volume series contains many of the papers that were presented at the symposium as well as volunteered chapters.

v

Although it could be argued that many of the papers are not strictly chemistry, it was felt that both basic and applied topics would give the reader an appreciation for the breadth and complexity of the field. This is by no means a comprehensive coverage of the field. Perhaps this meager beginning will stimulate further work and inform the growing technical community in this field.

Representatives from around the world, various governmental agencies, private industry and universities, all provided timely information which should form the foundation of a better understanding and appreciation of the interrelationships involved in water reuse.

William J. Cooper

ACKNOWLEDGMENT

No effort of this type could be accomplished without the help of many people.

The symposium which served as the catalyst for these volumes was sponsored by the Environmental Chemistry Division of the American Chemical Society. Financial assistance was received from the Society through the Division for travel funds and guest registration for foreign visitors.

The session chairmen Georges Belfort, Ronald J. Davenport, Richard D. Heaton, J. Donald Johnson, Frederick C. Kopfler, Francis M. Middleton, James P. Mieure and John A. Winter gave considerable time during the symposium and throughout the preparation of the books. The technical reviewers, too numerous to list, spent many hours evaluating the various contributions and provided invaluable input.

The U.S. Army Medical Biogengineering Research and Development Laboratory and the Environmental Protection Research Division kindly consented to allow me to pursue this extracurricular activity during the initial phases of planning.

Thanks are due to the staff of the Drinking Water Research Center of the Florida International University for helping complete the work.

Most importantly, throughout the project the hours which my wife, Karen, put into typing and retyping abstracts and other necessary tasks made the job much more enjoyable.

William J. Cooper is Associate Professor, Drinking Water Quality Research Center, Florida International University, Miami, Florida. He received his MS in organic geochemistry from Pennsylvania State University and his BS in chemistry from Allegheny College.

Mr. Cooper's research projects include water treatment technology improvements. He is primarily responsible for design and implementation of a water reuse research program. Previous research conducted with the Army Medical Bioengineering Research and Development Laboratory involved chemical degradation of pesticides, characterization of trace organics in waters and wastewaters, identification of impurities in chemicals for toxicological testing, and analytical methodology of water intended for reuse. He has also done extensive work in the development of methods for the determination of chlorine residuals.

To My Wife and Children
Karen, Jonathan and Derek
For Their Patience and Endurance

CONTENTS

Section 4
Disinfection

Section 5
Swimming Pools

CHAPTER 1

EVALUATION OF THE PUBLISHED LITERATURE
IN THE WATER REUSE/RECYCLE AREA

E. F. Rissmann and E. F. Abrams

Versar Inc.
Springfield, Virginia

R. J. Turner

Industrial Environmental Research Laboratory
U.S. Environmental Protection Agency
Cincinnati, Ohio 45242

Water reuse and recycling is practiced in industry for two primary reasons: to reduce waterborne waste discharges and to conserve water. The former is of particular importance in some industrial segments which have a potential for discharging hazardous materials, and the latter is important to those industrial segments located in water-deficient areas. A literature search and evaluation effort was recently completed which examined the use of water reuse and recycling technology in 39 industry segments. The results of this investigation are presented here.

The technologies used to achieve water recycling have ranged from installation of simple systems such as filters and clarifiers to remove muds and other suspended materials to rather complex technologies which, in some cases, have involved the development of entirely new processes. This chapter will examine the degree to which water recycling is employed in the industry segments studied; examine the technologies needed to achieve the reuse and recycle of process and cooling water in selected industries; and pinpoint those areas in which increased use of recycle technology is likely to occur because of either environmental reasons or the local availability and cost of suitable water.

SEARCH METHODOLOGY

The literature search was accomplished using in-house computer capabilities at Versar Inc. The National Technical Information Service (NTIS), Chemical Abstracts Condensates/CASIA, Enviroline and Pollution Abstracts computer files were searched.

Keywords used in the computer search were industry names, water, water reuse, water recycling and water conservation. A manual search was also made of chemical abstracts and pollution abstracts from 1970 through 1979 to identify publications not listed in the computer indexes. File lists of EPA publications from 1970 through 1979 were also manually searched. The search resulted in over 2000 references dealing with one or more aspects of water reuse technology. From these, 73 discrete cases were identified of water reuse and recycling technology in current practice. Each of these cases was then examined in more detail. Most case histories applied to six industrial categories:

1. pulp and paper,
2. electroplating,
3. inorganic chemicals,
4. petroleum refining,
5. iron and steel, and
6. food products.

The other industrial categories had either far fewer or no case histories reported.

The list of the industries for which this search was conducted is given in Table I, which also shows the number of case histories found for each industrial category and the treatment or novel process technologies used to achieve waste water recycle. Table II pinpoints the major waste sources and waste constituents for some of the more important industries studied. The complexities of technology used to achieve recycle varies considerably from one industry category to another. Thus, simple filtration and settling techniques are more than adequate to achieve recycle of some laundry waste waters. However, completely new process developments were required in the pulp and paper and photographic processing areas.

Table I. List of the Industries Studied and Numbers of Relevant Reuse and Recycle Case Histories Identified

Industry Category	References	Case Histories	Articles	Type Technology Most Used	Recovered or Recycled Materials
Adhesives and Sealants		0	0		
Aluminum Forming	1	1	1	Settling, chemical treatment	Process water
Auto and other Laundries	2	9	1	Settling, filtration	Process water
Batteries	3	1	1	Settling, filtration	Process water
Carbon Black		0	0		
Coal Mining and Processing	4	1	1	Chemical treatment, settling	Process water
Coil Coating	5	1	1	Reverse Osmosis	Chemicals, process water
Copper Forming		0	0		
Electroplating	6-9	8	5	Reverse osmosis, filtration, evaporation	Chiefly chemicals
Electrical Products	10	2	2	Ion exchange, reverse osmosis	Process water, chemicals
Explosives		1	1	Chemical treatment	Process water
Foundries		0	0		
Gum and Wood Chemicals	11	2	1	Skimming	Process water
Inorganic Chemicals	12-13	10	7	Settling, chemical treatment, carbon adsorption of organic materials	Chemicals, process and cooling water
Iron and Steel	14	2	2	Oil separation, chemical treatment	Process water

Table I, continued

Industry Category	References	Case Histories	Articles	Type Technology Most Used	Recovered or Recycled Materials
Leather Tanning	15	2	2	Aeration, chemical treatment	Chiefly chemicals
Mechanical Products		1	1	Sedimentation and filtration	Process and cooling water
Nonferrous Metals		1	2	Electrolysis, chemical treatment	Chiefly chemicals
Ore Mining	16	2	2	Clarification, ion exchange	Process water and chemicals
Organic Chemicals	17	2	2	Carbon adsorption reverse osmosis	Chiefly chemicals, some process water recovered for cooling use
Paint and Ink Formulation		0	0		
Paving and Roofing		0	0		
Pesticides		0	0		
Petroleum Refining	18-20	4	4	Oil separation, biological treatment	Chiefly chemicals
Pharmaceuticals		0	0		
Photographic Equipment		0	0		
Photo Processing	21	4	2	Electrolytic recovery of silver	Chemicals, process water, precious metals
Plastics and Synthetics	22	4	4	Chemical treatment, ion exchange	Chiefly chemicals
Plastic Processing		0	0		

Porcelain and Enamel		0	0		
Printing and Publishing		0	0		
Pulp and Paper	23,24	7	20	Chlorine dioxide bleaching, chemicals recovery[a]	Chemicals, process water
Rubber		0	0		
Shipbuilding		0	0		
Soap and Detergents		0	0		
Steam Electric	25	1	1	Evaporation, settling	Cooling water
Textiles	26	1	1	Hyperfiltration	Process water
Timber Products		0	0		
Food Products	27-28	6	6	Filtration, chemical treatment	Process water

[a]A number of technologies, used in series, are required to effect significant water recycling in the pulp and paper industry.

Table II. Summary of Major Wastes Generated by Some of the Industries Studied

Industry	Waste Origin	Major Constituents	Most Common Disposal Methods
Food Products	Processing of agricultural products	High biochemical oxygen demand (BOD), suspended solids, colloidal and dissolved organic matter	Biological treatment, trickling filtration, settling
Pharmaceuticals	Spent filtrates, wash waters and mycelium	High suspended and dissolved organic matter	Biological treatment, recovery of some materials for agricultural feeds
Textiles	Preparation of fibers and desizing of fabric, dyeing operations	Alkalinity, dyes, BOD, suspended solids	Neutralization, chemical and biological treatment, aeration and/or trickling filtration
Leather Goods	Unhairing, soaking and bating of hides	High suspended solids, chromates, BOD, sulfides	Biological and chemical treatment
Laundries	Washing of materials	Alkalinity and organic solids	Chemical precipitation, settling
Soaps and Detergents	Product purification	High BOD and saponified soaps	Flotation, skimming, calcium salt treatment
Pesticides	Washing and purification of chlorinated pesticides	High organic matter, chiefly aromatic compounds and chlorinated derivatives	Evaporation
Pulp and Paper	Cooking, refining and washing of pulp fibers, screening of paper pulp	High salinity, dissolved colloidal and suspended solids, colored waterborne materials	Biological treatment, chemical treatment, settling, aeration and recovery of by-product materials
Oil	Drilling muds, salt oil, acid sludges, alkaline wastewaters	High BOD, sulfur compounds and phenols	Chemical and biological treatment, limited by-product recovery
Photographic Products	Spent developer and fixing solutions	Organic and inorganic reducing agents, silver salts, complex cyanides	Silver recovery and discharge to POTW

Steel	Coking of coal, steel pickling, scrubbing of blast furnace, gaseous emissions	Acids, phenols, oils, suspended materials	Neutralization, recovery and reuse, chemical treatment
Gum and Wood Chemicals	Solvent recovery operations and washing of raw materials	High BOD and acid contents	By-product recovery, chemical treatment and reuse of water
Explosives	TNT and nitrocellulose production	TNT, organic materials	Chemical treatment
Electroplating	Metal cleaning and plating operations	Acids, heavy metals, cleaning agents	Chemical treatment

EXAMINATION OF CASE HISTORIES

The case histories found in the literature involve the reuse of several types of process waste water. A few of these systems will now be discussed to illustrate the current state-of-the-art.

Waste Transport Water

Perhaps the most widely adopted system is the reuse of waste transport water. This waste water is generally used to convey process-generated solid wastes to a lagoon where they can be disposed of or treated further before landfilling. One example of this technology, taken from the literature, is shown in Figure 1. This example treats the reuse of transport water used to slurry gypsum waste generated in the manufacture of hydrofluoric acid. The

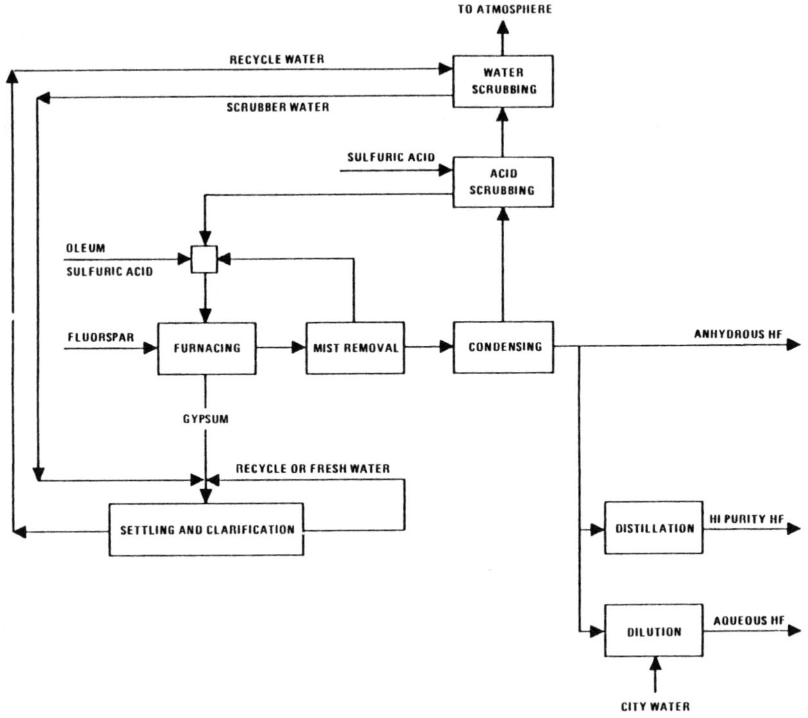

Figure 1. Settling pond recycling/reuse system in hydrofluoric acid production.

slurry waste water is pumped to the lagoon area, lime-treated to precipitate dissolved fluorides and lagooned to settle suspended materials, with the supernatant being reused to transport more waste to the lagoon. This system was originally installed at one facility in 1973. Since that time, it has partially been adopted by at least three other hydrofluoric acid facilities. With minor modifications, it has also found use in handling waste transport waters from the Bayer alumina process. Presently, all ten U.S. alumina plants reuse their waste transport water after chemical treatment to reduce alkalinity. Similar reuse of waste transport water has also been adopted in the alum and wet process phosphoric acid segments of the industry. Recycle of waste transport water has also found application in some segments of the mining industry, particularly in arid and semiarid locations.

Pulp and Paper Industry

A more complex reuse/recycle system was found in the pulp and paper industry. Since 1976, the Great Lakes Paper Company in Thunder Bay, Ontario, has been using novel technology to achieve zero discharge of process

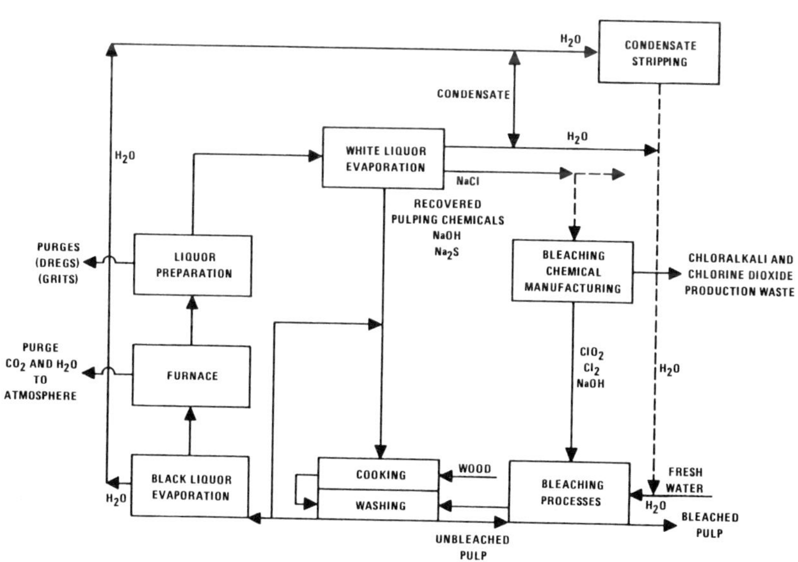

Figure 2. Closed-cycle bleached kraft pulp mill.

waste water and much recycle of process chemicals. This technology, known as the Rapson process, is shown in Figure 2. The novel features are as follows:

- Chlorine dioxide replaces 70% of the chlorine normally used for bleaching of pulp. This substitution effectively reduces levels of dissolved chlorides and chlorinated organic materials.
- Bleach plant effluents and evaporator condensates are used to wash unbleached pulp and are used in the chemical recovery system, eliminating the need for fresh water.
- Processes are used to recover sodium carbonate, sodium chloride and other chemicals. The recovered sodium chloride is converted electrolytically to sodium chlorate which in turn is converted to chlorine dioxide for use in the process.
- Dry debarking of the wood raw material is employed.

With the use of this technology, the only process wastes are solids and air-borne emissions purged from the process to avoid buildup of impurities, and a few waterborne waste streams generated from onsite production of chlorine, caustic soda and sodium chlorate. Purged gases consist mostly of carbon dioxide and water vapor, and the discarded solids are composed of some impure salt recovered from white liquor evaporation and various waste solids such as silica removed from the white liquor by filtration.

The technology developed to achieve a closed-cycle bleached kraft pulp mill has been partially adopted by many other facilities. The replacement of chlorine with chlorine dioxide for pulp bleaching is now occurring at many plants in the pulp and paper industry throughout the world. This new bleaching technology is currently causing an increase in production of sodium chlorate, the chemical from which chlorine dioxide is normally produced.

Several other published methods for reduction of wastewater quantity in the pulp and paper industry have also been reported. Most process modifications have involved substitution of sodium salts for the corresponding calcium compounds in the process. The Rapson process described above uses sodium salts entirely, and does not have any calcium salt disposal problems.

Photoprocessing Industry

Another industry where introduction of new technology has resulted in a reduction of waterborne waste discharges is the photoprocessing industry, which presently discharges a large fraction of its wastes to municipal wastewater systems. One firm has developed and is using a novel treatment system in which the following is accomplished:

1. Silver is recovered electrolytically from spent fixing solutions.
2. Aeration regenerates ethylene diamine tetraacetic acid (EDTA)-based bleach solutions which are recycled to the process.

3. Film and paper wash water is treated by reverse osmosis to remove residual silver and recover metal salt values as concentrated solutions. The permeates from the reverse osmosis units are reused. In all, 75,700 liter/day (20,000 gal/day) of waste water are treated and reused.

This treatment system (Figure 3) has been in operation since 1978 at the PCA International, Inc. plant in Matthews, NC. Of the treatments listed above, the recovery of silver is also widely used in the industry as a whole. This is partially the result of a continuing escalation of silver prices over the past decade.

The three examples described above were selected from the 73 case histories located during this investigation. As was shown in Table I, a wide variety of technologies has been employed to achieve the recycle of waste waters in a number of industries. Generally, most of this technology has beem employed to eliminate a number of troublesome wastewater discharges. However, water recycle technology has also been used as a method to reduce water consumption in areas where potable water is in short supply. A few examples of this application have appeared in the literature. Most of these have dealt with recycle of waste transport water, which has just been discussed. One case, however, has been reported which involves recovery of water from steam normally vented to the atmosphere. At one recently opened facility, both carbon dioxide and water vapor are recovered from the flue gases generated from calcination of sodium bicarbonate produced as an intermediate in the manufacture of soda ash from Searles Lake brines. At that facility, both the recovered carbon dioxide and water are reused.

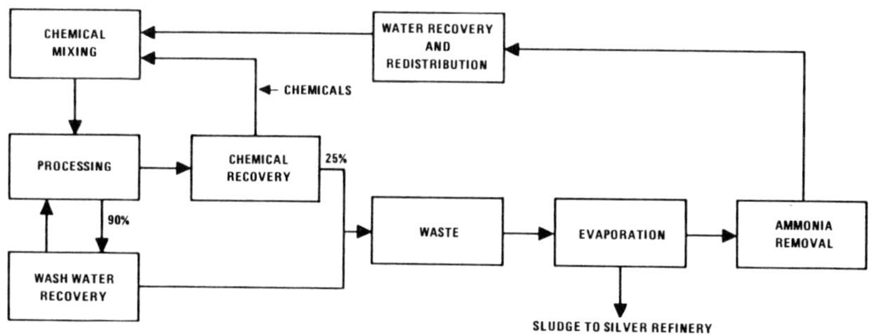

Figure 3. Zero-discharge system used in photographic processing industry.

CURRENT LIMITATIONS OF RECYCLE TECHNOLOGY

Most of the work reported in the literature dealt with the treatment and subsequent reuse of waste water containing inorganic contaminants. The only published work on removal of organic materials from reused process water has dealt with recovery of acetates and phenolic-type materials by such techniques as carbon adsorption. No case histories have been reported dealing with the removal of chlorinated organics from waste water before reuse. These are areas requiring attention as several chlorinated compounds are currently listed as priority pollutants by EPA.

Another problem encountered in the use of recycle technology is the removal of dissolved solids. Buildup of dissolved salts, such as sodium chloride, effectively limit the degree to which recycle technology can be applied economically to several process areas. Although processes do exist for the recovery of usable water from saline solutions, the technologies involved, such as distillation, are too expensive or energy-intensive to find widespread application.

A third potential problem involves finding uses or adequate disposal areas for the volumes of waste solids generated by some recycle technologies. Presently, millions of tons of waste gypsum, red muds and ore residues are being land-stored at plant sites for lack of other use or disposal options.

FUTURE USE OF REUSE AND RECYCLE TECHNOLOGY

Increased use of water reuse and recycle technology can be expected to occur in those industries where such technology is an economically achievable alternative to conventional wastewater treatment and discharge. Industries where this is currently the situation include the plating industry, some segments of the inorganic chemicals and mining industries, some segments of the steel industry, the pulp and paper industry, and the photoprocessing industry. In each of these areas, recently developed water reuse technology is already economically competitive with conventional treatment methods. Increased costs for potable water in some locations and further development and implementation of environmental regulations are likely to make the adoption of recycling technology more widely accepted in those industries where such technology has already been developed and demonstrated to be economically feasible.

Also, with further population growth, there is likely to be increased demands for limited water supplies in some arid areas. This may reflect itself locally in increased water costs and lead indirectly to the increased use of recycle technology in those industries centered in such areas.

CONCLUSIONS

As the result of an extensive literature search and evaluation, a significant number of case histories have been identified involving the reuse and recycle of process waste water in several industries. Most of the cases have involved removal of inorganic contaminants from waste waters before their reuse. Only a few case histories involving the removal of organic contaminants could be found and these dealt primarily with the removal and/or recovery of phenolic derivatives and acetates.

ACKNOWLEDGMENT

This work was supported by the Industrial Environmental Research Laboratory, Office of Research and Development, U.S. Environmental Protection Agency, under Contract No. 68-03-2604.

REFERENCES

1. Schwartz, S. M. *Ind. Water Eng.* 12(3):18 (1975).
2. Versar, Inc. "Study of Reuse of Water In Auto Laundries," Final report, Contract No. 68-01-3273, Effluent Guidelines Division, U.S. EPA (1976).
3. Versar, Inc. "Development of Data for Effluent Guidelines for the Batteries Manufacturing Segment of the Mechanical Products Point Source Category," Final report, Contract No. 68-01-3273, Task 2, Effluent Guidelines Division, U.S. EPA (1976).
4. Melios, P. *Chem. Eng. Prog.* 71(6):99 (1975).
5. Obrzut, J. J. *Iron Age* 221(43):139 (1978).
6. Bhatia, S., and R. Jump. *Environ. Sci. Technol.* 11(8):752 (1977).
7. Elicker, L. N., and R. W. Lacy. *Finishing Industries* 2(11):28 (1978).
8. Caprio, C., M. D. Beasley and L. Luttinger. *Ind. Water Eng.* (October 1977) pp. 24-30.
9. Warnke, J. E., K. G. Thomas and S. C. Creason. *Chem. Eng.* 84(7):75 (1977).
10. Ploos Van Amstel, J. J., and J. L. Frampton. *Environ. Sci. Technol.* 11(10):956 (1977).
11. Cresielski, L. F. "Tall Oil Refining Waste Water Treatment Systems," paper presented at American Oil Chemists Society 64th Annual Spring Meeting, New Orleans, Louisiana (1973).
12. Parmelle, C. S., and R. D. Fox. *Water Wastes Eng.* 9(11):F10-F12 (1972).
13. Reiter, W. M., and W. F. Stocker. *Chem. Eng. Prog.* 50(1):55 (1974).
14. Simon, R. *Iron Steel Eng.* 55(12):42 (1978).

15. Constantin, J. M., and G. B. Stockman. "Leather Tannery Waste Management Through Process Change, Reuse and Pretreatment," U.S. EPA Report EPA-600-2-77-034 (1977).
16. Gott, R. D., and J. M. Laferty. *Ind. Water Eng.* 15(2):6 (1978).
17. Baker, D. C., E. W. Clark, W. V. Jeserney and C. H. Heuther. *Chem. Eng. Prog.* 69(8):77 (1973).
18. Annessen, R. J., and G. D. Gould. *Chem. Eng.* 78(7):67 (1971).
19. Denbo, R. T., and F. W. Gowdy. *Environ. Sci. Technol.* 5(11):1098 (1971).
20. Hart, J. A. *Oil Gas J.* 71(24):93 (1973).
21. Daignault, L. G. *J. Appl. Photo. Eng.* 3(2):93 (1977).
22. "Putting the Closed Loop into Practice," *Environ. Sci. Technol.* 6(13): 1072 (1972).
23. Rapson, W. H. *Chem. Eng. Prog.* 72(6):69 (1976).
24. Romanschuk, H., and T. Voryolainin. *Chem. Eng.* 11(20):138 (1970).
25. Fosberg, T. M. *Ind. Water Eng.* 9(4):35 (1972).
26. Brandon, C. A. *Ind. Water Eng.* 12(4):14 (1976).
27. Esvelt, L. A. "Reuse of Treated Fruit Processing Wastewater in a Cannery," U.S. EPA Report EPA-600-12-78-203 (1978).
28. Tonka, W. C. *Poll. Eng.* (May 1979) p. 36.

SECTION 1

MONITORING AND CONTROL

CONTROL AND MEASUREMENT OF ORGANIC MICROPOLLUTANTS IN SOUTH AFRICAN WATER RECLAMATION PLANTS

J. F. J. van Rensburg, A. J. Hassett and S. J. Theron
National Institute for Water Research
Council for Scientific and Industrial Research
Pretoria, South Africa

By the year 2000, South Africa will have an estimated water deficit of about 30% [1]. With 80% of the water presently available being used for irrigation, this deficit necessitates extensive rethinking on the application of water supplies. Of the water used in urban areas, 60-70% is currently being discharged as waste water. The situation could therefore be improved by the extensive reuse of waste water for potable water.

The first, rather unsuccessful, attempt at reclaiming water for direct reuse at Chanute was described in 1956 [2]. The Windhoek water reclamation plant in South West Africa/Namibia was commissioned in January 1969, and marked the first deliberate step toward the reclamation of purified sewage effluents for unrestricted reuse [3]. In November 1970 the Stander water reclamation plant was inaugurated at Daspoort, Pretoria, to supplement research in this field (Figure 1). A third, much smaller, pilot-scale plant has since come into use at Athlone, Cape Town. The chemical and microbiological quality of the water produced by these plants has been studied intensively since 1966 [3-22], including biological testing with rats [20] and the Ames test for mutagenicity [21]. Organic micropollutants were to a large extent excluded

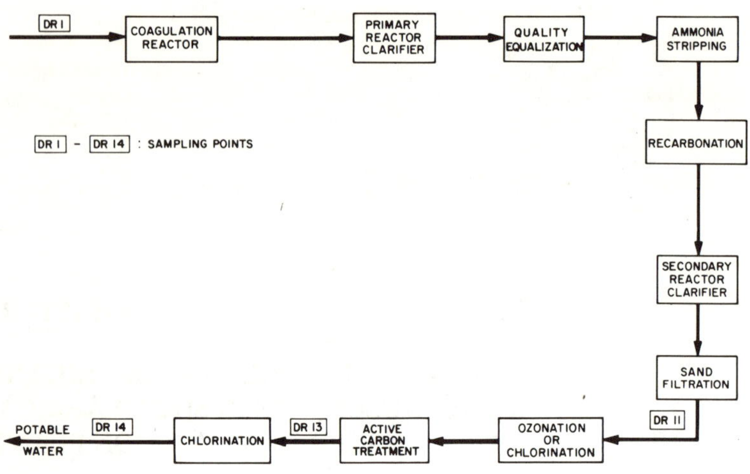

Figure 1. Flow diagram of the Stander water reclamation plant, Pretoria, South Africa, treating activated sludge effluent (4000 m^3/day).

from the initial chemical tests, because of the complex nature of these organics and the difficulties involved in devising suitable testing procedures. Since 1975, however, a number of chromatographic analytical techniques were developed by which potentially health-hazardous organic micropollution in potable water could be measured [23-26]. These techniques have also been applied to the measurement of organic micropollution in several water reclamation plants in South Africa. Although these results give insight into the ability of the plants to produce potable water from predominantly domestic sewage, the effect of industrial pollutants can only be gauged by inoculating pilot-plant scale systems with a number of organics, selected for their known or suspected health hazard implications, as well as their possible presence in effluents of various kinds. These experiments should also give insight into the removal and control of organic compounds by individual unit processes.

ANALYTICAL PROCEDURES

Selection of Micropollutants to be Measured

The compounds in Table I and the lower hazard limits (LHL) were chosen after consultation of available and suggested water quality standards, compounds found in water elsewhere, toxicological data and other relevant publi-

Table I. Organic Micropollutants Selected for Analysis

Compounds	Lower Hazard Limits (μg/l)
Volatile Halogenated Hydrocarbons (VHH)	
Trihalomethanes (chloroform, bromoform, bromodichloromethane, dibromochloromethane), tetrachloromethane	1 per compound
Chlorinated Hydrocarbons/Pesticides (CHP)	
Lindane, chlordane, dieldrin, endrin, (*bis*) chloroisopropylether, hexachloro-butadiene, hexachlorobenzene, PCB, DDT-complex, endosulfan	0.1 per compound
Dichlorobenzene, chloroethers	1 per compound
Chlorophenols (CPHEN)	
Di-, tri-, tetra- and pentachlorophenols	1 per compound
Polynuclear Aromatic Hydrocarbons (PAH)	
Benz[a]anthracene, benzo[b]- and [k]fluoranthene, benzo[a]pyrene, dibenz[a,h]anthracene, indeno[1,2,3-cd]pyrene	0.1 per compound
Anthracene, phenanthrene, fluoranthene, naphthalene, acenaphthene, fluorene, chrysene	1 per compound
Phenolic Compounds (PHEN)	
Phenol, cresols, xylenols, β-naphthol etc	1 per compound
Other Compounds (DIV)	
Dibutylphthalate, diphenylether, nitrotoluene	1 per compound
Unidentified compounds to be identified by GC/MS	If more than 1 per (0.1 for CPH and PAH) compound

cations [27]. The data in Table I form the basis of a set of analytical procedures used for measuring significant organic micropollution in South African potable waters.

Sample Preparation

Two water samples (4 liters and 250 ml) were drawn from each sampling point. Excess ascorbic acid was added to the 250-ml sample bottle before taking the sample, to reduce all the free chlorine in the water. The 4-liter sample was acidified to pH less than 2 with sulfuric acid. Both samples were stored at 4°C until the organic compounds could be extracted. The 4-liter sample (Figure 2) was saturated with Na_2SO_4 and extracted twice with 100 and 25 ml of ether/dichloromethane (7/3). After separation, the organic phase was dried through anhydrous Na_2SO_4 and concentrated to 0.4 ml using Kuderna-Danish and micro-Snyder evaporators [24]. Of this concentrate 0.1 ml was reserved for thin-layer (TLC) and liquid chromatography (LC). The remaining 0.3 ml was extracted twice with 1 ml of 0.1 M NaOH solution, and 5 μl of the organic phase was analyzed by gas chromatography, using a combined flame ionization-electron capture detection, for the general organic pollution index (OPI-GEN), as well as for chlorinated hydrocarbons (OPI-CH). The remaining organic phase was used for TLC identification of polynuclear aromatic hydrocarbons (PAH) by anticircular high performance thin layer chromatography (HPTLC) on caffeine-impregnated TLC plates [25] and for combined gas chromatography/mass spectrometry (GC/MS) studies when needed. Acetic anhydride was added to the NaOH combined extract to derivatize the extracted phenolic compounds to their acetates. These esters were then extracted and analyzed by gas chromatography with flame ionization detection.

Of the 250-ml sample, 10 ml was extracted with 200 μl of n-hexane for the determination of the trihalomethanes (THM) and related compounds [volatile halogenated hydrocarbons (VHH)]. Borax was added to a further 10 ml of sample to bring the pH to ±9.0 and this mixture was extracted with 200 μl of n-hexane. On separation, the organic phase was analyzed for chlorinated pesticides and other (chlorinated) hydrocarbons (CHP). The residual aqueous borax phase was shaken with acetic anhydride and 200 μl hexane to

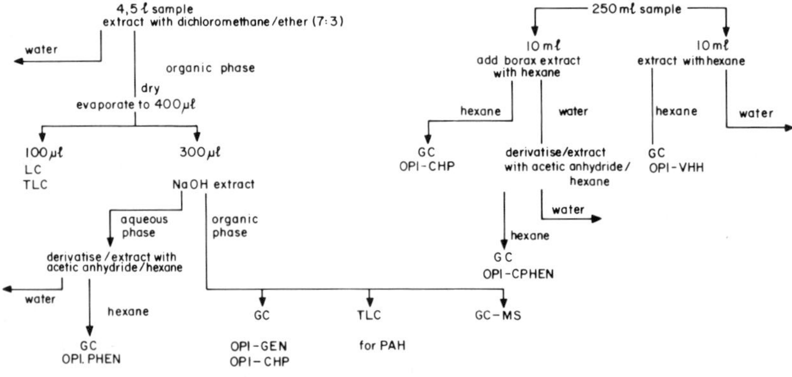

Figure 2. Schematic representation of analytical techniques used.

derivatize the chlorinated phenols to their acetates. The hexane phase was used for the gas chromatographic determination of these compounds.

The "syringe" extraction technique [23] was used throughout for the 10-ml water samples.

Chromatography

The general organic pollution profile was obtained by direct injection at 50°C onto a 30-m wallcoated open tubular (WCOT) column with SE-30 as stationary phase, and chromatographed by temperature programming from 70 to 250°C at 3°C/min. Flame ionization (FID) and electron capture detection (ECD) were used in parallel, with the ECD desensitized to produce a comparable response on the two detectors for aldrin. Analytical data were processed by a Hewlett Packard 3352 data system, which identifies predetermined compounds by relative response on the two detectors and retention indices. The compounds identified were quantified relative to the particular internal standards [24]. All peaks on a chromatogram were quantified and expressed as an organic pollution profile [27].

The phenolic compounds were gas chromatographed in the same way using only a FID; identification was done by retention time and quantification with an internal standard. All peaks were again expressed as an organic pollution profile.

The VHH and CHP were determined on a gas chromatograph equipped with an automatic sampler, ECD and a 60-m x 0.5-mm WCOT column coated with SE-30. Instrument control and data processing were done by the Hewlett Packard data system. Quantification was done by internal standard and identification by retention time. The analyses were done isothermally at 60°C for the VHH and 190°C for the CHP.

RESULTS AND DISCUSSION

The analytical procedures described provide a wealth of analytical data for the assessment of relevant water quality. By using suitable internal standards, a total concentration for each chromatographic procedure is obtained as the sum of the concentrations of known as well as analytically related unknown compounds. To express these data in a useful form use is made of an organic pollution index (OPI) [24]. For this purpose the number of peaks in each of four concentration groups (more than 10 μg/l, 1-10 μg/l, 0.1-1 μg/l and less than 0.1 μg/l) are determined and for the pollution profile these numbers are given consecutively after the total concentration obtained (e.g., 5-0-0-4-64). Since an OPI obviously also depends on the particular analytical technique employed, a suitable indication should be used (Tables II and III).

Table II. Volatile Halogenated Hydrocarbons and Chlorophenols Found in Water Samples (Average Values)

	Volatile Halogenated Hydrocarbons										Chlorophenols							
	Compounds Found (μg/l)						OPI-VHH				Compounds Found (μg/l)[a]				OPI-CPHEN[b]			
									Peak Distr.[b]								Peak Distr.[b]	
Sample	CCl_4	$CHCl_3$	C_2Cl_4	$CHBrCl_2$	$CHBrCl$	$CHBr_3$	Total (μg/l)	A	B	C	Cl_2	Cl_3	Cl_4	Cl_5	Total (μg/l)	A	B	C
Stander Reclamation Plant (Figure 1)																		
Activated sludge effluent—DRI																		
1977-1978	0.03	0.8	0.5	0.1	0.04	0	2	0	0	4	0.04	0.7	0.01	1.2	2	0	1	5
1979	0.05	0.9	0.4	0.2	0.2	0	2	0	0	5	ND[c]	ND	ND	ND				
After sand filtration—DR 11																		
1977-1978	0.3	4	0.6	0.1	0.2	0.1	6.3	0	1	4	0.2	0.7	0.3	1.1	3	0	1	4
1979	0.1	4.4	0.5	5.9	2	0.3	14	0	3	4	ND	ND	ND	ND				
After active carbon treatment—DR13																		
1977-1978	0.1	2.9	0.4	0.5	0.2	0	4.4	0	1	4	0.2	0.7	0.5	0.7	2.4	0	0	5
1979	0.1	8	0	9.8	2.1	0.2	21	1	3	3	ND	ND	ND	ND				
Final product—DR-14																		
1977-1978	0.3	4.1	0.5	0.8	0.6	0.4	6.9	0	2	4	0.2	0.7	0.1	0.6	2	0	0	4
6/78-2/79	0.2	3.8	0.4	0.7	0.7	0.3	6.4	0	2	4					2.7	0	0	6
3/79	0.04	3.9	0	3.6	4.1	2.7	14	0	3	1	ND	ND	ND	ND				

4/79-5/79	0.1	4.4	0	9.9	4.0	1.4	22	0	3	2					0.1	0	0	0
6/79	0.1	7.5	0	10.2	4.3	1.8	25	1	3	1	ND	ND	ND	ND				
LFB Pilot Plant (Figure 5)																		
Final Product	0	2	0	0.4	0	0	2.6	0	1	1	ND	ND	ND	ND				
Athlone Water Reclamation Plant																		
Final Product	0.03	3.3	0.02	2.9	2.1	1.2	12	0	4	4	ND	ND	ND	ND				
Municipal Water Supplies																		
Port Elizabeth	0.1	7.9	0.3	5.3	4.7	1.4	20	1	2	4	0.3	0.8	0.1	0.1	2.4	0	1	6
Pretoria[d]	0.2	8	0.2	20	12	1.5	45	2	3	4	0.1	0.6	0.05	1.7	2.8	0	0	4
Brits (a small rural town)	1	32	0.3	18	10		81	3	6	4	0.1	0.6	0.2	0.9	3.1	0	0	5
Cape Town[d]	0.04	9.4	0.03	5.1	3.3	0.4	21	1	2	3	0.2	0.8	0.1	0.5	2.4	0	0	5
Hazard Limits							50	0							10	0	0	

[a] Cl_2 = total dichlorophenols; Cl_3 = total trichlorophenols; Cl_4 = total tetrachlorophenol; Cl_5 = pentachlorophenol.
[b] Number of peaks in concentration ranges A: >10 μg/l; B: 10–1 μg/l; C: 0.1–1 μg/l; D: <0.1 μg/l (D is meaningless in this table and therefore left out.)
[c] ND = not determined.
[d] Representing one of the city's supplies.

Table III. Organic Pollution Indices (OPI) of Water Samples by Capillary Column Gas Chromatography (Average Values)

| | OPI-GEN | | | | | OPI-CHP | | | | | OPI-PHEN | | | | |
| | Total Conc. | Peak Distribution[a] | | | | Total Conc. | Peak Distribution[a] | | | | Total Conc. | Peak Distribution[a] | | | |
Sample	(μg/l)	A	B	C	D	(μg/l)	A	B	C	D	(μg/l)	A	B	C	D
Stander Reclamation Plant (Figure 1)															
Activated sludge effluent—DR 1															
1977-1978	13	0	1	39	89	0.3	0	0	1	6		ND[b]	ND	ND	ND
Final Product—DR 14															
1977-1978	3	0	0	7	37	1.2	0	0	3	20		ND	ND	ND	ND
1979	2.5	0	0	7	20	0.4	0	0	1	7		ND	ND	ND	ND
LFB Pilot Plant (Figure 6)															
Final Product	1.6	0	0	4	15	0.2	0	0	0	2	0.9	0	0	2	11
Athlone Reclamation Plant															
Final Product	8.2	0	0	24	56	0.9	0	0	1	19		ND	ND	ND	ND
Municipal Water Supplies															
Port Elizabeth[c]	3.4	0	1	2	21	0.4	0	0	0	3	4.6	0	1	9	12
Pretoria[c]	5.0	0	1	6	30	0.7	0	0	2	8		ND	ND	ND	ND
Brits (small rural town)	9	0	1	9	80	0.4	0	0	0	9	7.5	0	0	33	2
Cape Town[c]	2.3	0	0	6	22	0.1	0	0	0	4		ND	ND	ND	ND
Hazard Limits	10	0	0			1	0	0	0		10	0	0		

[a]Peak distribution; number of peaks found in the various concentration ranges: A: $>$10 μg/l; B:1-10 μg/l; C:0.1-1 μg/l; D: 0.01-0.1 μg/l.
[b]ND = not determined.
[c]Representing one of the city's supplies.

The OPI provides a simple way of comparing the organic micropollutant quality of various types of water. Thus a water with an OPI-GEN (Table III) of 13-0-3-18-59 is obviously "dirtier" than one with an index of 2.2-0-0-1-78; whereas an OPI of 9-0-2-15-64 could be more hazardous than 9-0-0-33-79, due to the compounds with concentration above 1 μg/l. By introducing hazard limits as indicated in Tables II and III, water samples can easily be divided into two categories, "safe" and "possibly unsafe—further investigation required," representing a considerable saving in time and costs, since only the latter group of samples would then require analytical treatment of a more complex and costly nature like GC/MS. In the authors' laboratories the costs of the routine procedures of Figure 1 amount to about $10 per sample, sampling and transportation excluded. If GC/MS work is required, the cost can easily be in the region of $30-300 per compound identified.

Results obtained from the water reclamation plants are presented in Tables II and III, together with comparable data from selected conventional potable water supplies in South Africa. Typical chromatograms are shown in Figures 3 and 4. These data indicate that a high quality water is being produced, especially at the Stander water reclamation plant, which formed the mainstay of the reclamation studies. When results in Tables II and III are compared, results are as good as or better than typical potable water supplies in South Africa. The higher quality produced by the LFB pilot plant (an integrated wastewater treatment-water reclamation plant, Figure 5) is probably due to

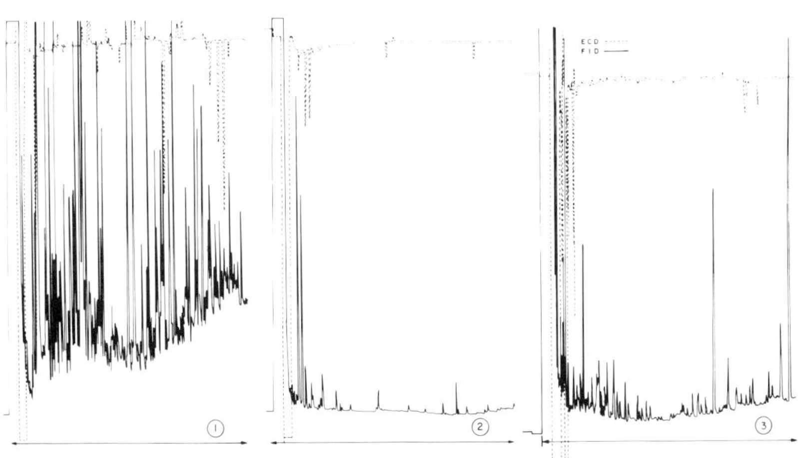

Figure 3. Capillary gas chromatograms obtained by electron capture (- - - -) and flame ionization detection (———). See Table III. (1) Feed water to LFB pilot plant, OPI-GEN: 76-0-19-123-58 [30]; (2) final water from LFB pilot plant, OPI-GEN: 1.6-0-0-4-15; (3) Brits municipal water supply, OPI-GEN: 9-0-1-9-80.

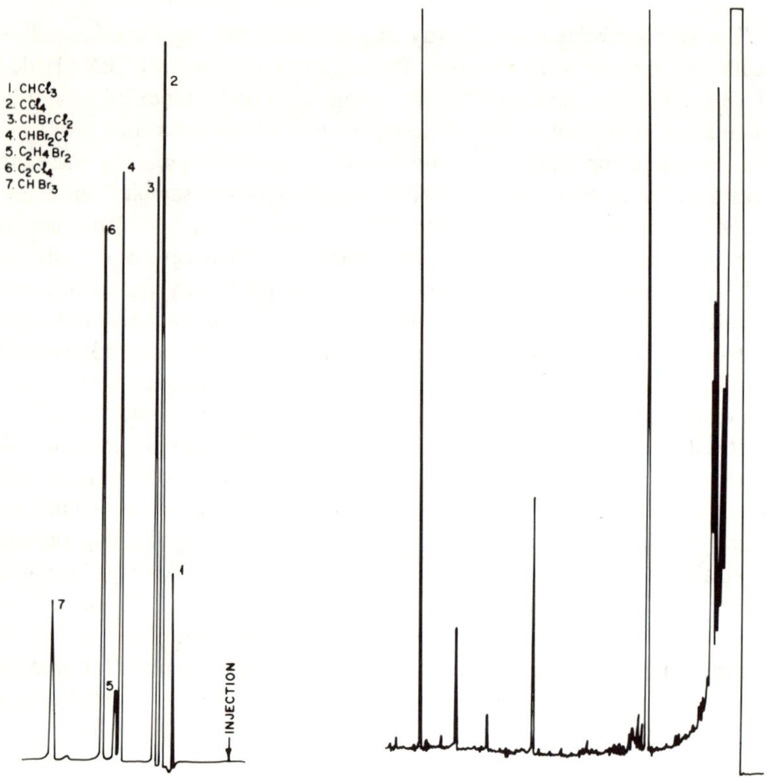

Figure 4. Typical chromatograms of analytical data from water samples; (left) volatile halogenated hydrocarbons (Table II); (right) phenols (Table III).

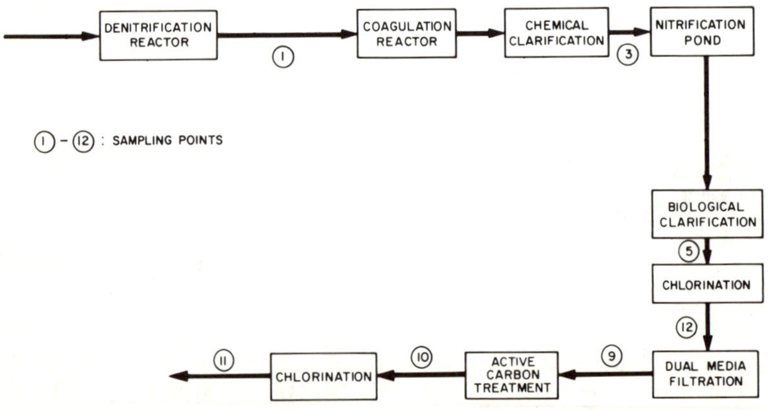

Figure 5. Flow diagram of the LFB pilot plant, Pretoria, South Aftica (60 m³/day).

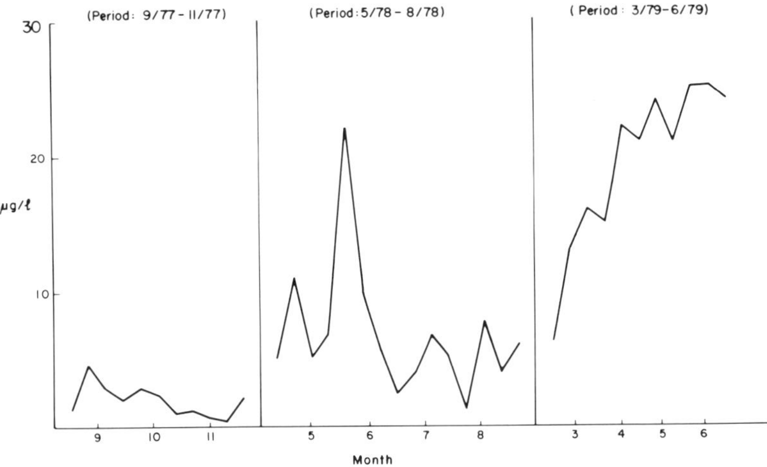

Figure 6. Total VHH concentrations in final water of Standard water reclamation plant. Monthly average values are indicated (Table II).

the superior state of its activated carbon as compared with that of the Stander plant. This is also indicated by the gradual increase in VHH values of the final water of the Stander plant (DR-14) over the final sampling period (Figure 6). While this increase was not apparent in the general profile, these results could indicate a gradual saturation of the activated carbon, allowing the smaller VHH molecules to pass in increasing concentrations. This observation could be of major importance when specifying carbon regeneration cycles. No relevant PAH have been found in the final water at the levels indicated, except naphthalene (noncarcinogenic) below 1 μg/l. Dibutylphthalate was found in all final waters at an average concentration of 0.4 μg/l.

GC/MS studies on the stronger peaks (see Table III) in the final water of the Stander plant indicate the additional presence of a number of saturated hydrocarbons, the methyl esters of the higher fatty acids such as myristic, palmitic and stearic acid, higher alcohols, a number of methylbenzoates, and aromatic compounds with an added hydrocarbon chain.

The presence of the esters from fatty acids seems to be typical of water reclaimed from domestic sewage and could be useful as a means of indicating the presence of this type of water in a water supply.

The OPI method is used to indicate the organic micropollutant quality of water also in water reclamation systems, and therefore is a simple method for measuring and controlling the quality of reclaimed water. Where the VHH seem to be one of the first indicators of granular active carbon (GAC) saturation and a pending breakthrough of organics, the simple and fast analytical technique used for their determination could act as an early warning system, provided that a prechlorination step is used in the purification procedures (Figures 1 and 5).

Recent instrumental developments for the determination of total halogenated organic compounds (TOH) [28] could provide a simpler and more economical means for the suggested monitoring procedure, even more so if the instrument could be automated to run unattended.

Inoculation Experiments

A scaled down version of the Stander water reclamation plant [19] was used to investigate the plant's and unit processes' abilities to remove a number of toxic organic pollutants of possible industrial, agricultural and domestic origin. The 17 compounds (Table IV) were continuously fed into the feed water of the plant for a period long enough to allow complete equilibrium in each unit process to be reached. Analyses of the final product, as well as the effluents from individual process units, demonstrated the stepwise and complete removal of these compounds through the plant to better than 99%.

The effects of the individual unit processes were, however, often inconclusive where particular pollutants had already been completely removed by preceding process units. A multiple barrier effect [20] could consequently not always be demonstrated.

While the higher concentrations used in this experiment proved the plant's ability to handle shock loads, solubility effects could have had an influence on the results. Consequently, in a later experiment with the LFB pilot plant (Figure 5) which is fully described elsewhere [30], the compounds in Table V were continuously fed into the feed water for 30 hr at lower concentrations (40 μg/l) and in followup experiments also into the feed waters to the individual process units. Analyses of the final product and the effluents of the various process units not only demonstrate the plant's ability to handle and completely remove shock loads of these toxic substances, but also produced the necessary insight into the effect of the various process units.

Dentrification

Results summarized in Table V indicate significant removal of up to 80% for 12 of the 19 compounds, with an average removal of 37%. No significant buildup of the compounds through sludge recycling, as found for inorganic anions [31], was encountered.

Table IV. Efficacy of a Pilot Plant in Removing Toxic Organic Compounds [19]

		Unit Process									
		High Lime Treatment		Secondary Clarification		Sand Filtration		Chlorination		Activated Carbon Treatment	
Compound	A (μg/l)	B (%)	C[a] (%)	B (%)	C (%)	B (%)	C (%)	B (%)	C (%)	B (%)	C (%)
Lindane	20	0	0	0	0	0	0	80	80	98.3	99.5
Dieldrin	40	12.5	12.5	0	12	22	30	4	33	99	>99.3
Chlorodane	300	18	18	2	20	13	30	32	52	99	>99.5
Demeton-s-methyl	3000	67	67	0	67	21	74	99.6	>100		
Parathion	4000	19	19	0	19	18	33	99.9	100		
Fenitrolhion	4000	22	22	0	22	21	38	99.9	100		
Fenthion	4000	22	22	0	22	19	37	58	74	100	100
Phenol	600	5	5	30	33	0	33	82	88	98	99.7
Hexachlorobutadiene	100	72	72	97	>99						
Trichlorophenol	400	28	28	0	28	0	29	25	47	97	99
Hexachlorobenzene	80	94	94	5	100						
Acenaphthene	600	27	27	95	97	13	98	75	100		
Fluoranthene	500	79	79	80	96	33	98	22	98		
Pyrene	500	85	85	86	98	20	99	50	100		
Dibutylphthalate	400	87	87	0	53[b]	0	35[b]	28	53	99	100
o-Nitrotoluene	400	24	24	73	80	10	82	18	85	93	100
Tetradecane	200	96	96	89	100						

[a] A = concentration of substances in feed water; B = percentage removal by unit process; C = percentage overall removal.
[b] Increase in dibutylphthalate probably due to plastic piping used in pilot plant.

Table V. Removal of Micropollutants by Individual Processes at LFB Pilot Plant
(Figure 6) [30]

Compound	Percent Removal of Individual Compounds from the Feed Water to Each Process						
	LFB-1	LFB-3	LFB-5	LFB-12	LFB-9	LFB-10	LFB-11
(*bis*)Chloroethylether	0	20	30	0	20	88	0
Dichlorobenzene	50	20	80	40	20	0	0
Nitrotoluene	60	20	70	0	0	100	0
Naphthalene	30	25	100				
Dichlorvos	0	70	100				
Acenaphthene	50	40	97	0	100		
Hexadecane	80	85	70	0	100		
α-BHC	30	30	20	25	0	100	
Atrazine	20	25	0	20	0	100	
Anthracene	80	70	100				
Chlorfenvinphos	30	40	0	25	0	100	
Endosulfan	50	80	0	0	20	100	
Dieldrin	80	70	70	0	0	100	
Phenol	0	40	97	0	75	100	
m-Cresol	0	30	100				
Chlorophenol	50	0	100				
Trichlorophenol	0	8	90	100			
Pentachlorophenol	20	0	80	80	75	100	
β-Naphthol	70	0	100				
Average	37	35	69	21	34	89	

Chemical Clarification

Significant removals of up to 85% for 15 of the compounds in the process feed water (effluent from LFB-1), with an average removal of 35%, were obtained. Removal was probably affected by adsorption on the flocs [32].

Combination of Nitrification Pond and Biological Clarification

Removal of up to 100%, with an average of 69% for 16 of the compounds, was obtained. At this stage naphthalene, dichlorvos, anthracene, cresol, chlorophenol and β-naphthol were removed to below their detection limits. The nominal trace quantities of about 1% remaining for acenaphthene, hexadecane, dieldrin and phenol represent nominal quantities of 0.5 μg/l. Of these compounds, only the toxicity of dieldrin places the residue of 1 μg/l within a hazardous concentration level.

Prechlorination

The inoculation of the feed water to this process and the subsequent residuals indicated a removal or breakdown [32] of more than 99% for the PAH (naphthalene, acenaphthene and anthracene);98-100% of the phenols was also removed.

Roughing and Dual Media Filtration

Significant removal of only the PAH, saturated hydrocarbons and other compounds of low solubility in water was obtained, consistent with the other clarification processes, through the removal of suspended matter on which these compounds were probably adsorbed.

Activated Carbon Adsorption

The ability of this unit process to remove organic material is well known [33]. In the present plant it acts mostly as an additional safety barrier because of the effective removal by the preceding processes of most of the relevant organic material to acceptable levels. The average removal of remaining material was 98%. The inefficient removal by the preceding processes of chloroethylether, nitrotoluene, α-BHC, atrazine and chlorfenvinphos demonstrated the necessity of this safety barrier. The 100% removal of these compounds, as well as remaining traces of the other compounds, confirms the efficacy of GAC described earlier.

Final Chlorination

Except for the removal of the final trace of phenol, no other effects were found, because of the high quality of the activated carbon effluent.

The overall results obtained through these inoculation experiments seem to be fairly consistent with those found elsewhere [34].

CONCLUSIONS

Organic pollution indices provide a simple means for indicating the quality of a water in terms of organic micropollution, and thus also for indicating and controlling the relevant quality of reclaimed water. Additionally, OPI-VHH or TOH may serve as an early warning against a pending breakthrough in GAC systems.

A better-than-conventional water quality can be and is produced by water reclamation systems because they are designed to handle "dirty" water, while conventional water purification systems are designed to handle water low in organic micropollutants (e.g., GAC process units are normally absent).

The ability of a water reclamation system (or any other water purification system) to effectively handle industrial and other pollutants at shock-load levels could only be gauged by inoculation of relevant chemicals into the feed water and not by routine analytical procedures.

The reclamation plants investigated were capable of effectively removing organic industrial and other pollutants in shock-load quantities. This was found to be true also for inorganic anions [31] and toxic trace metals [35].

ACKNOWLEDGMENT

This paper is presented with the approval of the Director of the National Institute for Water Research.

REFERENCES

1. "Water—75," Department of Water Affairs, Pretoria, South Africa (1975).
2. Metzler, D. J. *Am. Water Works Assoc.* 50:1021 (1958).
3. Van Vuuren, L. R. J., M. R. Henzen and G. J. Stander. "The Full-Scale Reclamation for the Augmentation of the Domestic Supplies of the City of Windhoek," paper presented at the 5th International Conference of the IAWPR, San Francisco, CA (1970).
4. Henzen, M. R., G. J. Stander and L. R. J. van Vuuren. *Prog. Water Technol.* 3:307 (1973).
5. Cillé, G. G., L. R. J. van Vuuren, G. J. Stander and F. F. Kolbe. In: *Advances in Water Pollution Research*, Vol. 2 (Washington, DC: Water Pollution Control Federation, 1966), pp. 1-19.
6. Funke, J. W., and P. Coombs. "Project Report No. 4, 6201/6434," National Institute for Water Research, CSIR, Pretoria, South Africa (1971).
7. Funke, J. W., and P. Coombs. "Project Report No. 5, 6201/6434," National Institute for Water Research, SCIR, Pretoria, South Africa (1972).
8. Grabow, W. O. K., N. A. Grabow and J. S. Burger. *Water Res.* 3:943-953 (1969).
9. Grové, S. S. "The Epidemiology of Reclaimed Water," paper presented at the 51st Conference of the Institute of Municipal Engineers of SA, Windhoek, June, 1974.
10. Hattingh, W. H. J., and E. M. Nupen. *Water SA* 2:33 (1976).
11. Nupen, E. M. *Water Res.* 4:661-672 (1970).
12. Nupen, E. M., B. W. Bateman and N. C. McKenny. "The Reduction of Virus by the Various Unit Processes Used in the Reclamation of Sewage to Potable Waters," paper presented at a Conference of Viruses in Water and Wastewater Systems, Austin, TX, April, 1974.

13. Nupen, E. M., and W. H. J. Hattingh. "Health Aspects of Reusing Wastewater for Potable Purposes—South African Experience," paper presented at a Workshop on Research Needs for the Potable Reuse of Municipal Wastewater, Boulder, CO, March 17-20, 1975.
14. Nupen, E. M., and G. J. Stander. "The Virus Problem in the Windhoek Wastewater Reclamation Project," paper presented at the 6th International Conference of the IAWPR, Jerusalem, June 1972.
15. Stander, G. J., and L. R. J. van Vuuren. *J. Water Poll. Control Fed.* 41: 355-367 (1969).
16. Van Vuuren, L. R. J., and M. R. Henzen. "Process Selection and Cost of Advanced Wastewater Treatment in Relation to the Quality of Secondary Effluents and Quality Requirements for Various Uses," paper presented at the IAWPR Specialized Conference on Application of New Concepts of Physical-Chemical Wastewater Treatment, Vanderbilt University, Nashville, TN, September 18-22, 1972.
17. Van Vuuren, L. R. J., G. J. Stander, M. R. Henzen, P. G. J. Meiring and S. H. V. van Blerk. *Water Res.* 1:463-470 (1967).
18. Grabow, W. O. K., and M. Isaäcson. "Microbiological Quality and Epidemiological Aspects of Reclaimed Water," *Prog. Water Technol.* 10:329-335 (1978).
19. Van Rensburg, J. F. J., P. G. van Rossum and W. H. J. Hattingh. "The Occurrence and Fate of Organic Micro-Pollutants in a Water Reclaimed for Potable Reuse," *Prog. Water Technol.* 10:41-48 (1978).
20. Van Rensburg, S. J., W. H. J. Hattingh, M. L. Siebert and N. P. J. Kriek. "Biological Testing of Water Reclaimed from Purified Sewage Effluents," *Prog. Water Technol.* 10:347-356 (1978).
21. Denkhaus, R., W. O. K. Grabow and O. W. Prozesky. "Removal of Mutagenic Compounds in a Wastewater Reclamation System Evaluated by Means of the Ames *Salmonella* Microsome Assay," paper presented at the Tenth International Conference of the IAWPR, Toronto, Canada, June 23-27, 1980.
22. Grabow, W. O. K., J. S. Burger and E. M. Nupen. "Evaluation of Acid Fast Bacteria, *Candida albicans*, Enteric Viruses and Conventional Indicators for Monitoring Wastewater Reclamation Systems," paper presented at the Tenth International Conference of the IAWPR, Toronto, Canada, June 23-27, 1980.
23. Van Rensburg, J. F. J., J. J. van Huyssteen and A. J. Hassett. "A Semi-automated Technique for the Routine Analysis of Volatile Organohalogens in Water Purification Processes," *Water Res.* 12:127-131 (1978).
24. Van Rensburg, J. F. J., S. J. Theron and A. Kühn. "Screening for Organic Pollutants in South African Fresh Waters. 1. Pollution Profiles," (in preparation).
25. Van Rensburg, J. F. J., and A. Kühn. "The Determination of Polynuclear Aromatic Hydrocarbons in Potable Water by Thin Layer Chromatography," (in preparation).
26. Hassett, A. J., J. F. J. van Rensburg and J. Coetzee. "A Method for the Routine Screening of Halogenated Organic Compounds in Water," (in preparation).
27. Van Rensburg, J. F. J. "Health Aspects of Organic Substances in South African Waters—Opinions and Realities," (in preparation).

28. Hart, O. O. "Operational and Control Aspects in Water Reclamation," *Prog. Water Technol.* 10:65-80 (1978).

29. Van Steenderen, R. A. "The Construction of a Total Organohalogen Analyser System," *Lab. Practice* (May 1980).

30. Van Rensburg, J. F. J., A. Hassett, S. Theron and S. G. Wiechers. "The Fate of Organic Micropollutants Through an Integrated Waste Water Treatment/Water Reclamation System," paper presented at the Tenth International Conference of the IAWPR, Toronto, Canada, June 23-27, 1980.

31. Siebert, M. L. "The Fate of Phenol, Cyanide, Fluoride and Nitrate Through an Integrated Physical-Chemical-Biological Water Reclamation System," (in preparation).

32. Liu, D., N. K. Chawla and A. S. Y. Chau. "Chlorinated Hydrocarbon Pesticides in Chemical Sewage Sludges," *Trace Substances Environ. Health* 9:189-196 (1975).

33. Henzen, M. R., G. J. Stander, and L. R. J. van Vuuren. "The Current Status of Technological Developments in Water Reclamation," *Prog. Water Technol.* 3:307-318 (1973).

34. Haberer, K., and Wendling-Edler. "Zur Schadstoffeleliminierung bei der Wasseraufbereitung. II. Eliminierungs-untersuchungen und ihre Ergebnisse," *Zeit. Wasser Abwasser-Forsch.* 12:230-236 (1979).

35. Smith, R. and S. G. Wiechers. "Elimination of Toxic Metals from Wastewater by an Integrated Wastewater Treatment/Water Reclamation System," paper presented at the Biennial Conference of the IWPC (SA Branch), Pretoria, South Africa, June 2-5, 1980.

CHAPTER 3

GROSS ORGANICS MEASUREMENTS FOR MONITORING OF WASTEWATER TREATMENT AND REUSE

Medy Michail and Emanuel Idelovitch

Tahal, Water Planning for Israel Ltd.
Tel Aviv, Israel

A major concern in supplying high-quality water to domestic consumers is the presence of trace concentrations of organic compounds, a few of which are known to have toxic effects on man, others of which have been associated with adverse long-term effects on human health. Organics have also become the main substances of concern in wastewater schemes which involve either potable reuse or groundwater recharge of potable aquifers. A comparative study of treated effluents and public drinking water supplies demonstrated that the same organic compounds are found in both natural and purified waste water [1]. The qualitative and quantitative determination of organic substances in water is thus a matter of paramount importance and wide applicability.

The major problems related with the monitoring of organic compounds in water and waste water are caused by (1) their large number and great variety and (2) the analytical difficulties involved in their identification and measurement.

Two approaches presently available for measuring organics are usually employed:

1. Gross measurement of organics by means of various general parameters, some of which can be easily and rapidly determined.
2. Identification and measurement of specific organic compounds by means of methods which require sophisticated equipment, such as gas chromatographs, high-pressure liquid chromatographs and mass spectrometers.

While the second approach can be used for sporadic analyses or for research purposes, it is only the first approach—measurement of gross organics by means of general parameters—that is suitable for routine monitoring of water and waste water quality.

Over the years a number of tests have been developed to measure the gross organics content of water, but no single one has so far proved entirely satisfactory. Consequently, a large number of general parameters is used, each providing partial information on the type and amount of organics present in water.

Because of the large number of methods available for gross organics measurement, the cost of a routine monitoring program which includes all these parameters and the logistics involved in carrying out these analyses (which are usually spread among different laboratories) make this approach more sophisticated than the information provided by the results of the analyses themselves. The need for devising a more rational organics monitoring program became particularly evident in the Dan Region Sewage Project in Israel, because of the complexity of the scheme, which provides for groundwater recharge to a potable aquifer as an intermediary step between treatment and reuse of the effluent. The large number of sampling points and the stringent requirements of professional and public authorities to ensure to the greatest possible extent that no contamination of potable supply wells occurs dictate the need for simple but reliable methods of measuring the organics content of the water.

The Dan Region Sewage Reclamation Project—the largest and most advanced wastewater reclamation scheme in Israel—comprises facilities for advanced treatment, groundwater recharge and reuse of municipal waste water from the Tel Aviv metropolitan area. Stage One of the project, which has been in operation since 1976 south of Tel Aviv, serves at present the southern parts of the metropolitan area, with a total connected population estimated at 400,000 by the end of 1979.

The treatment scheme of Stage One (Figure 1) consists of:

1. biological treatment in recirculated oxidation ponds;
2. high lime-magnesium treatment in a sludge-blanket reactor-clarifier;

3. free ammonia stripping and recarbonation in polishing ponds:
4. infiltration to ground water by spreading basins; and
5. long-time detention in the aquifer before pumping for reuse.

Stage Two of the project, now in the design and construction stage, will serve (in parallel with Stage One facilities) the whole metropolitan area, which has a current population over one million.

A pilot plant located in the north of Tel Aviv has **been** in operation to study the basic treatment system incorporated **in Stage** Two of the project, which is a modified activated sludge **process with** nitrification-denitrification (Figure 1).

Within the study presented in this chapter, some of the parameters used for gross organics evaluation were thoroughly examined to determine: (1) the advantages and disadvantages of each; (2) if relationships exist between various parameters; and, especially, (3) the suitability of using ultraviolet (UV) absorbance measurements as a major parameter for routine monitoring of gross organics.

LITERATURE REVIEW

Organic compounds usually found in municipal waste water include mainly proteins, carbohydrates, fats and oils, but also a wide variety of synthetic molecules, such as detergents, phenols and pesticides [2]. Only about

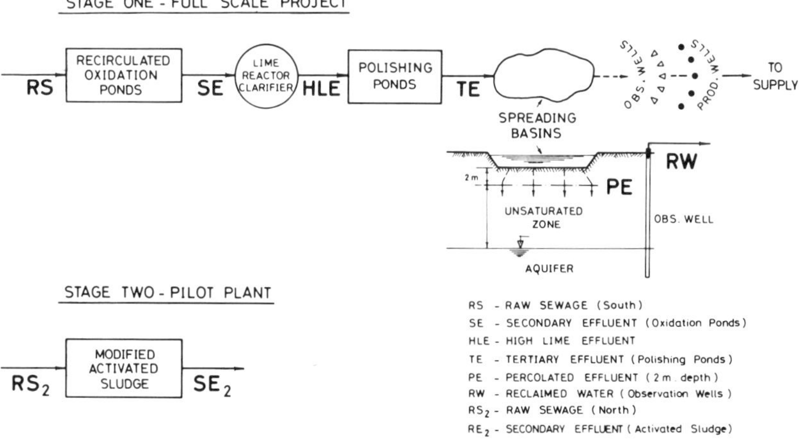

Figure 1. Dan Region Sewage Reclamation Project: flow diagram and sampling points.

500 of the 2 million organic compounds known to man have been identified in water to date [3]; many of these compounds are found only in trace concentrations, which makes their determination extremely difficult.

One of the oldest methods of organics analysis consists in determining the volatile fraction of the total solids by the difference between residual solids at 105 and 550°C. Other methods used in the past as a measure of organic matter include total, albuminoid, organic and ammonia nitrogen [2]. While the albuminoid nitrogen analysis is obsolete, the other nitrogen analyses are still used but for other purposes, e.g., in eutrophication control or as an indicator of nitrogen availability in biological wastewater treatment.

Carbon chloroform extract (CCE) is another parameter that was used in the recent past and is still included in the international and some national drinking water standards [4, 5]; however, because of its limited significance and the difficulty sometimes involved in analysis, it is not used any more and it has been excluded from the latest EPA potable water quality standards [6]. To complement the information provided by the CCE test, the carbon alcohol extract (CAE) analysis was also suggested [7].

Two methods most commonly used in the past as well as today are biochemical oxygen demand (BOD) and chemical oxygen demand (COD). In Europe as well as in Israel, permanganate ($KMnO_4$) consumption is used, especially for water and high-quality effluent, in place of or in addition to COD.

In the United States, the threshold odor number (TON) is also used as an indicator of the organic content of water, since most odors in water are caused by organics [8].

With the increasing importance attached in the last two decades to organic substances in water, improved analytical techniques have been developed for measuring their concentration. These include total and dissolved organic carbon (TOC and DOC), total oxygen demand (TOD), total and dissolved organic chlorine (TOCl and DOCl) and UV absorbance.

In addition to analyses for gross organics, techniques have been developed or improved for identification and analysis of specific groups of organic substances of high importance, such as organopesticides, polyaromatic hydrocarbons (PAH) and volatile halogenated organics or trihalomethanes (THM).

A brief description of each of the major parameters measuring gross organics follows [2, 9-14].

Biochemical Oxygen Demand

BOD is the amount of oxygen required for the degradation of organic substances present in waste water by a heterogenous microbial population. It measures the biodegradable organic carbon as well as the oxidizable nitrogen present. The latter can be excluded by adding to the sample inhibitors for the nitrifying bacteria.

The equipment required for the BOD test includes an incubator, incubation bottles and titration equipment for dissolved oxygen analyses. The advantage of the BOD test is that it measures only organics that can be oxidized by bacteria to carbon dioxide and water. Its disadvantages are the long duration of the test (usually 5 days) and the poor reproducibility of the results. The BOD values after 5 days (BOD_5) represent only a portion of the total BOD—usually about 70-80%.

The BOD test is applied mainly as a criterion for the selection of biological treatment processes and in the design of biological reactors for domestic and industrial waste waters. It is also used in stream pollution control, where organic loading must be limited to maintain certain dissolved oxygen levels required for protection of aquatic life.

Chemical Oxygen Demand

COD is the amount of oxygen required for the oxidation of organic substances present in waste water to carbon dioxide and water by a strong chemical reagent such as potassium dichromate ($K_2Cr_2O_7$).

The COD test requires routine laboratory reflux apparatus and titration equipment and takes a relatively short time (2-3 hr) in comparison with the BOD test. Its main disadvantages are: (1) the inability to distinguish between biologically degradable and inert inorganic matter because of the oxidation of some inorganics such as ferrous ion, nitrogen, sulfites and sulfide, and (2) the great variety of organic classes it includes; however, aromatic hydrocarbons and pyridine, which are not oxidized by dichromate are not included.

The method is accurate for COD values above 50 mg/l with concentrated dichromate. With dilute dichromate, values below 10 mg/l can also be determined, but are less accurate [9].

Permanganate ($KMnO_4$) Consumption

The old method of determining COD by oxidation with potassium permanganate—a weaker oxidant than potassium dichromate—is still in use in some countries, especially for clean effluents or slightly polluted waters.

In most European countries (with the exception of the U.K.), as well as in Israel, the Kubel method is generally used [15], which is simple, requires regular laboratory equipment and takes even a shorter time (about half an hour) than the COD (dichromate) method.

The main disadvantage of the permanganate consumption test lies in the incomplete and nonselective oxidation of organic compounds.

Total Oxygen Demand

The TOD test measures the amount of oxygen needed to oxidize all the impurities found in waste water by a special instrument (TOD analyzer). This is a sophisticated method whereby oxidation occurs in a platinum catalyzed combustion chamber, and the oxygen concentration of a nitrogen carrier gas is continuously monitored [16]. The method gives closer results to the theoretical oxygen demand than the COD analysis. Most of the common anions found in waste water do not interfere with the TOD analysis, but nitrates which decompose under acidic conditions may release oxygen to the combustion system, thus introducing an error in the results. TOD analyses take only about five minutes to perform but require sophisticated and expensive instrumentation.

Total Organic Carbon

The measurement of organic carbon has been used for many years in water pollution control. The old methods were based on titrimetric determination of the amount of CO_2 produced during oxidation of organic substances (which was trapped in a standard caustic solution). With the recent development of TOC analyzers, the pyrolysis technique is used for complete oxidation of the sample; the sample is either purged of inorganic carbon prior to analysis or the inorganic carbon is separately determined by low-temperature combustion.

The analysis takes only several minutes to perform, but the usual instrument (such as the Beckman analyzer) can accommodate only small samples (several microliters), and thus lacks the sensivity required for analysis of potable water. It is ususally accurate for values above 10mg/l.

Another technique employed for converting organic matter to carbon dioxide is the wet chemical method that uses acid persulfate with either heat or UV catalysis [17]. Greater sensitivity is achieved with this technique since samples are as large as 10 ml, but the oxidation is weaker than by pyrolysis, and some refractory organics may not be included.

By techniques which have recently become available, accurate TOC analyses in the range of less than 1 mg/l can also be performed for potable water [17].

With the instrumentation presently available, a wide range of TOC concentrations may be determined (from less than 0.05 to more than 1000 mg/l). The main disadvantage of the method lies in the expensive instruments required.

Relationships between BOD, COD, TOC and TOD

Correlations between various parameters measuring the organic content can be established for a particular waste water. Such correlations can be used for various purposes, e.g., BOD is a good indicator of organic load, but because of the long incubation time, it can be replaced by COD or TOC for routine monitoring purposes, if reliable correlations are available.

The BOD/COD ratio reflects what fraction of the dichromate oxidizable organics is amenable to biological degradation: the higher the ratio the higher the content of biodegradable organics. Correlations between BOD or COD and TOC can be made considering the following [16]:

- Part of COD is attributed to dichromate oxidation of some inorganics, whereas in the TOC analysis, these compounds are not oxidized.
- Many organic compounds are partially or totally resistant to biochemical or chemical oxidation and are thus excluded from BOD and COD values, whereas their organic carbon is included in the TOC analysis.
- BOD results are affected by factors extraneous to the sample, such as dilution, temperature, seed acclimation, pH and substances toxic to bacteria, whereas COD and TOC are not affected by such factors.

Theoretically, the COD/TOC ratio should approximate the molecular ratio of oxygen to carbon (32/12 or approx. 2.67), but could range from 0— when the organic matter is totally resistant to dichromate—to 5.33 for methane or even more when part of the COD is attributable to inorganic oxidation [8].

Correlation between BOD and TOC has been found to be fairly good for domestic waste water, but not for industrial wastes, which may include a variety of biodegradable and nonbiodegradable organic compounds. BOD_5/ TOC ratios ranging between 1.35 and 2.62 have been reported for domestic wastes [12]. The calculated ratio BOD_5/TOC was estimated at 1.85 by assuming that the BOD_5 is 77% of the ultimate BOD and that the ultimate BOD is 90% of the total oxygen demand [12].

TOD should correlate well with both COD and BOD_5 since all these parameters measure oxygen demand. The levels of correlation, depending on the nature of the wastes, range as follows [18]: BOD_5/TOD = 0.1-0.6; COD/TOD = 0.5-0.9.

UV Absorption

In recent years there has been increasing interest in the use of UV spectrophotometry for the evaluation of organic matter in water and waste waters. The relatively high cost of the TOC analyzer has prompted attempts to find an alternative, low-cost method for monitoring organics in water, and UV absorbance analysis would seem to provide the answer, since many significant organics commonly found in waste water show strong absorption of UV radiation.

The UV spectra of most water-soluble compounds are well known. UV absorption in the range of 200-300 nm is primarily due to organic compounds having aromatic or conjugated unsaturated molecular configurations such as C=C or C=0, since absorption is a function of electronic mobility [19, 20]. The common inorganic salts, with the exception of transition metal ions, do not have significant absorbances above 250 nm.

UV absorbance is subject to interference from several sources. The main interference in fresh water is from nitrate, which absorbs strongly around 210 nm [20]. Bromide also interferes at the lower end of the spectrum, but can be ignored at wavelengths above 250 nm [21].

Some organic components of waste waters do not absorb in the UV, e.g., sugars, simple aliphatic acids and alcohols, and simple amino acids such as glycine. However, these compounds are most rapidly consumed by microorganisms in biological treatment plants, and therefore they will not have a long-term polluting effect [22]. Paraffins are not absorbed in the UV either, but their volatility or insolubility limit the extent to which they can be present as soluble organics in waste waters [22].

A wide range of wavelengths has been employed in UV absorbance studies. It was demonstrated [22, 23] that the optical density of samples at 275 nm was closely proportional to the permanganate consumption value. Dornbush and Ryckman recommended in 1962 absorbance measurements at 250 nm as a useful parameter for evaluating removal of soluble trace organics by physico-chemical treatment [22]. In 1966 Bramer et al. [19] reported the use of UV_{254} (UV absorbance at about 254 nm) to monitor coke oven plant wastes and water from Lake Michigan. The use of UV absorbance measurements to assess the organic carbon content of waters and waste waters has since been used by several researchers. Ogura and Hanya [21] suggested that the dissolved organic substances in natural waters can be classified according to the absorbance ratio at 250 and 220 nm. Balch et al. [24] demonstrated the usefulness of measuring dissolved organic matter in effluent and receiving waters by UV absorption at 250 nm. Briggs and Melbourne [23] studied the relationship between UV_{254} and the organic carbon content of water from six different rivers and three different sewage plant effluents. Fahnrich and Soukup [25] determined the UV absorption of river waters at 260 and 280 nm; from these values the two main organic components (humic and lignin substances) could be distinguished quantitatively. Mrkva [26] in a study of river water indicated the possibility of using the absorbance at 280 nm as a measure of the oxygen demand. Later Mrkva [27] measured UV_{254} of various types of water and waste water and obtained good correlations with permanganate consumption and dichromate oxygen demands.

Dobbs et al. [22] described the correlation obtained between UV_{254} and the TOC content of a variety of treated and untreated water samples, ranging from secondary sewage effluent to raw and treated river water. To select the most suitable wavelengths for measuring the UV absorbance they scanned a sample of secondary effluent before and after granular carbon treatment. UV absorbance increased with decreasing wavelength but showed almost a "plateau" in the range 240-270 nm. Since this plateau included the sharp 253.7-nm spectral line of the low-pressure mercury lamp, that particular wavelength was chosen for the absorbance measurements.

Tambo and Kamei [28] used for their treatability investigations the gelchromatographic grouping method combined with the TOC/UV_{260} ratio.

Briggs et al. [17] proposed an on-line organic pollution monitor based on measurements of UV_{254}. They obtained a good correlation with BOD, COD and TOC for a wide range of water, waste water and treated effluents. Kölle and Sontheimer [29] reported that UV absorbance together with DOC and COD are the most unspecific organics analyses.

Schalekamp et al. [30] proposed UV_{254} as a means of monitoring surface waters to detect immediately any accidental organic pollution. Gurguis et al. [31] also used UV_{254} in an investigation of the performance of activated carbon.

UV absorbance analysis has many advantages over other methods of gross organics measurement. It is fast, requiring only 2-5 min, the equipment required is minimal, it includes a UV spectrophotometer, which can also be used in the visible range for all colorimetric analyses, and it may be used for a wide range of organics concentrations.

A disadvantage of UV analysis is the significant interference from turbidity. The extent to which turbidity affects the UV absorbance depends on the number and size of the suspended particles. UV radiation is scattered to a much greater degree by particles of "colloidal" size (i.e., less than 500 nm diam) than is radiation of longer wavelengths [22]. To overcome this interference in their organic monitors Mrkva [27] proposed to adjust UV_{254} values by substracting Vis_{545} (the absorbance in the visible range at 545 nm), while Briggs et al. [17] proposed Vis_{510} (510 nm in the visible range) for the same purpose.

In Israel, the use of UV absorbance for measuring organics was first suggested by Helfgott, who recommended the wavelength of 274 nm [32].

The application of UV absorbance to monitor the organics content of wastewater effluents was included in the present study and will be discussed further.

METHODOLOGY OF STUDY

The study was carried out on eight types of water and waste water from the Dan Region Reclamation Project including: six from Stage One and two from Stage Two. Tap water (TW) was included for reference purposes; it was sampled from Tahal laboratory (Azur) which is located south of Tel Aviv, in an area supplied from the same water source (wells) as most of the southern Tel Aviv metropolitan area, which is connected to the Stage One project.

The samples from the Dan Region Project were taken as follows (Figure 1).

Stage One Project (Full-Scale)

1. raw waste water (RS) from south Tel Aviv taken at the entrance to the oxidation ponds;
2. secondary effluent (SE) from oxidation ponds taken at the entrance to the lime reactor-clarifier;
3. high-lime effluent (HLE) taken from the outlet of the lime reactor-clarifier;
4. tertiary effluent (TE) taken from the last polishing pond, prior to pumping for groundwater recharge;
5. percolated effluent (PE) taken below the spreading basins, after 2-m vertical percolation through the upper layer of the unsaturated zone; and
6. reclaimed water (RW) taken from observation wells surrounding the recharge basins, after vertical percolation through the whole unsaturated zone (20-30 m) and horizontal flow in the aquifer (60-500 m).

Stage Two Project (Pilot Plant)

1. Raw waste water from north Tel Aviv taken at the entrance to the pilot plant (RS_2); and
2. Secondary effluent from the modified activated sludge plant taken at the outlet of the secondary clarifier (SE_2).

Batch samples were taken from all sampling points, except for RS_2 where composite samples were taken. The study was carried out in two steps: a basic investigation and a complementary investigation.

Basic Investigation

The basic investigation, which covered all sampling points enumerated above, was carried out between November 1977 and February 1978, i.e., during the winter, when rains occur and water temperatures are generally in the range 10-15°C. Six major gross organics parameters were analyzed: BOD, COD, $KMnO_4$ consumption, TOC, DOC and UV absorbance at 254 nm (UV_{254}). Volatile dissolved solids and organic nitrogen were also determined.

The analyses were carried out on 10-13 samples for raw waste waters, secondary effluents, high-lime effluent, tertiary effluent and tap water, and on 3-5 samples for percolated effluent and reclaimed water, the quality of which does not show wide fluctuations. The reclaimed water was sampled only from one observation well (No. 61) which is located some 60 m from the nearest recharge basin. During this period, a new microbial population developed in the aquifer. This population was not acclimated, and the concentrations of organics were higher than that recorded later [33].

The objectives of the basic investigation were to determine:

1. the range of values for each parameter in each type of water (especially for UV_{254}, which had never been analyzed previously);
2. the ratios between various parameters;
3. whether simple mathematical correlations exist between various parameters; and
4. whether UV_{254} was a promising parameter for routine monitoring purposes so that the acquisition of a spectrophotometer for further analyses was justified.

Complementary Investigation

After the preliminary findings of the basic investigations indicated the sutibility of UV_{254} for routine monitoring analyses, additional UV_{254} analyses together with some other gross organic parameters that showed good correlation with UV_{254} were carried out only for some types of water. The complementary investigation, which was carried out in the period June 1979-February 1980, concentrated mainly on raw waste water and on reclaimed water from the Dan Region Project Stage One, which was sampled from four additional observation wells—Nos. 54, 60, 62, 63 (in addition to well 61, which was included in the basic investigation).

The main objective of the complementary investigation was to consolidate and widen the findings of the basic investigation especially with respect to UV_{254}.

ANALYTICAL METHODS

BOD_5, COD, volatile dissolved solids and organic nitrogen measurements were performed in accordance with *Standard Methods* [9]. Permanganate consumption was determined by the Kubel Method [15].

TOC was determined with a Beckman analyzer. In this apparatus 10-20 μl of sample are injected into a combustion tube which is maintained at 900°C and contains a small plug of cobalt catalyst. The carbon dioxide produced is measured by a nondispersive infrared absorptiometer.

DOC was determined with an Oceanography International Corporation analyzer, in which the oxidation part was replaced by that of the Maihak UV-DOC-UNOR apparatus. The sample (1-20 ml) is injected into a vessel and irradiated with UV light which, in the presence of oxygen, converts the organic material to carbon dioxide [34]. The carbon dioxide is detected by means of a nondispersive infrared absorptiometer. This analyzer allows determination of concentrations as low as 0.01 mg/l.

UV absorbance was determined with a double beam UV-visible spectrophotometer (Varian Technotron-Model 635 and Model 634 S[digital]). A 1-cm cell was used for absorbance measurements. To correct for the possible interference of turbidity, the absorbance in the visible at wavelength 545 nm was subtracted from the UV absorbance for most of the samples, as proposed by Mrkva [27]. For convenience of presentation, absorbance results were multiplied by 10^3 and given as ($UV_{254} \times 10^3$).

All samples were filtered through GFC glass fiber filter paper to remove the suspended solids and colloidal particles greater than 1.2 μ which interfere with measurements of some parameters (e.g. DOC and UV absorbance). All analytical results therefore refer to the filtrable fraction of the sample.

The residual turbidity of the samples was determined with a Hach Turbidimeter. TOC analyses were carried out by the Weizmann Institute of Science, Department of Plastics Research. DOC analyses were performed by the Hebrew University, Environmental Health Laboratory.

UV absorbance measurements included in the basic investigation were carried out at the Tel Aviv University, Department of Zoology on a UV-visible spectrophotometer model 635 (Varian). All other analyses, including UV absorbance measurements carried out in the complementary investigation on a UV-visible spectrophotometer model 634 S (digital) were performed at Tahal Laboratory (Azur).

SCANNING FOR WAVE LENGTH SELECTION

All samples taken within the basic investigation were scanned between 210 and 275 nm to select the most suitable wavelength for UV absorbance measurements.

The scannings carried out indicated a decrease of the absorbance values with increasing wavelength, with a "plateau" recorded for all samples in the wavelength range 250-270 nm. The 254-nm wavelength has been used by several researchers [19, 22, 27, 29-31]; thus, it appeared to be suitable for gross organics evaluation for all types of water covered by this investigation. Figure 2 illustrates the scanning results obtained for raw waste water.

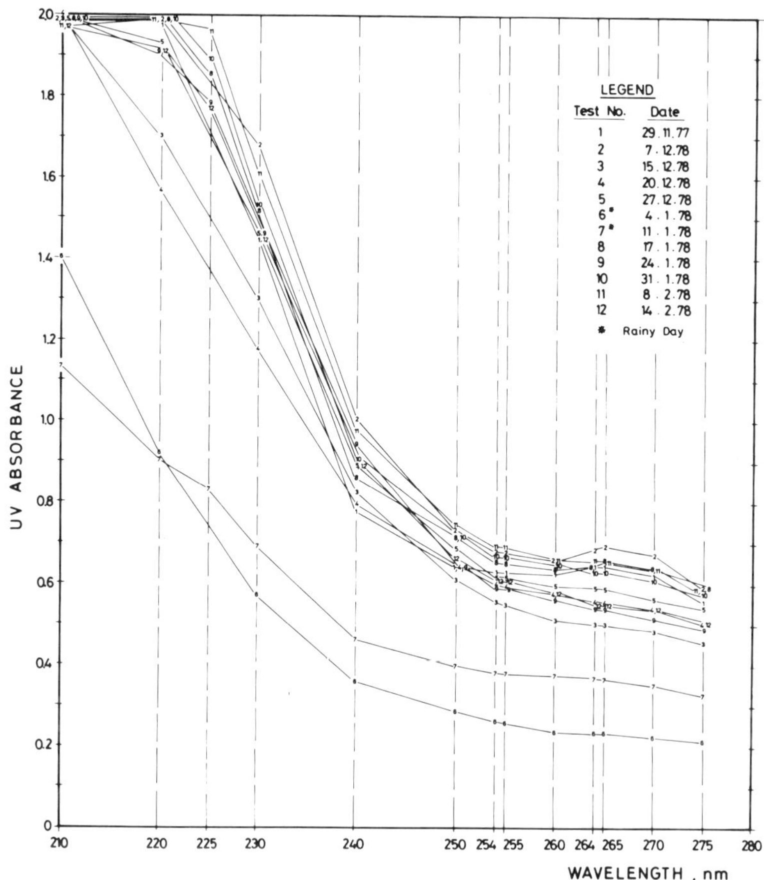

Figure 2. Wavelength scanning of raw waste water.

RESULTS AND GENERAL DISCUSSION

The results of the analyses carried out within the basic investigation on six gross organic parameters and nine types of water are summarized in Table I, which shows mean values and standard deviations.

Rainfall occurred on two of the days when sampling was carried out; consequently two results were consistently lower (because of the dilution effect of rain water), especially for raw waste water and secondary effluent from the activated sludge pilot plant. These results were included in the over-all analysis carried out, since the effect of dilution was equal for all param-

Table I. Gross Organics Parameters: Results of Basic Investigation

	RS[a]	SE	HLE	TE	PE	RW	TW	RS$_2$	SE$_2$
No. of Data	12	11	10	12	4	5	11	11	13
Mean Values									
BOD$_5$ (mg/l)	102	7.3	9.4	2.3	1.2	<0.5	<0.5	197	1.7
COD (mg/l)	199	84.7	58.2	52.1	40.0	20.6	2.8	386	54.8
KMnO$_4$ Consump. as O$_2$									
(mg/l)	40	21.0	14.7	12.0	8.6	3.1	0.5	110	17.5
TOC (mg/l)	64	23.5	21.4	18.3	13.7	5.3	1.8	137	21.8
DOC (mg/l)	56	18.7	15.7	14.3	6.9	3.1	0.4	122	15.4
VDS (mg/l)	144	79.6	58.7	45	48	142	48	235	56.9
Organic N (mg/l)	5.7	4.9	4.9	1.5	1.2	1.6	1.0	6.0	1.8
UV$_{254}$(x10^3)	580	490	270	220	197	58	2	850	370
UV$'_{254}$(x10^3)[b]	534	420	258	199	185	61	2	778	331
Standard Deviations									
BOD$_5$	43.5	2.9	2.4	0.6	0.8	0.2	0.3	77.8	1.4
COD	63.4	8.1	8.4	5.0	6.8	6.4	0.7	101.4	8.5
KMnO$_4$ Consump. as O$_2$	11.5	2.2	2.2	1.8	1.7	0.4	0.1	47.2	2.9
TOC	22.6	4.1	4.8	2.1	0.6	1.9	0.7	37.8	5.5
DOC	27.1	3.9	3.2	5.0	2.1	1.1	0.3	44.7	2.5
VDS	58.8	24.4	23.6	22.1	26.1	75.5	24.0	130.9	29.4
Organic N	6.0	5.3	5.1	1.8	0.6	0.9	0.7	6.0	1.1
UV$_{254}$(x10^3)	130	80	50	20	20	10	2	180	70
UV$'_{254}$(x10^3)	148	41	42	14	27	10	2	205	56

[a]RS = raw sewage (south); SE = secondary effluent (oxidation ponds); HLE = high-lime effluent; TE = tertiary effluent; PE = percolated effluent; RW = reclaimed water; TW = tap water; RS$_2$ = raw sewage (north); SE$_2$ = secondary effluent (activated sludge).

[b]UV$'_{254}$ = UV$_{254}$ - Vis$_{545}$. For UV$'_{254}$, the number of data was smaller, because measurements of Vis$_{254}$ (absorbance in the visible range at 545 nm) started later; thus, UV$'_{254}$ and UV$_{254}$ values cannot be compared directly.

eters; however when taking into account the absolute values of various gross organics parameters for these types of water, it should be borne in mind that they reflect the average for the whole winter season (including rainfall effect); averages for periods without rain should be slightly higher. For the other types of effluents the diluting effect of the rain was much smaller because of the buffering capacity of the ponds and the groundwater aquifer.

Figure 3, which shows the UV absorption spectra of various types of water studied, demonstrates the reduction in gross organics content at various treatment steps, and confirms the suitability of the selected wavelength (254 nm) for all types of water studied.

The values of UV absorbance as well as those of other parameters measuring gross organics showed that the filterable fraction of raw sewage in the north of the metropolitan area (RS_2) is much stronger than in the south (RS); this is attributable to light industries (beverages, food processing, etc) which carry heavy organic loads.

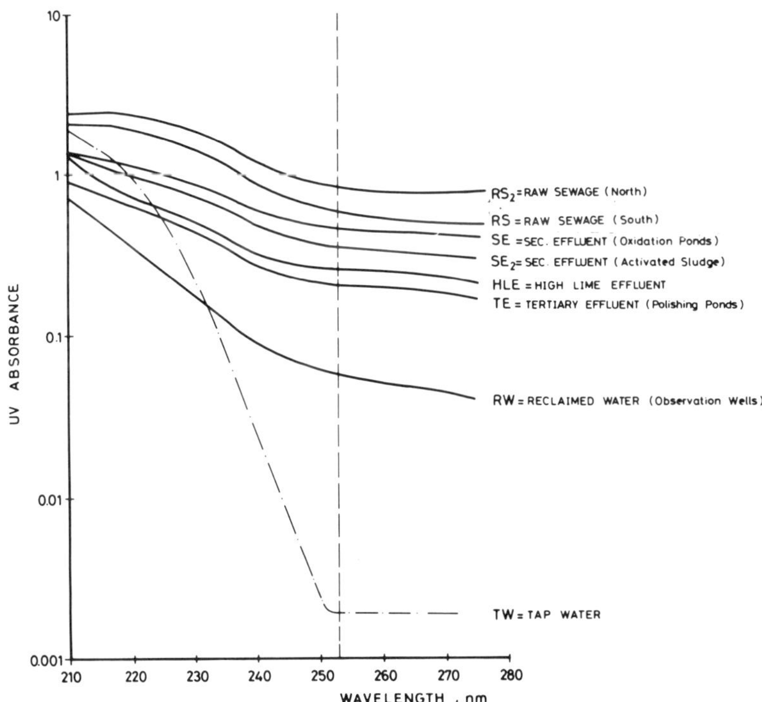

Figure 3. Ultraviolet spectra of various types of water.

A comparison of the secondary effluent (SE_2) from the modified activated sludge process, which is an advanced biological treatment process, with the various effluents from the treatment sequence of Stage One, which includes simple biological treatment in oxidation ponds and advanced chemical treatment showed the following:

- SE_2 was of higher quality than the secondary effluent from oxidation ponds (SE) but lower than the high-lime effluent (HLE) with respect to UV absorbance and permanganate consumption.
- It was approximately equal to the HLE with respect to COD, TOC and DOC, and to the tertiary effluent (TE) with respect to BOD.

The values obtained for the well known gross organics parameters such as BOD, COD and TOC were in accordance with the expected performance of the treatment processes and with previous analyses carried out. The range of UV_{254} values obtained for the various types of water studied are shown in Table II.

RATIOS BETWEEN MAJOR PARAMETERS

The common ratios between gross organic parameters (filtrable fraction) were calculated from the mean values of the data obtained in the basic investigation and are shown in Table III.

BOD/COD ratio for raw sewage (both south and north) was 0.51. As a result of the considerable removal of soluble biodegradable substances, the BOD/COD ratio in the secondary effluents decreased considerably (0.09 for oxidation ponds and 0.03 for the activated sludge process). In the subsequent

Table II. UV Absorbance for Various Types of Water

Type of Water	Sample Symbol	Range	
		$UV_{254}(x10^3)$	$UV'_{254}(x10^3)^a$
Raw Sewage (South)	RS	257-693	249-652
Secondary Effluent (Oxid. Ponds)	SE	376-669	348-460
High-Lime Effluent	HLE	220-350	211-344
Tertiary Effluent	TE	179-251	172-214
Percolated Effluent	PE	172-221	165-204
Reclaimed Water	RW	43-74	52-72
Tap Water	TW	0-9	0-8
Raw Sewage (North)	RS_2	538-1095	519-1065
Secondary Effluent (Act. Sludge)	SE_2	286-476	278-440

$^a UV'_{254} = UV_{254} - Vis_{545}$.

Table III. Ratios Between Gross Organics Parameters

	RS	SE	HLE	TE	PE	RW	RS$_2$	SE$_2$
BOD/COD	0.51	0.09	0.16	0.04	0.03	0.005	0.51	0.03
BOD/KMnO$_4$	2.55	0.35	0.64	0.19	0.14	0.03	1.79	0.10
BOD/TOC	1.59	0.31	0.44	0.13	0.09	0.02	1.44	0.08
BOD/DOC	1.81	0.39	0.59	0.16	0.14	0.03	1.62	0.11
COD/KMnO$_4$	4.97	4.03	3.96	4.34	4.65	6.64	3.51	3.13
COD/TOC	3.1	3.61	2.72	2.85	2.92	3.89	2.81	2.51
COD/DOC	3.53	4.52	3.70	3.63	5.80	6.65	3.16	3.56
TOC/DOC	1.14	1.25	1.36	1.27	1.99	1.71	1.13	1.42
UV/BOD[a]	5.69	67.12	28.72	95.65	164.17	580	4.31	217.65
UV/COD	2.91	5.78	4.64	4.22	4.92	2.81	2.20	6.75
UV/KMnO$_4$	14.5	23.30	18.37	18.33	22.91	18.71	7.73	21.14
UV/TOC	9.06	20.85	12.62	12.02	14.38	10.94	6.20	16.97
UV/DOC	10.36	26.20	17.20	15.38	28.55	18.71	6.97	24.02

[a]$UV = UV_{254} \times 10^3$

treatment steps of Stage One, the BOD/COD ratio decreased further with the exception of the high-lime effluent, due to the slight rise in BOD values, presumably caused by hydrolysis of refractory organic substances, making them more amenable to biodegradation [35, 36].

BOD/KMnO$_4$ ratio for raw sewage was 2.55 and 1.79 for south and north, respectively; it followed the same pattern as the BOD/COD ratio for the other types of water.

BOD/TOC ratio for raw sewage was 1.59 and 1.44 for south and north, respectively. These values lie in the range reported in literature for municipal waste water [12]. In the treated effluents, the BOD/TOC ratio followed the same pattern as the BOD/COD ratio, i.e., gradual decrease with the exception of the high-lime effluent. The BOD/DOC ratio was slightly higher than BOD/TOC ratios, as DOC values were slightly lower than TOC.

COD/TOC ratios ranged between 2.5 and 3.9, which correspond to values expected from the molecular ratio of oxygen to carbon. COD/DOC ratios were higher than COD/TOC ratios because of the lower values of DOC than TOC, especially for percolated effluent and reclaimed water.

COD/KMnO$_4$ ratios were in the range 3-5, which indicates the degree by which potassium dichromate is a stronger oxidant than permanganate.

TOC/DOC ratios were in the range 1.1-2.0, although similar values for both parameters should have been obtained, because the analyses were carried out on filtered samples. The highest ratios were obtained for the cleaner effluents (percolated effluent and reclaimed water), which reflects the poorer accuracy of the TOC analyzer at low values.

Ratios between UV ($UV_{254} \times 10^3$) and other parameters were determined for the various types of water studied.

The UV/BOD ratio for raw sewage was 5.69 and 4.31 for south and north, respectively; for other types of water the ratio UV/BOD was much higher—from about 67 in oxidation pond effluent to 580 in reclaimed water, indicating the increasing ratio between refractory and biodegradable organics at higher treatment levels.

The ratios UV/COD, $UV/KMnO_4$, UV/TOC and UV/DOC followed approximately the same pattern, i.e., an increase from raw sewage to secondary effluent and stable values (approximately in the same range) for the other types of water.

As no data are found in literature on ratios between UV and other gross organics parameters, the information provided by the ratios determined in this study should be considered of a preliminary nature. These ratios, which are based on mean values of a relatively limited set of data, need to be confirmed by additional investigations before they can be used for any practical purpose.

MATHEMATICAL CORRELATIONS

From the data collected within the framework of the basic investigation, correlation coefficients and equations between pairs of major gross organic parameters were calculated for all types of water by a best-curve fit computer program based on the least-square method. The program gave the equation in the linear form $y = ax + b$, and the value of the correlation coefficient (r) which indicates the degree to which the data fit the curve. In most cases, good ($r > 0.5$) to very high ($r > 0.9$) correlation coefficients were obtained; however, for some parameters in some types of water, poor correlations or even negative correlations ($r < 0$) were obtained.

Of all types of water studied, the best correlations for most parameters were obtained for raw sewage (south), raw sewage (north) and secondary effluent from the activated sludge process (Tables IV and V). UV absorbance showed good to very good correlation with all other parameters; the correlation was even improved when UV absorbance was corrected for turbidity interference by subtraction of absorbance in the visible.

For reclaimed water some correlations were very good, e.g., UV-TOC and UV-DOC, but others were poor (UV-COD and $UV-KMnO_4$), which was attributed to the limited number of samples collected in the basic investigation (from observation well 61). Consequently, in the complementary investigation more data were collected on reclaimed water from well 61 and from four additional observation wells incorporated in the current monitoring program of Stage One full-scale plant, as well as from raw waste water and tertiary effluent.

Table IV. Simple Correlation Coefficients (r) Between Gross Organic Parameters

	RS					RS$_2$				
	BOD	COD	KMnO$_4$	TOC	DOC	BOD	COD	KMnO$_4$	TOC	DOC
BOD	1.000					1.000				
COD	0.634	1.000				0.710	1.000			
KMnO$_4$	0.554	0.862	1.000			0.632	0.888	1.000		
TOC	0.654	0.846	0.829	1.000		0.736	0.885	0.936	1.000	
DOC	0.545	0.414	0.633	0.579	1.000	0.622	0.671	0.822	0.906	1.000
UV	0.790	0.891	0.919	0.838	0.666	0.596	0.832	0.656	0.735	0.519
UV$'$	0.959	0.896	0.969	0.911	0.685	0.707	0.940	0.945	0.993	0.984

Table V. Simple Correlation Coefficients (r) Between Gross Organic Parameters:SE$_2$

	BOD	COD	KMnO$_4$	TOC	DOC
BOD	1.000				
COD	0.573	1.000			
KMnO$_4$	0.682	0.711	1.000		
TOC	0.481	0.781	0.787	1.000	
DOC	0.445	0.729	0.548	0.681	1.000
UV	0.776	0.676	0.930	0.681	0.578
UV$'$	0.743	0.695	0.906	0.873	0.897

The results, which are presented below, demonstrated the feasibility of using UV absorbance as a general parameter for routine monitoring of gross organics in a wastewater treatment and reclamation scheme.

Raw Sewage

All data collected for raw sewage (both south and north) in the basic, as well as in the complementary investigation, were plotted and computed to obtain correlations and equations between UV$'$ (corrected UV absorbance) and other gross organics parameters. The correlations obtained were even better than those obtained from the data of the basic investigation only (Figure 4). The equations and correlation coefficients obtained are as follows:

$$BOD = -89.5 + 359.2 \; (UV') \tag{1}$$
$$r = 0.85$$

$$COD = -28.7 + 452.1 \; (UV') \tag{2}$$
$$r = 0.90$$

$$TOD = -60 + 246.2 \; (UV') \tag{3}$$
$$r = 0.95$$

UV absorbance can thus be used for general monitoring of organic matter in the raw waste water feeding the treatment plant. It has the great advantage of being capable of easy and quick identification of strong organic industrial wastes entering the municipal sewage network.

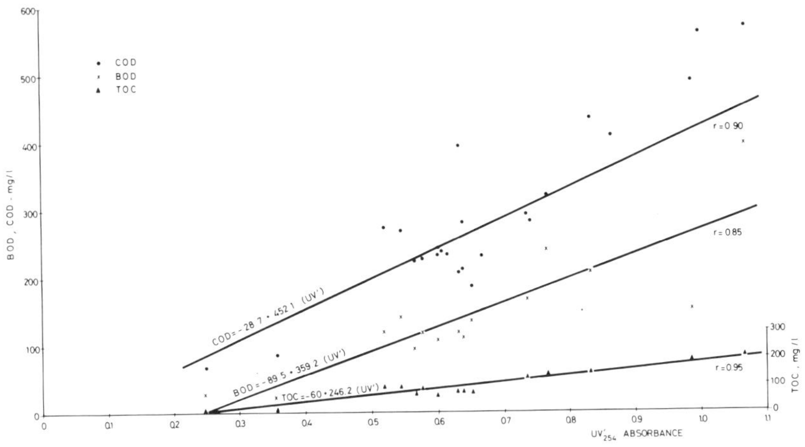

Figure 4. Correlations between gross organic parameters in raw wastewaters.

Secondary Effluent (Activated Sludge)

From the data collected in the basic investigation, the correlations between UV and the other parameters were computed and plotted (Figure 5). The equations are given below and the correlation coefficients are shown in Table V.

$$BOD = -4.2 + 16.0 \ (UV) \qquad (4)$$

$$COD = 23.3 + 85.7 \ (UV) \qquad (5)$$

$$TOC = \ \ 0.8 + 58.3 \ (UV) \qquad (6)$$

$$KMnO_4 = \ \ 2.6 + 40.5 \ (UV) \qquad (7)$$

Satisfactory correlations were obtained, especially between UV and BOD, indicating the feasibility of using UV absorbance for routine monitoring of activated sludge treatment plant performance with respect to carbonaceous matter.

Reclaimed Water

From the large amount of data collected during the complementary investigation, correlation coefficients and equations were computed and plotted

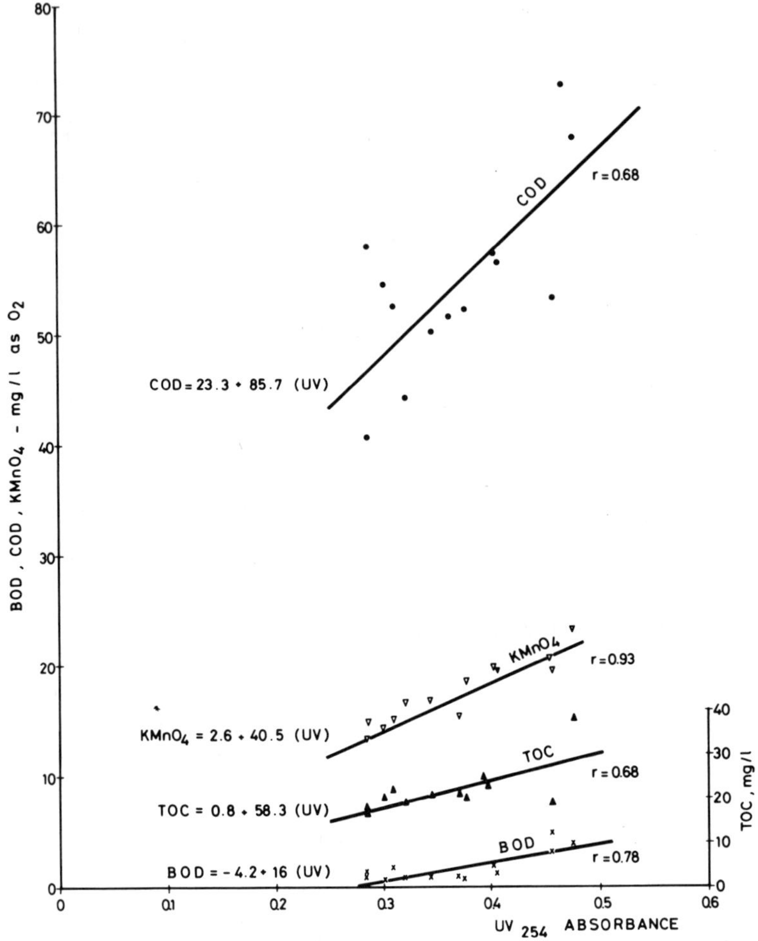

Figure 5. Correlations between gross organic parameters in activated sludge effluent.

for UV$'$–COD and UV$'$–KMnO$_4$ (Figures 6 and 7, respectively). The results obtained are as follows (data collected during the basic investigation on well 61 were excluded):

$$COD = 3.6 + 144.4 \ (UV') \tag{8}$$
$$r = 0.82$$

$$KMnO_4 = 0.5 + \ 24.9 \ (UV') \tag{9}$$
$$r = 0.85$$

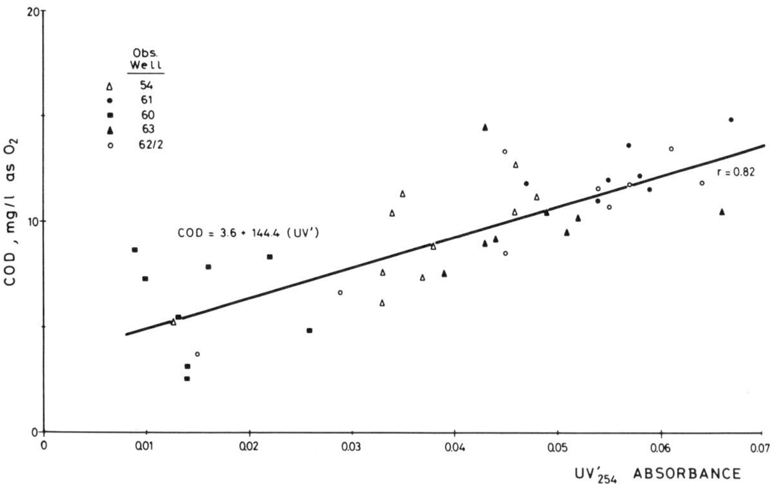

Figure 6. Correlations between UV absorbance and COD in reclaimed water.

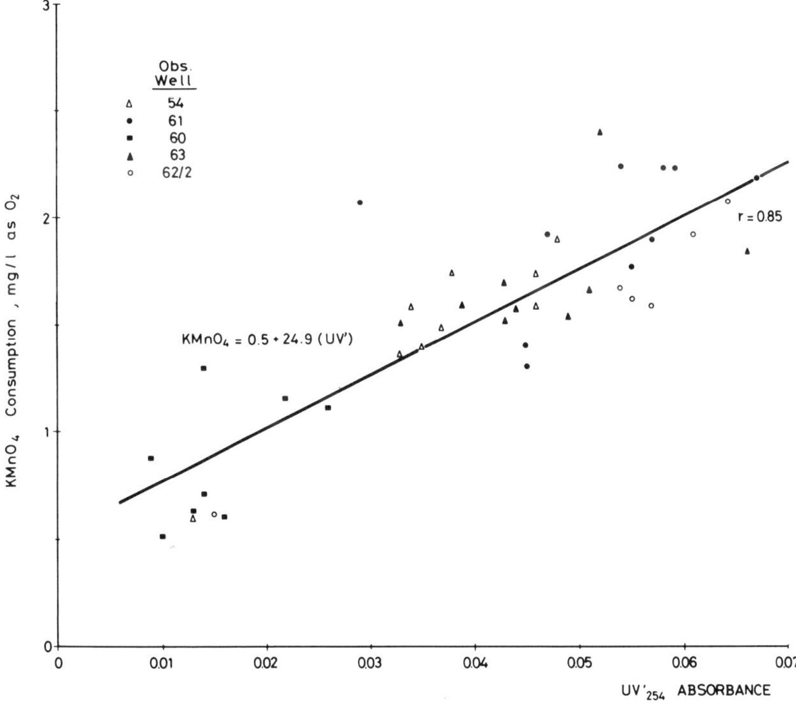

Figure 7. Correlations between UV absorbance and permanganate consumption in reclaimed water.

As good correlations were obtained, it appears that UV absorbance can be used as a simple and easy indicator of reclaimed water organic quality, as well as of effluent movement in the ground water aquifer.

AWT Influent and Effluents

Correlations obtained in the basic investigation for secondary effluent (oxidation ponds), for high-lime effluent and for tertiary effluent were poor for most gross organics parameters, including UV absorbance. However, when plotted together, the points of correlation between UV'_{254} and other parameters fitted well a straight line (Figure 8). The following equations and correlation coefficients were obtained:

$$COD = 18.9 + 144.7 \ (UV')$$
$$r = 0.89 \qquad (10)$$

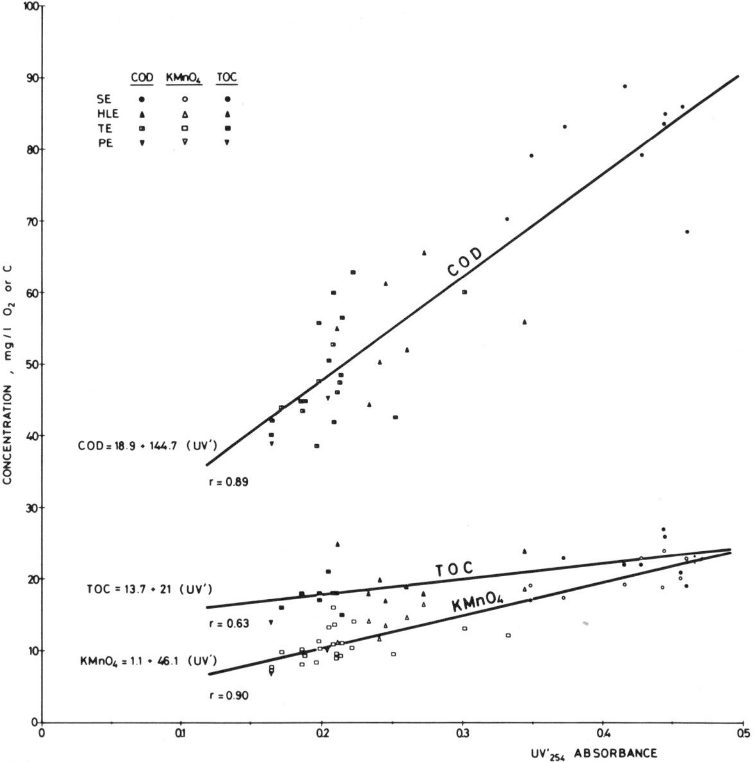

Figure 8. Correlations between gross organics parameters in various effluents of stage one project.

$$KMnO_4 = 1.1 + 46.1 \, (UV') \qquad (11)$$
$$r = 0.90$$

$$TOC = 13.7 + 21.0 \, (UV') \qquad (12)$$
$$r = 0.63$$

It thus appears that UV absorbance can also be used for general monitoring of the advanced wastewater treatment plant with respect to organics.

GROSS ORGANICS AT VARIOUS TREATMENT STEPS

From the values of various parameters (including UV') at each treatment step, the removal efficiency of filtrable organics by biological and chemical treatment was also calculated (Table VI). The great discrepancies in the results seem to reflect the limited, and sometimes even misleading, information provided by the percentage removal efficiency related to gross organics parameters (with the exception of BOD, which reflects the removal of a defined group—biodegradable organics).

The reduction of the organic content of waste water by biological treatment, chemical treatment and groundwater recharge is much better illustrated by means of the levels of gross organic parameters at each treatment step (Figure 9). With the exception of BOD, which reflects the biodegradability of organics at each treatment step, all other lines are approximately parallel. It appears that UV absorbance can replace any of the other gross organic parameters measuring gross organics, after it becomes more "popular" and correlations are established with the more common gross organic parameters.

SUMMARY AND CONCLUSIONS

Organics are the main substances of concern in water and wastewater treatment, as well as in wastewater reuse. Measurement of concentration of organic substances in water and waste water is usually carried out by use of gross parameters such as BOD, COD, TOC, DOC and $KMnO_4$ consumption.

From the literature reviewed and from experience, the advantages and disadvantages of each gross organic parameter were determined, as well as the relationships between various parameters.

A study was carried out on nine types of water from the Dan Region Sewage Reclamation Project which provides biological treatment, chemical treatment, groundwater recharge and reuse of municipal waste water from the Tel Aviv metropolitan area. The samples studied included raw waste water,

Table VI. Mean Removal Efficiency (%) of Filtrable Gross Organics by Biological and Chemical Wastewater Treatment

Parameter	Biological Treatment		Chemical Treatment		
	Oxidation Ponds	Activated Sludge	High-Lime Treatment	Polishing Ponds	Tertiary Treatment (High-Lime + Polishing Ponds)
BOD	92.8	99.1	−28.8	75.5	68.5
COD	57.4	85.8	31.3	10.5	38.5
TOC	63.3	84.1	8.9	14.5	22.1
DOC	66.6	87.4	16.0	8.9	23.5
$KMnO_4$ Cons.	47.5	84.1	30.0	18.4	42.9
UV	15.5	56.6	44.9	18.5	55.1
UV′	21.3	57.5	38.6	22.9	52.6

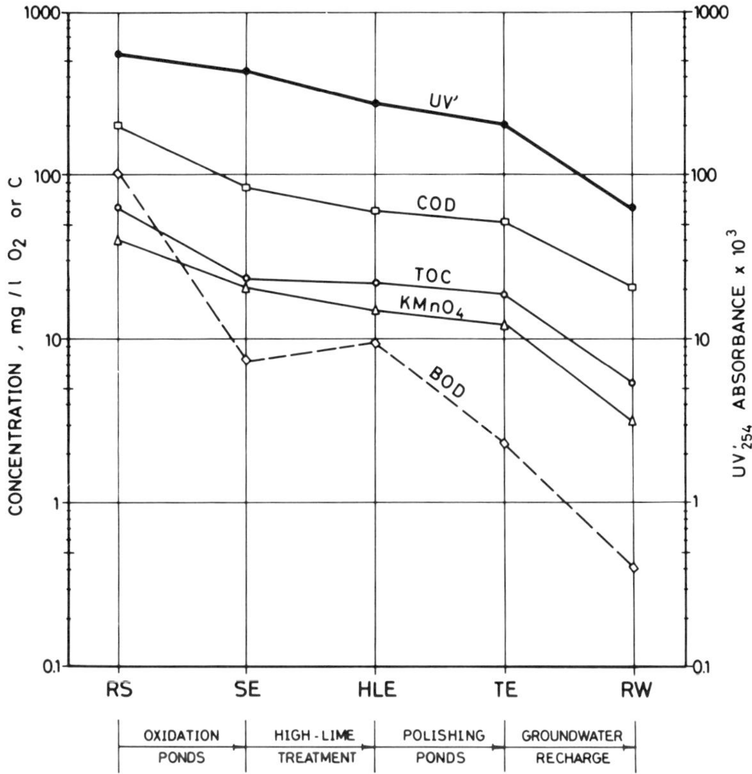

Figure 9. UV absorbance and other gross organic parameters at various treatment steps.

secondary effluent, tertiary effluent, reclaimed water and tap water. Six major gross organics parameters were included in the study: BOD, COD, TOC, DOC, $KMnO_4$ consumption and UV absorbance.

The wavelength of 254 nm was selected for UV absorbance measurements as the most suitable for all types of water studied. The corrected UV absorbance was used, which is the UV absorbance at 254 nm from which the absorbance in the visible at 545 nm was deducted to compensate for turbidity interference.

The range of values for each parameter was determined for each type of water studied. The ratios between the mean values of pairs of parameters were also determined.

Good to very high correlation coefficients were obtained for most parameters, including UV absorbance. The results obtained indicated the feasibility of using UV absorbance for routine monitoring of gross organics in wastewater treatment and reuse, and its particular suitability to:

1. general monitoring of organic matter in municipal raw waste water feeding the treatment plant, for easy and quick detection of organic-laden industrial wastes.
2. monitoring of activated sludge plant performance with respect to carbonaceous matter;
3. monitoring of organic quality of reclaimed water and of effluent movement in the groundwater aquifer; and
4. general monitoring of advanced wastewater treatment plant performance with respect to organics.

UV absorbance, which can be measured easily and rapidly and does not require sophisticated instrumentation, is as good as any other parameter for measuring gross organics, provided good correlations are established for each type of water and after its values become "familiar" to water and wastewater treatment utilities.

REFERENCES

1. Warner, H. P., J. N. English and I. J. Kugelman. "The Influence of Municipal Wastewater Treatment on Used Water Withdrawn for Domestic Supplies," paper presented at the American Water Works Association Annual Convention, Anaheim, CA, 1977.
2. Metcalf and Eddy, Inc. *Wastewater Engineering* (New York: McGraw-Hill Book Company, 1972).
3. Trussel, A. R., and M. D. Umphres. "An Overview of the Analysis of Trace Organics in Water," *J. Am. Water Works Assoc.* (November 1978).
4. *International Standards for Drinking Water*, 3rd ed. (Geneva, Switzerland; World Health Organization, 1971).
5. "Israeli Standards for Drinking Water Water," 3117, Tel Aviv (December 1974).
6. "National Interim Primary Drinking Water Regulations," *J. Am. Water Works Assoc.* (February 1976).
7. *Standard Methods for the Examination of Water and Wastewaters*, 13th ed. (New York: American Public Health Association, 1971).
8. "Measurement and Control of Organic Contaminants by Utilities," *J. Am. Water Works Assoc.* (May 1977).
9. *Standard Methods for the Examination of Water and Wastewaters*, 14th ed. (New York: American Public Health Association, 1976).
10. "Methods for Chemical Analysis of Water and Wastes," Environmental Monitoring and Support Laboratory, ORD, U.S. EPA-600/4-79-020 (March 1979).
11. Sawyer, C. N., and P. C. McCarty. *Chemistry for Environmental Engineering* (New York: McGraw-Hill Book Company, 1978).

12. Eckenfelder, W. W., and D. I. Ford. *Water Pollution Control* (New York: The Pemberton Press, 1970).
13. "Analysis of Raw, Potable and Waste Waters," Department of the Environment, London, England (1972).
14. Callely, A. G., C. F. Forester and D. A. Stafford. *Treatment of Industrial Effluents*, (London: Hodder and Stoughton, 1977).
15. Busse, H. J. "Instrumentelle Bestimmung der organischen Stoffe in Wässern," *Zelt. Wasser Abwasser–Forsch.* (6) (1975).
16. "Handbook for Monitoring Industrial Wastewater," U.S. EPA (August 1973).
17. Briggs, R., J. W. Schofield and P. A. Gorton. "Instrumental Methods of Monitoring Organic Pollution," *Water Poll. Control* (1976).
18. Ford, D. L., J. M. Eller and E. F. Gloyna. "Analytical Parameters of Petrochemical and Refinery Wastewaters," *J. Water Poll. Control Fed.* (August 1971).
19. Bramer, H. C., M. J. Walsh and S. C. Caruso. "Instrument for Monitoring Trace Organic Compounds in Water," *Water Sewage Works* (August 1966).
20. Sheppard, C. R. C. "Problems with the Use of Ultra-Violet Absorption for Measuring Carbon Compounds in a River System," *Water Res.* Vol. 11 (1977).
21. Ogura. N., and T. Hanya. "Ultraviolet Absorbance as an Index of the Pollution of Seawater," *J. Water Poll. Control Fed.* (March 1968).
22. Dobbs, R. A., R. H. Wise and R. B. Dean. "The Use of Ultra-Violet Absorbance for Monitoring the Total Organic Content of Water and Wastewater," *Water Res.* Vol. 6 (1972).
23. Briggs, R., and K. V. Melbourne. "Recent Advances in Water Quality Monitoring," *Water Treat. Exam.* 17 (1968).
24. Balch, N., D. Brown, R. Pym, E. Marles, D. Ellis and J. Littlepage. "Monitoring Marine Outfalls by Using Ultraviolet Absorbance," *J. Water Poll. Control Fed.* (January 1975).
25. Fähnrich, V., and M. Soukup. "UV-Spectrophotometric Determination of Organic Compounds in Water," *Wasserwirtsch.-Wassertech.* (1964).
26. Mrkva, M. "Investigation of Organic Pollution of Surface Waters by Ultraviolet Spectrophotometry," *J. Water Poll. Control Fed.* (November 1969).
27. Mrkva, M. "Automatic UV-Control System for Relative Evaluation of Organic Water Pollution," *Water Res.* Vol. 9 (1975).
28. Tambo, N., and T. Kamei. "Treatability Evaluation of General Organic Matter, Matrix Conception and Its Application for a Regional Water and Wastewater System," *Water Res.* Vol. 12 (1978).
29. Kölle, W., and H. Sontheimer. "Experience with Activated Carbon in West Germany," paper presented at Activated Carbon in Water Treatment Symposium, University of Reading, April 3-5, 1973.
30. Schalekamp, M., K. Dietlicher, J. Valenta, R. Wattenhofer and U. Zimmerman. "L'Évolution des Polluants Organiques Spécifiques dans les Eaux de Surface et au Cours de la Préparation de l'Eau Potable," *Gas Wasser Abwasser* No. 11 (1978).
31. Gurguis, W., T. Cooper, J. Harris and A. Ungar. "Improved Performance of Activated Carbon by Pre-Ozonation," *J. Water Poll. Control Fed.* (February 1978).

32. Helfgott, T. B. "Parameters for Measuring Organics in Waters," (Tel Aviv: Tahal—Water Planning for Israel Ltd., 1977).
33. Idelovitch, E., R. Terkeltoub, M. Butbul, M. Michail and R. Friedman. "Dan Region Project, Groundwater Recharge with Municipal Effluent, Second Recharge Year-1978, Annual Report," (Tel Aviv: Tahal—Water Planning for Israel Ltd., 1979).
34. Peleg, M. Personal communication (1977).
35. Zuckerman, M. M., and A. H. Molof. "High Quality Reuse Water by Chemical-Physical Wastewater Treatment," *J. Water Poll Control Fed.* (March 1970).
36. Rebhun, M., and J. Manka. "The Fate of Soluble Organics in Effluents from Chemical Treatment and Recharge," Project No. 013-568, Annual Report, Technion—Israel Institute of Technology, Environmental Engineering Laboratories, Haifa (August 1976).

CHAPTER 4

DETERMINATION OF CARBOHYDRATES AND PRIMARY AMINES IN RIVER WATER

Lawrence E. Conroy

Department of Chemistry

Walter J. Maier

Environmental Engineering Program
Department of Civil and Mineral Engineering

Young-Tzung Shih

Department of Chemistry
University of Minnesota
Minneapolis, Minnesota

Surface waters are host to a bewildering variety of organic compounds. Myriad organic chemical constituents are continuously being added to the aquatic environment from industrial point sources as well as from diffuse sources such as agriculture, silviculture and natural drainage areas. While the bulk of this influx represents natural organic decay products, the rapidly rising influx of anthropogenic compounds that resist decomposition and accumulate in the environment is of paramount concern. The potential hazards for human health and the ecosystem that are posed by manmade compounds depend on their chemical composition. The effects of many of

65

these substances have not yet been defined; nevertheless, government agencies are committed to monitor and regulate their influx and to limit the concentrations of organic compounds in drinking water. Analytical procedures for the detection and estimation of organic constituents in the aquatic environment are therefore urgently demanded.

Decay products of biomass include materials related to humic and fulvic acids that are leached from soil organics. These are biochemically refractory materials that are structurally related to lignin and tannin. In contrast, the lower-molecular-weight constituents, such as amino acids, proteins, carbohydrates and lipids, are for the most part rapidly metabolized by aquatic microbes and must be regarded as transients. Their concentrations are likely to be very low except in the immediate vicinity of a source, but the total flux of these transient components through an aquatic system may be appreciable if the rate of degradation of biomass and the rate of metabolization are high.

Many techniques for concentrating, separating and identifying organic matter in aquatic media have been evaluated during the past 75 years. In recent times, membrane ultrafiltration has proven to be one of the most successful and practical procedures for separating and classifying the "solution" fraction of surface water samples. A study by Chian [1] on the main classes of soluble organics that are present in leachate samples showed that the largest fraction consisted of free volatile fatty acids capable of penetrating a low-molecular-weight ultrafiltration membrane. The second largest fraction comprised a fulvic acidlike material having a high density of carboxyl and aromatic hydroxyl groups. A small percentage of the organic compounds consisted of a high-molecular-weight fraction with properties characteristic of humic acid carbohydrates but also containing a significant quantity of hydrolyzable amino acids.

The concentration of dissolved free amino acids (DFAA) and carbohydrates was found to be generally detectable, but low. However, because of the ability to penetrate virtually all filtration media, these soluble components are among the more chemically significant species affecting the quality of effluents and drinking water. Methods for detection and measurement of the dissolved carbohydrate and amino acid contents of water are of significant importance. This study has utilized analytical techniques that had been devised for biochemical and industrial solutions and has tested their applicability to analysis of river water. Carbohydrates were assayed by use of the anthrone reagent, and primary amines were assayed with fluorescamine. The utility of these techniques was evaluated on carefully fractionated samples of Mississippi River water (MRW). The fractionation procedure will be described briefly, and then the analytical techniques will be discussed.

Molecular Size Fractionation of Aquatic Organics by Membrane Ultrafiltration

This research was a part of a larger study of aquatic organic species in the Mississippi River. Sampling was performed at four sites along the Mississippi, chosen to represent (1) natural background, (2) agricultural drainage, (3) urban runoff and (4) the influence of a major wastewater discharge. Site 1 is located near the Mississippi River headwaters 175 miles northwest of Minneapolis-St. Paul (Twin Cities) near Bemidji (BEM) (Figure 1). Site 2 is 70 miles upstream from the Twin Cities in an agricultural area near Royalton (ROY). Site 3 is at the St. Anthony Falls (SAF) in the center of Minneapolis. Site 4 is located south of Minneapolis-St. Paul, 3.5 miles downstream of the major metropolitan treatment facility near the Inver Grove Bridge (IGB). In addition, seasonal fluctuations were observed at the Minneapolis site. Samples of MRW were obtained in May, September and February from the SAF site, and each sample was treated identically. Each sample was prefiltered with a pleated depth filter (Millipore CP-20) and a 0.22-μm filter (Millipore EGWP-PHV-20). The ultrafiltration fractionation was accomplished with a batch system using 90-mm agitated high flux ultrafiltration cells (Millipore XX42-090-50). In addition to the 100,000, 10,000 and 1000 nominal molecular weight (NMW) membranes, a 500-NMW membrane (Amicon Type UM05) was used. Operating pressures were 25 psig for the 100K and 10K membranes and 40 psig for the 1 K- and 500-NMW (0.5K) membranes.

Each cell was fitted with a precleaned ultrafiltration membrane, filled with 500 ml of SAF river water, and the retentate volume was reduced by 350 ml. More water was added and filtered, with the process repeated until all the sample had been added to the cell. The sample was filtered until the retentate volume was 300 ml. Next, the 300 ml of retentate was washed with three 200-ml portions of pH 8.3 wash water (NaOH in deionized water) to remove all organics smaller than the NMW cutoff of the membrane. The filtrate was combined with that from the washes and used as feed for the next ultrafiltration with a smaller NMW cutoff membrane. Figure 2 indicates the volumes used for the ultrafiltration and the scheme followed.

Following ultrafiltration, the membrane was soaked overnight in 1.0 M NaCl for cleaning and then stored in 0.1% NaN_3 to prevent bacterial growth. The ultrafiltration fractions were analyzed for dissolved organic carbon (DOC) content.

To determine the molecular size distribution of organics in the river water, samples were subjected to batch ultrafiltration fractionations. The results appear in Table I. DOC concentration in the river shows considerable seasonal variation, which is probably the result of both streamflow conditions and

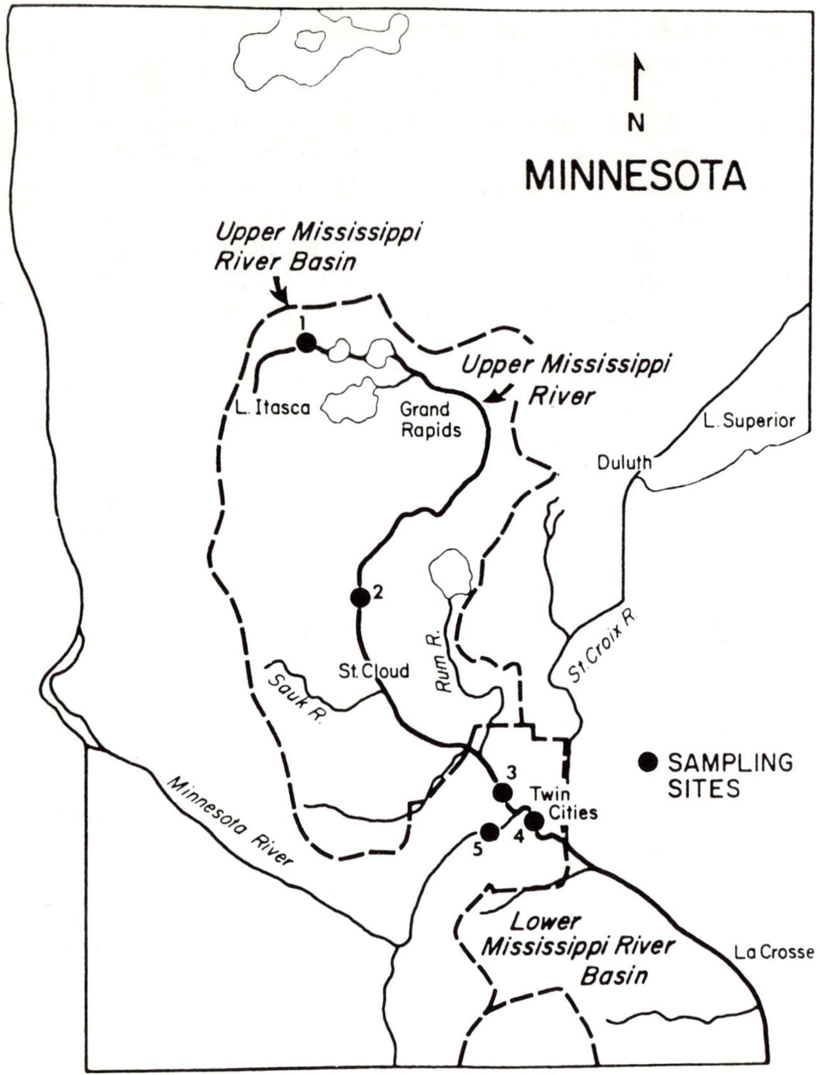

Figure 1. Water sampling sites on the Mississippi River in Minnesota.

source inputs. The May concentration (9.7 mg/l) can be attributed to a high streamflow coupled with a large surface input of carbon from runoff, while the fall concentration is the result of low streamflow. The low value of 7.70 mg/l in winter can be ascribed to groundwater inputs since the streamflow is low at that time of the year and the surface inputs are nonexistent due to frozen conditions.

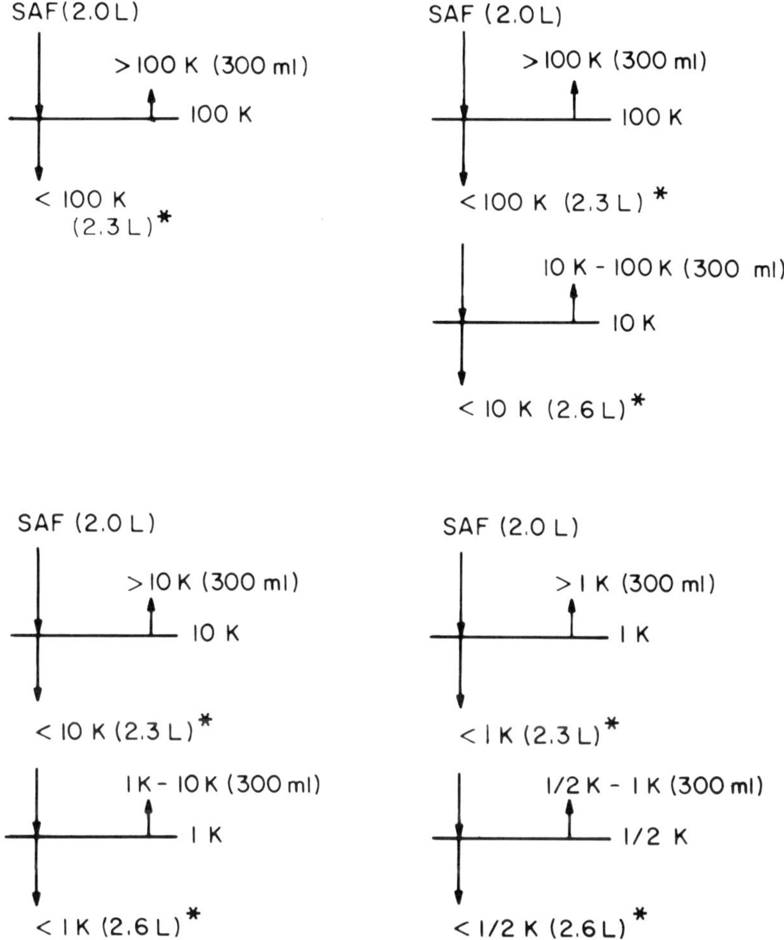

* Volume includes washes

Figure 2. Fractionation scheme for Mississippi River water using ultrafiltration.

The DOC molecular size distribution indicated that the majority of aquatic organics are of molecular size in the range of 1K-10K (59-68%). In addition, the <1K fraction accounts for 29-36% while the >10K fraction accounts for only 4-13%. This shows that the vast majority of aquatic organics are less

Table I. Seasonal Distributions of Dissolved Organic Carbon Among Molecular Weight Fractions of Organic Compounds

Fraction (MW)	Dissolved Organic Carbon[a] (mg/l)			Distribution (%)		
	5/78	9/78	2/79	5/78	9/78	2/79
$<$1,000	2.80	3.57	2.80	28.8	22.6	36.3
1,000-10,000	5.65	10.7	4.57	58.1	67.7	59.3
10,000-100,000	0.71	1.01	0.15	7.3	6.4	2.0
100,000-0.22-μ	0.57	0.54	0.19	5.9	3.4	2.5
Total $<$0.22-μ	9.73	15.8	7.71	100.1	100.1	100.1

[a] St. Anthony Falls site, passing 0.22-μ filter

than 10,000 NMW in size. In addition to the molecular size distribution of organics in river water, a seasonal variation was also noted. The percentage of small molecular size (less than 1K) increased in the winter (36%) compared to fall (23%). While the percentage of small organics increased, the percentage of large organics (greater than 10K) decreased. Both changes are attributed to the change in the source of organics (i.e., ground water) previously mentioned. The largest fraction (1K-10K), which constitutes the "core aquatic organics," remained relatively constant from fall through winter but decreased in percent of the total during the high runoff spring conditions when other organics were being flushed into the river.

Although seasonal variations were noted, the distribution of DOC in the river followed the same pattern. The 1K-10K fraction contained the majority of the organics with the less than 1K fractions a close second in importance. Very little of the organic carbon was greater than 10K in molecular size.

ASSAY OF CARBOHYDRATES IN AQUEOUS MEDIA

The Anthrone Method

A qualitative method for the detection of carbohydrates by the use of anthrone (9,10-dihydro-9-ketoanthracene) in concentrated sulfuric acid was first described by Dreywood [2] in 1946. The formation of a green-colored complex constituted a positive test. The use of this reagent for quantitative estimation of carbohydrates was evaluated by Morse [3]. Viles and Silverman [4] showed that anthrone in concentrated sulfuric acid could be used to detect starch or cellulose spectrophotometrically in the range of 10-200 μg with a sensitivity of 2 μg. Figure 3 reproduces their spectral transmittance curves for starch and cellulose using a 0.1% anthrone solution. In both cases

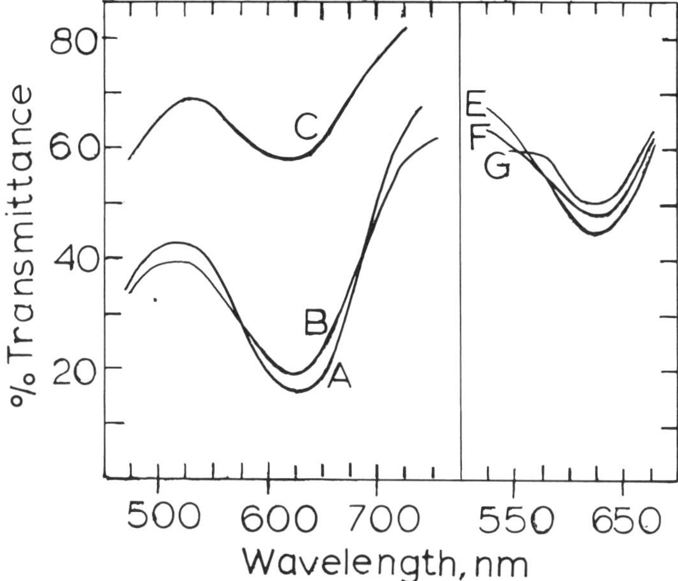

Figure 3. Spectral transmittance of complexes of anthrone [4] (A) 90 μg cellulose; (B) 90 μg cellulose (reagent 1 day old); (C) 25 μg cellulose; (E) 50 μg starch; (F) 50 μg starch (reagent 1 day old); (G) sample E after several hours.

the absorption maximum is located approximately at 625 nm. Dreywood [2] also demonstrated that successful detection was possible not only with carbohydrates, but also with certain carbohydrate derivatives. Samsel and DeLap [5] adapted the method to the determination of methylcellulose, and Black [6] extended it to sodium carboxymethylcellulose. Up to this time the technique had exhibited some difficult features. Maximum color development occurs at the time when the rates of formation of the chromogen and its destruction in hot acid are in balance, a situation that differs for different carbohydrates. In addition, the effect of temperature on the rate of formation of the chromagen varies widely; for example, it is much slower for aldoses than for ketoses. In 1975 Jermyn [7] significantly improved the technique with his discovery that the addition of formic acid and hydrochloric acid increases the rate of formation of color and thereby greatly enhances the sensitivity of the test. All known saccharides, polysaccharides and glycosides give a positive reaction with anthrone. Furfural is the only noncarbohydrate that has been reported to give a positive test.

The mechanism of the reaction between anthrone and carbohydrates has never been elucidated definitely.

Anthrone (II) is the tautomer of 9-hydroxyanthracene (I).

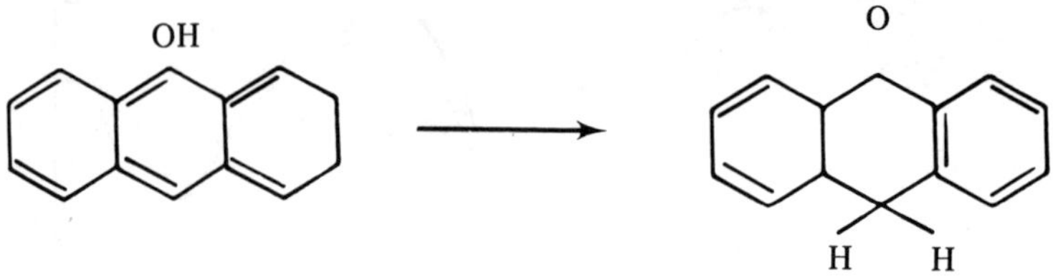

$$OH \qquad\qquad\longrightarrow\qquad\qquad O$$

$$\text{I} \qquad\qquad\qquad\qquad\qquad \text{H} \quad \text{H}$$

$$\text{II}$$

The reactivity of anthrone with aldehydes and ketones has long been known [8] as a condensation reaction, presumably with the keto form (II) above, for example:

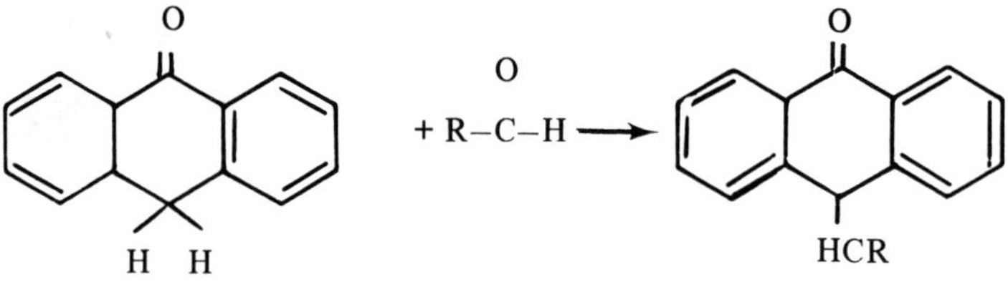

$$O \qquad\qquad + R{-}\overset{\displaystyle O}{\overset{\|}{C}}{-}H \longrightarrow \qquad O$$

$$\text{H} \quad \text{H} \qquad\qquad\qquad\qquad\qquad \text{HCR}$$

$$\text{III}$$

The positive reaction of furfural with the anthrone reagent led to the postulate that the primary reaction with sugars involves the conversion of the sugar to furfural or a furfural derivative by dehydration and ring formation. Wolfrom et al. [9] demonstrated the formation of 5-hydroxymethylfurfural (IV) from glucose by refluxing in neutral or acid aqueous solution.

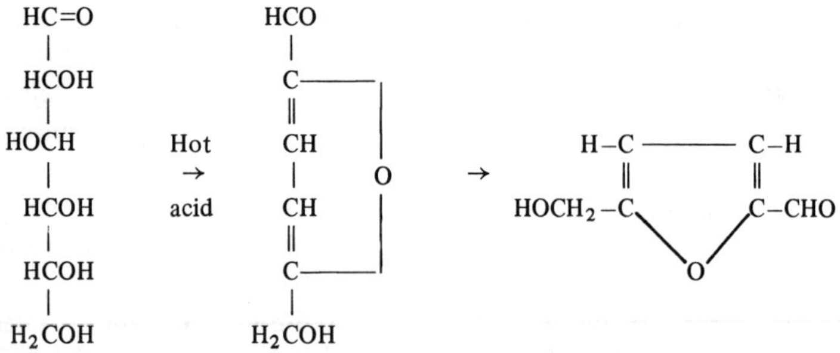

$$
\begin{array}{c}
HC{=}O \\
| \\
HCOH \\
| \\
HOCH \\
| \\
HCOH \\
| \\
HCOH \\
| \\
H_2COH
\end{array}
\quad\xrightarrow[\text{acid}]{\text{Hot}}\quad
\begin{array}{c}
HCO \\
| \\
C \\
\| \\
CH \\
| \\
CH \\
\| \\
C \\
| \\
H_2COH
\end{array}
\quad\longrightarrow\quad IV
$$

IV

Sattler and Zerban [10] confirmed this mechanism by applying the anthrone test to a diverse group of carbohydrates. Methylated glucose, di-heterolevulosan, levulose tetraacetate, maltose octaacetate, hydroxymethyl-furfural and the diphenylacethydrazide of hydroxymethylfurfural all produced a positive test with anthrone. In addition, the phenylosazones of glucose and galactose and the phenylosotriazole of glucose do not give the test, whereas mannose phenylhydrazone gave a positive test. Since the osazones and osotriazoles differ from the other compounds by forming osones (V) in strong acids:

$$
\begin{array}{l}
HC=NNHC_6H_5 \\
| \\
C=NNHC_6H_5 \\
| \\
(CHOH)_3 \\
| \\
CH_2OH
\end{array}
\quad + 2H_2O + 2HCl \rightarrow \quad
\begin{array}{l}
HC=O \\
| \\
C=O \\
| \\
(CHOH)_3 \\
| \\
CH_2OH
\end{array}
\quad + 2C_6H_5NHNH_3{}^{+}Cl^{-}
$$

V

whereas the phenylhydrazones are reconverted to the original sugars

$$
\begin{array}{l}
HC=NNHC_6H_5 \\
| \\
HOCH \\
| \\
HOCH \\
| \\
HCOH \\
| \\
HCOH \\
| \\
CH_2OH
\end{array}
\quad + H^{+} \rightarrow \quad
\begin{array}{l}
HCO \\
| \\
HOCH \\
| \\
HOCH \\
| \\
HCOH \\
| \\
HCOH \\
| \\
CH_2OH
\end{array}
\quad + C_6H_5NHNH_3{}^{+}Cl^{-}
$$

only the latter phenylhydrazine derivative could condense to a furfural ring. Black [6] has provided further evidence in favor of the furfural mechanism by applying the anthrone test to solutions of cellulose and carboxymethyl cellulose. The ultraviolet (UV) absorption spectra obtained from those tests and the spectrum obtained from a test on 5-hydroxymethylfurfural were all very similar in shape and absorption maximum. As shown in Figure 4, the indications are that cellulose was hydrolyzed to glucose by the sulfuric acid. Glucose yields 5-hydroxymethylfurfural by dehydration and ring formation, and the 5-hydroxymethylfurfural then reacts with the anthrone to produce the color.

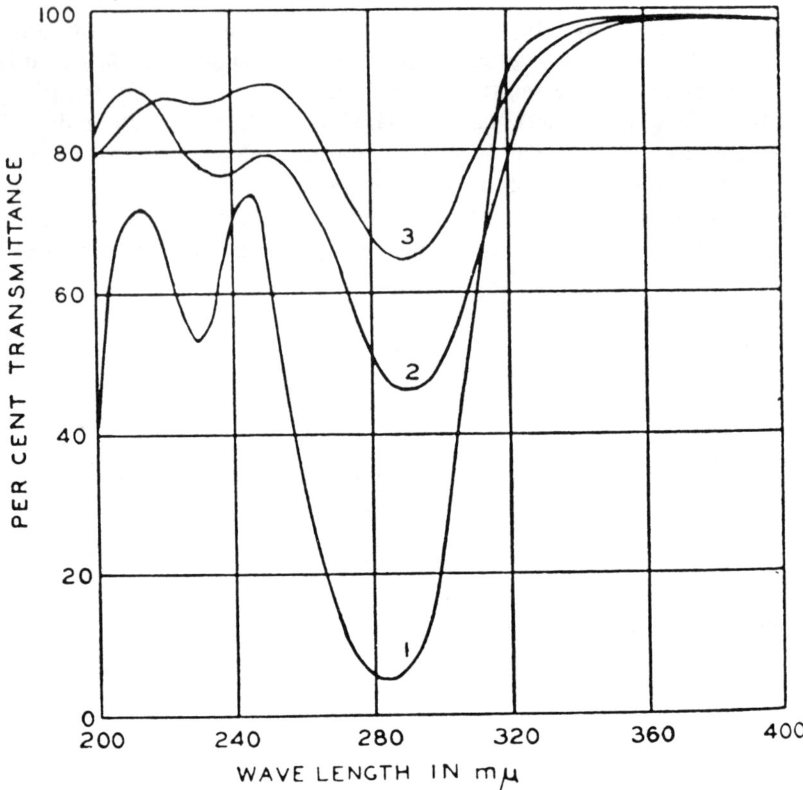

Figure 4. UV spectral transmittance of solutions [6]. (1) Hydroxymethylfurfural; (2) solution of cellulose in 60% sulfuric acid; (3) solution of carboxymethylcellulose in 60% sulfuric acid.

Experimental

Reagent-grade anthrone and anhydrous granular glucose (dextrose) were obtained from the J. T. Baker Chemical Company. The formic acid was "certified" grade from Fisher Scientific Company. Inorganic acids were all of reagent grade.

The anthrone reagent must be prepared freshly (within four hours) for use by dissolving 20 mg of anthrone in 100 ml of 80% (v/v) sulfuric acid at room temperature. To a 1-ml riverwater sample in a 25-ml test tube were added 1 ml of conc hydrochloric acid and 0.1 ml of 88% formic acid. The anthrone reagent (8 ml) was added slowly to avoid excessive foaming. After thorough mixing, the sample tube, along with similar tubes containing glucose standards and pure H_2O blanks, were heated 12 min in a boiling water bath, followed by immediate quenching in a cold water bath. Each sample was thoroughly stirred on a Vortex mixer, then allowed to stand for 5 min to disperse bubbles. Absorbance was read at 627.5 nm on a Beckman Model 26 spectrophotometer.

Results

Experiments with known samples containing glucose as the carbohydrate show that the color development with anthrone is linear with concentration for quantities up to 20 mg/liter and that the sensitivity is 1 mg/liter or better. Figure 5 displays a typical plot of absorbance at 627 nm vs concentration of glucose in the presence of anthrone reagent. Incremental additions of pure glucose solution to a sample of MRW also showed the same linearity. Samples of MRW, using the four ultrafiltration fractions that had been separated by the procedure described above, were then evaluated by the same anthrone reagent technique. Table II summarizes typical data from the anthrone test applied to a sample of MRW fractionated by membrane ultrafiltration.

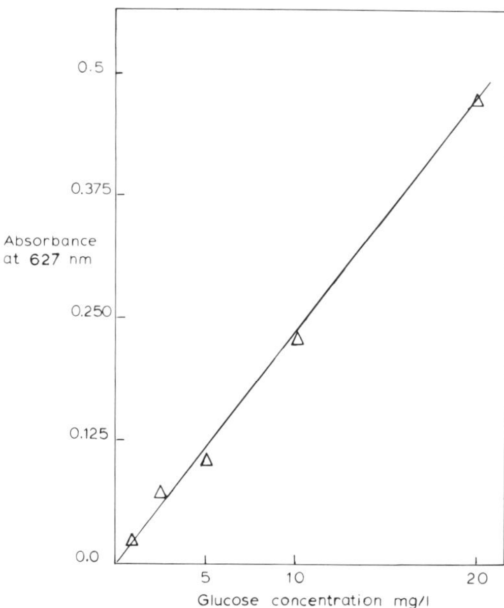

Figure 5. Plot of absorbance vs glucose concentration.

**Table II. Carbohydrate Determinations in Mississippi River Water
Using Anthrone Reagent**

Filtration Fraction (MW)	Absorbance at 627.5 nm			Average Absorbance	Carbohydrate (mg/l)[a]
	Sample 1	Sample 2	Sample 3		
<1000	0.009	0.000	0.000	0.003	0.003
1000-10,000	0.000	0.012	0.012	0.008	0.005
10,000-100,000	0.035	0.037	0.032	0.035	0.16
>100,000	0.067	0.062	0.065	0.065	0.24

[a] As glucose.

Column 1 lists the nominal molecular weight range of each fraction. Columns 2 and 3 show the absorbances at 627.5 nm of the anthrone test solutions, and column 4 records the carbohydrate concentrations (as glucose) calculated from columns 2 and 3. An explanation of this distribution is discussed below.

ASSAY OF DISSOLVED AMINES IN RIVER WATER

Fluoresence techniques for determination of primary amines and the related amino acids, peptides, and proteins have been in use for well over a decade. In 1962 McCaman and Robins [11] described a fluorometric method, in wide use today, for assay of serum phenylalanine which utilizes the interaction with ninhydrin and peptides. Samejima et al. [12,13] reported that the reaction of phenylalanine with ninhydrin first formed phenylacetaldehyde which then combined with additional ninhydrin and peptide, or any other primary amine, to yield highly fluorescent products. The structure of these products was then elucidated by Weigele et al. [14] who proceeded to synthesize a new and novel reagent [15] to replace the fluorogenic ninhydrin reaction. The Weigele reagent is 4-phenylspiro[furan-2(3H), 1′-phthalan]3,3′-dione which is given the trivial name fluorescamine and which reacts with primary amines to form the same fluorophores (390 nm excitation, 475 nm emission) as were produced in the original ninhydrin-phenylacetaldehyde reaction. A number of factors favor fluorescamine as a reagent for assaying primary amines, including amino acids, peptides and proteins. Reactions are rapid; Udenfriend et al. [16] showed that at pH 9 primary amines react with half times of fractions of seconds. Excess reagent is eliminated with a half time of a few seconds [16]. These reactions are summarized in Figure 6. Fluorescamine and its hydrolysis products do not fluoresce. The fluorogenic reaction proceeds to near completion (80-95%) with small peptides, even in

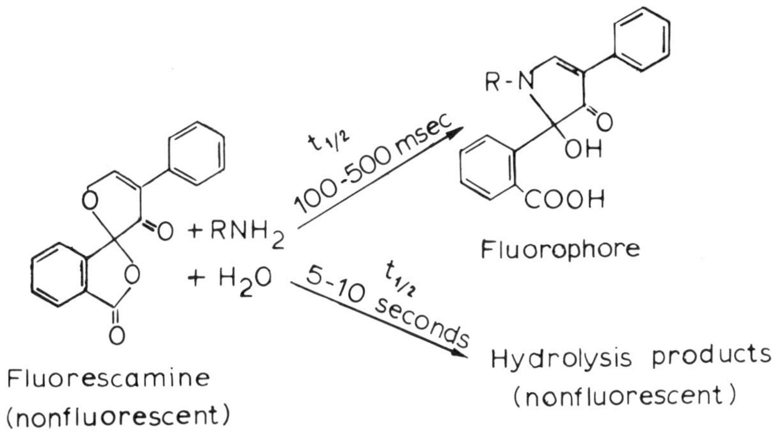

Figure 6. Reactions of fluorescamine reagent.

the absence of large excess of fluorescamine [16]. Primary amine assay is carried out in buffered (pH >7) solution, and the fluorescamine, dissolved in an appropriate water-miscible nonhydroxylic solvent such as acetone, is added. At room temperature the reaction is complete in less than a second and the excess reagent is depleted in less than one minute, yielding the amine fluorophore as the only fluorescent product. The resulting solution fluorescence is proportional to the total primary amine concentration; this fluorescence remains stable over several hours.

Ring nitrogen amines, such as proline, will not react with fluorescamine, but such molecules may be hydrolyzed or otherwise converted to primary amines for successful assay. A significant advantage of the fluorescamine assay over the colorimetric ninhydrin procedure is that negligible fluorescence arises from reaction with ammonia or urea [17,18]. In uncomplicated situations, such as solutions that contain only a single amine component, the fluorescamine assay is highly sensitive. Quantities as small as 50 pmol of amino acid have been determined by Undenfriend et al. [16]. In mixtures of amine molecules some limitations to the quantification of components will occur. Because peptides generally yield greater fluorescence than their component amino acids, it is not feasible to quantify such mixtures or partially hydrolyzed peptides on a molar basis. Furthermore, the maximum fluorescence occurs at different pH values for peptides (pH ~7) and amino acids (pH ~9). Since the relative fluorescence of individual fluorescamine derivatives is constant from pH 4 to 10, the differences in fluorescence intensity is due to the reactivity of individual amines and differences in quantum yields for the various fluorophores. It is noted that fluorescamine will function successfully as an amine assay reagent not only in aqueous solution, but in organic solvents or on solids as well. It can be used as a spray reagent to detect amino acids and peptides on thin layer chromatograms with as little as 20 pmol of amino acid [19].

Experimental

Fluorescamine was obtained from Pierce Chemical Co. as Fluram, Roche. Glycine and glycosamine were obtained from Sigma Chemical Co. Spectrograde acetone (Fisher Scientific Co.) was used. A solution of 20 mg of fluorescamine in 100 ml of acetone was prepared. A 3-ml sample of standard or river water was buffered by adding 2 ml of sodium borate buffer solution (pH 8.9). To this 5-ml sample of buffered water was added a 1-ml aliquot of the fluorescamine solution. The solution was mixed on a Vortex mixer for at least 2 min. The fluorescence intensity was read on an Aminco-Bowman Spectrophotofluorometer using 390 nm excitation and 480 nm emission. The fluorophore was stable over several hours.

Results

Optimum concentrations of fluorescamine, acetone and hydrogen ion were established from experiments with standard glycine solutions. Typical data relating the quantities of reagent and pure glycine are shown in Table III.

Table III. Fluorescence Intensity for Fluorescamine Assay of Aqueous Glycine Solution[a]

Sample	Buffer Volume[b] (ml)	Fluorescamine[c] Solution Volume (ml)	Relative Intensity Sample 1	Relative Intensity Sample 2
6 ml H_2O			0.00	0.00
6 ml acetone			0.00	0.00
		6 ml	30.5	34.0
5 ml H_2O		1 ml	5.8	5.9
	5 ml	1 ml	5.9	6.0
2.5 ml H_2O		2.5 ml	7.1	6.6
	2.5 ml	2.5 ml	7.9	8.2
3 ml glycine[d]	2 ml	1 ml	9.6	9.8
3 ml glycine[d]	1 ml	2 ml	12.5	12.6
2 ml glycine[d]	2 ml	2 ml	10.6	11.0

[a] 390 nm excitation, 480 nm emission.
[b] pH 8-9.
[c] 0.02 mg/l in acetone.
[d] 1.00 mg/l.

These data permitted the choice of optimum ratios for solution, buffer and fluorescamine solution. The linear relationship between relative fluorescence intensity and glycine concentration was found to hold for the concentration range of 0.5 to 5 μM (Figure 7). The limit of detection was 0.2 μM in glycine.

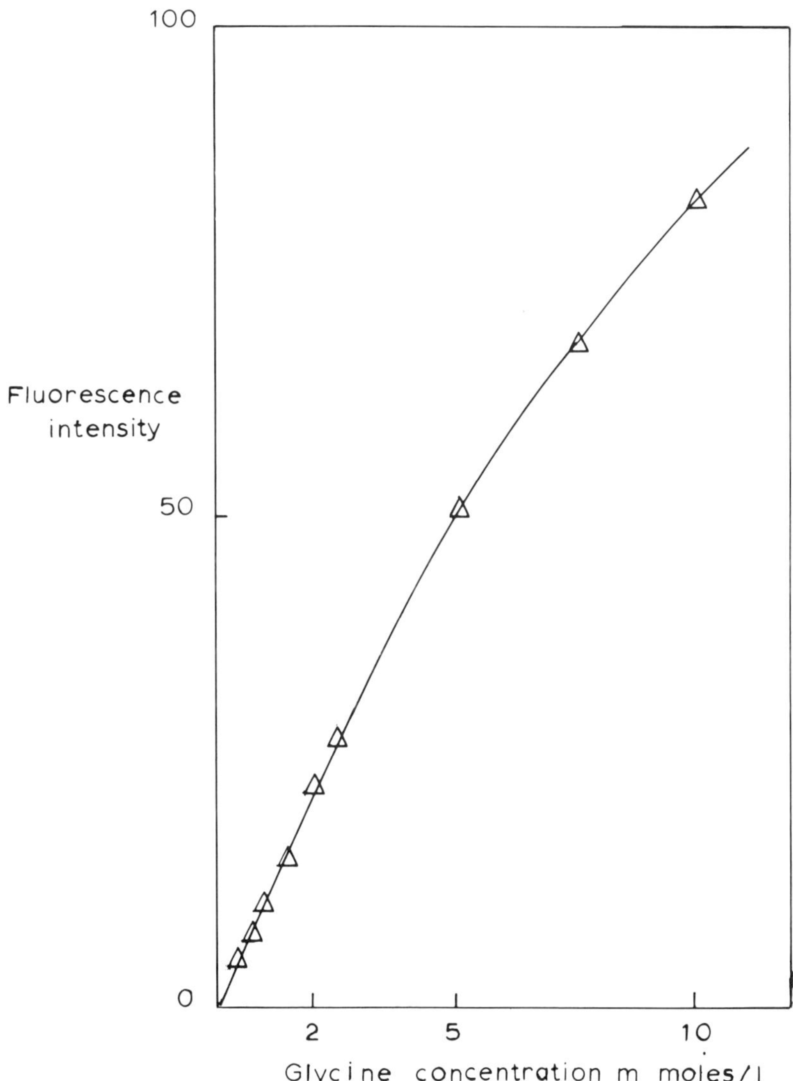

Figure 7. Plot of fluorescence intensity vs glycine concentration.

To determine possible influence on the fluorescence of the other components of river water, the glycine calibration was also carried out on MRW samples and the linear relationship between glycine concentration and fluorescence was found still valid within our limits of detection.

Table IV summarizes data from the fluorescamine assay of primary amine in a typical MRW sample that was fractionated by ultrafiltration according to the procedure described earlier.

DISCUSSION

Among the various methods that were investigated for the assaying of carbohydrate and amine functionalities occurring in soluble riverwater organic species, the anthrone test and the fluorescamine reagent, respectively, have proven to be the most sensitive and reliable. Both techniques were applied to the soluble ($<0.22 \mu$) organic matter in MRW samples taken in May 1978 at the four sites described earlier. Examination of these results permits some inferences concerning the nature of the organic constituency in the river water as well as the variations with geographic location. All of the following discussion assumes that the carbohydrate and/or amine functional groups are equally accessible for reaction with the appropriate reagent throughout the range of molecular sizes. The validity of this assumption has not been directly established and awaits further study.

Carbohydrates

Table V is a summary of the concentrations of carbohydrates and amines in each molecular weight fraction for samples taken at the four Mississippi River sites: Bemidji (BEM), Royalton (ROY), St. Anthony Falls (SAF) and the Inver Grove Bridge (IGB). See Introduction for description of these sites.

Table IV. Determination of Primary Amines in Mississippi River Water with Fluorescamine Reagent

Ultrafiltration Fraction (MW)	Relative Intensity Average	Amine Concentration as Glycine (μg/l)
$<$1000	5.0	38.0
1000-10,000	7.9	59.0
10,000-100,000	12.3	90.0
$>$100,000	4.3	30.0

Table V. Geographic Variation of Carbohydrate and Primary Amine Distributions
in Mississippi River Water, May 1978

Filtration Fraction (MW)	Carbohydrate Concentration[a] (μg/l)				Primary Amine Concentration[b] (μg/l)			
	BEM	ROY	SAF	IGB	BEM	ROY	SAF	IGB
<1K	< 10	< 10	53	282	16.25	17.48	8.98	13.02
1K-10K	175	149	355	263	8.57	18.80	21.16	16.35
10K-100K	66	103	118	158	0.61	2.39	2.21	2.24
100K-0.22 μ	48	55	68	89	0.29	1.01	1.49	1.14
Total, <0.22 μ	289	307	594	792	25.72	39.68	33.84	32.75

[a] As glucose.
[b] As glycine.

The same data are represented in the form of a bar graph in Figure 8. Some significant distinctions between the carbohydrate amine geographical distributions are apparent. The total concentration of soluble carbohydrates is seen to increase from the headwaters to the area south of the Twin Cities, by approximately 165% from BEM to IGB (based on BEM) during the change from a rural, forested area to a location downstream from a large wastewater facility. To further illuminate this observation, the concentrations of carbohydrates were compared to the concentrations of total organic carbon (TOC) in the same MRW samples. These data are presented, as ratios of carbohydrate to TOC, in Table VI. That ratio is seen to be relatively constant within the fluctuations of the data, indicating that the geographical distribution of carbohydrate roughly follows that of the total organic content. If the distribution of carbohydrate among the various molecular weight functions is considered, it is seen that the bulk of the carbohydrates occur in the 1-10K fraction (33-61%) and the 10K-100K fraction (20-33%), and that these two fractions together comprise 53-84% of the carbohydrates. This is not the same pattern of distribution as TOC, which is distributed in roughly equal percentages between the <1K and 1-10K fractions. The explanation is found in the carbohydrate/TOC ratios (Table VI) which increase with molecular weight while the TOC values decrease, so that the maximum carbohydrate content occurs in the intermediate-molecular-weight range. The remarkably low carbohydrate content of the low-weight fractions would be the consequence if the smaller carbohydrate species were being metabolized rapidly by aquatic organisms. Moreover, if hydrolysis, which is usually a slow process, were the rate-determining step in the metabolism of the carbohydrate in larger molecules, a higher content of carbohydrate would be present in those fractions. Finally, it should be noted that the overall carbohydrate content of

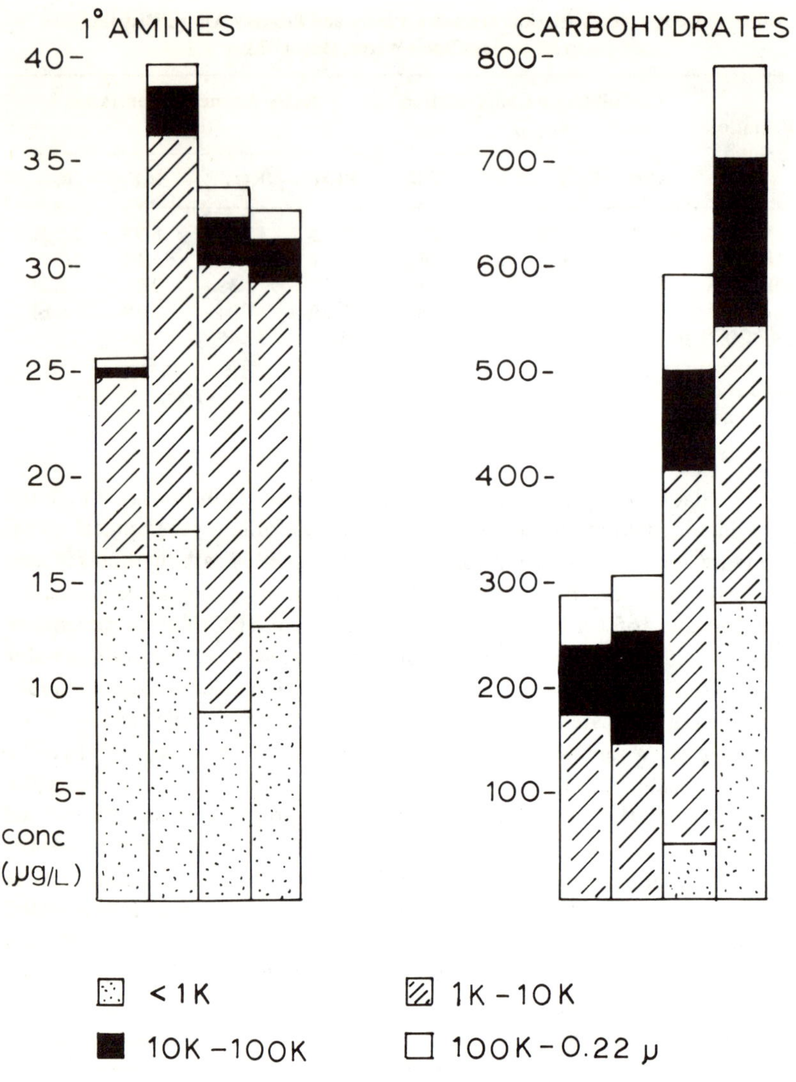

Figure 8. Bar graph representations of concentrations of primary amines and carbohydrates in Mississippi River water at four sites in Minnesota.

MRW is low, less than 0.8 mg/l in all samples, indicating that carbohydrates are readily being assimilated.

Table VI. Distributions of Organics Among Molecular Weight Fractions[a]
and Geographical Locations in Mississippi River Water

Fraction (MW)	Carbohydrate[b]/TOC Ratio				Primary Amine[c]/TOC Ratio x 10^3			
	BEM	ROY	SAF	IGB	BEM	ROY	SAF	IGB
$<$1K	0.002	0.003	0.019	0.071	3.8	5.4	3.2	3.3
1K-10K	0.072	0.030	0.063	0.063	3.5	3.8	3.8	3.9
10K-100K	0.41	0.10	0.17	0.23	3.8	2.4	3.1	3.2
100K-0.22 μ	2.9	1.0	1.0	2.2	2.9	3.4	2.6	3.2

[a] Passing 0.22-μ filter.
[b] As glucose.
[c] As glycine.

Amines

Considering the data on the distribution of primary amines in MRW it is seen in Table V that the total amine content is even lower than that of carbohydrates, typically by factors in the range of 7-20. The total amine content fluctuates from upstream to downstream by 27% from BEM to IGB, a much smaller increase than that observed for the carbohydrates. Comparison for the amine distributions with the distributions of TOC show that the variation in amine concentrations parallels rather closely the distributions of TOC. This parallel is obvious in the amine/TOC ratios in Table VI, where it is seen that the amines make up a fairly constant proportion of the total organic content of MRW. Furthermore, this relative uniformity of the amine/TOC ratio suggests that the amine functional groups are present in the molecules of all sizes and that any consumption of amines by aquatic microorganisms proceeds at the same rate on all molecules, regardless of size. A less likely possibility is that the rate of hydrolysis of larger species is equal to the rate of consumption.

ACKNOWLEDGMENT

We are pleased to acknowledge funding of this research by the National Science Foundation under Grant No. NSF/ENV 77-04496.

REFERENCES

1. Chian, E. S. K. *Environ. Sci. Technol.* 11:158 (1971).
2. Dreywood, R. *Ind. Eng. Chem. Anal. Ed.* 18:499 (1946).
3. Morse, E. E. *Anal. Chem.* 19:1012 (1947).
4. Viles, F. J., and L. Silverman. *Ind. Eng. Chem. Anal. Ed.* 21:950 (1949).
5. Samsel, E. P., and R. A. DeLap. *Anal. Chem.* 23:1795 (1951).
6. Black, H. C. *Anal. Chem.* 23:1792 (1951).
7. Jermyn, M. A. *Anal. Biochem.* 68:332 (1975).
8. Karrer, P. *Organic Chemistry* (New York: Nordeman Publishing Co., 1938), p.405.
9. Wolfrom, M. L., R. D. Schuetz and L. F. Cavalieri. *J. Am. Chem. Soc.* 70:514 (1948).
10. Sattler, L., and F. W. Zerban. *Science* 108:207 (1948).
11. McCaman, M. W., and E. Robins. *J. Lab. Clin. Med.* 59:885 (1962).
12. Samejima, K., W. Dairman and S. Udenfriend. *Anal. Biochem.* 42:222 (1971).
13. Samejima, K., W. Dairman, J. Stone and S. Udenfriend. *Anal. Biochem.* 42:237 (1971).
14. Weigele, M., J. F. Blount, J. P. Tengi, R. C. Czaikjowski and W. Leimgruber. *J. Am. Chem. Soc.* 94:4052 (1972).
15. Weigele, M., S. L. DeBernado, J. P. Tengi and W. Leimgruber. *J. Am. Chem. Soc.* 94:5927 (1972).
16. Udenfriend, S., S. Stein, P. Böhlen, and W. Dairman. *Science* 178:871 (1972).
17. Stein, S., P. Böhlen, J. Stone, W. Dairman and S. Udenfriend. *Arch. Biochem. Biophys.* 155:202 (1973).
18. Böhlen, P., S. Stein, W. Dairman and S. Udenfriend. *Arch. Biochem. Biophys.* 155:213 (1973).
19. Felix, A. M., and M. Jimenez. *Anal. Biochem.* 48:417 (1972).

CHAPTER 5

ULTRAVIOLET MULTIWAVELENGTH ABSORBANCE MEASUREMENTS FOR MONITORING TRACE ORGANICS IN WATER

Walter J. Maier and Lawrence E. Conroy

University of Minnesota
Minneapolis, Minnesota

Myriad organic chemical constituents are continuously being added to the aquatic environment from industrial point sources and nonpoint sources such as agriculture, silviculture and runoff from land. The bulk of this influx represents natural organic decay products, but there is an increasing influx of manmade compounds that resist biological decomposition and therefore tend to accumulate in the aquatic environment. The potential hazards of some manmade compounds to human health and the ecosystem are recognized, and government agencies have taken steps to monitor and regulate their inflow and limit concentrations in drinking waters. However, the detrimental effects of naturally occurring aquatic organics are still not fully understood. They have been linked to the transport-solubilization of anthropogenic chemicals and are also implicated as precursors of carcinogens (halomethane formation during chlorination). There is, therefore, an urgent need for new analytical methods that can be used routinely, rapidly, simply, continuously and at low cost to measure naturally occurring organics as well as specific pollutants. The available nonspecific tests for biochemical oxygen demand (BOD), chemical oxygen demand (COD) and total organic carbon (TOC) are inadequate because they give no information about the composition of pollutants.

Ultraviolet (UV) absorbance measurements are used extensively for analyzing pure compounds and mixtures of naturally occurring organic materials [1,2]. Interest centers on the 200- to 300-nm region because most solvents, including water, are transparent, and instruments and equipment are readily available. Organic materials found in natural waters (aquatic organics) generally have strong absorbance bands in this region and are therefore well suited to UV spectral analysis.

Aquatic organic matter has not been studied as extensively as the related organic materials found in soils. However, elemental compositions are similar, as are their absorbance spectra. The latter are relatively featureless, showing progressively increasing absorbance from 350 to 200 nm [3,4]. Absorbance of soil organics is ascribed to the presence of polymerlike molecules such as fulvic and humic acids comprised of substituted aromatic rings, carboxyl groups, saturated and unsaturated alkyl groups, and heterocyclic compounds, derived from decay of biomass. Specific absorbance maxima of single chromophores are usually not observed. The absence of absorbance maxima is ascribed to the presence of a large variety of different chromophores, which masks the contribution of single compounds. However, the presence of unique and characteristic absorbance bands is readily apparent in differential spectrophotometry. The featurelessness of UV absorbance spectra of aquatic organics may be an inherent characteristic of large, multichromophore molecules found in water; peak broadening and formation of subsidiary absorbance bands are known to occur in large molecules through interaction of conjugated chromophores, steric effects of substituent groups, the presence of polar substituents and ring size. Featurelessness may, therefore, be viewed as an identifying characteristic but does not imply that all absorbance spectra are indistinguishable. It follows that absorbance spectra of aquatic organics should be analyzed in terms of groups of chromophores in order to characterize their composition. Utilization of UV absorbance measurements for characterizing aquatic organics is potentially attractive because:

1. absorbance measurements are simple, require very little time, small samples are needed, results are reproducible and high sensitivity allows measuring low concentrations;
2. instrumentation and equipment are readily available in most laboratories at reasonable cost; and
3. it is well suited to continuous monitoring applications as, for example, in process control of water and wastewater treatment, treatment for water reuse, or monitoring of ambient water quality in lakes or rivers.

One promising technique that has received relatively little attention is the application of multiwavelength UV absorbance measurements. Absorbance measurements at a series of wavelengths can be used to characterize chemical structure and determine concentrations of specific chromophores or classes of chromophores. In principle, the analysis involves superposition of absorbance spectra of pure compounds or classes of compounds to match the absorbance spectrum of the mixture and to estimate their respective concentrations. This chapter describes the initial results of research aimed at utilizing multiwavelength UV absorbance as a nonspecific measure of organic pollutant concentrations in natural waters.

The overall objective of the research was to study the UV absorbance characteristics of organic matter in Mississippi River waters to obtain insight on their physico-chemical properties. This basin was chosen because of its diversity; it includes upstream reaches that are unpopulated and unstressed by pollution, agricultural watersheds and a population- and pollution-stressed region. The first phase objective was to determine the absorbance characteristics of organic constituents from each of four geographical locations representing progressively more pollution-stressed regions and to determine seasonal changes. The second phase objective focused on analyzing the multiwavelength absorbance data in the context of absorbance spectra of pure model compounds and naturally occurring chromophores to characterize aquatic organics. A corollary objective was to develop methodologies for matching the absorbance spectra of aquatic organics by superposing the absorbance contribution of chromophoric groups and/or pure model compounds suspected to be present in aquatic organics. This approach allows calculating the respective concentrations of chromophores and therefore gives quantitative information about the mixture.

A state of the art review will be presented; pertinent theory of electronic absorption is summarized in the context of chromophores found in naturally occurring organics. This information provides the basis for choosing appropriate model compounds. Absorbance characteristics of model compounds will be described. The mathematical approach used for matching riverwater data will be developed in terms of model compounds. Preparation of samples and conventional analysis of riverwater absorbance data are discussed, and the last section describes results in terms of the absorbance contributions of model compounds as representative chromophoric groups.

ULTRAVIOLET SPECTROPHOTOMETRY

State of the Art

Applications of UV measurements in the water pollution control field have been reviewed by Dobbs et al. [5]. Absorbance measurements at 254 nm have been correlated with organic carbon concentration [5,6], potential oxygen demand [7] and used for monitoring industrial effluents [8]. Concentrations of TOC and A_{254} are strongly correlated for specific rivers [5,9] but finished drinking waters from 80 different cities and hence different raw water sources exhibited considerable scatter [6].

Applications of UV spectrophotometry for analyzing natural products are well documented [1,2]; it is probably the most widely used analytical measurement. Quantitative measurements, however, are usually made at a single wavelength at or near maximum absorbance. Analysis of multiwavelength measurements has not been widely used. Arends et al. [10] analyzed mixtures of arylsulfonic acids with reference to the pure compound spectra of each constituent. Metzler et al. [11] described a mathematical analysis for resolving spectra of multicomponent mixtures and showed how this approach could be applied to evaluating pK values. Superposition of absorbance spectra and measurement of difference spectra relative to known model compounds has been used to characterize lignin and its degradation products [2]. Kankare [12] calculated molar absorptivities and equilibrium constants from multicomponent spectrophotometric data. It is, therefore, apparent that multiwavelength analysis of mixtures is a potentially useful technique for obtaining quantitive information about their constituents.

Theory of Electronic Absorbtion

Ultraviolet light absorption is coincident with excitation of electrons from their ground state. The energy of transition and hence wavelength of absorbed light depends on the structure of the atoms that constitute the chromophore and to some extent on solvent environment. Intensity of absorption is a measure of the probability of electronic transition and therefore a measure of the concentration of chromophores. In dilute solutions of nonpolymeric compounds Beer's law applies, and absorbance is directly proportional to chromophore concentration; nonlinearity is likely at high concentrations and/or with aggregated molecules due to shading. Theoretical considerations of molecular structure on electronic absorbance have been used to correlate absorbance maxima and intensities of most organic compounds and empirical correlations are available that describe the effects of substituent groups on single chromophores (e.g., olefins, double bonds of carbonyl groups) and conjugated chromophores (e.g., aromatic ring structures and heterocyclic rings) [1,2]. Band shifts induced by substituent groups are correlated in terms of electron donating and withdrawing properties. These correlations are useful for the insight they give on absorbance characteristics of substituted aromatic ring structures typically ascribed to soil organics as illustrated below.

The conjugated ethylenic bonds of the benzene ring have principal absorption bands ascribed to local excitation (LE) at 202 and 255 nm, with maximum extinction coefficients (ϵ_{max}) of 7000 and 300 cm^2/mol, respectively. Intro-

Table I. Bathochromic Effects of Oxygen-Containing Substituents [2]

Benzene		Benzoic Acid		Benzaldehyde		Acetophenone	
λmax (nm)	ϵ (mol/cm^2)	λmax (nm)	ϵ (mol/cm^2)	λmax (nm)	ϵ (mol/cm^2)	λmax (nm)	ϵ (mol/cm^2)
202	7,000	202		200		200	
255	300	232	12,320	244	15,880	242	12,040
		275	1,050	280	1,529	278	1,100
		284	890	317[a]	20-280	319	80

[a] In cyclohexane absorbance ranges from 317 to 353 nm.

duction of substituents into aromatic rings tends to shift absorbance to longer wavelengths (bathochromic shift) and frequently increases maximum absorption (ϵ_{max}). Alkyl substituents such as CH_3, C_2H_5 and halogens exert minor effects: large shifts to 240 nm and also to the 320- to 350-nm range. The shifts have large bathochromic effects (Table I).

The carboxyl group in benzaldehyde and acetophenone have similar effects: large shifts to 240 nm and also to the 320- 350-nm range. The shifts in benzoic acids are less pronounced and show no absorbance above 300 nm.

The hydroxyl group in phenol induces a bathochromic shift with a strong peak at 210 nm (ϵ = 10,000) and a second maximum at 272 nm (ϵ = 2000); similar shifts to the 210- to 220 and 270- to 280-nm regions are reported for the dihydroxybenzenes. However, there is no absorption above 300 nm at neutral pH. Hydroxy substitutions into benzaldehydes or acetophenones give rise to two bands in the 230- to 300-nm regions for *ortho* or *meta* positions, and a single intense band (200-280 nm) for *para* groups [1]. Acetophenones exhibit an additional band near 320 nm from the carboxyl group. The hydroxy and alkoxy substituted compounds of benzaldehyde and acetophenone retain the general absorbance pattern as illustrated in Table I.

Quinone structures have intense absorption bands at 240-290 nm, with weak absorbance at 330-360 nm and into the visible range. Condensed ring aromatic structures also exhibit intense absorbance bands as exemplified by naphthalene at 220, 280, 312 and 320 nm. These four bonded spectra persist throughout its derivatives with minor shifts [1]. Heterocyclic compounds such as pyrrole and thiophene structures have similar absorption characteristics with extinction coefficients ranging from 4000 to 12,000 depending on the α and β substituents and solvent. Furan absorbance maxima occur at somewhat lower wavelengths. The six-membered pyridine derivatives absorb in two bands at 195-225 and 260-290 nm. The 3-hydroxy, methoxy and carboxy derivatives exhibit absorbance in the 260-280 region and a second absorbance band at lower wavelength (216-246). 3-hydroxypyridine has a third absorbance band at 315 nm (ϵ = 3060) and typifies the spectral absorbance of pyridoxine (Vitamin B_6) and related compounds. There are, of course, many other naturally occurring organic compounds in surface waters that absorb in the UV spectral range and it is, therefore, appropriate to consider characterizations in terms of classes of chromophores rather than pure compounds.

Absorbance of Mixtures, pH Effects

Review of the literature shows that structural information can be deduced from UV measurements in terms of the location of absorbance bands and their extinction coefficients. However, the overlap between absorbance bands due to band broadening induced by substituent groups precludes utilization of single wavelength absorbance measurements for quantitive measurement of specific compounds in a mixture of chromophores. This has prompted utilization of multiwavelength measurements for quantitative analysis of mixtures [2,10-13]. Systematic analysis of the whole absorbance spectrum (200-360 nm) has been used for quantitative determinations of specific chromophores in solutions containing mixtures of chromophores; it is conceptually related to the measurement of difference spectra where the pure solvent reference solution is replaced by a matrix solution that contains other chromophores. However, by analyzing the absorbance characteristics at a series of discrete wavelengths, additional information is obtained that identifies specific chromophores. Theoretical considerations indicate that multiwavelength analysis should be capable of identifying as many different chromophores in a mixture as there are discrete wavelength absorbance measurements; the assumption is made that no two chromophores have identical spectra. This assumption is reasonable in terms of the molecular structure that relates to electronic absorption behavior but does not apply to those molecular structures that are UV-transparent and do not influence electron absorbance.

Bathochromic shifts occur in phenolic chromophores when pH is increased from neutral to alkaline domains. Smaller shifts are reported for benzoic acids; amine-substituted chromophores also change with pH. It is, therefore, necessary to consider pH as a potential variable in absorbance matching calculations or to carry out all measurements at the same pH.

MODEL COMPOUNDS

Chemical Structures of Soil Organics

Aquatic organics, soil organics and lignin have many similarities in terms of origin, occurrence, environmental cross connections and elemental composition [3,4]. All three materials are known to be mixtures of natural constituents that have undergone physical-chemical-biological transformations. Lignins and soil organics have been studied extensively, and the available information gives insight on the chemical properties of aquatic organics. Comparison of the absorbance spectra of aquatic organics, lignin and soil organics (Figure 1) illustrates similarities and some differences; absorbance spectra are relatively featureless, but the absorbance increases with decreasing wavelength; similarity in spectra suggests the presence of related chromophores in all three samples. However, significant maxima and minima are observed when lignin is used instead of water as a reference solution. The

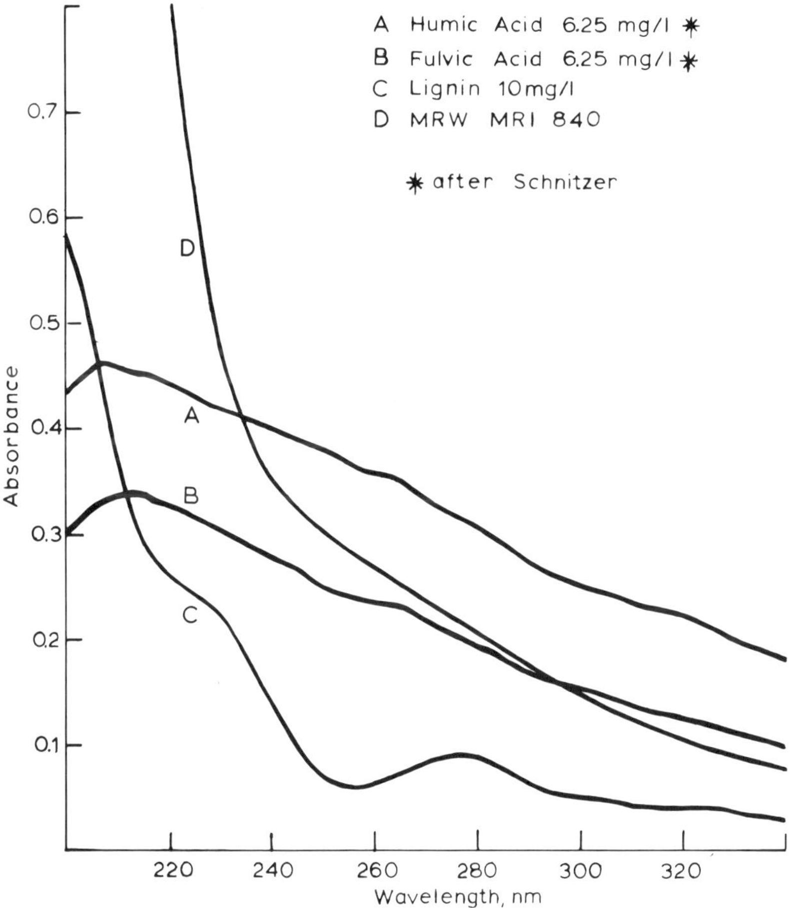

Figure 1. Aquatic organics absorbance features resemble humics and ligins [3].

positive and negative peaks show that there are significant differences in UV absorbance of aquatic organics compared to lignin, even though both of the overall spectra appear featureless. Negative absorbance bands indicate that certain chromophores are absent or present at lower concentrations, whereas positive peaks indicate the presence of chromophores with large extinction coefficients and/or high concentrations even though the TOC contents of the solutions are essentially the same. Similar results were obtained on a number of riverwater samples. The positive absorbance bands identified by differential spectrophotometry were used to guide the selection of pure model compounds whose absorbance characteristics could be used to match those of aquatic organics.

A structural scheme of lignin is reproduced in Figure 2; lignin is shown as a mixture of polymerized units made up of hydroxy-, carboxy- and alkoxy-substituted ring structures bonded by ether linkages and some carbon-carbon bonds. Figure 3 lists major degradation products of lignin. The structure of fulvic acids has been described as a polymeric, hydrogen-bonded, open structure [3]. Compounds isolated from fulvic acid after exhaustive methylation and polarity fractionation are illustrated in Figure 4. Although the concentrations of each of the identified compounds was very low, the structural variety is an indication of the composition of the structural units that make up the bulk of fulvic and humic acids. Working on the premise that aquatic organics are of similar origin, pure model compounds were chosen on the basis of chemical and structural similarity to breakdown products of lignin and soil organics and availability of pure compounds. A few model compounds were included because they are representative of transient decay products of biomass, e.g., amino acids, ring nitrogen compounds and long chain fatty acids. The presence of such compounds in aquatic organics is indicative of ongoing or recent decomposition.

Analytical Procedures

Some 35 model compounds have been incorporated into the program. Absorbance spectra were measured on a Beckman Model 26 using 10-mg/l solutions in Milli-Q water buffered with phosphate to pH 7. Milli-Q water was used as a reference; buffer addition had no effect.

Elemental composition data for pure compounds were obtained from suppliers. Data on lignin and the river water samples were measured in the laboratory. TOC was measured on a Beckman 915A carbon analyzer [15]. Organic nitrogen was measured by a micro-Kjeldhal technique [15]. Reproducibility of measurements is discussed in a following section. Organically combined hydrogen and oxygen were measured with a CDS 1200 Elemental Composition Analyzer but the results were not used because they were not reproducible. The absorbance values were digitized in two nanometer increments from 200 to 358 nm except in earlier work which was limited to the 230- to 348-nm ranges. The digitized absorbance values were inventoried into the computer program as discussed in subsequent sections.

Analysis of Absorbance Spectra of Model Compound

Absorbance characteristics of model compound were studied to give insight on the effects of structural and chemical composition differences. The objective was to identify chromophores with characteristic absorbance bands that resemble or piecewise match the absorbance spectra of river water, keeping in mind that the latter is a mixture of many compounds.

Absorbance spectra of model compounds were categorized into i absorption bands centered on each absorbance peak to facilitate comparisons between

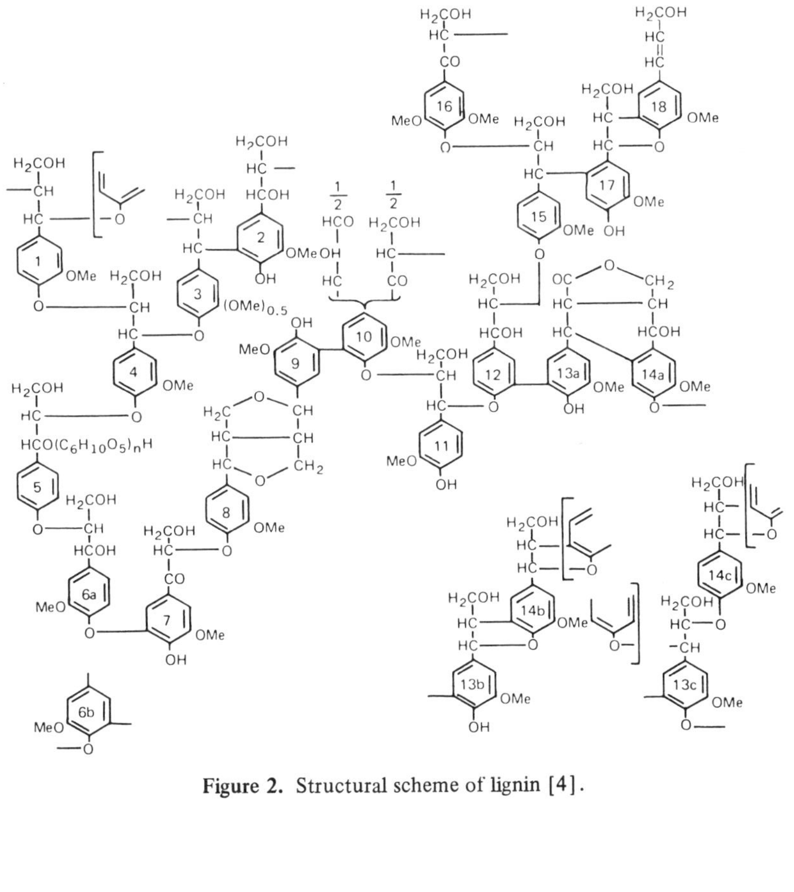

Figure 2. Structural scheme of lignin [4].

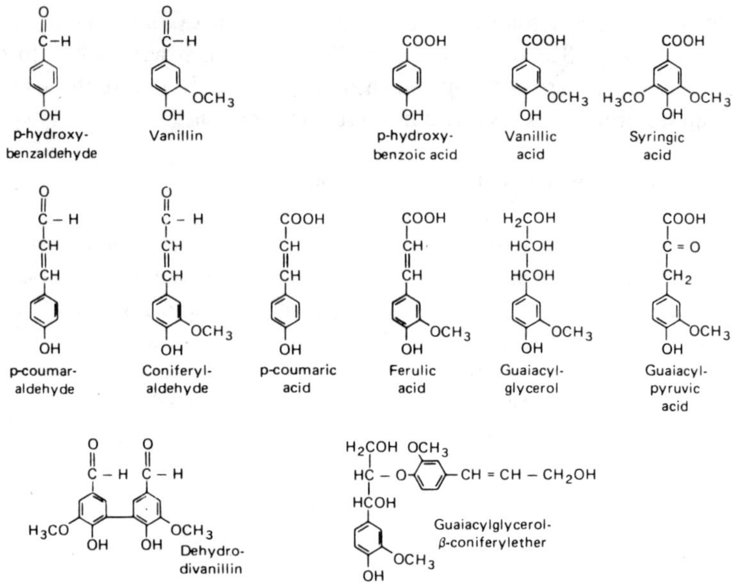

Figure 3. Lignin degradation products [4].

compounds and to document the effects of substituent groups on absorbance characteristics of ring structures. Average absorbance values (A_i) were calculated as the integral under a truncated curve

$$A_i = a_i \frac{\int_{a_i}^{b_i} f(\lambda)}{|a_i - b_1|}$$

where
$f(\lambda)$ = the absorbance at wavelength λ
a_i and b_i = the values of λ at which absorbance is half the maximum absorbance of the ith waveband

Model compounds with λ_{max} near 200 nm were truncated at 200 nm. Averaged absorbance bands are shown in Figures 5 to 9. This stylized but simplified method of data presentation facilitates comparison of absorbance band shifts associated with structural differences. Total absorbance (A_T) defined by

$$A_T = \int_{200}^{360} f(\lambda)$$

was calculated for each model compound as an overall indicator of absorbance.

$1R_1 = R_2 = CO_2CH_3; R_3 = R_4 = R_5 = R_6 = H$

$2R_1 = R_4 = CO_2CH_3; R_2 = R_3 = R_5 = R_6 = H$

$3R_1 = R_2 = R_3 = CO_2CH_3; R_4 = R_t = R_6 = H$

$4R_1 = R_2 = R_4 = CO_2CH_3; R_3 = R_5 = R_6 - H$

$5R_1 = R_3 = R_5 = CO_2CH_3; R_2 = R_4 = R_6 = H$

$6R_1 = R_3 = R_5 = CO_2CH_3; R_4 = R_6 = H$

$R_4 = R_6 = H$

$7R_1 = R_2 = R_3 = R_4 = CO_2CH_3; R_5 = R_6 = H$

$8R_1 = R_2 = R_4 = R_5 = CO_2CH_3; R_3 = R_6 = H$

$9R_1 = R_2 = R_3 = R_5 - CO_2CH_3; R_4 = R_6 = H$

$10R_1 = R_2 = R_3 = R_4 = CO_2CH_3; R_5 = OCH_3; R_6 = H$

$11R_1 = R_2 = R_3 = R_4 = R_5 = CO_2CH_3; R_6 = H$

$12R_1 = R_2 = R_3 = R_4 = R_5 = CO_2CH_3; R_6 = OCH_3$

$13R_1 = R_2 = R_3 = R_4 = R_5 = R_6 = CO_2CH_3$

$14R_1 - R_3 = R_4 = R_5 = CO_2CH_3; R_2 = OCH_3; R_6 = H$

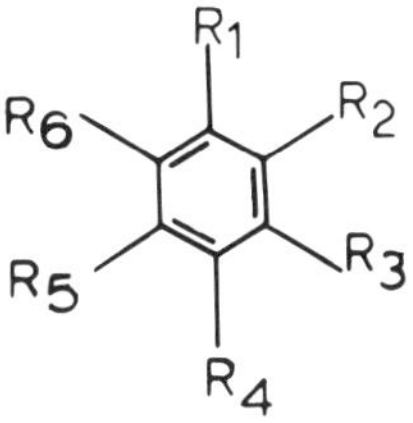

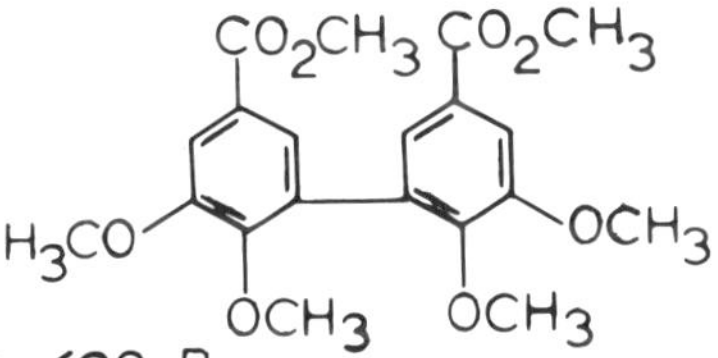

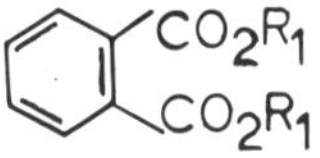

$16\ R_1 = R_2 = C_4H_9$

$17\ R_1 = R_2 =$

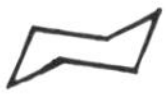

$18\ R_1 = R_2 = CH_2CH\text{--}C_4H_9$ (C_2H_5)

$19\ R_1 = R_2 = C_8H_{17}$

Figure 4. Chemical structures of compounds isolated from fulvic acids [14].

Figure 5 compares the three methoxy-substituted isomers of benzoic acid. They represent structural analogues of naturally occurring compounds related to anisic acid. An intense band at 200 nm is present in all three isomers with secondary bands centered near 220 and 280 nm for the *ortho* and *meta* isomers, whereas, the *para* isomer has a single intense peak at 244 nm. *Ortho* and *meta* isomers have similar spectra, but *meta* substitution results in a slight shift to longer wavelengths.

The marked effect of *para* substitution is confirmed in Figure 6 where both the 3,4,5-trimethoxybenzoic acid and 2,4-dimethoxybenzoic acid exhibit strong absorbance bands near 240 nm, whereas the 2-β-dimethoxybenzoic acid peaks resemble the *ortho* and *meta* isomers in Figure 5. Total absorbance, A_T of the para substituted compounds is consistently larger, 36-41 vs 25-33. All seven compounds are UV-transparent above 300 nm.

The comparison of dimethoxybenzoic acids in Figure 6 shows that the symmetric attachment of methoxy groups at the number 2 and 6 positions eliminates one secondary absorption band completely and gives a weak band at 271 nm. It appears that the presence of the methoxy groups at the 2 and 6 positions reduces the absorbance associated with the carboxyl group, whereas shifting one methoxy group to the para position leads to a bathochromic shift

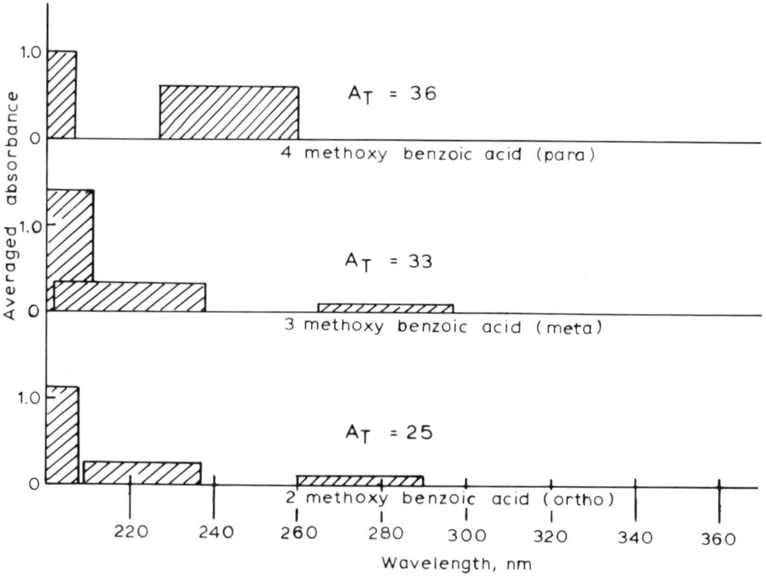

Figure 5. Comparison of methoxy-substituted organics.

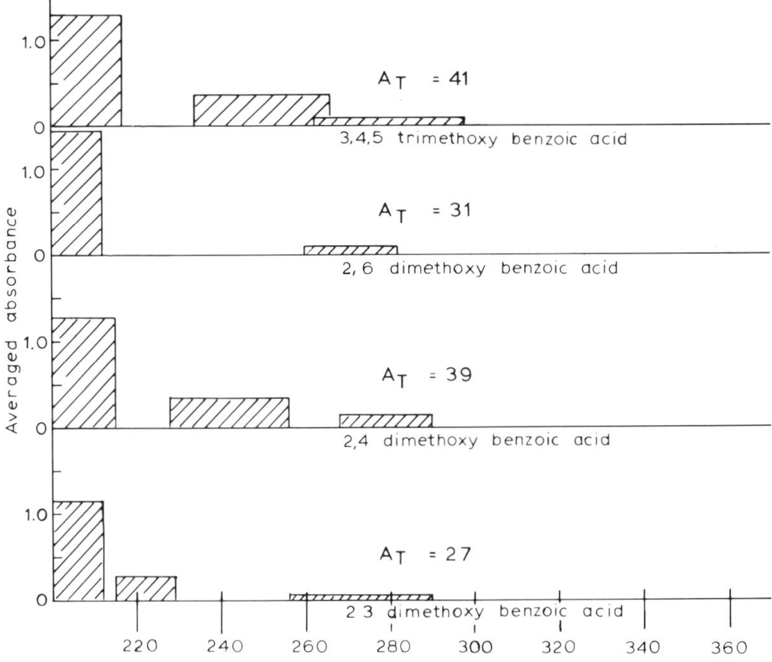

Figure 6. Comparison of multimethoxy substitutions.

of the secondary peak from 220 to 242 nm. The same bathochromic shift was noted for 4 methoxybenzoic acid (Figure 5). The absorbance waveband near 200 nm is remarkably similar in all the mono- and dimethoxybenzoic acids; it appears that this waveband is associated with the benzene ring structure per se and is not as susceptible to substituent group effects.

The spectra of catechol (1,2-dihydroxybenzene) and resorcinol (1,3-dihydroxybenzene) are essentially alike. As shown in Figure 7, the three absorption bands of catechol at 200, 215 and 270 nm are closely related to the corresponding 2 and 3-methoxy substituted benzoic acids except that the second peak at 215 nm is more intense and displaced towards shorter wavelengths.

Comparisons of compounds with different oxygen-containing substituent groups are illustrated in Figure 7. The presence of two carboxyl groups as opposed to hydroxyl groups broadens the absorbance bands but reduces intensity as shown by 1,2-benzenedicarboxylic acid vs 1,2-dihydroxybenzene; substitution by -OCH$_3$ resembles the dihydroxy compound. However, salicylic acid, which has adjacent -OH and -COOH groups, shows a markedly higher

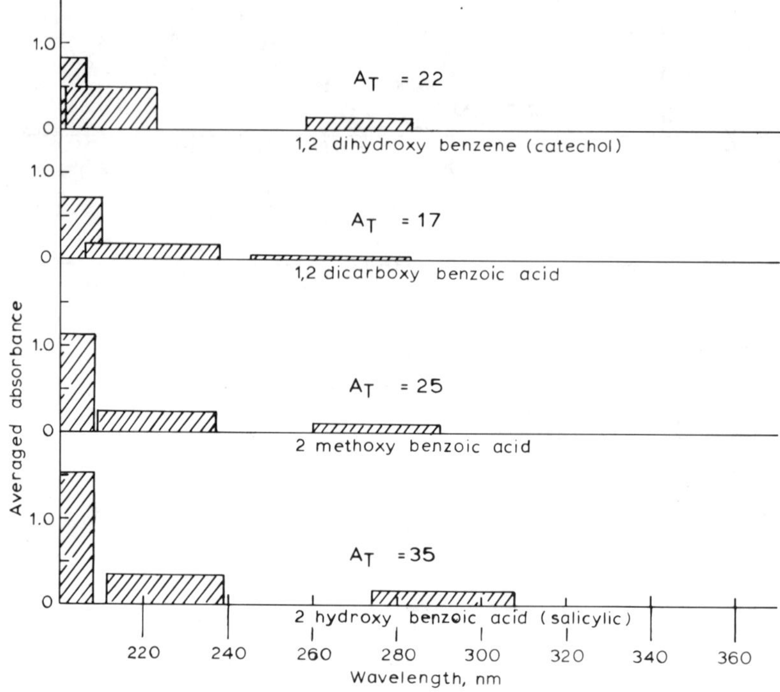

Figure 7. Comparison of different substituent groups.

intensity at 200 nm and a shift of absorbance to the 275- to 308-nm waveband. The A_T values of salicylic acid are also substantially larger than the corresponding dicarboxylic acid.

Comparison of 2-hydroxy-, 2-methoxy- and 2-carboxybenzoic acids (Figure 7) shows that hydroxyl group attachment results in more intense absorbance and a shift to longer wavelengths. The 200-nm waveband intensity is twofold larger for OH over COOH and methoxy is in between; the 220 waveband is shifted to longer wavelengths by OH substitution; absorbance at longer wavelengths shows a marked shift from 264 to 276 to 292 for -COOH, $-OCH_3$ and -OH, respectively, and total absorbance increases in the same order. The absorbance band shifts due to -OH substitution is also evident from comparison of the benzoic acids with 3,4,5-trimethoxy (Figure 6) and 3,4,5-trihydroxy (gallic acid in Figure 9) groups, respectively. Absorbance peaks of the hydroxylated compound (gallic acid) are shifted to longer wavebands and a low intensity band extends into the 320-nm region; A_T values are also higher.

The effects of location and multisubstitution of carboxylic acids on the absorbance characteristics of benzene rings are illustrated in Figure 8. Absorbance bands near 200 nm are common to all compounds and increase with the number of carboxyl substituents except that the presence of *para* substituents also induces band broadening and a shift to longer wavelengths; the higher intensity of 1,3 vs 1,2-dicarboxylic acid is consistent with the isomer effects observed with methoxy substituents (Figure 5). The A_T values follow the same trends, pentacarboxylic acid is highest but the symmetrical hexacarboxylic acid is an exception with low intensity bands at 200 nm and low overall absorbance. Comparison of secondary absorbance bands shows a consistent shift towards longer wavebands with increasing numbers of substituent groups per ring; the tetra- and penta-substituted compounds have the highest absorbance in the 260- to 300-nm region. However, all compounds are transparent above 300 nm.

Ketone and aldehyde chromophores are widely distributed in naturally occurring organic materials and have strong absorption bands [1,2]. Acetophenones and aldehydes differ from their carboxylic acid analogs by replacement of -OH by -CH_3 and -H.

As shown in Figure 9, all four compounds with ketone or aldehyde groups (acetophenones, tannic acid and vanillin) have characteristic absorption bands in the 260- to 300-nm range. Comparison of the 2,4-dimethoxyacetophenone with its 2,4-benzoic acid analog shows a secondary absorbance band centered around 225 nm, whereas the 230- to 255-nm absorbance band of substituted benzoic acids is absent; the acetophenone compound also has medium-intensity dual absorbance bands ranging from 255 to 320 nm, whereas benzoic acid has a low intensity band near 280 nm. Comparison of the absorbance characteristics of 3,4,5-trimethoxyacetophenone (Figure 9) with its benzoic acid analog (Figure 6) shows the same differences.

The absorbance of vanillin is the most unusual of all the single-ring structures tested (Figure 9). The medium intensity absorption band centered around 280 nm is typical of acetophenone groups; however, appearance of relatively high intensity absorbance bands from 300 to 355 nm is unique. By contrast, the spectrum of vanillic acid (Figure 9) is typical of substituted benzoic acid spectra, although its total absorbance is higher (A_T = 61 vs about 40 for the dimethoxy compounds). Vanillin has the highest total absorbance (A_T = 84).

Tannic acid and lignin are examples of naturally occurring materials consisting of polymeric combinations of substituted ring structures. Analysis of their UV spectra gives insight on the influence of molecular size and configuration on the absorbance characteristics of subgroups. The structure of tannic acid is usually described in terms of corilagin, which consists of three gallic acid groups esterified to a six carbon sugar [16].

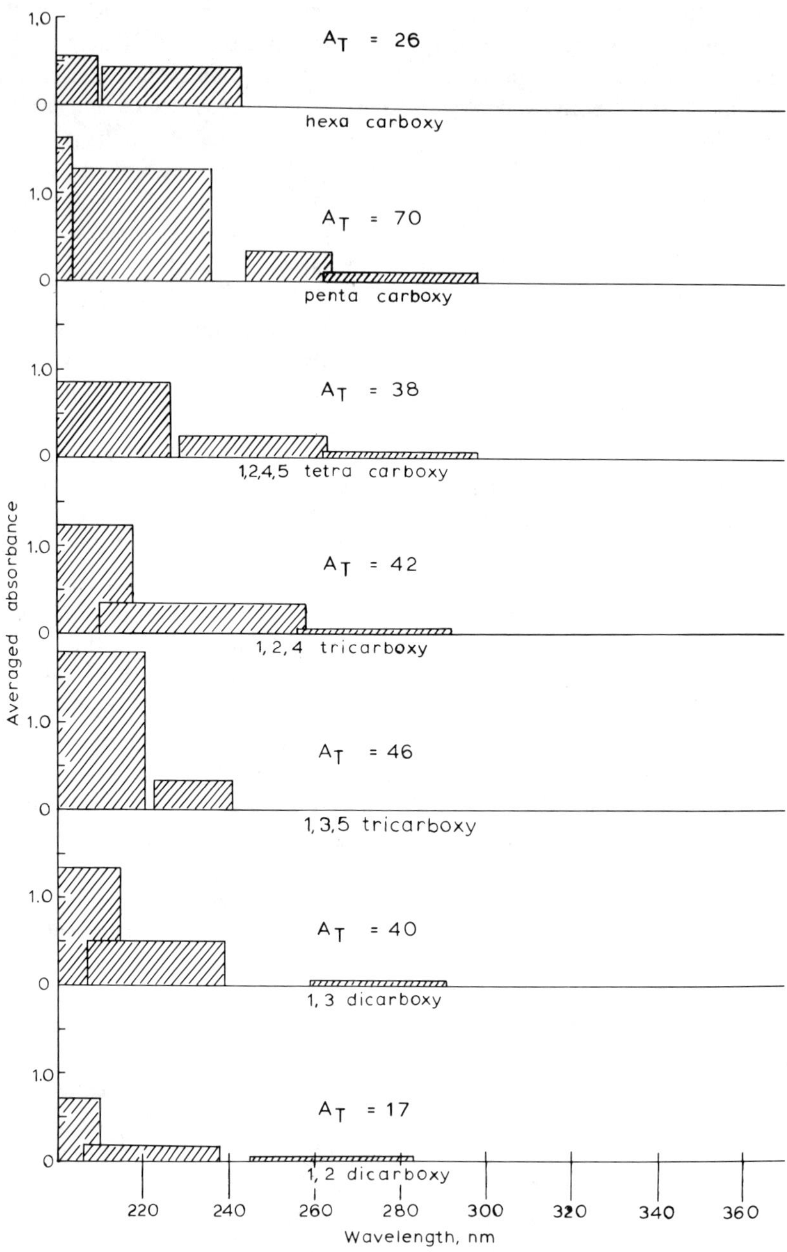

Figure 8. Comparison of multicarboxy substitutions of benzene.

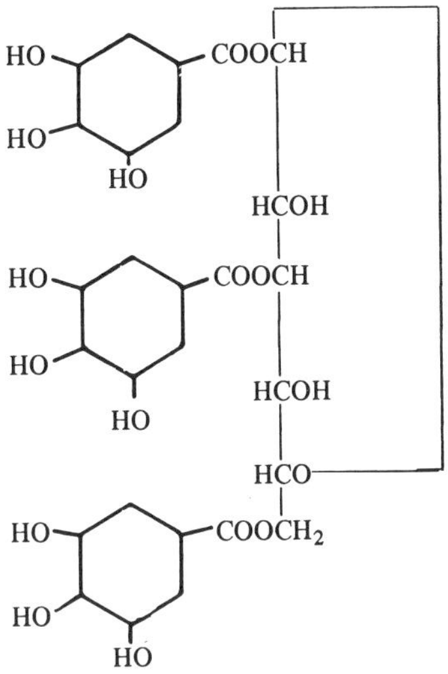

HO — COOCH
HO
HO

HCOH

HO — COOCH
HO
HO

HCOH

HCO

HO — COOCH$_2$
HO
HO

CORILAGIN

The absorbance spectrum of gallic acid shown in Figure 9 is similar to its 3,4,5-trimethoxybenzoic acid analog (Figure 6). However, the abosrbance characteristics of tannic acid are much more closely related to the aceto-phenones. The ring structures of tannic acid have the same 3,4,5-trihydroxy substituents but differ from gallic in that the carboxylic acid group is esteri-fied. The tannic acid structure shifts the ring related absorbance maximum from 200 to 209 nm and the secondary band from 250 to 274 nm. This shift is ascribed to the extended structure of tannic acid, which comprises three gallic acid rings esterified to a six-carbon sugar. The esterified carboxyl groups appear to have the same absorbance characteristics as the keto group of acetophenone. The absorbance bands are somewhat broader but other-wise similar to 3,4,5-methoxyacetophenone and are ascribed to the inter-action of the C=O group (conjugation) with the unsaturated ring structure combined with the steric effects of the extended molecular structure.

The lignin absorption spectrum as shown in Figure 9 is less featured and has low intensity bands extending into the 300- to 350-nm waveband. This is qualitatively consistent with the proposed structures (Figure 2) which include a variety of methoxy-substituted rings connected by ether linkages and keto acids that are analogs of acetophenone structure. The diffuse nature of the major bands is probably indicative of steric effects associated with the ex-tended structure of lignin molecules. Quantitative information on these effects is lacking at this time. It should also be noted that lignin samples from different sources have different absorbance spectra.

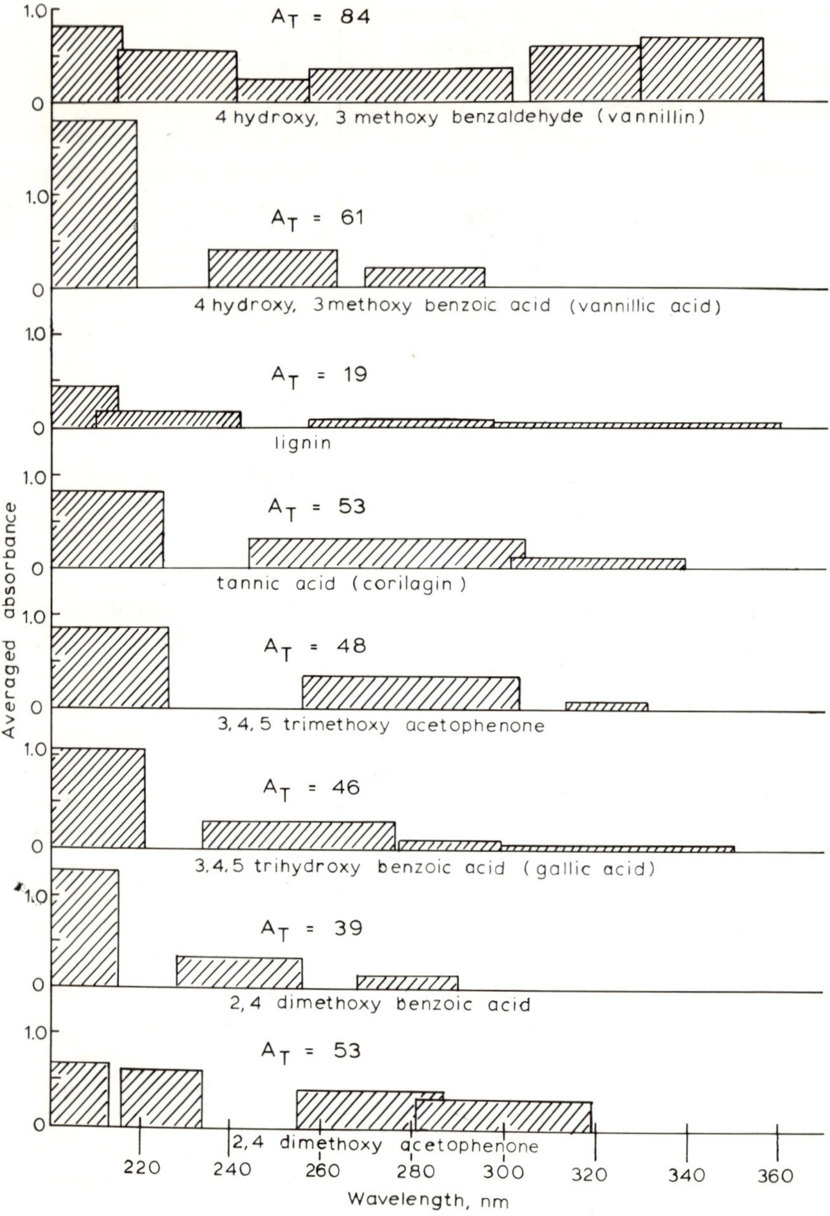

Figure 9. Ketone and aldehyde groups shift absorbance to longer wavelengths.

Summary of Structure-Absorbance Relationships of Model Compounds

The absorbance bands of a variety of substituted aromatic ring compounds, related to or identified as major constituents of natural decay products in soils and water, have been analyzed to give insight on the effects of molecular structure. The ultimate objective is to match absorbance spectra of aquatic organics by identifying chromophores that resemble or piecewise match riverwater spectra.

1. Characteristic absorbance bands at 200 nm were observed in all ring compounds and are ascribed to the benzene ring structure.
2. Ring structures have characteristic absorbance maxima in terms of wavelength and intensity; *para* substitution gives more intense and featured spectra than *ortho* or *meta* substitution; multisubstitution leads to band broadening, shifting to longer wavelength and higher overall absorbance; hydroxyl groups result in more intense absorbance and a shift to longer wavebands compared to carboxyl and methoxy groups.
3. Aldehyde and keto groups directly attached to the ring give unique strong absorbance bands at long wave lengths; absorbance in the 300- to 360-nm waveband is typical, whereas this spectral region is completely transparent with other substituents. The long wavelength absorption bands are probably due to interaction (conjugation) of the double bonds of the C=O group with the ring double bond system. Typically this band centers around 320 nm, but is frequently masked or shifted by further substitution of the molecule.
4. Analysis of tannic acid and lignin indicates that steric effects due to aggregation of ring structures into large molecules result in band broadening; however, quantitative information is lacking. The similarity of tannic acid absorbance with acetophenone shows that ester linkages involved in formation of extended molecules correspond to the absorbance behavior of keto groups.
5. Comparison of 35 model compounds shows that their absorbance spectra are in all cases uniquely different; the differences are small but measurable. Chromophores with large UV transparent attachments were not tested and it is conceivable that such compounds could have identical absorbance spectra although their extinction coefficients would be different. This measurable dissimilarity of the 200- to 360-nm wavelength-absorbance characteristics of organics is, therefore, a useful measurement for characterization.

As regards analysis of mixtures of components in aquatic organics, the uniqueness of the UV spectra of pure compounds implies that the concentrations of individual chromophores could be calculated by matching the absorbance spectrum of the mixture in terms of the pure compound spectra. This matching calculation assumes that Beer's law applies, that the absorbances of individual chromophores are additive, and that chemical interactions between compounds do not change absorbances. One problem with this approach is that all specific compounds in aquatic organics have not been identified. Only the breakdown products after hydrolysis, oxidation or reduction have been partly identified for fulvic acid and soil organics. However as indicated above, the similarity in spectral shifts induced by different substituent groups shows that chromophores can be categorized into classes of compounds with equivalent absorbance patterns. It is therefore proposed to use absorbance matching to calculate the relative concentration of chromophore groups using the available pure compounds as characteristic chromophoric units. Theoretically it should be possible to obtain precise estimates of the concentration of each chromophore in a mixture, provided that the spectra of all constituents are known and the abovementioned assumptions are valid. By using the absorbance spectra of a limited number of characteristic chromophoric groups to match aquatic organic mixtures the calculated composition reflects a certain degree of approximation. The calculated composition therefore should be regarded as "quantitative characterization" which can be used to compare aquatic organics from different sources, give insight on the concentration of major chromophores, and in some cases identify the presence of specific compounds. The latter is obtainable by using the equivalent of a standard addition technique to determine the response of the absorbance matching calculations to a specific chromophore. This technique has been evaluated and will be discussed in a subsequent publication.

METHODOLOGY FOR ANALYZING MULTIWAVELENGTH DATA

Mathematical Model

Analysis of the absorbance spectrum of mixtures to elucidate their composition is based on the assumptions that absorbance follows Beer's law, that there is no interference between chromophores and absorbance is additive; the latter implies that the absorbance of mixtures is equal to the sum of absorbance contributions of each chromophore. The procedure is conceptually simple; absorbance spectra of pure compounds are measured independently and used to simulate the measured absorbance spectrum of the mixture by summation of the constituent spectra at the appropriate concentration.

Matrix algebra is ideally applicable to the treatment of spectrophotometric data. The generalized form of Beer's law can be represented by the matrix equation [1]:

$$AX = B \tag{1}$$

where   A = the m by n absorbance matrix of n pure compounds
                 B = the m by p absorbance matrix of the mixture
                 X = the n by p matrix of contributing pure compounds in the mixtures

The rank of the matrix A is important as a measure of linear dependency between its columns. In principle the rank of the matrix A can be determined by computing the corresponding second moment matrix and determining its eigenvalues and eigenvectors.

For a specific mixture of j compounds Equation 1 is:

$$\sum_j A(i,j) \cdot X(j) = B(i) \tag{2}$$

The summation is made at any arbitrary number of wavelengths (i = m). The elemental composition data are analyzed in a similar manner because mass of carbon, hydrogen, oxygen and nitrogen are additive. Material balances on elemental carbon, hydrogen, oxygen and nitrogen are calculated as shown below and are added as four additional rows to the matrix A making it m + 4 by n matrix and an m + 4 column vector in Equation 2.

$$\sum_j TOC(j) \cdot X(j) = B_{toc}$$

$$\sum_j TOH(j) \cdot X(j) = B_{toh}$$

$$\sum_j TOO(j) \cdot X(j) = B_{too} \tag{3}$$

$$\sum_j TON(j) \cdot X(j) = B_{ton}$$

TOC, TOH, TOO and TON represent the concentration of organic carbon, hydrogen, oxygen and nitrogen per unit of concentration for each of the n constituents and the mixture B. The objective is to compute values of the column vector X so as to minimize the summation

$$\sum [B(i) - \sum A(i,j) \cdot X(j)]^2 + [B_{toc} - \sum TOC(j) \cdot X(j)]^2 + \ldots \tag{4}$$

subject to the constraint $X \geqslant 0$. The constraint $X \geqslant 0$ precludes negative values that would imply emission of light energy or removal of elements. Analytical methods were studied to find algorithms that converge on a reasonable solution in a finite number of iterations. Two algorithms are used, the Marquardt algorithm for solving nonlinear least squares problems, which is available as a subroutine in IMSL library and, the nonnegative least squares algorithm described by Lawson [17]. The algorithms of the computer program UVSA is attached as Appendix A.

Model Verification

The effectiveness of the UVSA program for calculating the concentration of constituents that best simulate the absorbance of a mixture was tested using simple linear absorbance-wave number correlations illustrated in Figure 10. Simple absorbance-wave number relationships were used to facilitate interpretation; absorbance of models A and B increase with wave number and are linearly related; C and D are linearly related but are inverses of A and B. Similarly, E, F and G are linearly related to one another. H and J are step functions designed to evalueate the algorithm's effectiveness for handling constituents with narrow absorption bands. The absorbance-wave number data of A through K were provided as input to the UVSA program and represent the inventory of possible constituents. Absorbance spectra of arbitrary mixtures of A through J were calculated and used to verify the effectiveness of the program for calculating constituent concentrations and their respective contribution to the combined spectrum.

The results in Table II show that the program calculates an exact match of the absorbance values as evidenced by the low values of the Euclidian norm (EN). EN is a measure of the deviations of the calculated absorbance spectrum compared to the input spectrum. However, the concentrations were not matched in mixtures that contain constituents whose absorbance spectra are

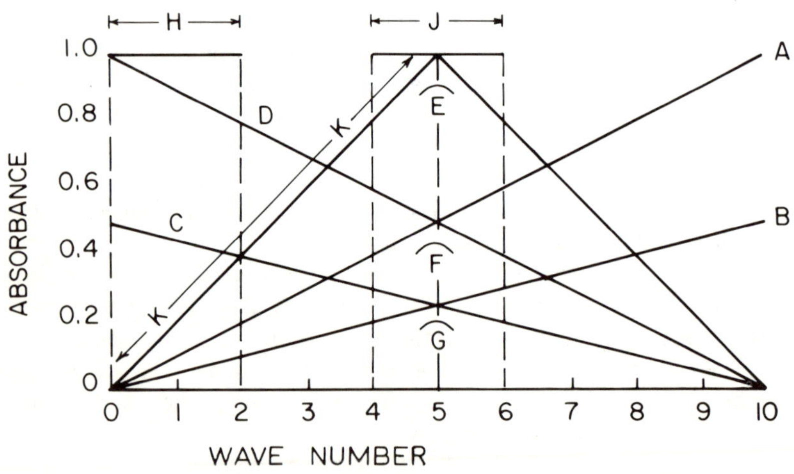

Figure 10. Linear absorbance models.

Table II. UVSA Program Calculations of Linear Model Mixtures

Mixture	Input Mixture Compositions[b]	Constituent Concentrations[a]										
		A	B	C	D	E	F	G	H	J	K	EN
M1	$A + B \Rightarrow 1.5A$	1.5	0	0	0	0	0	0	0.134×10^{-14}	0	0	10^{-14}
M2	$1/6A + 1/6B \Rightarrow 0.25A$	0.25	0	0.119×10^{-15}	0	0	0	0	0.995×10^{-16}	0	0	10^{-15}
M3	$1/2A + 1/2B \Rightarrow 0.75A$	0.75	0	0	0	0	0	0	0.695×10^{-15}	0	0	10^{-14}
M4	$1/4(A + B + C + D) \Rightarrow 0.375A + 0.375C$	0.375	0	0.375	0	0	0	0	0	0.289×10^{-16}	0	10^{-14}
M5	$A + C + D + H \Rightarrow A + 1.5C + H$	1.0	0	1.5	0	0	0	0	1.0	0	0	10^{-13}
M6	$C + G + H \Rightarrow C + 0.25E + H$	0.36×10^{-15}	0	1.0	0	0.25	0	0	1.0	0.721×10^{-15}	0.413×10^{-14}	10^{-14}
M7	$D + E + K \Rightarrow 0.5C + E + K$	0.334×10^{-14}	0	0.5	0	1.0	0	0	0.548×10^{-14}	0.67×10^{-14}	1.0	10^{-14}
M8	$0.4A + 0.5C + 0.3F + 0.55H + 0.66J + 0.45K \Rightarrow$ $0.4A + 0.5C + 0.15E + 0.55H + 0.66J + 0.43K$	0.4	0	0.5	0	0.15	0	0	0.55	0.66	0.43	10^{-14}
M9	$0.5A + 0.5C$	0.5	0	0.5	0	0	0	0	0	0	0.943×10^{-16}	10^{-13}
M10	$0.55H + 0.66J + 0.43K$	0	0	0	0	0	0	0	0.55	0.66	0.43	10^{-13}

[a] Numbers represent unit concentrations.
[b] Absence of a coefficient in front of letter means unit concentration.

linearly related; the linearly related model constituents are $A = 2B$, $C = 2D$, $E = 2F = 4G$. Mixtures M1 through M8 all contain linearly related constituents and only constituents with the highest absorbance value are identified. By contrast, mixtures M9 and M10 which do not contain linearly related constituents are correctly identified. The reason for not identifying low absorbance constituents of linearly related compounds is inherent in the mathematical logic.

Analysis of the simulation tests leads to the following conclusions:

1. The algorithm calculates an exact match of the whole absorbance spectrum for multicomponent mixtures. Values of the EN are 10^{-13} or less.

2. Concentrations of constituents are calculated exactly for mixtures of constituents whose absorbance spectra are not linearly related.

3. Mixtures containing constituents whose absorbance-wave number spectra are linearly related over the whole range are lumped together and only the constituent with the highest absorbance is identified.

4. The shape of the absorbance-wave number correlation of the mixture has no effect on the calculations; mixtures with several maxima, minima or completely featureless and uniform spectra are matched exactly.

5. If the input values of the absorbance of the mixture are in error, the algorithm calculates a best fit. This may result in identification of spurious constituents. The subroutine is sensitive to small errors. EN values of 10^{-2} are indicative of small mismatches that may result from errors in the third significant figure of one of the absorbance values; the calculated error in concentration is less than 0.3%. EN values in the 10^{-1}-10^{-2} range result from an error in the second significant figure of one of the absorbance values (1-10%) and give errors up to 6% in calculated concentrations. Incorrect values of absorbance values of the order of 50% may result in substantial mismatching and concommittantly large errors in calculated concentrations and identification of spurious compounds. EN values may, therefore, be used as one measure of the reliability of the fit.

6. If input errors lead to mismatching in a specific waveband, spurious constituents with the highest absorbance are brought into the solution.

7. If mismatching is evident, a series of tests with progressive elimination of inventory constituents may be useful for characterizing the absorbance "errors." The absorbance "error" may reflect the fact that a critical constituent is not available in the data bank inventory or that the absorbance input data are erroneous.

MISSISSIPPI RIVER WATER ANALYSIS

Four critical sampling locations were identified in the upper Mississippi River for detailed and intensive characterization of aquatic organics and metal complexes, namely, the unpolluted headwaters at Bemidji (BEM), the agriculturally developed midregion near Royalton (ROY), the population-stressed region near Minneapolis-St. Paul (SAF) and the Inver Grove Bridge (IGB) site which reflects the full effects of the industrialized regions of Minnesota. Samples were taken in spring, fall and winter to determine seasonal variations.

Sample Preparation and Analytical Methods

Water taken from the Mississippi River (MRW) was filtered through a pleated depth filter to remove large solids followed by 0.22-μm membrane filtration to remove suspended materials. Ultrafiltration membranes (Table III) were used in a Millipore 90-mm High Flux Ultrafiltration Cell to size-fractionate the organics. Operating pressures of 25 psig were used with the 100K and 10K membranes and 40 psig for the 0.5K and 1K membranes. A 300-ml sample of 100 K$^+$ concentrate was prepared by filtering 2 liters of MRW to 300 ml retentate and washing with 600 ml of pH-adjusted megapure water to reduce the concentration of filtrate organics in the retentate. The corresponding 300 ml of 10/100K organic concentrate was obtained by filtering the 2.3 liters of filtrate from 100K filtration through a 10K membrane followed by a 600-ml wash with pH-adjusted megapure water. The corresponding 1/10K and 0.5/1K cuts were prepared in a similar manner, starting with 2-liter aliquots of MRW to avoid excessive dilution. The less than 0.5K fraction was collected as filtrate from the 0.5/1K concentrate and was therefore diluted by a factor of 2.6/2.0. Washing of the retentate with two 300-ml or three 200-ml volumes of organic-free water should reduce the quantity of filtrable organics retained in the concentrates to a nominal 22-25% of its original concentration but is not sufficient to obtain complete separations.

Membrane filtration is operationally defined as a molecular size separation process with nominal molecular weight cut-off shown in Table III. In practice, separations are influenced by surface charge characteristics as well as molecular size [18-20]. The effect of charge is particularly noticeable with the low-molecular-weight cutoff membranes (0.5 and 1K) when filtering ionizable organic molecules; pH control is required to ensure reproducible results.

All samples were analyzed for dissolved organic carbon (DOC), organic nitrogen and UV absorbance as described in the section on physical-chemical analysis of model compounds. Accuracy of analytical measurements and reproducibility of molecular size fractionations were evaluated using pure compounds and MRW. Two or three replicate DOC measurements were made on all samples and average values are ±0.1 mg/l; comparisons of data from different time periods are less reproducible. Repeat UV absorbance measurements showed a maximum deviation of 5.8% at one specific wavelength, but the average deviation of the absorbance values over the whole spectral range (200-360 nm in 10-nm increments) is 2.2%. Reproducibility could be im-

Table III. Characteristics of Ultrafiltration Membranes

Membrane	Nominal Molecular Weight Cutoff	pH Range	Membrane Composition	Manufacturer
UMO5	500 (0.5K)	<12	Polyelectrolyte	Amicon
PSAC	1,000 (1K)	2-10	Polyelectrolyte	Millipore
DM5	5,000 (5K)			Amicon
PTGC	10,000 (10K)	1-14	Aromatic polymer	Millipore
PSED	25,000 (25K)	2-10	Polyelectrolyte	Millipore
PTHK	100,000 (100K)	1-14	Aromatic polymer	Millipore

proved by digitized readout in place of strip chart recording and reading of charts. Reproducibility of size separation of organics by membrane filtration depends on membrane characteristics, pretretment procedures and prior membrane use. All membranes were pretested with pure water to determine permeabilities (a number of membrane batches were rejected or returned to the manufacturer). Pretest procedures and membrane preparation have been published [18,21]. Repeat filtrations using pure compounds gave essentially identical separations, retentate concentrations differed by less than 1% when the separations were carried out at the same pH and operating procedures [18].

Interpretation of Absorbance Spectra

UV measurements of MRW as illustrated in Figure 11 typically show a featureless gradually increasing absorbance with decreasing wavelength. Figure 11 shows the corresponding absorbance values (unadjusted) of molecular size fractionated samples from SAF.

The less than 0.5K sample is a 2.6 to 2.0 dilution relative to Mississippi River water (MRW) because of the presence of washwater from the filtration; all the other fractions represent 6.7-fold concentrates prepared by filtering 2 liters of MRW to 300 ml and washing as described in the previous section. DOC concentrations are shown in parentheses. The less than 0.5K fraction exhibits relatively high absorbance at 200-230 nm despite the low DOC concentration (0.8 mg/l); absorbance trails off to very low values at 300 nm. This resembles the classical spectral characteristics of minimally substituted ring structures which show primary absorbance bands near 200 nm and weak absorbance bands near 250 nm. By contrast, the less than 0.5K fractions of spring samples taken at BEM and ROY showed very low absorbance at 200-220 nm which indicates that low-molecular-weight, minimally substituted ring structures were essentially absent. Because the less than 0.5K fraction contains almost all the inorganic ions originally present in the river water, their contribution to UV absorbance must also be considered. Inorganic ion contributions in MRW have been studied and are generally quite small and

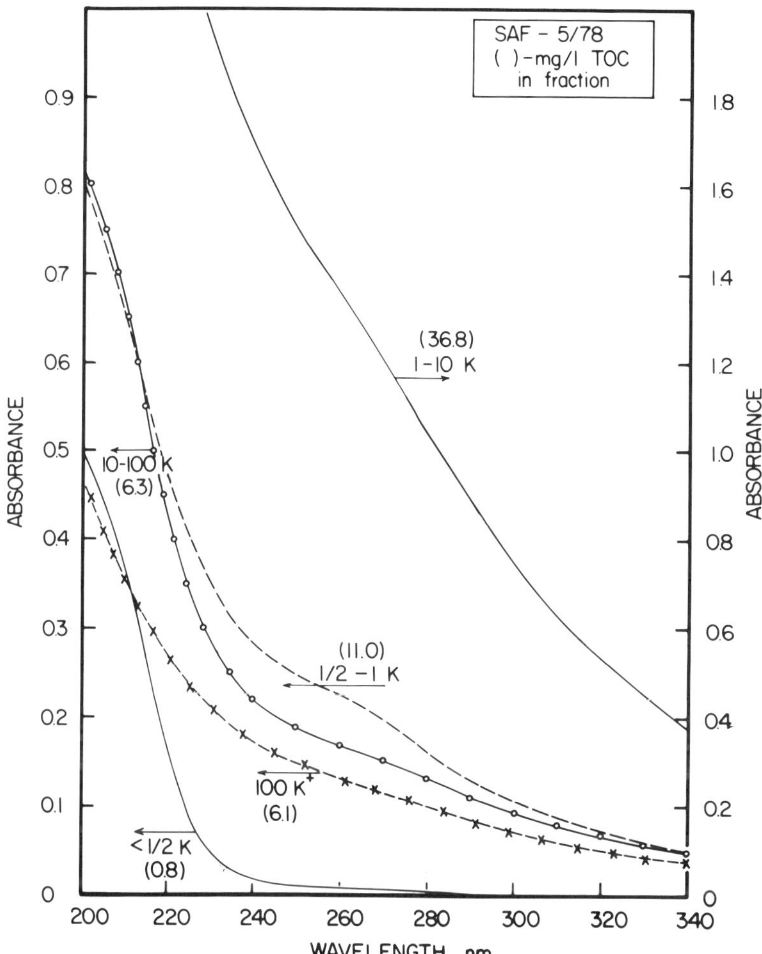

Figure 11. Absorbance of Mississippi River fractions.

can be neglected in the larger molecular size fractions. The effect of inorganic ions in the less than 0.5K fraction will be discussed in a subsequent paper.

The 0.5-1K fraction shows a distinct absorbance band near 260 nm. The 1-10K fraction has the highest absorbance but also the highest DOC concentration; absorbance spectra of the other fractions are lower but have the same general shape, relatively featureless. All the larger molecular size fractions exhibit relatively strong absorbance at longer wavelengths (250-340 nm) contrasted with that of the less than 0.5K sample. This bathochromic shift may be the result of increasing substitution of the larger molecular weight constituents.

Extinction Values of Aquatic Organics

Extinction values rather than molar absorptivities have been calculated to give insight on the relative intensity of absorbance. Extinction values (ϵV) are calculated by dividing measured absorbance by DOC concentrations in mg/l because the molecular weights are not known; in effect, this tends to normalize the absorbance data to a common organic carbon concentration (absorbance/mg/l DOC). Data for the less than 0.5K fractions are shown in Figures 12 to 14 for the spring, fall and winter samplings, respectively. Unusually high ϵV values occur in spring (Figure 12) at the downstream, pollution-stressed, urban sampling points (SAF and IGB). By contrast, winter (Figure 14) samples which contain minimal urban runoff contribution, have relatively low ϵV values. Fall samples (Figure 13) are intermediate. The low ϵV values of the less than 0.5K fraction, particularly at 200-220 nm, at both BEM and ROY in spring (Figure 12) indicate the absence of minimally substituted benzene ring structures even though DOC levels are relatively high (2.0 and 1.0 mg/l, respectively). Absorbance contributions from inorganic ions have not been accounted for but are unlikely to be major factors.

Comparison of the less than 0.5K fall samples from all four locations shows that there is relatively little difference in either DOC or ϵV (Figure 13). One possible exploration is that fall runoff is primarily from surface drainage, and surface organics have been exposed to similar environmental conditions and therefore the same degree of biochemical transformations throughout the whole watershed. As shown in Figure 13 the ϵV above 280 nm are very low at all four locations. This is consistent with the explanation that aquatic organics have been exposed to similar biodegradation environments in which these chromophores have been removed. By contrast, the winter samples (Figure 14) at BEM and ROY have high ϵV values throughout the whole spectral range. The source of this material is ascribed to hydrolysis of humified sediments and biomass in ice covered impoundments. The presence of non-UV-absorbing material in the downstream reaches is postulated because DOC is relatively high but absorbance is low; this material probably represents primary decomposition products of biomass not humified due to low temperatures.

The ϵV and DOC data for the 0.5-1K fractions are summarized in Figures 15-17. ϵV values are highest at short wavelengths and decrease at longer wavelengths. The similarity in ϵV values despite the rather substantial differences in TOC of the spring samples suggest that the organics are of similar composition. Comparison of the four sampling sites in fall and winter show that upstream samples have slightly higher ϵV values in winter and the reverse in fall.

The ϵV and DOC values for the 1-10K size fraction are shown in Figures 18-20; the ϵV spectra over the 200- to 340-nm range are remarkably similar for the spring and fall samples despite the fact that DOC concentrations differ

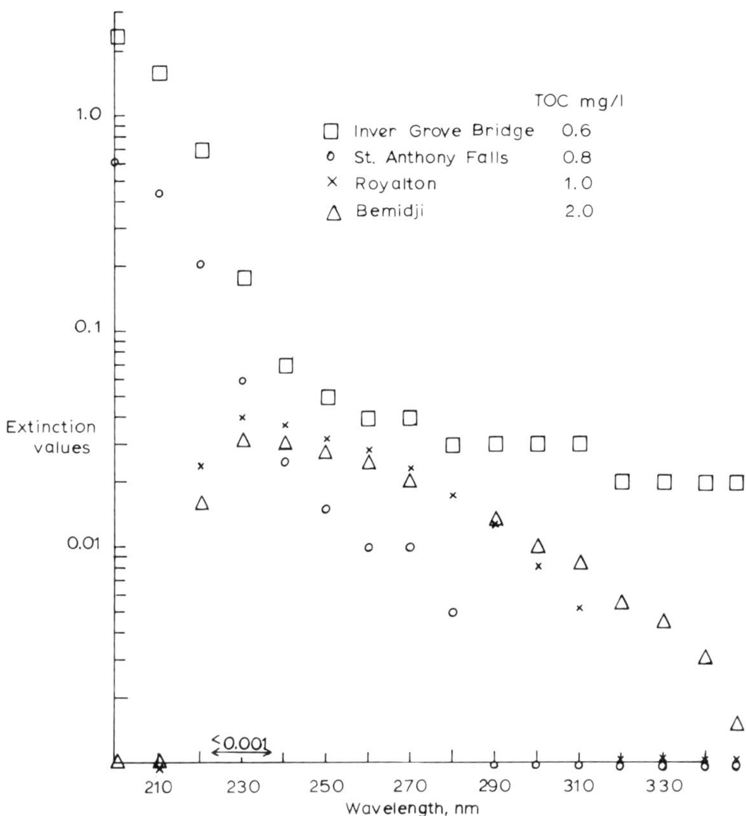

Figure 12. Less than 0.5 fractions, spring MRW.

by a factor of five. The winter samples show the same trend but have a lower extinction value at all wavelengths. DOC concentrations in winter and spring are lower than in fall. The lower DOC values in spring coincide with high volumetric flows and can be ascribed to a dilution effect. However, the ϵV values are not affected, suggesting that the aquatic organics in the 1-10K fractions during spring and fall have the same characteristics. The lower ϵV values of winter samples are probably the result of two interrelated effects, namely the low flow conditions and reduced biodegradation rates during cold weather. The low flow conditions represent base flow from under-ice drainage and ground water inflows. The former probably contains relatively high concentrations of incompletely biodegraded organics (carbohydrates and proteins) that are weak absorbers. The inflow of ground water dilutes overall DOC

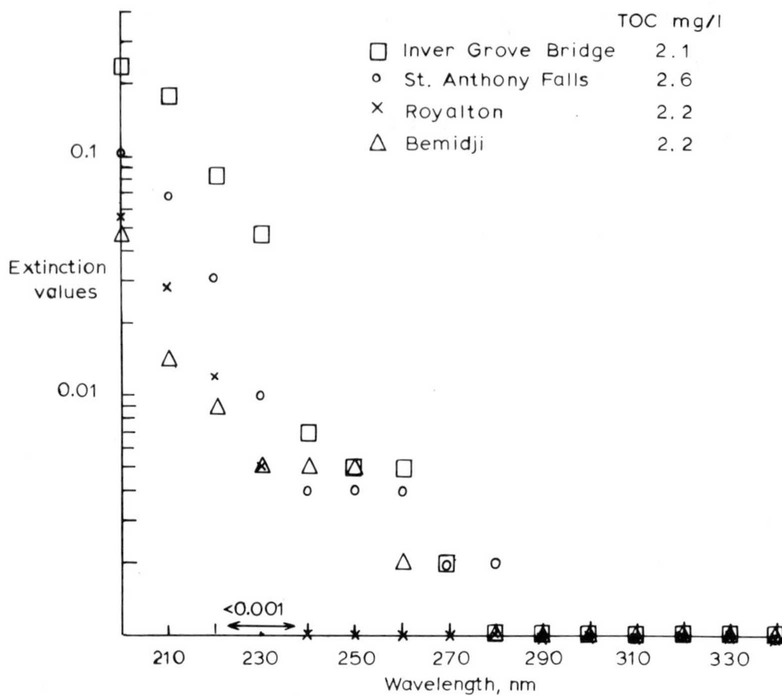

Figure 13. Less than 0.5 fractions, fall MRW.

concentrations because ground waters contain very little DOC [22]. As a result, the DOC concentrations are similar to spring samples, but ϵV is lower due to lower absorbance.

Analysis of the 10-100K fractions shows that the absorbance character-istics of larger molecules differ both seasonally and geographically. As shown in Figures 21-23, the trend of increasing ϵV values with decreasing wavelength is predominant except in spring samples which exhibit low values at 200-210 nm, most notably at BEM. There is some similarity between ϵV features, compared to the ϵV spectra of polymeric molecules such as lignin and tannic acid which show distinctive absorbance featuring at 210, 240 and 280 nm (Figure 18); the 1-10K organics have completely linear $ln(\epsilon V)$ vs wavelength spectra.

The linear characteristics of the $ln(\epsilon V)$ spectra are of interest for two reasons. The linearity and similarity of the spectra shows that all Mississippi River waters contain a characteristic aquatic organic constituent regardless of

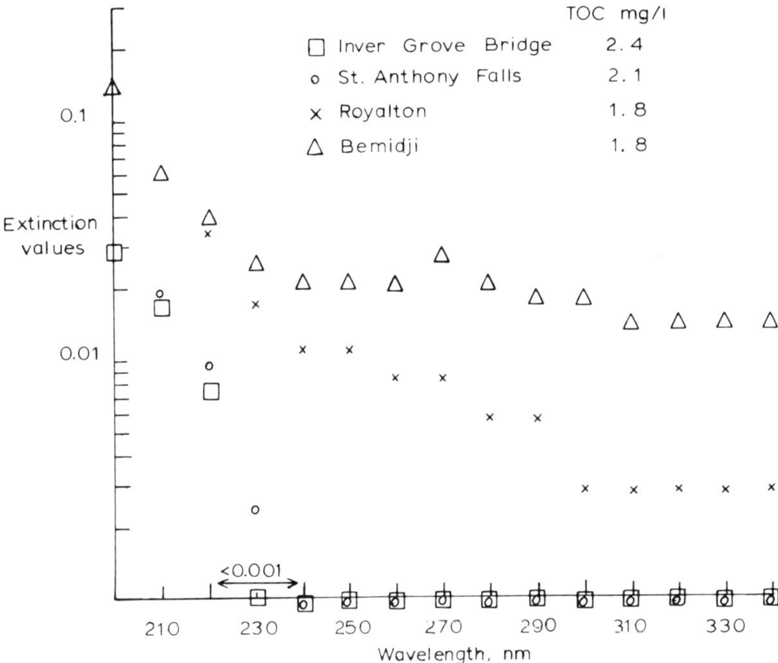

Figure 14. Less than 0.5 fractions, winter MRW.

geographical location or time of year. It should be noted that the 1-10K fraction accounts for over 50% of the TOC. The ϵV characteristics are, therefore, potentially useful as a correlating parameter or as a reference point for comparing the relative concentrations of other constituents. (Interpretation of the UV absorbance spectrum of the 1-10K fraction in terms of model compound by matching absorbance spectra will be discussed in later sections.) The linearity and magnitude of the 1-10K ϵV can also give insight on the physicochemical properties of this material. Qualitatively, the inverse relationship between $ln(\epsilon V)$ and λ is consistent with the absorbance characteristics of a variety of aromatic ring structures, e.g., high absorbance in the 200- to 220-nm wave band and lesser extinction values at longer wavebands. However, linearity is not observed in any of more than 36 model compounds tested. The ϵV values of lignin and tannic acid are shown in Figure 18 and serve to illustrate the uniqueness of the 1-10K materials. Tannic acid and lignin, as discussed above, are polymeric in nature and illustrate the effects of polymer bonding on the absorbance characteristics of chromophore subgroups. The

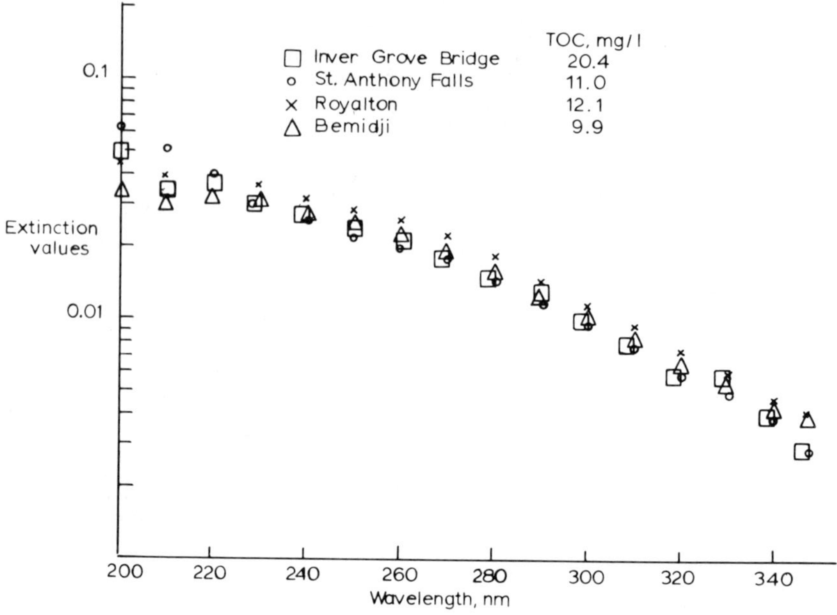

Figure 15. 0.5K-1K fractions, spring MRW.

strong bathochromic shift induced in gallic acid, which is a subunit of tannic acid (corilagin), has been described in a previous section. By analogy, it is postulated that the absorbance contributions of chromophore subunits in the 1-10K fraction is composed of a specific molecular arrangement or polymer that contains a consistent combination of chromophore subgroups; band broadening and shifting of absorbance bands occur as a result of chemical bondings. The presence of ketoaldehyde structures whose C=O groups are conjugated with the double bond system of aromatic rings are inferred because these structures have been shown to result in absorbance band broadening and relatively strong absorbance bands at long wavelengths (Figure 9). The presence of keto and aldehyde structures is also consistent with the reported elemental composition of soil and aquatic organics [3]. Weak physical bonds are ruled out because they would not be expected to cause sufficient band broadening and shifting of absorbance to long wavelengths.

The ϵV values and hence the concentration of chromophores of the 1-10K organics is slightly lower than that of tannic acid but greater than that of lignin (Figure 18). This indicates that the concentration of ring structures is high. The lower ϵV values in winter samples (Figure 20) can be interpreted as

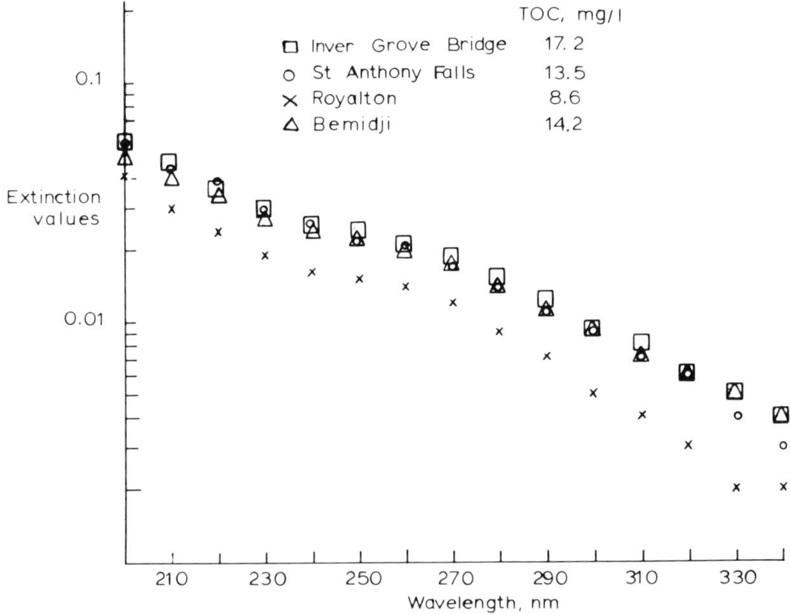

Figure 16. 0.5K-1K fractions, fall MRW.

showing that dilution of polymer by non-UV-absorbing organics occurs in winter when biological and chemical rate processes are slow.

Comparison of absorbance characteristics of the low-molecular-weight size fractions indicate these materials to be more transient; less than 0.5K fractions show distinct seasonal and geographic differences in absorbance. High extinction values at low wavelengths in pollution-stressed waters in spring in contrast to low values in winter is consistent with seasonal variations in fluxes from urban drainage. By contrast, high ϵV values in the upstream unpolluted regions in the winter samples may indicate in situ transients or intermediate decay products. Microbial decay of biomass and humified organics proceeds via hydrolysis of large molecules (polymers) into soluble fragments which are subsequently utilized by microorganisms. When environmental conditions are favorable (ambient temperatures above freezing), overall rates of biodegradation are rate limited by hydrolysis; conversely, in winter, the rate of utilization of hydrolyzed fragments may be rate-limiting and results in transient accumulations of low-molecular-weight materials. It is therefore anticipated that the less than 0.5K materials consist of different chemical species, depending on geographical location and time of year. Contributions of ab-

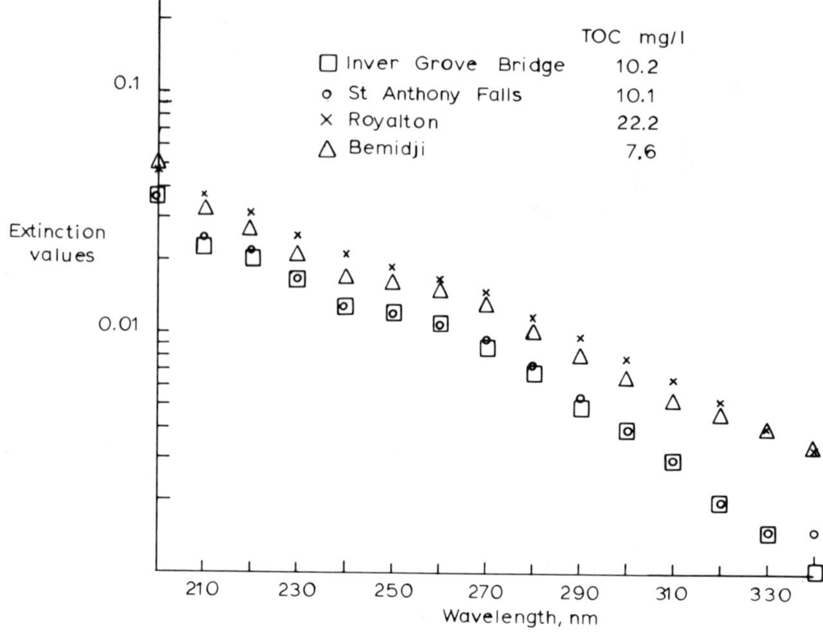

Figure 17. 0.5K-1K fractions, winter MRW.

sorbance from inorganic ions at the low wavelengths may also be a factor and will be reported on in the future.

The ϵV spectra and hence chemical characteristics of the other fractions are similar to those of the 1-10K fractions and will be discussed in more detail in a subsequent section.

APPLICATION OF UVSA PROGRAM

Absorbance Matched Model Compound Analysis of Mississippi River Water

The absorbance spectra of Mississippi River water samples were matched using the previously described algorithm (UVSA) and an inventory of the 29 model compounds shown in Figures 5-9. Absorbance data from 200 to 350 nm were used in 2-nm increments, but elemental composition data were not included. Absorbance Matched Model Compound Concentrations (AMMCC) are summarized in Table IV for fall, winter and spring at each of the four sampling sites. Vanillin was identified in each sample at concentrations ranging

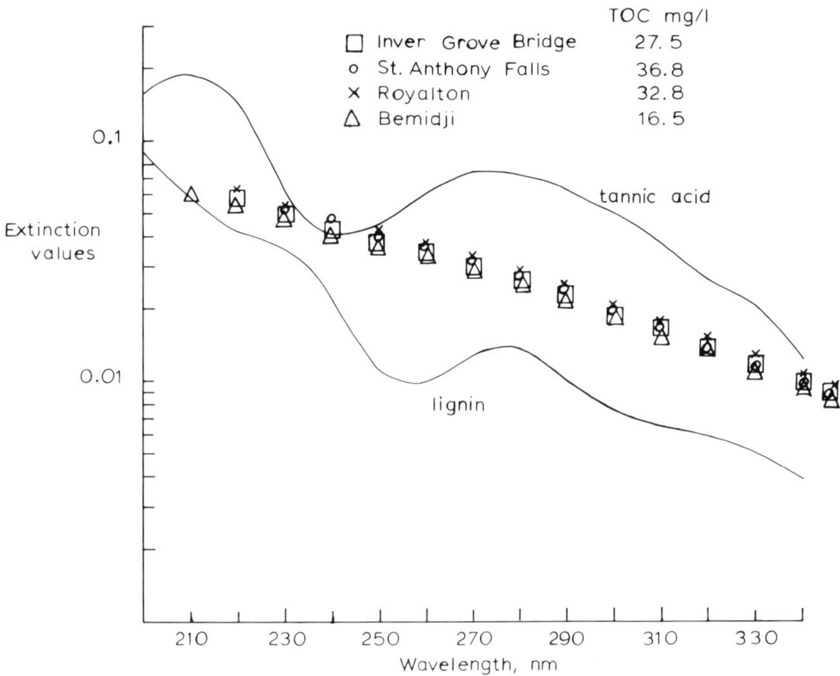

Figure 18. 1K-10K fractions, spring MRW.

from 0.2 to 1.8 mg/l; tannic acid, 3,4-dimethoxyacetophenone and *p*-anisic were identified in all but one sample at concentrations ranging from 0.4 to 4.7, 0.1 to 3.0, and 0.4 to 2.9 mg/l, respectively; gallic, mellitic (hexabenzoic acid) and syringic acid were identified in more than half the samples but at lower concentrations. The winter samples had the largest number of model compounds, including salicylic and pentacarboxybenzoic acid and catechol. Lignin was only identified in two fall samples. In all, some 20 compounds were identified at least once. Comparison of AMMCC for the three seasons shows the distribution in fall and spring are similar, whereas winter samples have a greater diversity of model compounds.

Identification of vanillin, tannic acid and 3,4-dimethoxyacetophenone reflects the relatively high absorbance of aquatic organics in the 280- to 350-nm range; these three model compounds have relatively strong absorption bands at long wavelengths due to the presence of keto or aldehyde groups as shown in Figure 9. It may, therefore, be inferred that the chromophores of MRW

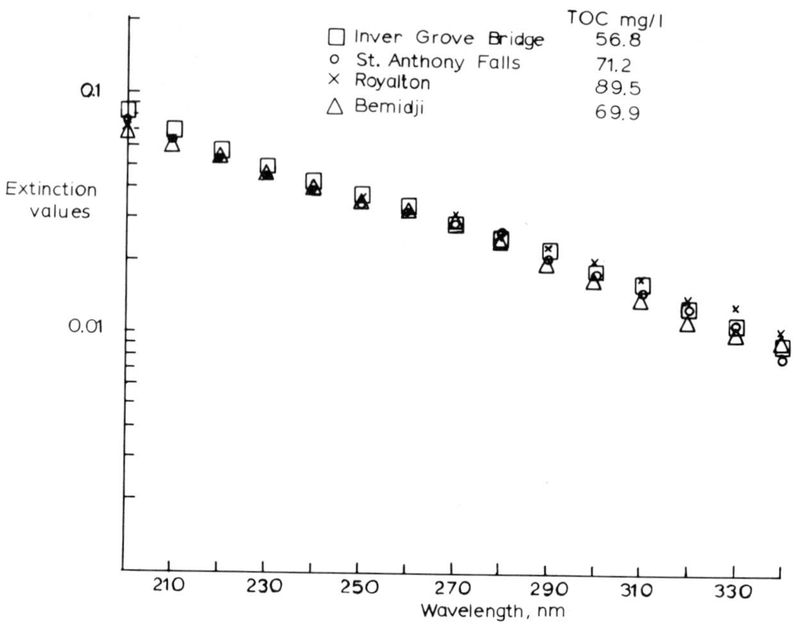

Figure 19. 1K-10K fractions, fall MRW.

organics have similar structures, namely aromatic structures with carbonyl substituents. Scott [1] has presented a comprehensive summary of data on keto- and aldehyde-substituted ring compounds; data on the effects of different substituent groups (-H, -OH, alkyl, alkoxy) on the wavelength and intensity of absorption have been correlated in terms of ring position and it is shown that steric interactions have important effects on absorbance bands in the 300- to 350-nm range. AMMCC identifications should therefore be interpreted as indicative of the presence of a class of ketoaldehyde aromatic ring structures.

Identification of *p*-anisic acid indicates the presence of chromophores with relatively strong absorption bands at short wavelengths (200 nm and 240 nm) as shown in Figure 5. Scott [1] states that "non-interacting chromophores give rise to absorption corresponding to the simple summation of the contributing absorptions, whereas conjugation of chromophores gives rise to new and intense absorption bands." It may, therefore, be inferred that MRW organics contain noninteracting chromophores containing para substituted aromatic ring structures with absorbance characteristics related to *p*-anisic acid.

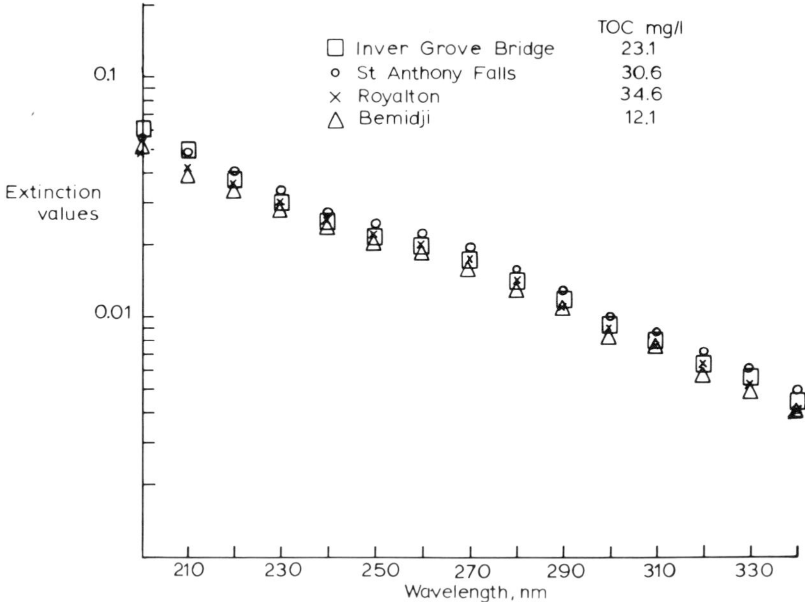

Figure 20. 1K-10K fractions, winter MRW.

Although the molecular size of aquatic organics is not precisely known, information from physical separation processes such as ultrafiltration [21] and gel filtration [19] indicate that the major fraction is in the 1000- to 10,000-MW size range; lower molecular weights have been reported after chemical treatment [23]. It is widely believed that a major fraction of aquatic organic molecules comprise polymeric structures of substituted aromatic rings [14,19,24], and it was anticipated that lignin, being a polymeric model compound, would be identified as a major constituent. The absence of lignin in all except two samples is, therefore, somewhat surprising. Deletion of lignin from the model compound inventory did not have a major effect on AMMCC as shown for the fall IGB sample (Table IV). The fact that lignin was not identified as a major contributing chromophore is probably due to the presence of alternative model compounds that give as good or better fit than lignin. The algorithm tends to preferentially identify those model compounds that make the largest contribution to the summed absorbance values. Lignin as a model compound was therefore not identified because other model compounds made larger absorbance contributions. This is illustrated in the last column of Table IV which lists the AMMCC calculated for lignin in terms of the 28 model compound inventory and shows that its overall absorbance

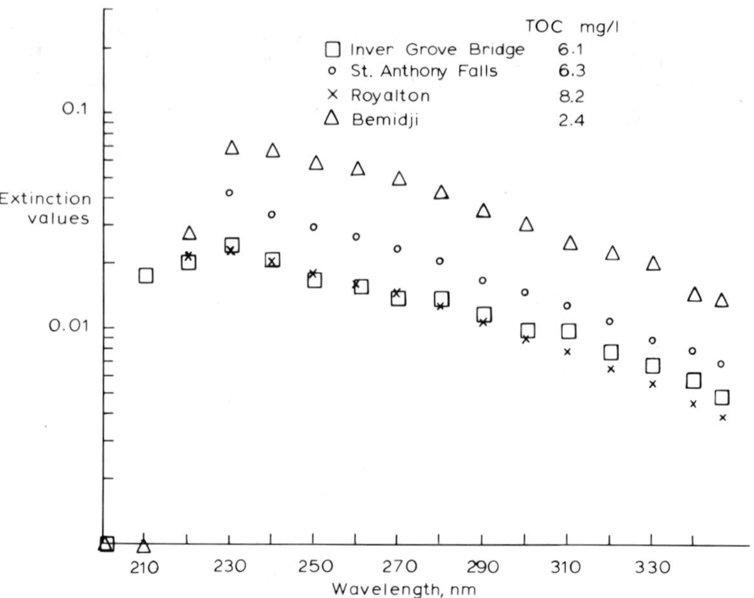

Figure 21. 10K-100K fractions, spring MRW.

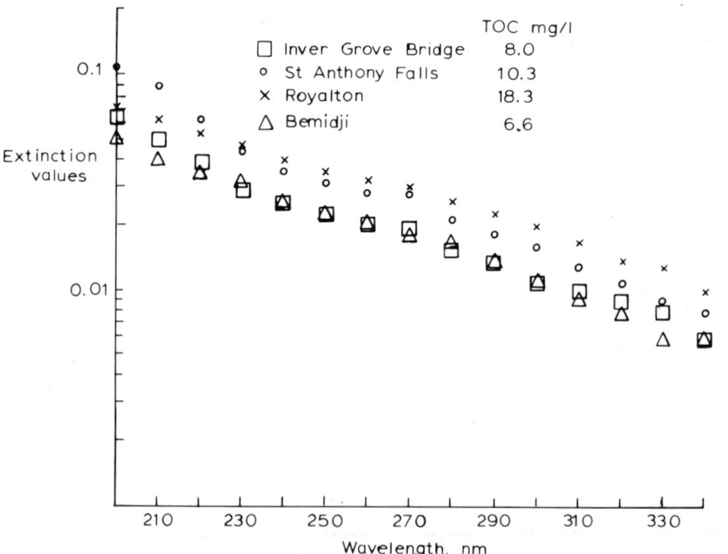

Figure 22. 10K-100K fractions, fall MRW.

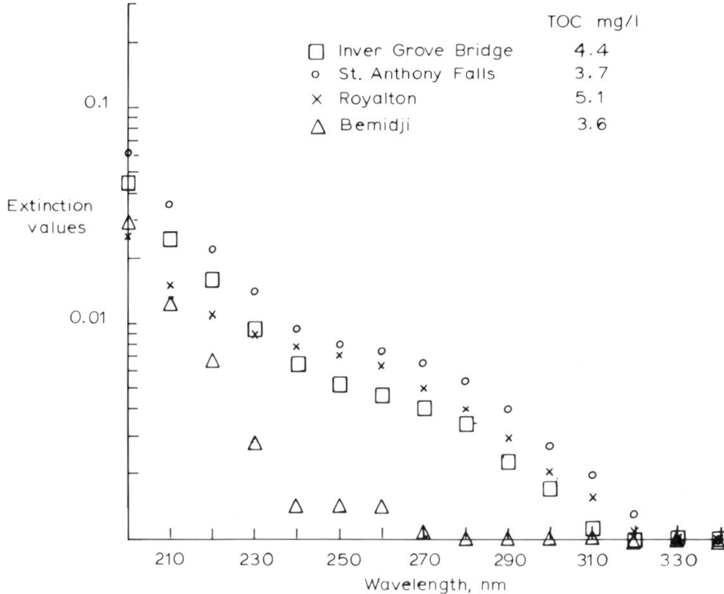

Figure 23. 10K-100K fractions, winter MRW.

characteristics can be decomposed into 8 model compound contributions. Although the specific chromophores are not the same, multicomponent identification is consistent with the proposed structural diversity of lignin illustrated in Figure 2 and lends credence to the idea that the absorbance characteristics of extended molecular structures can be approximated by summation of noninteracting chromophores. However, information about secondary effects from extended structures is needed to refine the analysis.

Identification of mellitic, gallic, syringic and 3,4,5-trimethoxybenzoic acids in a majority of samples from all locations and seasons indicates that these chromophores or their analogs are widely distributed in aquatic organics and hence are of natural origin. However, it is not apparent from these data whether they are present as noninteracting chromophores in extended molecular structures or whether they represent relatively small individual molecules. The greater diversity of chromophores and concomitant reduction in vanillin and acetophenone chromophores in the winter samples is noteworthy because it shows a consistent seasonal pattern related to environmental factors. Winter MRW flows are low and usually contain low concentrations of organics because of ice cover. Groundwater inflow and under-ice drainage from lakes and bogs are the major sources of water in winter; ground waters contain very

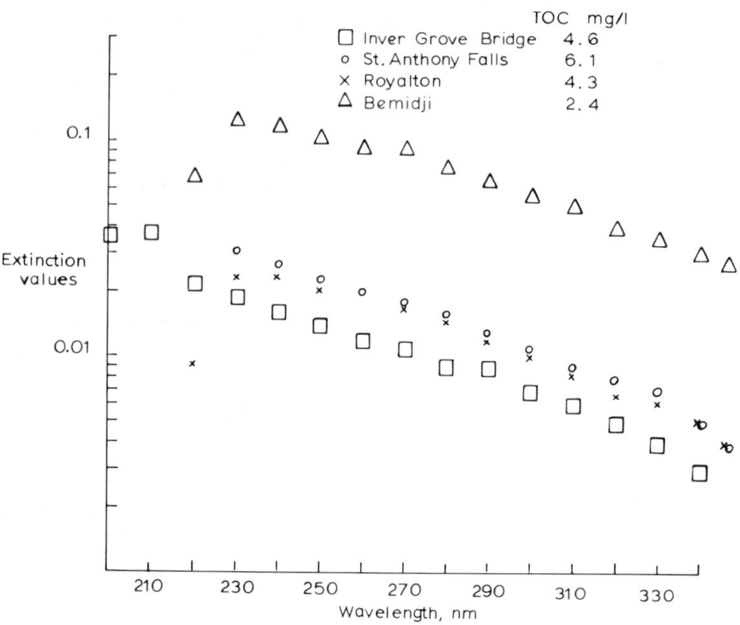

Figure 24. >100K fractions, spring MRW.

little organic matter [22]; furthermore, rates of biological-chemical decay processes in flowing waters are minimal. This suggests that the organic constituents in winter MRW originate as under-ice leachates from bogs and residual organics in underice lake waters supplemented by products of hydrolytic breakdown of humified organics from bottom deposits. Accordingly, when winter samples are compared with other seasons, they will characteristically (1) have relatively low organic carbon concentrations (groundwater dilution); (2) contain some of the same chromophores (leachates from bogs); (3) contain a larger diversity of chromophores and non-UV-absorbing organics (slow rates of decomposition); and (4) show uniform compositions at all sites (water originates in lake region). These effects have been observed.

Absorbance Matched Model Compound Analysis of Ultrafiltered Fraction

Molecular size fractionated samples were prepared from each of the four sampling sites at three different times of the year (procedures outlined above). UV absorbance measurements (some concentrates had to be diluted) were analyzed using the UVSA program and the complete absorbance matching

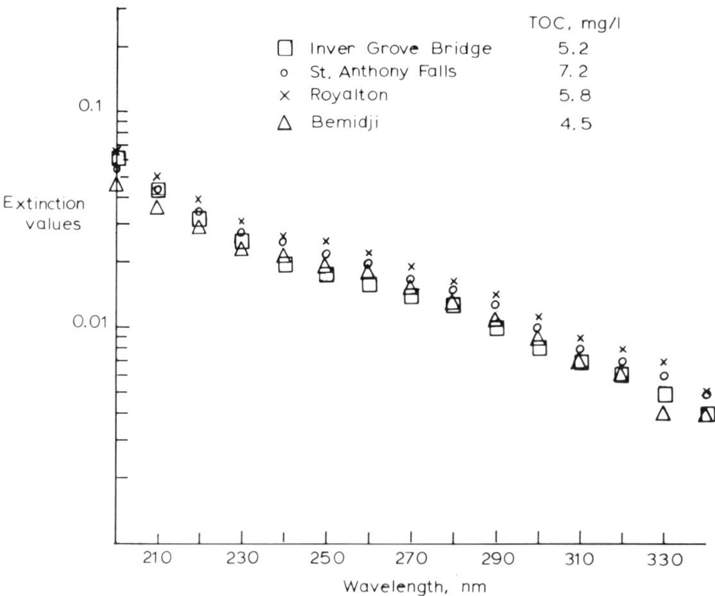

Figure 25. >100K fractions, fall MRW.

compound concentrations (AMMCC) are listed in Appendix B for the winter, spring and fall samples; concentrations are expressed as mg/l based on the original volume of river water. Vanillin, 3,4-dimethoxyacetophenone, tannic, *p*-anisic and mellitic acid (VATAM) spectra were identified as major constituents in almost all the 0.5-1K and 1-10K fractions regardless of location or season. Table V focuses on the 1-10K fractions; concentrations vary seasonally and geographically. Comparison by seasons shows a consistent pattern; winter samples have the lowest concentrations, spring samples are intermediate, while fall samples are highest. Comparison of geographical locations shows a mixed trend. The upstream site (BEM) has the lowest concentration in winter and spring but in fall minimum concentrations occur at ROY; maximum values occur at ROY and SAF in winter and spring but shift to IGB in fall. Compositions at ROY and SAF are remarkably consistent in winter and spring even though river flows differ substantially (Table V). The similarity of AMMCC in winter for ROY, SAF and IGB is ascribed to the relatively small change in river flow, which indicates that drainage occurs largely in the upper reaches of the basin. The composition of IGB differs from the other sites, lower concentrations are observed in winter and spring (except for tannic). This trend is probably due to inflows from the Minnesota River, which is a

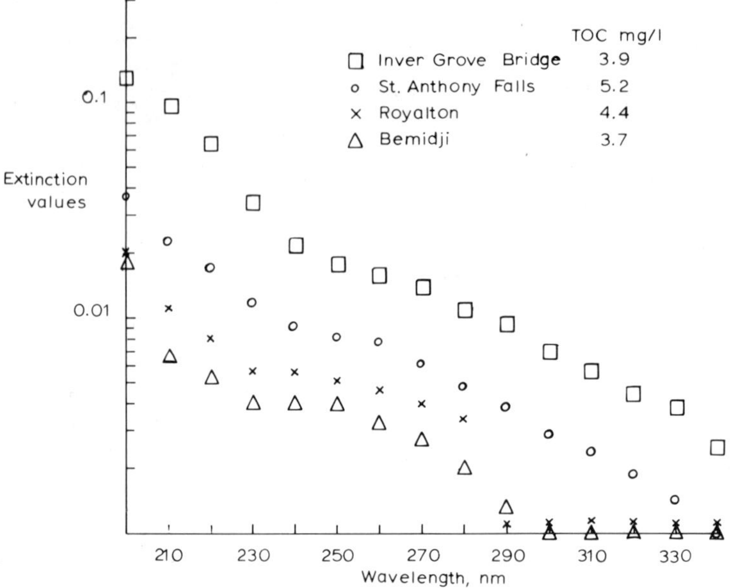

Figure 26. >100K fractions, winter MRW.

major tributary. By contrast, IGB concentrations are higher in fall which coincides with low tributary inflows as evidenced by the fact that river flows are nearly the same as at SAF. The increase in VATAM concentrations probably reflects the influx of treated wastewaters which have been shown to have similar composition as aquatic organics [21]. The 1-10K fractions are unique in that the program identified the same five model compounds in all samples; furthermore, no other model compounds were identified. These results are consistent with the previously discussed similarity in extinction value spectra of the 1-10K fractions (Figures 18-20). Identification of the same chromophores can be interpreted as indicating the presence of polymers consisting of noninteracting chromophores characteristic of VATAM. The fact that VATAM are found in unfiltered river water rules out the explanation that they are artifacts of the membrane filtration process. The fact that VATAM are also identified in the smaller molecular size fraction 0.5-1K rules out the explanation that these spectral characteristics represent a single chromophore.

Absorbance matching calculations of the less than 0.5K fractions show that the AMMCC include a diversity of substituted ring structures; seven model compounds were identified more than once or at high concentrations as shown in Table VI. Resorcinol, salicylic and vanillic acid are ubiquitous constituents of naturally occurring organics. However, the presence of bases

Table IV. Absorbance Matched Model Compound Concentrations (mg/l)—Mississippi River Water

	Fall 1978				a	Winter				Spring 1978				b
	BEM	ROY	SAF	IGB		BEM	ROY	SAF	IGB	BEM	ROY	SAF	IGB	
EN	0.37	0.35	0.05	0.06	(0.06)	0.03	0.02	0.03	0.02	0.37	0.49	0.77	0.10	0.04
Vanillin	1.03	1.32	1.12	1.03	(1.06)	0.19	0.24	0.36	0.43	0.89	1.11	1.77	0.81	0.38
Tannic Acid	3.08	4.07	4.73	4.33	(4.55)	0.38	0.89	1.16	0.62	0.41	0.39		1.61	0.69
3,4-Dimethoxyacetophenone	2.01	1.80	0.78	0.61	(0.58)		0.17	0.26	0.14	2.20	2.64	2.98		0.14
p-Anisic	2.71	2.89	2.50	2.11	(2.18)	0.43	0.71	0.86	0.76	1.81	1.99	1.18		
Gallic Acid			0.93	0.74	(0.68)	0.22	0.19	0.08	0.33			2.94	1.46	
Mellitic Acid	1.89	2.44	2.91	2.81	(2.83)		0.81		1.24	0.87	1.22			1.61
Syringic Acid			0.42	0.22	(0.29)		0.41	0.10	0.06			0.04		
3,4,5-Trimethoxybenzoic Acid			0.33	0.80	(0.63)		0.08	0.04	0.18					
Pentacarboxybenzoic Acid						0.61	0.09	1.02				0.90		
Salicylic Acid						0.08	0.16	0.05	0.30					0.07
1-7 Dimethyl Xanthine		0.04						0.19	0.25				0.18	
Catechol						0.46	0.50	0.28	0.10					
Resorcinol			0.45		(0.23)	0.21			0.37				1.47	
3,4,5-Trimethoxy-acetophenone						0.19			0.48					
1,3,5-Tricarboxybenzoic Acid									0.20				2.09	
Lignin		1.42	1.17						0.47					
o-Methoxybenzoic Acid					(0.06)		0.40							1.14
o-Phthalic Acid				0.07	(0.93)									
m-Anisic Acid														1.0
1-Methyl Inosine														0.14
2,4-Dimethoxybenzoic Acid														
1,2,4,5-Tetracarboxybenzoic Acid													1.60	
2,4-Dicarboxylic Acid													0.46	

[a] 28 model compounds; lignin was not included as a model compound.
[b] 10 mg/l lignin solution matched in terms of 28 model compounds.

Table V. AMMCC of 1-10K Fractions (mg/l)

Vanillin	Winter				Spring				Fall			
	BEM	ROY	SAF	IGB	BEM	ROY	SAF	IGB	BEM	ROY	SAF	IGB
Tannic	0.06	0.22	0.21	0.16	0.31	0.69	0.76	0.46	1.05	0.31	1.13	1.18
Acetophenone	0.39	1.02	1.14	0.86	0.35	1.37	1.45	1.73	2.58	0.60	2.80	0.88
Anisic	0.26	0.71	0.70	0.45	0.68	1.15	1.34	0.92	2.16	0.58	2.29	1.58
Mellitic	0.25	0.81	0.87	0.68	0.30	1.03	1.12	0.89	1.80	0.46	1.96	3.64
Flow (m^3/sec)		86	125	138		216	475	838		213	329	342

Table VI. AMCCC of Less Than 0.5K Fractions (mg/l)

	Winter				Spring				Fall			
	BEM	ROY	SAF	IGB	BEM	ROY	SAF	IGB	BEM	ROY	SAF	IGB
Catechol	0.78											
Resorcinol		0.73	0.66	0.34				2.72	0.43	0.19		
Salicylic Acid	0.16	0.15	0.01					1.15	0.08			
1,7-Dimethyl Xanthine		0.05		0.22			0.50	0.50		0.04	0.24	0.53
2,4-Dimethoxy-benzoic Acid			0.15	0.17				2.53		0.32	0.21	
Vanillic Acid		0.05						1.53		0.02	0.03	
Isophthalic Acid							0.11					0.14
		a						a	b			b
	c	c			c	c			c		c	

[a] Low concentrations of two other substituted benzoic acid analogs were identified.
[b] Low concentrations of one other substituted benzoic acid analogs were identified.
[c] One or more VATAM chromophores was identified.

such as 1,7-dimethylxanthine indicate active metabolism; the latter appears mostly in the downstream pollution-stressed areas (SAF and IGB) and may indicate waste waters. Identification of these specific model compounds is consistent with the absence of absorbance bands in the 300- to 350-nm region and the presence of absorbance at short wavelengths (200-220 nm) as noted in the extinction value spectra. The presence of hydroxy-, methoxy- or carboxy-substituted single ring structures is consistent with the molecular size range (0.5K membrane gives a nominal 500 mol wt cutoff). It is noteworthy that the VATAM which account for all the absorbance of the 1-10K fractions are absent from most of the less than 0.5K fractions. This difference in absorbance characteristics between size fractions gives further support to the argument that the 1-10K organics consist of a unique molecular arrangement or polymer comprised of VATAM as noninteracting chromophores. Further studies of the less than 0.5K fractions to evaluate the absorbance contributions of inorganic ions are in progress and will be reported in subsequent papers.

Absorbance matching calculations of 0.5-1K fractions identify the same model compounds as the 1-10K fractions (VATAM); however, the concentrations are generally lower and the proportions of vanillin, tannic, 3,4-dimethoxyacetophenone, p-anisic and mellitic acids are not the same; tannic acid is absent from three winter samples (Appendix Table B-III). Identification of the same groups of model compounds in both 0.5-1K and 1-10K fractions suggests that there is some overlap in size separations. This could be the result of variations in size of polymeric molecules and/or an artifact of membrane separations. Membrane separations do not achieve completely quantitative size separations of molecules whose dimensions are close to the nominal cutoff of the membrane [18]. Analysis of the 0.5-1K fractions identifies two model compounds, syringic acid and 3,4,5-trimethoxyacetophenone in the downstream (SAF and IGB) sites in winter. Model compound concentrations of these two samples are identical (Appendix Table B-III) and indicate that water composition is uniform during winter months (ice-covered and low urban runoff).

Mississippi River waters contain relatively small concentrations of high molecular weight organics as measured by TOC of the 10-100K and > 100K fractions. Furthermore, these fractions have relatively low-intensity UV absorbance and their calculated model compounds are correspondingly low (Appendix B). The same five model compounds (VATAM) as found in the 1-10K fraction were identified in almost all samples. Table VII summarizes the concentrations in the 10-100K fraction and shows that they are higher than the corresponding > 100K fractions (Table VIII). This is consistent with measured TOC values. As with the 1-10K fractions, tannic acid is identified at the highest concentrations and vanillin and 3,4-dimethoxyacetophenone are

Table VII. AMMCC of 10-100K Fraction (mg/l)

	Fall				Spring			
	BEM	ROY	SAF	IGB	BEM	ROY	SAF	IGB
Vanillin	0.07	0.39	0.15	0.07	0.05	0.10	0.08	0.04
Tannic Acid	0.18	0.70	0.54	0.14			0.34	0.11
3,4-Dimethoxy-acetophenone	0.10	0.42	0.13	0.09	0.07	0.23	0.01	0.02
p-Anisic Acid	0.15	0.52	0.31	0.16	0.03	0.16	0.16	0.08
Mellitic Acid	0.11	0.47	0.22	0.22		0.08	0.07	

Table VIII. AMMCC of $100K^+$ Fractions (mg/l)

	Fall				Spring			
	BEM	ROY	SAF	IGB	BEM	ROY	SAF	IGB
Vanillin	0.03	0.05	0.06	0.03	0.05	0.05	0.06	0.04
Tannic Acid	0.11	0.25	0.27	0.13			0.24	0.11
3,4-Dimethoxy-acetophenone	0.05	0.03	0.05	0.02	0.07	0.13	0.03	0.02
p-Anisic Acid	0.09	0.14	0.16	0.09	0.04	0.09	0.14	0.08
Mellitic Acid	0.07	0.11	0.13	0.09			0.17	

lowest; overall distribution ratios are similar. However, the high-molecular-weight fractions of the winter samples are different; absorbance calculated concentrations of VATAM range from 0 to 0.03 mg/l. These traces of UV-absorbing material are believed to be an artifact of the size separation procedure and represent incompletely washed 1-10K materials rather than large-molecular-weight UV-absorbing compounds. Membrane filtration followed by two volumes of organic-free wash water does not completely remove low-molecular-weight materials from the retentate. Theoretical considerations predict that two 300-ml washes of the 300-ml retentate leave approximately 25% of the small molecular size (filtrable) organics in the retentate, and the measured concentration shoud be adjusted accordingly.

Absorbance matching calculations of the winter samples identified low concentrations of several substituted ring structures in the 10-100K and > 100K fractions, namely syringic acid, salicylic acid, resorcinol, catechol, pentacarboxybenzoic acid, 3,4,5-trimethoxyacetophenone, gallic acid and 1,7-dimethylxanthine. These compounds were also identified in the corresponding unfractionated river which suggests that they are noninteracting chromophoric constituents of large nonfiltrable molecules.

SUMMARY

Comparison of the model compounds identified by absorbance matching shows the same chromophoric units, indicating that the organic constituents of the 1-10K molecular size fractionates of Mississippi River water have similar absorbance characteristics regardless of season or geographical location. Five major model compounds, namely vanillin, 3,4-dimethoxyacetophenone, tannic, p-anisic and mellitic acid (VATAM), and no others, are consistently identified by UV absorbance matching calculations. By contrast, AMMCC of the less than 0.5K materials is indicative of hydroxy-, methoxy- or carboxy-substituted single ring structures with major absorbance bands at short wavelengths. VATAM are not identified in the less than 0.5K fractions.

Absorbance matching calculations on the 0.5-1K fractions identify the same five model compounds but at lower concentrations and in different proportions. The same applies to the larger size fractions; absorbance of the 10-100K and $100K^+$ fraction is characterized by the same five model compounds in spring and fall, but the winter samples have a distinctly lower absorbance level and a greater variety of model compound chromophores. The greater diversity of model compounds in the fractionated winter samples is consistent with the raw water analysis in which a greater variety of model compounds are identified, albeit at very low concentrations.

Identification of the same five recurring chromophores shows that a large fraction of aquatic organics in the Mississippi River consist of a unique molecular arrangement or polymer that occurs in the unpolluted headwaters regions as well as the more pollution-stressed reaches at all times of the year. Information on size characteristics deduced from ultrafiltration indicates that the molecules are predominantly in the 1000- to 10,000-mol wt size range. Percentage distributions of vanillin, 3,4-dimethoxyacetophenone, and tannic, p-anisic and mellitic acids are shown in Table IX.

The distribution patterns for 1-10K spring and fall are similar, whereas the 1-10K winter distribution is more comparable with the 0.5-1K fraction. Most of the VATAM constituents measured in the 10-100K and $> 100K$ fractions can be ascribed to incomplete washing of the 1-10K materials as discussed in previous paragraphs. This explanation is consistent with the observed similarity in the percentage distributions of VATAM chromophores in the larger molecular size fractions compared to the 1-10K fractions.

The similarities in UV spectral and molecular size characteristics of organics in the 800-km stretch of the Mississippi River (from BEM to IGB) with total

Table IX. Averaged Percentage Distribution of Major Model Compounds

	1-10K Fractions			Fall		
	Winter	Spring	Fall	0.5-1K	10-100K	>100K
Vanillin	7	13	12	5	12	9
Tannic Acid	36	27	22	33	30	38
3,4-Dimethoxy- acetophenone	9	18	20	13	14	8
p-Anisic Acid	22	24	22	27	23	25
Mellitic Acid	26	18	24	23	21	21

flow times of 20-30 days, encompassing flow variations from 86 to 883 m^3/sec, led to the following hypothesis:

1. VATAM characterized polymers are formed throughout the whole watershed;
2. these polymers are chemically and biologically relativly stable;
3. they predominate the 1000-10,000 molecular size range;
4. they account for most of the TOC but are less significant in winter; and
5. the presence of a greater diversity of chromophores and a correspondingly smaller concentration of VATAM polymer in winter is ascribed to slower chemical-biological reaction rates.

Practical Implications

The results of this study have interesting potential applications in water pollution control and municipal water supply technology. The high sensitivity of multiwavelength UV measurements combined with relatively low cost, reliability and simplicity, makes this an ideal continuous flow monitoring tool. Continuous monitoring is desperately needed to indicate the inflow of increasing concentrations of organics in municipal water treatment plants that obtain their raw water supplies from surface sources. Semicontinuous UV absorbance scans over the 200- to 350-nm waveband would serve as an early warning system to change processing conditions, for example, increase alum and/or activated carbon dosage to help remove organics.

UV monitoring coupled with absorbance matching calculations to identify specific compounds is particularly well suited to industrial situations where known organics are likely to be present and remedial actions depend on compound identification.

Continuous UV absorbance monitoring of sewage treatment plant effluents is particularly interesting in the light of ongoing studies that show an excellent correlation between BOD and a combination of 200- to 350-nm UV absorbance and TOC measurements. The advantages of the combined UV-TOC measurements, which can be obtained instantly, compared to the 5-day BOD test for process control purposes are obvious. It appears that multiwavelength absorbance analysis is a good measure of biochemically refractory organics, whereas TOC measures total organics. For specific industrial wastes, absorbance matching calculations to identify pollutants would also be of interest.

The potential usefulness of multiwavelength UV absorbance analysis for monitoring environmental water quality as, for example, in continuous monitoring systems on major river basins is evident in view of the sensitivity of the measurement for detecting specific pollutants in a background of naturally occuring chromophores. Such monitoring could be used to detect spills from specific industries or sewage treatment wastes.

The advantages of multiwavelength absorbance measurements in research studies where more than one chromophore is present have been described in the chemistry literature. There is an even greater potential for application in biological studies that have to contend with mixtures of chromophores.

Future Work

The model compounds identified in these computations differ from those reported in an earlier publication [13] which presented preliminary results on Mississippi River water and waste waters. The major difference is that additional model compounds were incorporated and the absorbance range was extended to 200-358 nm. Further improvements could be achieved by incorporating elemental composition and spectrophotometric measurements at more extended wavelengths and different pH values. Studies to evaluate the absorbance contributions of inorganic ions are in progress and will be reported in subsequent publications.

ACKNOWLEDGMENTS

The help of Christine Macko, P. D. Nath and Nancy Pelt Lyche in carrying out the experimental program are gratefully acknowledged. Many thanks are due to Mrs. M. Henderson and Mrs. S. Maier for preparing this manuscript. This work was supported by funds from the National Science Foundation and the Office of Water Resources Research.

REFERENCES

1. Scott, A. I. *Ultraviolet Spectra of Natural Products* (New York: The MacMillan Co., 1964).
2. Morton, R. A. *Biochemical Spectroscopy, Vol. 2* (New York: John Wiley & Sons, 1975), pp. 696-702.
3. Schnitzer, M., and S. U. Khan. *Humic Substances in the Environment* (New York: Marcel Dekker, Inc., 1972).
4. Gieseking, J. E. *Soil Components, Vol. I., Organic Components* (New York: Verlag, 1975).
5. Dobbs. R. A., R. H. Wise and R. B. Dean. "The Use of Ultraviolet Absorbance for Measuring the Total Organic Carbon Content of Water and Wastewater," *Water Research*, 6:1173-1180 (1972).
6. Symons, J. M., T. A. Bellar, J. K. Carswell, J. DeMarlo, K. L. Kropp, G. R. Robeck, D. R. Seeger, C. J. Slocum, B. L. Smith, and A. A. Stevens. "National Organics Reconnaissance Survey for Halogenated Organics," *J. Am. Water Works Assoc.* 67:634-647 (1975).
7. Mrkva, M. "Investigation of Organic Pollution in Surface Waters by Ultraviolet Spectroscopy," *J. Water Poll. Control Fed.* 41:1923-1931 (1969).
8. Bramer, H. C., M. J. Walsh and S. C. Caruso. "Instrument for Monitoring Trace Organic Compounds in Water," *Water Sew. Works*, 113:275-278 (1966).
9. Briggs, R., and K. W. Melbourne. "Recent Advances in Water Quality Monitoring," *Water Treat. Exam.* 17:107-120 (1968).
10. Arends, J. M., H. Cerfontain, I. S. Herschberg, A. J. Prinsen, and A. C. M. Wanders. "Ultraviolet Spectrometric Determination of Mixtures of Arylsulfuric Acids," *Anal. Chem.* 36(9):1802 (1964).
11. Metzler, D. E., C. M. Harris, R. L. Reeves, W. H. Lawton and M. S. Maggio. "Digital Analysis of Electronic Absorbtion Spectra," *Anal. Chem.* 49(11): 864-874 (1977).
12. Kankare, J. J. "Computation of Equilibrium Constants for Multicomponent Systems from Spectrophotometric Data," *Anal. Chem.* 42:1322-1325 (1970).
13. Maier, W. J., L. E. Conroy, S. J. Eisenreich, M. J. Hoffmann, C. A. Macko and P. D. Nath. "Multicomponent UV Spectral Analysis of Aquatic Organics," in *Proceedings of the 1978 International Congress on Analytical Techniques in Environmental Chemistry*, J. Albaiges, Ed. (Barcelona, Spain, 1980).
14. Khan, S. U., and M. Schnitzer. "Further Investigations of Fulvic Acid, a Soil Humic Fraction," *Can. J. Chem.* 49:2302 (1971).

15. *Standard Methods for the Examination of Water and Wastewater*, 13th ed. (New York: American Public Health Association, 1971).
16. *The Merck Index*, 9th ed. (Merck & Co., 1976).
17. Lawson, C. L., and R. J. Hanson. "Jet Propulsion Laboratory 1973," in *Solving Least Squares Problems* (Englewood Cliffs, NJ: Prentice-Hall, 1974).
18. Macko, C., W. J. Maier, S. J. Eisenreich and M. R. Hoffman. *AIChE Symp. Ser.–Water*, 75:162-169 (1979).
19. Gjessing, E. T. *Physical and Chemical Characteristics of Aquatic Humus* (Ann Arbor, MI: Ann Arbor Science Publishers, Inc., 1975).
20. Wheeler, J. R. *Limnol. Oceanog.* 2(16):846 (1976).
21. Macko, C. "The Removal of Organic Matter from Surface Water Supplies by Anion Exchange Resins," PhD Thesis, University of Minnesota (1980).
22. Maier, W. J., R. G. Gast, C. T. Anderson and W. W. Nelson. "Carbon Content of Surface and Ground Waters in South Central Minnesota," *J. Environ. Qual.* 4(2):124 (1976).
23. Shapiro, J. "The Relation of Humic Acid Color with Iron in Natural Waters," *Int. Ver. Theor. Agnew. Berh.* 16:477 (1966).
24. "Organics Removal by Coagulation," *J. Am. Water Works Assoc.* 71(10): 588 (1979).

APPENDIX A

```
 1.    000000B           PROGRAM UVSA(INPUT,CUTPUT,TAPE5=INPUT,TAPE6=OUTPUT)
 2.    002131B           DIMENSION A(80,50),A1(80,50),B(80),B1(80),D(80,50),C(80,50),E(80),
                        1MTOV(4,50),STOV(4,50),TOV(4),W(50),ZZ(80),X(50)
 3.    002131B           REAL MTOV
 4.    002131B           INTEGER INDEX(50),FLOW
 5.    002131B           READ(5,100) FLOW,M,N,ND,NS,L2
 6.    043115B       100 FORMAT(6I2)
 7.    043115B           WRITE(6,8)
 8.    043120B           READ(5,120)   ((A(I,J),I=1,M),J=1,N)
 9.    043137B           READ(5,120)   ((D(I,J),I=1,M),J=1,NS)
10.    043156B           WRITE(6,130) ((A(I,J),J=1,N),I=1,M)
11.    043200B           WRITE(6,180)
12.    043203B           WRITE(6,150) ((D(I,J),J=1,NS),I=1,M)
13.    043225B           WRITE(6,8)
14.    043230B           IF(L2 .EQ. 0) GO TO 140
15.    043231B           WRITE(6,180)
16.    043234B           READ(5,160)   ((MTOV(I,J),I=1, 4),J=1,N)
17.    043251B           READ(5,160)   ((STOV(I,J),I=1, 4),J=1,NS)
18.    043266B           WRITE(6,170)
19.    043271B       170 FORMAT(14X,3HTOC,15X,3HTOH,15X,3HTOO,15X,3HTON)
20.    043271B           WRITE(6,165) ((MTOV(I,J),I=1,4),J=1,N)
21.    043306B           WRITE(6,180)
22.    043311B           WRITE(6,165) ((STOV(I,J),I=1,4),J=1,NS)
23.    043326B           WRITE(6,6)
24.    043331B         8 FORMAT(1H1)
25.    043331B       120 FORMAT(10F8.3)
26.    043331B       130 FORMAT(2X,14F8.4)
27.    043331B       150 FORMAT(2X, 4F7.3)
28.    043331B       160 FORMAT(4F8.3)
29.    043331B       165 FORMAT(2X,2(10X,F8.3,10X,F6.3,10X),/)
30.    043331B       180 FORMAT(///)
31.    043331B       140 L1 = ND - M
32.    043332B       400 FORMAT(//,2X,* TOC CONTRIBUTION OF THE MODELS :- *,//)
33.    043332B       410 FORMAT(//,2X,* TOH CONTRIBUTION OF THE MODELS :- *,//)
34.    043332B       420 FORMAT(//,2X,* TOO CONTRIBUTION OF THE MODELS :- *,//)
35.    043332B       430 FORMAT(//,2X,* TON CONTRIBUTION OF THE MODELS :- *,//)
36.    043332B           L = M+1
37.    043333B           IF((L1 .NE. L2) .OR. (FLOW .LT. 1) .OR. (FLOW .GT. 3)) GO TO 1
38.    043343B           IF((FLOW .EQ. 2) .OR. (FLOW .EQ. 3)) GO TO 3
39.    043351B           MM = M
40.    043352B           GO TO 4
```

```
 41.   043353B      1 WRITE(6,5)
 42.   043357B      5 FORMAT(2X, * EITHER FLOW OR L2 WAS DECLARED WRONG *)
 43.   043357B        RETURN
 44.   043360B      3 MM = ND
 45.   043360B        IF(FLOW .EQ. 3) IND = 3
 46.   043365B        DO 6 I = L,ND
 47.   043366B        K = I - M
 48.   043367B        DO 6 J = 1,N
 49.   043372B        A(I,J) = MTOV(K,J)
 50.   043376B        DO 6 J1 = 1,NS
 51.   043402B      6 D(I,J1) = STOV(K,J1)
 52.   043417B      4 DO 7 II = 1,NS
 53.   043421B     11 WRITE(6,8)
 54.   043425B        WRITE(6,9) II
 55.   043431B      9 FORMAT(/////,58X,7HSAMPLE ,I2,/)
 56.   043431B        IF(FLOW .EQ. 1) WRITE(6,440)
 57.   043436B        IF(FLOW .GT. 1) WRITE(6,450)
 58.   043443B    440 FORMAT(50X,28HANALYSIS NOT CONSIDERING TOV,//////)
 59.   043443B    450 FORMAT(52X,24HANALYSIS CONSIDERING TOV,/////)
 60.   043443B        DO 10 I = 1,MM
 61.   043445B        B(I) = D(I,IT)
 62.   043447B        B1(I) = B(I)
 63.   043447B        DO 10 J = 1,N
 64.   043454B     10 A1(I,J) = A(I,J)
 65.   043464B        MDA = ND
 66.   043464B        CALL NNLS(A1,MDA,MM,N,B1,X,RNORM,W,Z7,INDEX,MODE)
 67.   043467B        IF(MODE .EQ. 1) WRITE(6,12)
 68.   043474B        IF(MODE .EQ. 2) WRITE(6,13)
 69.   043501B     12 FORMAT(2X,* THE SOLUTION HAS BEEN COMPUTED SUCCESSFULLY *,//)
 70.   043501B     13 FORMAT(2X,* THE DIMENSIONS OF THE PROBLEM ARE BAD *,//)
 71.   043501B        WRITE(6,14) RNORM
 72.   043505B     14 FORMAT(2X,* THE EUCLIDEAN NORM = *,E15.8,//)
 73.   043505B        WRITE (6,15)
 74.   043510B     15 FORMAT(2X,* THE VECTOR (X) :- *,//)
 75.   043510B        WRITE(6,16) (X(I),I=1,N)
 76.   043520B     16 FORMAT(2X,14F8.4)
 77.   043520B        WRITE(6,180)
 78.   043523B        DO 17 I = 1,MM
 79.   043525B        E(I) = 0.
 80.   043525B        DO 17 J = 1,N
 81.   043530B        C(I,J) = A(I,J) * X(J)
 82.   043533B     17 E(I) = E(I) + C(I,J)
 83.   043540B        DO 18 I = 1,MM
 84.   043541B        IF(I .EQ. L) WRITE(6,400)
 85.   043547B        IF(I .EQ. L+1) WRITE(6,410)
 86.   043555B        IF(I .EQ.L+2) WRITE(6,420)
 87.   043563B        IF(I .EQ. L+3) WRITE(6,430)
 88.   043571B        WRITE(6,19) (C(I,J),J=1,16)
 89.   043604B     19 FORMAT(2X,16F8.4)
 90.   043604B     18 CONTINUE
 91.   043606B        DO 26 I = 1,MM
 92.   043607B        IF(I .EQ. L) WRITE(6,400)
 93.   043615B        IF(I .EQ. L+1) WRITE(6,410)
 94.   043623B        IF(I .EQ.L+2) WRITE(6,420)
 95.   043631B        IF(I .EQ. L+3) WRITE(6,430)
 96.   043637B        WRITE(6,19) (C(I,J),J=17,N),E(I),B(I)
 97.   043657B     26 CONTINUE
 98.   043661B        IF(FLOW .EQ. 1) GO TO 20
 99.   043663B        DO 21 I = 1,L2
100.   043665B        TOV(I) = 0.
101.   043665B        DO 21 J = 1,N
102.   043670B     21 TOV(I) = TOV(I) + MTOV(I,J) * X(J)
103.   043702B        WRITE(6,180)
104.   043705B        DO 22 I = 1,L2
105.   043707B        WRITE(6,23) TOV(I),STOV(I,II)
106.   043721B     23 FORMAT(2X,2E15.8)
107.   043721B     22 CONTINUE
108.   043723B     20 IF(FLOW .EQ. 3) GO TO 24
109.   043725B        IF(IND .EQ. 3) FLOW = 3
110.   043731B        IF(IND .EQ. 3) MM = ND
111.   043734B        GO TO 7
112.   043734B     24 FLOW = IND - 2
113.   043736B        MM = M
114.   043737B        GO TO 11
115.   043737B      7 CONTINUE
116.   043743B        RETURN
117.   043744B        END
               C      SUBROUTINE NNLS (A, MDA,M,N,B,X,RNORM,W,Z7,INDEX,MODE)
               C      GIVEN AN M BY N MATRIX, A AND AN M-VECTOR, B, COMPUTE AN
               C      N-VECTOR,X,WHICH SOLVES THE LEAST SQUARES PROBLEM
               C             A*X=B SUBJECT TO X.GE.0
               C      A(),MDA,M,N,   MDA IS THE FIRST DIMENSIONING PARAMRTER
```

```
C     A(),MDA,M,N,     MDA IS THE FIRST DIMENSIONING PARAMETER FOR THE
C                      ARRAY,A().ON ENTRY A() CONTAINS THE M BY N
C                      MATRIX,A. ON EXIT A() CONTAINS
C                      THE PRODUCT MATRIX, Q*A, WHERE Q IS AN M BY M
C         ORTHOGONAL MATRIX GENERATED IMPLICITELY BY THIS ROUTINE
C     B()              ON ENTRY B() CONTAINS THE M-VECTOR,B. ON EXIT B() CONTAIN
C     X()              ON ENTRY X() NEED NOT BE INITIALIZED. ON EXIT X() WILL
C                      CONTAIN THE SOLUTION VECTOR.
C     RNORM            ON EXIT RNORM CONTAINS THE EUCLIDEAN NORM OF THE
C                      RESIDUAL VECTOR.
C     W()              AN N-ARRAY OF WORKING SPACE. ON EXIR
C     W()              AN N-ARRAY OF WORKING SPACE. ON EXIT W() WILL CONTAIN
C                      THE DUAL SOLUTION VECTOR. W WILL SATISFY W(I)=0.
C                      FOR ALL I IN SET P AND W(I) .LE.0.FOR ALL I IN SET Z
C     ZZ()             AN M-ARRAY OF WORKING SPACE.
C     INDEX()          AN INTEGER WORKING ARRAY OF LENGTH AT LEAST N.
C                      ON EXIT THE CONTENTS OF THIS ARRAY DEFINE THE SETS
C                      P AND Z AS FOLLOWS..
C                      INDEX(1) THRU INDEX(NSETP) = SET P.
C                      INDEX(IZ1) THRU INDEX(IZ2) = SET Z.
C                      IZ1 = NSETP + 1 = NPP1
C                      IZ2 = N
C     MODE  THIS IS A SUCCESS-FAILURE FLAG WITH THE FOLLOWING MEANINGS.
C           1     THE SOLUTION HAS BEEN COMPUTED SUCCESSFULLY.
C           2     THE DIMENSIONS OF THE PROBLEM ARE BAD.
C                 EITHER M.LE. 0 OR N .LE. 0.
C           3     ITERATION COUNT EXCEEDED. MORE THAN 3*N ITERATIONS.
```

```
  1.    000000B          SUBROUTINE NNLS (A,MDA,M,N,B,X,RNORM,W,ZZ,INDEX,MODE)
  2.    000000B          DIMENSION A(80,50),B(80),X(50),W(50),ZZ(80),DUMMY(1)
  3.    000000B          INTEGER INDEX(50)
  4.    000000B          DIFF(X,Y) = X - Y
  5.    000006B          ZERO=0.
  6.    000006B          ONE=1.
  7.    000007B          TWO=2.
  8.    000011B          FACTOR=0.01
  9.    000012B          MODE=1
 10.    000014B          IF (M.GT.0.AND.N.GT.0) GO TO 10
 11.    000020B          MODE=2
 12.    000020B          RETURN
 13.    000023B       10 ITER=0
 14.    000023B          ITMAX=3*N
                 C          INITIALIZE THE ARRAYS INDEX() AND X().
 15.    000025B          DO 20 I=1,N
 16.    000027B          X(I)=ZERO
 17.    000027B       20 INDEX(I)=I
 18.    000032B          IZ2=N
 19.    000032B          IZ1=1
 20.    000034B          NSETP=0
 21.    000034B          NPP1=1
                 C            ****** MAIN LOOP BEGINS HERE ******
 22.    000036B       30 CONTINUE
                 C       QUIT IF ALL COEFFICIENTS ARE ALREADY IN THE SOLUTION.
                 C       OR IF M COLS OF A HAVE BEEN TRIANGULARIZED.
 23.    000037B          IF (IZ1.GT.IZ2.OR.NSETP.GE.M) GO TO 350
                 C       COMPUTE COMPONENTS OF THE DUAL (NEGATIVE GRADIENT) VECTOR W().
 24.    000043B          DO 50 IZ = IZ1,IZ2
 25.    000046B          J = INDEX(IZ)
 26.    000046B          SM = ZERO
 27.    000050B          DO 40 L=NPP1,M
 28.    000053B       40 SM=SM+A(L,J)*B(L)
 29.    000061B       50 W(J)=SM
                 C       FIND LARGEST POSITIVE W(J).
 30.    000066B       60 WMAX=ZERO
 31.    000067B          DO 70 IZ=IZ1,IZ2
 32.    000072B          J=INDEX(IZ)
 33.    000072B          IF (W(J).LE.WMAX) GO TO 70
 34.    000076B          WMAX = W(J)
 35.    000077B          IZMAX=IZ
 36.    000100B       70     CONTINUE
                 C              IF WMAX .LE. 0.GO TO TERMINATION.
                 C              THIS INDICATES SATISFACTION OF THE KUHN-TUCKER CONDITIONS.
 37.    000104B          IF (WMAX) 350,350,80
 38.    000106B       80 IZ=IZMAX
 39.    000106B          J=INDEX(IZ)
                 C       THE SIGN OF W(J) IS OK FOR J TO BE MOVED TO SET P.
                 C       BEGIN THE TRANSFORMATION AND CHECK NEW DIAGONAL ELEMENT TO AVOID
                 C       NEAR LINEAR DEPENDENCE.
 40.    000110B          ASAVE=A(NPP1,J)
 41.    000113B          CALL H12(1,NPP1,NPP1+1, M,A(1,J),1,UP,DUMMY,1,1,0)
 42.    000123B          UNORM = ZERO
 43.    000123B          IF (NSETP.EQ.0) GO TO 100
 44.    000125B          DO 90 L=1, NSETP
 45.    000127B       90 UNORM = UNORM + A(L,J)**2
```

```
 46.   000134B      100 UNORM=SQRT(UNORM)
 47.   000137B          IF (DIFF(UNORM+ABS(A(NPP1,J))*FACTOR,UNORM)) 130,130,110
       C   COL J IS SUFFICIENTLY INDEPENDENT. COPY B INTO ZZ,UPDATE ZZ AND
       C   SOLVE FOR ZTEST ( = PROPOSED NEW VALUE FOR X(J) ).
 48.   000152B      110 DO 120 L=1,M
 49.   000154B      120 ZZ(L)=B(L)
 50.   000157B          CALL H12 (2,NPP1,NPP1+1,M,A(1,J),1,UP,ZZ,1,1,1)
 51.   000170B          ZTEST=ZZ(NPP1)/A(NPP1,J)
       C   SEE IF ZTEST IS POSITIVE
 52.   000173B          IF (ZTEST) 130,130,140
       C   REJECT J AS A CANDIDATE TO BE MOVED FROM SET Z TO SET P.
       C   RESTORE A(NPP1,J),SET W(J)=0 ., AND LOOP BACK TO TEST DUAL COEFFS
 53.   000175B      130 A(NPP1,J)=ASAVE
 54.   000201B          W(J)=ZERO
 55.   000202B          GO TO 60
       C   THE INDEX J=INDEZ
       C   THE INDEX J=INDEX(IZ) HAS BEEN SELECTED TO BE MOVED FROM
       C   SET Z SO SET P. UPDATE B, UPDATE INDICES, APPLY HOUSEHOLDER TRANSF
       C   SET Z SO SET P. UPDATE B, UPDATE INDICES, APPLY HOUSEHOLDE
       C   TRANSFORMATIONS TO COLS IN NEW SET Z, ZERO SUBDIAGONAL ELTS IN
       C   COL J,SET W(J)=0.
 56.   000203B      140 DO 150 L=1,M
 57.   000206B      150 P(L)=ZZ(L)
 58.   000211B          INDEX(IZ)=INDEX(IZ1)
 59.   000212B          INDEX(IZ1)=J
 60.   000214B          IZ1=IZ1+1
 61.   000216B          NSETP=NPP1
 62.   000217B          NPP1 = NPP1 + 1
 63.   000220B          IF (IZ1.GT.IZ2) GO TO 170
 64.   000222B          DO 160 JZ=IZ1,IZ2
 65.   000225B          JJ=INDEX(JZ)
 66.   000225B      160 CALL H12 (2,NSETP,NPP1,M,A(1,J),1,UP,A(1,JJ),1,MDA,1)
 67.   000243B      170 CONTINUE
 68.   000244B          IF(NSETP .EQ. M) GO TO 190
 69.   000245B          DO 180 L = NPP1,M
 70.   000250B      180 A(L,J) = ZERO
 71.   000254B      190 CONTINUE
 72.   000255B          W(J)=ZERO
       C   SOLVE THE TRIANGULAR SYSTEM.
       C   STORE THE SOLUTION TEMPORARILY IN ZZ().
 73.   000255B          ASSIGN 200 TO NEXT
 74.   000260B          GO TO 400
 75.   000260B      200 CONTINUE
       C   ****** SECONDARY LOOP BEGINS HERE ******
       C   ITERATION COUNTER.
 76.   000261B      210 ITER=ITER+1
 77.   000262B          IF (ITER.LE.ITMAX) GO TO 220
 78.   000264B          MODE=3
 79.   000264B          WRITE (6,440)
 80.   000270B          GO TO 350
 81.   000270B      220 CONTINUE
       C   SEE IF ALL NEW CONSTRAINED COEFFS  ARE FEASIBLE.
       C   IF NOT COMPUTE ALPHA.
 82.   000271B          ALPHA=TWO
 83.   000271B          DO 240 IP=1,NSETP
 84.   000274B          L=INDEX(IP)
 85.   000274B          IF (ZZ(IP)) 230,230,240
 86.   000277B      230 T=-X(L)/(ZZ(IP)-X(L))
 87.   000303B          IF (ALPHA.LE.T) GO TO 240
 88.   000305B          ALPHA=T
 89.   000306B          JJ=IP
 90.   000307B      240 CONTINUE
       C   IF ALL NEW CONSTRAINED COEFFS ARE FEASIBLE THEN ALPHA WILL
       C   STILL = 2. IF SO EXIT FROM SECONDARY LOOP TO MAIN LOOP.
 91.   000313B          IF (ALPHA.EQ.TWO) GO TO 330
       C   OTHERWISE USE ALPHA WHICH WILL BE BETWEEN 0. AND 1. TO
       C   INTERPOLATE BETWEEN THE OLD X AND THE NEW ZZ.
 92.   000315B          DO 250 IP=1,NSETP
 93.   000317B          L=INDEX(IP)
 94.   000317B      250 X(L)=X(L)+ALPHA*(ZZ(IP)-X(L))
       C   MODIFY A AND B AND THE INDEX ARRAYS TO MOVE COEFFICIENT I
       C   FROM SET P TO SET Z.
 95.   000324B          I=INDEX(JJ)
 96.   000325B      260 X(I)=ZERO
 97.   000327B          IF (JJ.EQ.NSETP) GO TO 290
 98.   000332B          JJ=JJ+1
 99.   000334B          DO 280 J=JJ,NSETP
100.   000336B          II=INDEX (J)
101.   000336B          INDEX(J-1)=II
102.   000337B          CALL G1 (A(J-1,II),A(J,II),CC,SS,A(J-1,II))
103.   000350B          A(J,II)=ZERO
104.   000353B          DO 270 L=1,N
105.   000355B          IF (L.NE.II) CALL G2(CC,SS,A(J-1,L),A(J,L))
```

```
106.    000366B      270 CONTINUE
107.    000370B      280 CALL G2 (CC,SS,B(J-1),B(J))
108.    000377B      290 NPP1=NSETP
109.    000400B          NSETP=NSETP-1
110.    000401B          IZ1 = IZ1 - 1
111.    000403B          INDEX(IZ1)=I
                C        SEE IF THE REMAINING COEFFS IN SET P ARE FEASIBLE. THEY SHOULD
                C        BE BECAUSE OF THE WAY ALPHA WAS DETERMINED.
                C        IF ANY ARE INFEASIBLE IT IS DUE TO ROUND-OFF ERROR. ANY
                C        THAT ARE NONPOSITIVE WILL BE SET TO ZERO
                C        AND MOVED FROM SET P TO SET Z.
112.    000404B          DO 300 JJ=1,NSETP
113.    000406B          I=INDEX(JJ)
114.    000406B          IF (X(I)) 260,260,300
115.    000407B      300 CONTINUE
                C        COPY B() INTO ZZ( ). THEN SOLVE AGAIN AND LOOP BACK.
116.    000414B          DO 310 I=1,M
117.    000416B      310 ZZ(I)=B(I)
118.    000421B          ASSIGN 320 TO NEXT
119.    000422B          GO TO 400
120.    000422B      320 CONTINUE
121.    000423B          GO TO 210
                C    ******END OF SECONDARY LOOP ******
122.    000423B      330 DO 340 IP=1,NSETP
123.    000426B          I=INDEX(IP)
124.    000426B      340 X(I)=ZZ(IP)
                C        ALL NEW COEFFS ARE POSITIVE. LOOP BACK TO BEGINNING.
125.    000432B          GO TO 30
                C            ****** END OF MAIN LOOP ******
                C            COME TO HERE FOR TERMINATION .
                C            COMPUTE THE NORM OF THE FINAL RESIDUAL VECTOR.
126.    000432B      350 SM=ZERO
127.    000433B          IF (NPP1.GT.M) GO TO 370
128.    000436B          DO 360 I=NPP1,M
129.    000441B      360 SM=SM+P(I)**2
130.    000445B          GO TO 390
131.    000445B      370 DO 380 J=1,N
132.    000450B      380 W(J)=ZERO
133.    000452B      390 RNORM=SQRT(SM)
134.    000455B          RETURN
                C        THE FOLLOWING BLOCK OF CODE IS USED AS AN INTERNAL SUBROUTINE
                C        TO SOLVE THE TRIANGULAR SYSTEM,PUTTING THE SOLUTION IN ZZ( ).
135.    000457B      400 DO 430 L=1, NSETP
136.    000461B          IP=NSETP+1-L
137.    000462B          IF (L.EQ.1) GO TO 420
138.    000464B          DO 410 II=1,IP
139.    000467B      410 ZZ(II)=ZZ(II)-A(II,JJ)*ZZ(IP+1)
140.    000474B      420 JJ=INDEX(IP)
141.    000476B      430 ZZ(IP)=ZZ(IP)/A(IP,JJ)
142.    000506B          GO TO NEXT,(200,320)
144.    000506B      440 FORMAT (35H0 NNLS QUITTING ON ITERATION COUNT.)
144.    000507B          END
                C        SUBROUTINE H12 (MODE,LPIVOT,L1,M,U,IUE,UP,C,ICE,ICV,NCV)
                C        C.L.LAWSON AND R.J. HANSON,JET PROPULSION LABORATORY.1973 JUN 12
                C        TO APPEAR IN  SOLVING LEAST SQARES PROBLEMS , PRENTICE-HALL,1974
                C        CONSTRUCTION AND/OR APPLICATION OF A SINGLE
                C        HOUSEHOLDER TRANSFORMATION..   Q = I + U*(U**T)/B
                C        MODE  = 1 OR 2 TO SELECT ALGORITHM  H1 OR H2.
                C        LPIVOT IS THE INDEX OF THE PIVOT ELEMENT.
                C        L1,M IF L1.LE.M THE TRANSFORMATION WILL BE CONSTRUCTED TO
                C        ZERO ELEMENTS INDEXED
                C        ZERO ELEMENTS INDEXED FROM L1 THROUGH M. IF L1 GT. M
                C        THE SUBROUTINE DOES AN IDENTITY TRANSFORMATION.
                C        U( ),IUE,UP ON ENTRY TO H1 U() CONTAINS THE PIVOT VECTOR.
                C                    IUE IS THE STRORAGE INCREMENT BETWEEN ELEMENTS.
                C                       ON EXIT FROM H1 U() AND UP
                C                       CONTAINS QUALITIES DEFINING THE VECTOR U.OF THE
                C                       HOUSEHOLDER TRANSFORMATION.  ON ENTRY TO H2 U()
                C        AND UP SHOULD CONTAIN QUANTITIES PREVIOUSLY COMPUTED
                C                       BY H1. THESE WILL NOT BE MODIFIED BY H2.
                C        C()    ON ENTRY TO H1 OR H2 C() CONTAINS A MATRIX WHICH WILL BE
                C               REGARDED AS A SET OF VECTORS TO WHICH THE HOUSEHOLDER
                C               TRANSFORMATION IS TO BE APPLIED. ON EXIT C() CONTAINS
                C               THE SET OF TRANSFORMED VECTORS
                C        ICE    STRORAGE INCREMENT BETWEEN ELEMENTS OF VECTORS IN C().
                C        ICV    STORAGE INCREMENT BETWEEN VECTORS IN C().
                C        NCV    NUMBER OF VECTORS IN C() TO BE TRANSFORMED. IF NCV .LE. 0
                C               NO OPERATIONS WILL BE DONE ON C().
  1.    000000B          SUBROUTINE H12 (MODE,LPIVOT,L1,M,U,IUE,UP,C,ICE,ICV,NCV)
  2.    000000B          DIMENSION U(IUE,M),C(1)
  3.    000000B          DOUBLE PRECISION SM,B
  4.    000000B          ONE=1.
  5.    000004B          IF (0.GE.LPIVOT.OR.LPIVOT.GE.L1.OR.L1.GT.M) RETURN
  6.    000016B          CL=ABS(U(1,LPIVOT))
  7.    000021B          IF (MODE.EQ.2) GO TO 60
                         ****** CONSTRUCT THE TRANSFORMATION ******
```

```
  8.    000023B          DO 10 J=L1 ,M
  9.    000026B       10 CL=AMAX1(ABS(U(1,J)),CL)
 10.    000034B          IF (CL) 130,130,20
 11.    000035B       20 CLINV=ONE/CL
 12.    000036B          SM=(DBLE(U(1,LPIVOT))*CLINV)**2
 13.    000045B          DO 30 J=L1,M
 14.    000051B       30 SM=SM+(DBLE(U(1,J))*CLINV)**2
         C                        CONVERT DBLE. PREC. SM TO SNGL. PREC. SM1
 15.    000065B          SM1=SM
 16.    000065B          CL=CL*SQRT(SM1)
 17.    000070B          IF (U(1,LPIVOT)) 50,50,40
 18.    000075B       40 CL=-CL
 19.    000077B       50 UP = U(1,LPIVOT) - CL
 20.    000103B          U(1,LPIVOT)=CL
 21.    000106B          GO TO 70
         C              ******* APPLY THE TRANSFORMATION I+U*(U**T)/B TO C ******
 22.    000107B       60 IF (CL) 130,130,70
 23.    000110B       70 IF (NCV.LE.0) RETURN
 24.    000114B          B = DBLE(UP) * U(1,LPIVOT)
         C              B MUST BE POSITIVE HERE. IF B= / , RETURN.
 25.    000123B          IF (B) 80,130,130
 26.    000125B       80 B=ONE/B
 27.    000133B          I2=1-ICV+ICE*(LPIVOT-1)
 28.    000140B          INCR=ICE*(L1- LPIVOT)
 29.    000143B          DO 120 J=1,NCV
 30.    000145B          I2=I2+ICV
 31.    000146B          I3=I2+INCR
 32.    000147B          I4=I3
 33.    000147B          SM=C(I2)*DBLE(UP)
 34.    000154B          DO  90 I=L1,M
 35.    000160B          SM=SM+C(I3)*DBLE(U(1,I))
 36.    000171B       90 I3=I3+ICE
 37.    000174B          IF(SM) 100,120,100
 38.    000176B      100 SM=SM*B
 39.    000202B          C(I2)=C(I2)+SM*DBLE(UP)
 40.    000212B          DO 110 I=L1,M
 41.    000214B          C(I4)=C(I4)+SM*DBLE(U(1,I))
 42.    000225B      110 I4=I4+ICE
 43.    000230B      120 CONTINUE
 44.    000232B      130 RETURN
 45.    000235B          END


  1.    000000B          SUBROUTINE G1 (A,B,COS,SIN,SIG)
         C              C.L.LAWSON AND R.J.HANSON,JET PROPULSION LABORATORY,1973 JUN 12
         C              TO APPEAR IN  SOLVINGLEAST SQUARES PROBLEMS  , PRENTICE HALL,1974
         C              COMPUTE ORTHOGONAL ROTATION MATRIX..
         C              COMPUTE..MATRIX (C,S) SO THAT (C,S)(A)= (SQRT(A**2+B**2))
         C              COMPUTE SIG = SQRT(A**2+B**2)
         C              SIG IS COMPUTED LAST TO ALLOW FOR THE POSSIBILITY THAT
         C              SIG MAY BE IN THE SAME LOCATION AS A OR B .
  2.    000000B          ZERO=0.
  3.    000000B          ONE=1.
  4.    000001B          IF (ABS(A).LE.ABS(B)) GO TO 10
  5.    000007B          XR=B/A
  6.    000010B          YR=SQRT(ONE+XR**2)
  7.    000014B          COS=SIGN(ONE/YR,A)
  8.    000016B          SIN=COS*XR
  9.    000017B          SIG=ABS(A)*YR
 10.    000022B          RETURN
 11.    000024B       10 IF (B) 20,30,20
 12.    000025B       20 XR=A/B
 13.    000026B          YR=SQRT(ONE+XR**2)
 14.    000032B          SIN=SIGN(ONE/YR,B)
 15.    000034B          COS=SIN*XR
 16.    000035B          SIG=ABS(B)*YR
 17.    000040B          RETURN
 18.    000042B       30 SIG=ZERO
 19.    000042B          COS=ZERO
 20.    000043B          SIN=ONE
 21.    000045B          RETURN
 22.    000050B          END


  1.    000000B          SUBROUTINE G2 (COS,SIN,X,Y)
         C              C.L.LAWSON AND R.J.HANSON,JET PROPULSION LABORATORY,1972 DEC 15
         C              TO APPEAR IN  SOLVING LEAST SQARES PROBLEMS , PRENTICE-HALL,1974
         C              APPLY THE ROTATION COMPUTED BY G1 TO (X,Y).
  2.    000000B          XR=COS*X+SIN*Y
  3.    000003B          Y=-SIN*X+COS*Y
  4.    000004B          X=XR
  5.    000005B          RETURN
  6.    000010B          END
```

APPENDIX B

Table B-I. Absorbance-Matched Model Compound Concentrations (mg/l based on total river water)—Spring Samples of Mississippi River Water and Ultrafiltered Constituents

	BEM						ROY, 216 m^3/sec					
	<0.5K	0.5-1K	1-10K	10-100K	>100K	MRW	<0.5K	0.5-1K	1-10K	10-100K	>100K	MRW
EN	0.21	0.42	0.59	0.30	0.25	0.37	0.11	0.53	1.44	0.32	0.28	0.49
Vanillin	0.09	0.09	0.31	0.05	0.05	0.89		0.11	0.69	0.10	0.05	1.11
Tannic Acid			0.35			0.41		0.11	1.37			0.39
3,4-Dimethoxyacetophenone	0.50	0.31	0.63	0.07	0.07	2.20	0.26	0.39	0.98	0.23	0.13	2.64
p-Anisic Acid	0.30	0.22	0.68	0.03	0.04	1.81	0.23	0.32	1.15	0.16	0.09	1.99
Gallic Acid												
Mellitic Acid		0.16	0.30			1.22		0.27	1.03	0.08		1.22

	SAF, 475 m^3/sec						IGB, 838 m^3/sec					
	<0.5K	0.5-1K	1.10K	10-100K	>100K	MRW	<0.5K	0.5-1K	1-10K	10-100K	>100K	MRW
EN	0.4	0.3	1.48	0.02	0.02	0.77	0.36	0.31	0.58	0.03	0.02	0.10
Vanillin		0.30	0.76	0.08	0.06	1.77		0.14	0.46	0.07	0.04	0.81
Tannic Acid		0.30	1.45	0.34	0.24			0.71	1.73	0.20	0.11	1.61
3,4-Dimethoxyacetophenone		0.07	1.07	0.01	0.03	2.98		0.21	0.37	0.02	0.02	
p-Anisic Acid		0.19	1.34	0.16	0.14	1.18		0.46	0.92	0.09	0.08	
Gallic Acid		0.13		0.04		2.94					0.03	1.46
Mellitic Acid		0.29	1.12	0.07	0.17			0.53	0.89	0.09		
Syringic Acid		0.06			0.04	0.04				0.01		
3,4,5-Trimethoxybenzoic Acid				0.01							0.02	
Pentacarboxybenzoic Acid				0.08		0.90				0.10		
Salicylic Acid		0.02		0.02	0.01	1.15					0.04	

1,7-Dimethoxyxanthine	0.50		0.02		0.50	0.01	0.03	0.18
Catechol		0.02						
Resorcinol		0.02		0.01	2.72	0.01		1.47
1,3,5-Tricarboxybenzoic Acid	0.14		0.08			0.02	0.10	2.09
Lignin			0.04				0.06	
2,4-Dimethoxybenzoic Acid					2.53			
1,2,4,5-Tetracarboxybenzoic Acid								1.60
Vanillic Acid					1.53	0.5		
iso-Phthalic Acid	0.11		0.04					
2,4-Dicarboxybenzoic Acid	0.04							0.46

Table B-II. Absorbance-Matched Model Compound Concentrations (mg/l based on total river water)—Fall Samples of Mississippi River Water and Ultrafiltered Fractions

	BEM						ROY, 213 m³/sec						SAF, 329 m³/sec						IGB, 342 m³/sec					
	$<$1/2K	1/2-1K	1-10K	10-100K	$>$100K	MRW	$<$1/2K	1/2-1K	1-10K	10-100K	$>$100K	MRW	$<$1/2K	1/2-1K	1-10K	10-100K	$>$100K	MRW	$<$1/2K	1/2-1K	1-10K	10-100K	$>$100K	MRW
EN	0.03	0.22	2.58	0.12	0.05	0.37	0.03	0.05	0.71	0.67	0.04	0.35	0.02	0.14	2.46	0.14	0.06	0.05	0.04	0.19	4.62	0.07	0.01	0.06
Vanillin		0.08	1.05	0.07	0.03	1.03		0.02	0.31	0.39	0.05	1.32		0.05	1.13	0.15	0.06	1.12		0.08	1.18	0.07	0.03	1.03
Tannic Acid		0.38	2.58	0.18	0.11	3.08		0.11	0.60	0.70	0.25	4.07		0.47	2.80	0.54	0.27	4.73		0.63	0.88	0.14	0.13	4.33
3,4-Dimethoxy-acetophenone		0.20	1.82	0.10	0.05	2.01		0.08	0.52	0.42	0.03	1.80		0.12	1.76	0.13	0.05	0.76		0.16	2.25	0.09	0.02	0.61
p-Ansic Acid	0.10	0.32	2.16	0.15	0.09	2.71		0.15	0.58	0.52	0.14	2.89	0.07	0.31	2.29	0.31	0.16	2.50		0.40	1.58	0.16	0.09	2.11
Gallic Acid																		0.93					0.01	0.74
Mellitic Acid		0.24	1.80	0.11	0.07	1.89		0.12	0.46	0.47	0.11	2.44		0.30	1.96	0.22	0.13	2.91		0.41	3.64	0.22	0.09	2.81
Syringic Acid								0.04										0.42				0.13	0.04	0.22
3,4,5-Trimethoxy-benzoic Acid																		0.33						0.80
Salicylic Acid	0.08																						0.01	
1,7-Dimethoxy-xanthine							0.04						0.24						0.53				0.01	
Catechol																								0.45
Resorcinol	0.43						0.19																	
3,4,5-Trimethoxy-acetophenone								0.03																
1,3,5-Tricarboxy-benzoic Acid																		1.42	0.08				0.03	1.17
Lignin																								
o-Methoxybenzoic Acid	0.09																						0.01	
o-Phthalic Acid																								0.07
2,4-Dimethoxy-benzoic Acid							0.32						0.21											
Vanillic Acid							0.02						0.03											
iso-Phthalic Acid																			0.14					

Table B-III. Absorbance-Matched Model Compound Concentrations (mg/l based on total river water)—Winter Samples of
Mississippi River Water and Ultrafiltered Fractions

	BEM						ROY, 86 m^3/sec						SAF, 125 m^3/sec						IGB, 138 m^3/sec					
	<1/2K	1/2-1K	1-10K	10-100K	>100K	MRW	<1/2K	1/2-1K	1-10K	10-100K	>100K	MRW	<1/2K	1/2-1K	1-10K	10-100K	>100K	MRW	<1/2K	1/2-1K	1-10K	10-100K	>100K	MRW
EN	0.04	0.04	0.14	0.02	0.02	0.03	0.02	0.02	0.59	0.01	0.01	0.02	0.04	0.05	0.53	0.01	0.01	0.03	0.07	0.05	0.11	0.01	0.01	0.02
Vanillin	0.41	0.05	0.06			0.19	0.06	0.10	0.22			0.24		0.02	0.21			0.36		0.02	0.16		0.01	0.43
Tannic Acid		0.02	0.39			0.38		0.64	1.02	0.01	0.01	0.89			1.14		0.02	1.16			0.86		0.03	0.62
3,4-Dimethoxy-acetophenone	0.06	0.07	0.10		0.01			0.17	0.40	0.01	0.01	0.17		0.08	0.34			0.26		0.08	0.11			0.14
p-Anisic Acid	0.26	0.15	0.26		0.01	0.43	0.10	0.42	0.71	0.03	0.02	0.71		0.15	0.70	0.02	0.02	0.86		0.15	0.45	0.01	0.02	0.76
Gallic Acid						0.22						0.19					0.01	0.08					0.01	0.33
Mellitic Acid		0.11	0.25					0.46	0.81	0.02		0.81		0.16	0.87		0.01			0.16	0.68	0.01	0.01	1.24
Syringic Acid		0.04										0.41		0.03		0.02		0.10		0.03	0.07			0.07
3,4,5-Trimethoxy-benzoic Acid												0.08						0.04						0.18
Pentacarboxy-benzoic Acid						0.61						0.09				0.01	0.02	1.02						0.05
Salicylic Acid	0.16					0.08	0.15					0.16	0.01			0.02	0.01	0.05					0.01	0.30
1,7-Dimethoxy-Xanthine							0.05											0.19	0.22				0.01	0.25
Catechol	0.78	0.01			0.02	0.46			0.02	0.02		0.50					0.04	0.28					0.02	0.10
Resorcinol			0.03			0.21	0.73						0.66							0.34				0.01	0.37
3,4,5-Trimethoxy-acetophenone		0.09				0.19			0.01					0.10						0.10			0.03	0.48
1,3,5-Tricarboxy-benzoic Acid																								0.20
Lignin																		0.47						
o-Methoxybenzoic Acid																		0.40						
2,4-Dimethoxy-benzoic Acid													0.15						0.17					
Vanillic Acid	0.05						0.06																	
iso-Phthalic Acid																								
p-Cresol													0.03						0.04					
2,6-Dimethoxy-benzoic Acid							0.17																	

CHAPTER 6

DEVELOPMENT OF A TOC/COD ANALYZER FOR PROCESS APPLICATIONS

J. B. Lantz, R. J. Davenport and R. A. Wynveen

Life Systems, Inc.
Cleveland, Ohio

W. J. Cooper

Drinking Water Research Center
Florida International University
Miami, Florida

Environmental and public health considerations are increasing our awareness of the importance of detecting and monitoring organic contaminants in waters and waste waters. Many laboratory instruments have been developed to determine specific organic solutes and general indices of organic contamination. However, development of instrumentation for monitoring and control applications has lagged behind the development of laboratory instruments. This is probably due to the added sophistication and development time required to (1) reliably automate the analytical functions; (2) develop special operator interfaces to simplify control of the instrument, eliminating the need for skilled chemists or technicians; (3) develop special electrical interfaces, required in certain applications, to permit receiving commands and transmitting status indicators from the instrument to remote operator control panels or to centralized control/monitoring instrumentation; and (4) develop special features to decrease the frequency of operator maintenance to make it practical to use the instrument in remote locations or where skilled maintenance personnel are not available.

A total organic carbon (TOC)/chemical oxygen demand (COD) analyzer has been developed to an advanced breadboard level to provide these characteristics for general organic solute concentration monitoring and control. TOC is an organic solute measurement based on oxidation of the organics in a portion of the sample and measurement of the resulting carbon dioxide, either directly or after conversion to methane [1]. COD is a measure of the oxidizability of solutes in the sample (generally organic), and COD is determined by measuring the amount of a strong oxidizing agent required to oxidize the solutes in a portion of the sample [2].

The TOC/COD analyzer uses the ultraviolet light (UV)/persulfate technique. Organics are oxidized to carbon dioxide, and persulfate is reduced to bisulfate:

$$CH_3OH + 3S_2O_8{}^{2-} + H_2O \Rightarrow CO_2 + 6HSO_4{}^- \tag{1}$$

$$HCOOH + S_2O_8{}^{2-} \Rightarrow CO_2 + 2HSO_4{}^- \tag{2}$$

Inorganic carbon (IC) is automatically removed from the sample prior to the persulfate oxidation. After the oxidation, the carbon dioxide is measured to determine the TOC concentration, while the amount of persulfate consumed determines the COD concentration. The analyzer can be operated without IC removal to measure total carbon (TC). The sample can also be filtered in the analyzer through a 0.45-μm filter. When this is done, the measured parameter is dissolved organic carbon (DOC).

Equations 1 and 2 illustrate how two samples containing the same TOC concentration can have different COD concentrations, depending upon the oxidation state of the organic solutes. The ratio of COD to TOC values (or to TC or DOC values) can therefore be used to indicate the type of organics in the sample.

The TOC/COD analyzer combines both measurements in a single instrument. The advanced breadboard version totally automates these analyses, displays COD/carbon ratios and provides for the special requirements of on-line organic solute monitoring and control.

EARLY DEVELOPMENTS

The concept of the TOC/COD analyzer originated in response to the need for a monitor of residual organics dissolved in the ozonated effluent of the Water Processing Element (WPE) developed for the Medical Unit, Self-Contained, Transportable (MUST) Army field hospital [3]. Water Processing Element specifications called for monitoring effluents having only 0.1-10 mg/l TOC and 0.3-30 mg/l COD. The WPE specifications also required automated, on-line operation with minimal expendables to reduce the logistic burden of operating in remote areas. Compact size and reliable performance were required because of the mobile nature of the WPE and the labor-short nature of the application.

The concept of the TOC/COD analyzer was derived and refined to meet these requirements. Sparging the sample to remove IC prior to measuring TOC was rejected, for example, because it requires use of compressed gases that are a logistic burden. Carbon dioxide (CO_2) and persulfate concentration measurement techniques that consume reagents were rejected for the same reason.

Life Systems, Inc. (LSI) subsequently assembled a continuous chemical TOC analyzer to demonstrate some of the concepts selected for the TOC/COD analyzer. An IC stripper, using a gas-permeable membrane to extract CO_2 from the sample, was assembled and proven. A UV reactor was also assembled and tested, and a commercially available CO_2 sensor was adapted for measuring TOC in the analyzer.

LABORATORY BREADBOARD TOC/COD ANALYZER

The complete TOC/COD analyzer concept was tested and evaluated. Key analyzer components were first assembled and individually tested, then integrated and tested as part of a laboratory breadboard TOC/COD analyzer (Figures 1 and 2). The laboratory breadboard TOC measurements agreed well with measurements made using a commercial laboratory TOC analyzer (Figure 3 and Table I).

A somewhat lower-than-theoretical COD response to many organics (typically 60%) was observed, and this was shown experimentally to be due to oxygen in the water furnishing some of the oxidation power (instead of persulfate) under the influence of UV light. The response to the refractory acetic acid (not air oxidizable) agreed with the theoretical value. After calibrating the analyzer with potassium hydrogen phthalate (KHP), the COD measurement for all but acetic acid produced relatively close agreement to the theoretical values (Figure 4).

Calibrated COD values (vs KHP) were determined for several types of natural water samples (Table I). Some values were somewhat higher than for COD values determined by the dichromate method, indicating that in general the organics were more refractory (less susceptible to air oxidation) than the standard, KHP.

The laboratory breadboard demonstrated that the TOC/COD analyzer could provide the characteristics necessary for process monitoring/control applications and remote water quality monitoring.

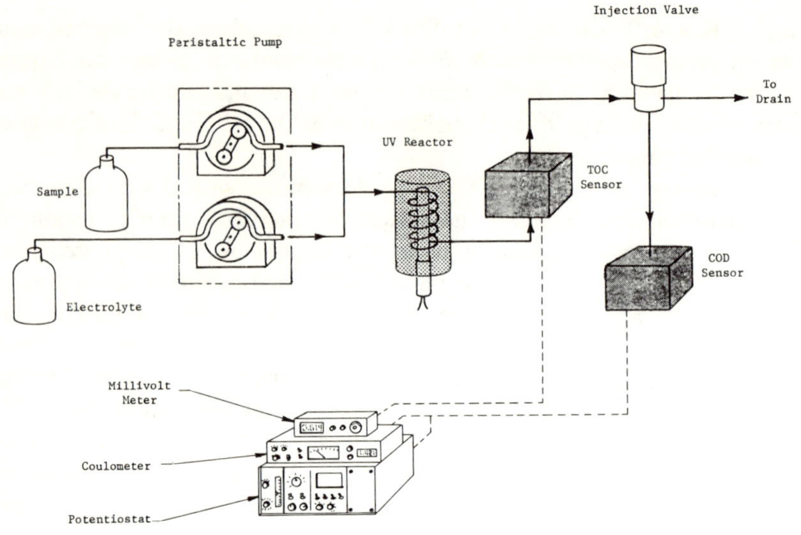

Figure 1. Schematic of laboratory breadboard TOC/COD analyzer.

ADVANCED BREADBOARD TOC/COD ANALYZER

At this point in the development, applications other than monitoring ozonated effluents were proposed for the TOC/COD analyzer. Some of these required measurements of samples having more than 100 mg/l TOC or 300 mg/l COD. Samples containing particulates were also considered for the first time. As a result, the design specifications of the TOC/COD analyzer underwent changes to satisfy the needs of these new applications. The design requirements and specifications of the current stage of development, the advanced breadboard TOC/COD analyzer, reflect these changes and the requirements developed during testing of the laboratory breadboard.

The advanced breadboard TOC/COD analyzer was designed with the goal of providing all the special features desirable for on-line monitoring and control of wastewater and water processes (Table II). The design of the advanced breadboard incorporates a microcomputer to automate the analyzer's startup, calibration and shutdown procedures. This simplifies operation and reduces the skill required by the operator. Electrical interfaces are provided to output the analyzer's status and to receive remote commands. In situ hydroxide and persulfate generation, and the use of an electrolyte recycle loop were designed to extend the period between reagent resupply activities to 30 days. The range and sampling capability goals of the advanced bread-

Figure 2. Laboratory breadboard TOC/COD analyzer.

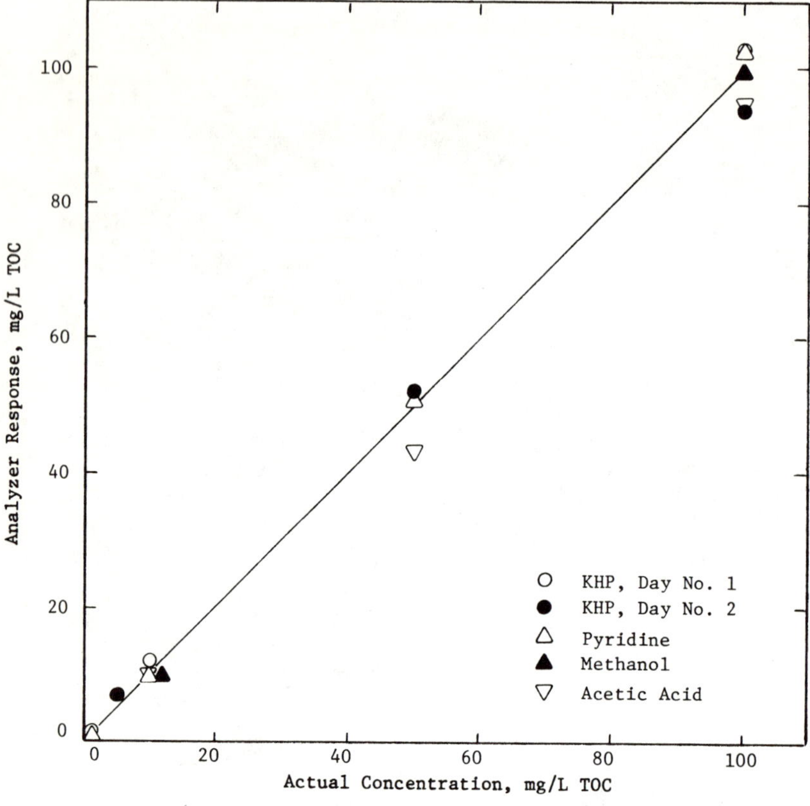

Figure 3. TOC response of laboratory breadboard analyzer.

board were extended from monitoring only ozonated effluents to also monitoring natural waters and treated water and sewage. Therefore, the advanced breadboard TOC/COD analyzer was designed to achieve a level of performance that heretofore has not been achieved.

The advanced breadboard is composed of two major packages: the analytical package and the instrumentation package. These are shown mounted on the TOC/COD analyzer characterization test stand in Figure 5.

Analytical Package

The analytical package (right side of Figure 5) contains all actuators, sensors, fluid handling components and the other mechanical and electrochemical devices in the analyzer. The components are located on both sides

Table I. Laboratory Breadboard Analyzer Results

Sample	TOC Concentration (mg/l)			COD Concentration (mg/l)		
	TOC/COD Analyzer	Dohrmann DC-50	EPA Value	TOC/COD Analyzer	Cr(VI) Method	EPA Value
EPA Demand Sample, No.1	6.0	5.8	6.1	16.2		15.4
Tap Water, No. 1	2.6	2.7		11.1	6.5	
Tap Water, No. 2	0.4	1.0				
Well Water	3.0	3.0		6.4	4.1	
Chagrin River Water	5.0	4.8				
Domestic Wastewater, Secondary Effluent						
Sample No. 1 (filtered)	4.3	4.5		17.5	11.2	
Sample No. 2	9.8	9.8		20.4	16.8	
Sample No. 3	9.5	9.4		30.4	23.0	
Sample No. 4	7.4	7.6				
Domestic Wastewater, Untreated	31.1	27.3		166.8	164.5	

of a vertical mounting panel, rather than in an enclosure, to facilitate assembling, debugging and maintenance of the analyzer.

The analytical package incorporates six basic processes:

1. electrochemical generation of the persulfate from an acidic electrolyte;
2. continuous sampling and mixing of the sample with the electrolyte and persulfate;
3. removal of IC (HCO_3^-, CO_3^{2-} and dissolved CO_2) from the sample;
4. oxidation of the organics;
5. measurement of the CO_2 produced and persulfate consumed in the oxidation; and
6. elimination of water from the electrolyte that was introduced during sampling.

Schematic

These processes are integrated as shown in Figure 6. During normal operation, the sample is continuously drawn into the analyzer through a 10-μm filter that removes particles that might otherwise eventually restrict liquid flow through the analyzer. If DOC measurements are made, the sample also passes through a 0.45-μm membrane filter.

For IC removal during TOC or DOC measurements, the IC stripper is used. The sample is divided into two streams. One is mixed with the acidic electrolyte (containing persulfate), and IC is immediately converted to CO_2 by the

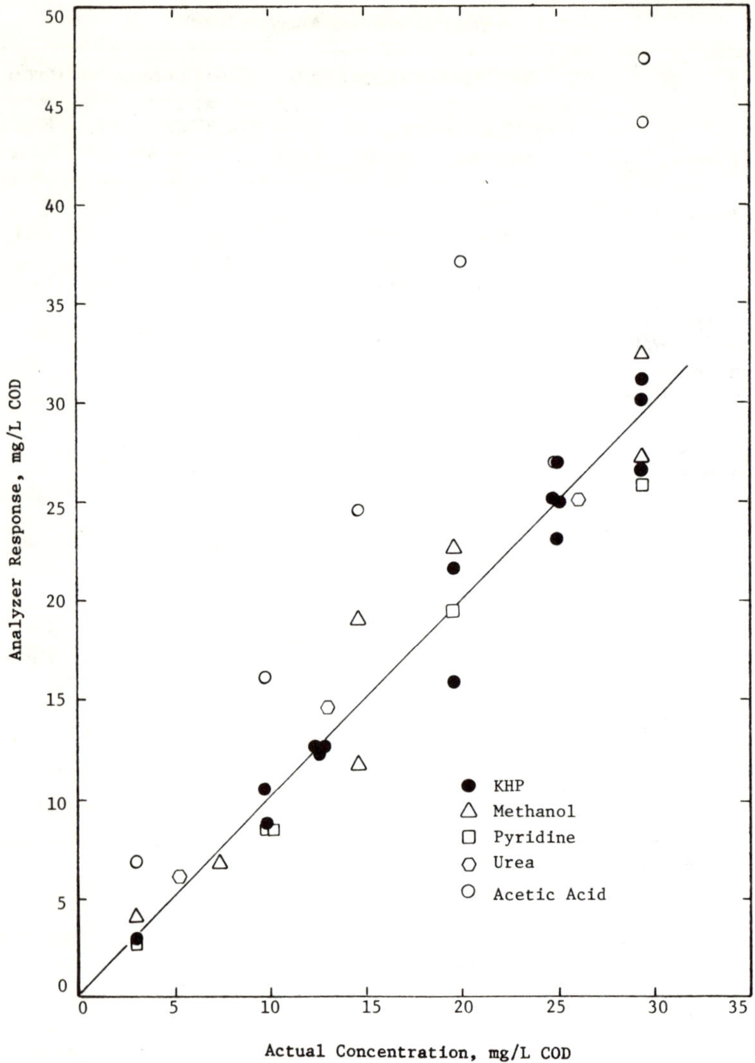

Figure 4. COD response of KHP-calibrated laboratory breadboard analyzer.

acid. The second sample stream passes through the hydroxide generator, which injects sodium hydroxide (NaOH) into the sample to make it basic.

When the acidic and basic sample streams flow through the IC stripper, CO_2 diffuses from the acidic stream, through a gas-permeable membrane, to the basic stream. Effective IC removal is thereby achieved. If TC measurements are made, the acidified sample stream bypasses the stripper.

Table II. Design Goals

Self-Contained, Packaged Design

Concentration Range
- 0.1-500 mg/l TOC
- 100-1500 mg/l COD

Continuous, On-Line Sampling

Direct Concentration Readout

Analog Outputs Compatible with Strip Chart
  Recorders and Data Acquisition System

In Situ Persulfate Generation

Acceptance of up to 300 mg/l Cl^- [a]

Automated Calibration
- Autozero
- Autospan

Measurement of TOC, DOC and TC

Acceptance of up to 0.3-mm diameter particles
  with homogenizer accessory [a]

Unattended Operation for 30 Days

Electrolyte Recycle Loop

---

[a]Design pending as of this writing.

The effluent from the stripper is also divided into two streams. Both flow through the UV reactor, but only one flows through the quartz coil in the reactor, which is exposed to UV radiation. This is called the "analytical stream." It exits the UV reactor and flows through the TOC sensor, where CO_2 produced in the oxidation of organic solutes (and from acidification of IC, if TC measurements are made) is measured. The analytical stream continues to the analytical cell of the COD sensor.

The COD sensor consists of two identical electrochemical cells that measure the persulfate concentration in each of the two sample/electrolyte streams. The concentration of persulfate in the analytical stream is reduced by the amount of persulfate consumed in the oxidation. However, the other stream (the "reference stream," which is not exposed to UV radiation in the UV reactor) contains the persulfate concentration that would exist if the sample contained no organics.

The difference in persulfate concentrations between the reference and analytical streams is measured by the COD sensor. The concentration difference is proportional to the COD concentration in the sample. This differential technique is designed to compensate automatically for variations in the

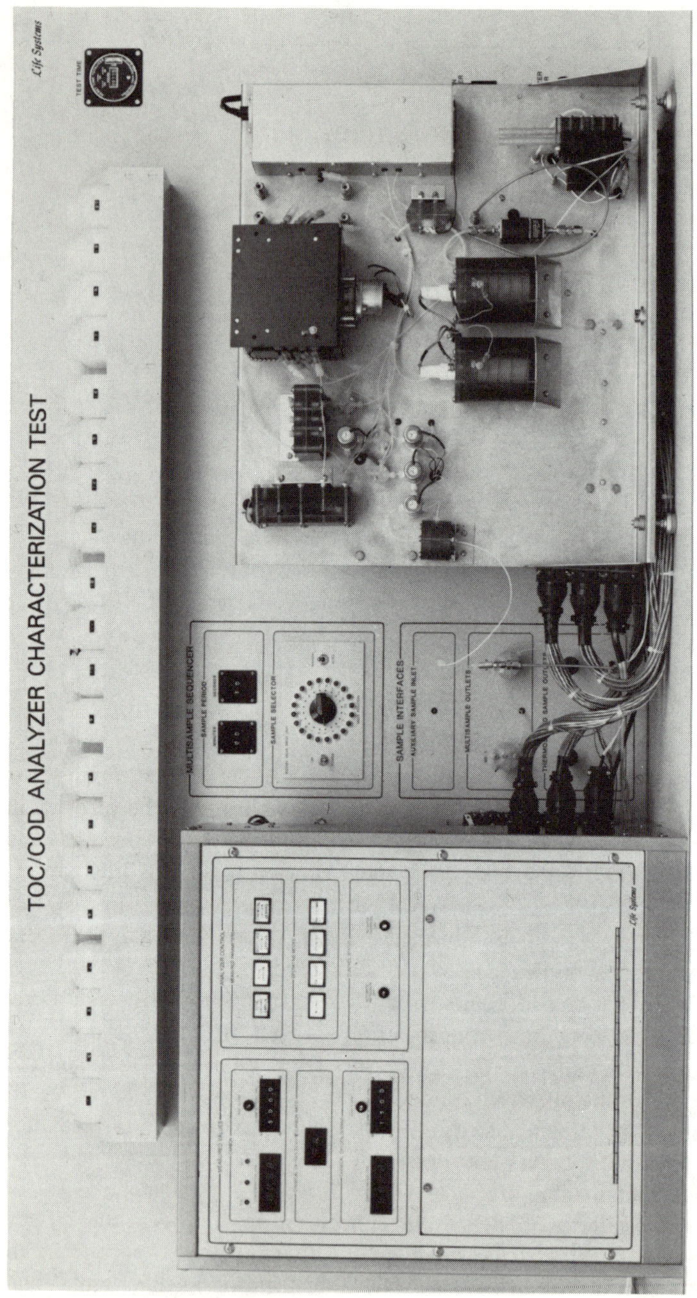

Figure 5. Advanced breadboard TOC/COD analyzer, front view.

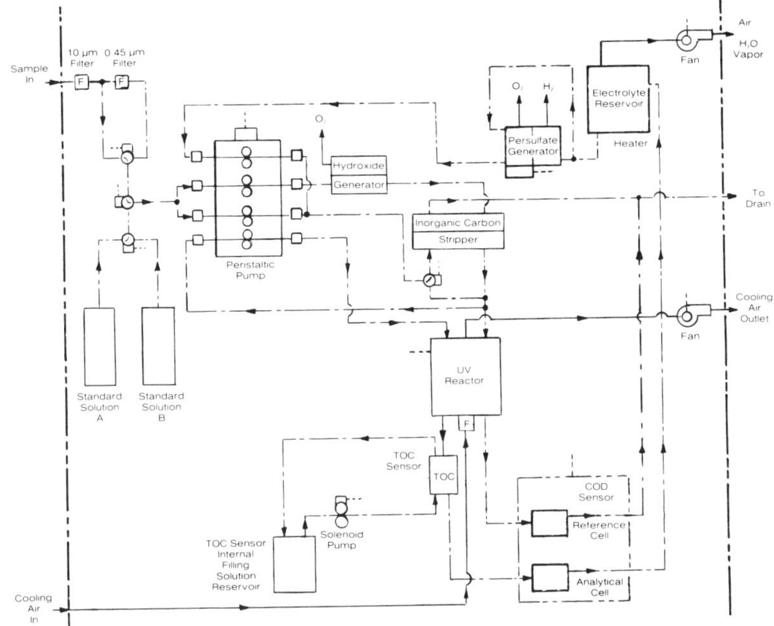

Figure 6. Advanced breadboard TOC/COD analyzer schematic.

persulfate generation efficiency in the persulfate generator and for changes in the rate of spontaneous thermal decomposition of persulfate.

The analytical stream is returned to the electrolyte reservoir, where it is heated, and water vapor is removed to remove water added during sampling. Electrolyte is therefore brought to its original concentration and recycled to the persulfate generator. This device electrochemically oxidizes sulfate $(SO_4{}^{2-})$ in the electrolyte to generate persulfate.

The major mechanical analyzer components are discussed in more detail below.

Major Components

Many of the features of the advanced breadboard analyzer are the result of the unique or innovative components used in the analyzer.

The hydroxide generator operation is based on the LSI patented electrochemical valve principle [6]. The hydroxyl ions (OH^-) are electrochemically generated in the sample stream, while sodium ions (Na^+) migrate through a cation exchange membrane from a slurry of NaOH pellets in water. The concentration of NaOH injected into the sample stream is controlled by controlling the current flowing between the electrodes in the generator.

In the IC stripper, CO_2 passes from the acidic solution to the basic solution through a gas-permeable membrane. The IC in the sample is removed effectively because of the large CO_2 partial pressure difference that exists between the acidic solution (in which IC exists as CO_2) and the basic solution (in which the CO_2 is converted to the nonvolatile $CO_3{}^{2-}$). This minimizes the loss of volatile organics and avoids use of compressed gas attendant with use of a sparger.

The persulfate generator avoids frequent replacement of a persulfate solution with fresh reagent by continuously generating the persulfate required for operation in an electrochemical cell. Persulfate is extremely reactive, and solutions containing persulfate are not stable for more than 24 hours. With the persulfate generator, however, sulfate anions $(SO_4{}^-)$ are electrochemically oxidized to form persulfate as required. The persulfate concentration is feedback controlled, based on the COD reference sensor output, to compensate for generation condition and electrolyte composition variations.

The UV reactor consists of parallel coils of quartz and opaque Teflon$^{®}$ tubing (for the analytical and reference streams, respectively) surrounding a UV lamp. The UV lamp irradiates the sample/electrolyte mixture in the quartz coil within a residence time of about eight minutes. The reactor is forced-air-cooled to minimize spontaneous thermal decomposition of persulfate and maintain a high oxidation efficiency.

The TOC sensor is based on the membrane electrode concept. This type of sensor is responsive to CO_2 over the concentration range of the TOC analyzer, is simpler than most CO_2 measurement devices and eliminates the need for sparging of CO_2 from the sample. The TOC sensor determines the concentration of CO_2 contained in the sample/electrolyte mixture by detecting the pH change caused by the diffusion of CO_2 from the sample/electrolyte mixture, across a gas-permeable membrane, into an internal filling solution containing HCO_3. The pH change, logarithmically proportional to the CO_2 concentration, is automatically linearized by the analyzer instrumentation.

It was found that with conventional membrane CO_2 electrodes, reestablishment of equilibrium response to low CO_2 concentration samples (e.g., 1 mg/l TOC) could require more than an hour if the previous sample produced a high CO_2 concentration (e.g., 100 mg/l TOC). Restoration of low equilibrium CO_2 concentrations in the internal filling solutions is slow, due to slow diffusion of CO_2 from the internal filling solution to the sample (vs relatively rapid diffusion in the reverse direction). In contrast, the TOC sensor in the advanced breadboard analyzer pumps refreshed, low CO_2 concentration internal filling solution to the membrane electrode before each measurement. Therefore, regardless of the CO_2 concentration in the oxidized TOC/DOC or

TC sample, CO_2 diffusion is always from the sample to the filling solution—relatively fast—and the TOC sensor can recover quickly when the sample carbon concentration drops.

The COD sensors operate on an amperometric principle. The difference between electrical currents required to electrochemically reduce persulfate in the analytical and reference sensors is proportional to the COD—the amount of persulfate consumed by the sample components during oxidation in the analytical stream.

Instrumentation Package

The instrumentation package (left side of Figure 5) contains the operator interfaces, process monitoring/control interfaces and the other electrical components of the analyzer. A microcomputer in the instrumentation package performs calculations, monitors component status and controls the analyzer's operating modes. Features of the advanced breadboard analyzer, made possible by the analyzer's advanced instrumentation, are listed in Table III.

Table III. Advanced Breadboard Instrumentation Features

Automated Startup, Shutdown Procedures
- Manually initiated
- Automatically initiated every 24 hours

Automatic Calculation of COD/Carbon Concentration Ratio

Acceptance of Command Inputs for:
- Initiation of operating mode transitions
- Selection of measurement parameters

Transmittal of Status Indicators
- Parameters measured
- Operating mode
- Operating mode transition underway

Autoprotection
- Rejects incorrect commands
- Detects failure in major components—
  initiates automatic shutdown

Fault Detection
- Displays code identifying incorrect commands
- Displays code identifying component causing shutdown

Operating Modes

The advanced breadboard has four operating modes: NORMAL, CALI-BRATION, STANDBY and SHUTDOWN. Commands received by the analyzer directing it to undergo undesirable transitions between these modes are automatically rejected.

In the SHUTDOWN mode, the instrumentation package and sensors retain power, but the display is blank and other components are unpowered. Shutdown is called for by manual actuation, by detection of improper analytical package component operating conditions or by unsuccessful calibrations or mode transitions. On subsequent (manual) call for the NORMAL mode, displays are not reactivated for a preset interval to permit proper monitoring conditions to be restored. In the NORMAL mode the analyzer is performing cally on completion of a calibration. In the CALIBRATION mode, the analyzer automatically adjusts the zero and span of the COD and TOC sensors to conform to two standards. This mode is called for automatically every 24 hr and, optionally, manually at any time. The STANDBY mode is available on manual actuation and is used during brief interruptions in monitoring (there is no time delay on reactivation of NORMAL mode). All components except the peristaltic pump and hydroxide and persulfate generators remain operating (displays and analog outputs are disengaged, however).

Operator Interfaces

The primary operator interfaces are located on the upper front panel of the instrumentation package, Figure 7. Autocalibrated carbon concentration, COD concentration and COD/carbon ratio are displayed digitally (indicator lights show which carbon parameter is being displayed). Adjustable high alarm setpoints and alarm indicator lights provide warnings when the carbon concentration and COD exceed preset values.

By pushing the appropriate illuminated switches, the operator can select the parameters to be measured and can initiate operating mode transitions. Indicator lights also are located there to indicate when automatic protection (shutdown capability) of any component is off, or when automatic operation of any actuator is overridden.

Manual actuation overrides, autoprotection controls and actuator controls are located on the recessed panel located at the bottom of the front panel (Figure 7). These controls are not normally required for operation, but are provided for initial analyzer setup and debugging.

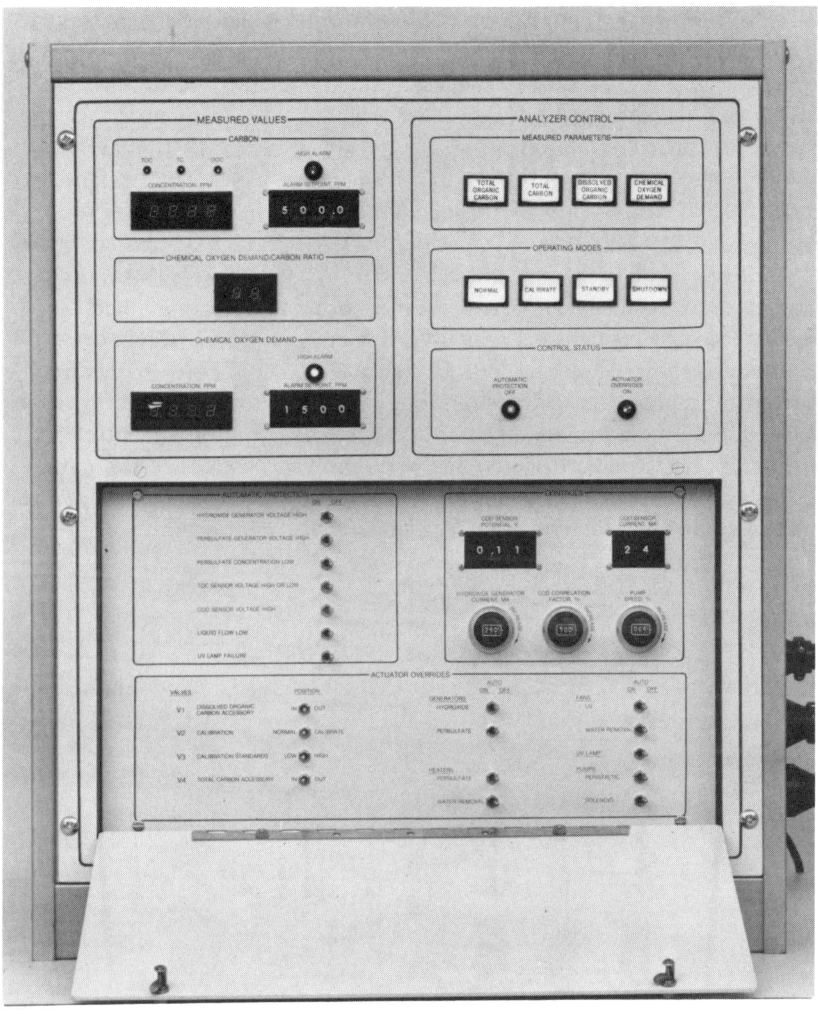

Figure 7. Instrumentation package, front panel.

Electrical Interfaces

The electrical interfaces of the advanced breadboard are illustrated schematically in Figure 8. Digital inputs are provided to receive commands (0-5 V dc) from water and wastewater treatment process control/monitoring instrumentation, or from remote operator control panels. These command inputs can be used to select parameters to be measured and to initiate operating mode transitions.

Analog electrical outputs (0-5 V dc) are provided to transmit carbon (TOC, TC or DOC) concentration, COD concentration and the COD/carbon ratio. Digital overrange alarm signals (0-5 V dc) are provided for both the carbon and COD measurements.

Additional digital electrical outputs (0-5 V dc) are provided to indicate the status of the analyzer (parameter measured and operating mode), and to indicate when transitions between operating modes are underway.

These electrical outputs will be valuable in monitoring or controlling treatment processes and in monitoring water quality in remote locations. Using these interfaces, the operator or the process monitoring/control instrumentation will be able to determine when problems occur and can initiate a calibration sequence or shut down the analyzer, as required. The analyzer can be operated only when desired and calibrated when convenient.

Evaluation

Evaluation of the advanced breadboard TOC/COD analyzer is in process as of this writing. The interim status of that evaluation is being reported elsewhere [7].

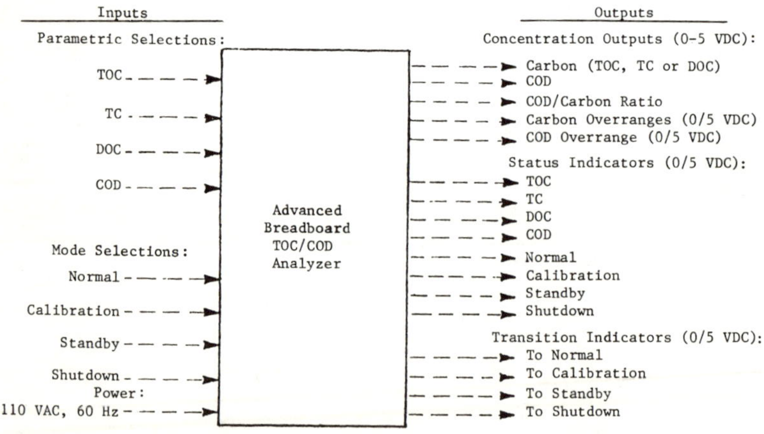

Figure 8. Electrical interfaces.

CONCLUSION

The TOC/COD analyzer under development possesses the potential to move general organic solute analysis from the laboratory to the water treatment process line and field. Its features will not only provide automated on-line monitoring, but will greatly simplify operation, reduce routine maintenance and enable interfacing with modern centralized plant instrumentation.

ACKNOWLEDGMENTS

The authors gratefully acknowledge the financial support of the U.S. Army Medical Bioengineering Research and Development Laboratory, Ft. Detrick, Frederick, MD; the Civil Engineering Laboratory, Port Hueneme, CA; the Civil and Environmental Engineering Development Office, Tyndall Air Force Base, FL; the Chemical Systems Laboratory, Aberdeen Proving Ground, MD; Life Systems, Inc., Cleveland, OH; and the U.S. Environmental Protection Agency, EMSL-Cincinnati, OH.

REFERENCES

1. "Methods for Chemical Analysis of Water and Wastes," U.S. EPA Report EPA-625-/6-74-003a (1974).
2. *1979 Annual Book of ASTM Standards, D 1252-78* (Philadelphia, PA: American Society for Testing and Materials, 1979).
3. Davenport, R. J., and R. A. Wynveen. "Development of Organic Solute and Total Organic Carbon Monitors," Annual Report, Contract DAMD17-75-C-5070, ER-285-3; Life Systems, Inc., Cleveland, OH (June 1976).
4. Davenport, R. J., and R. A. Wynveen. "Fundamental Studies for Development of Electrochemical COD and TOC Analyzers," Annual Report, Contract DAMD17-76-C-6077, ER-310-4; Life Systems, Inc., Cleveland, OH (September 1977).
5. Davenport, R. J., T. A. Berger and R. A. Wynveen. "Evaluation of a Breadboard Electrochemical TOC/COD Analyzer," Annual Report, Contract DAMD17-76-C-6077, ER-310-4-2; Life Systems, Inc., Cleveland, OH (October 1978).
6. Wynveen, R. A., J. D. Powell and F. H. Schubert. "Iodine Generator for Reclaimed Water Purification," U.S. Patent No. 4,061,570 (December 1977).
7. Lantz, J. B., R. J. Davenport and R. A. Wynveen. "Development of an Advanced Breadboard TOC/COD Analyzer," Annual Report, Contract DAMD17-76-C-6077, TR-310-4-3; Life Systems, Inc., Cleveland, OH (August 1980).

CHAPTER 7

CHEMICAL AND MICROBIOLOGICAL MONITORING OF A SOLE-SOURCE AQUIFER INTENDED FOR ARTIFICIAL RECHARGE, NASSAU COUNTY, NEW YORK

Brian G. Katz
Water Resources Division
U.S. Geological Survey
Towson, Maryland

Gail E. Mallard
Water Resources Division
U.S. Geological Survey
Syosset, New York

Ground water is the sole source of fresh water for the nearly 1.5 million residents of Nassau County, one of four counties on Long Island, New York (Figure 1). Rapid population growth in Nassau County over the past 30 years has increased both the demand for fresh water and the amount of domestic waste water discharged to the ground water through cesspools and septic tanks. Because ground water is the sole source of public supply for Nassau and Suffolk Counties, the U. S. Environmental Protection Agency (EPA) in 1978 assigned "sole-source" status to the Long Island groundwater reservoir [1]. This status prohibits federally assisted projects from discharging waters that would contaminate the groundwater reservoir.

As of 1978, about 65% of the land in Nassau County was residential, and only 1.8% of the area remaining was vacant land. Approximately 60% of the population, or 850,000, resided in unsewered areas and discharged 60 Mgal/day of domestic wastes through 150,000 cesspools into the shallow zone of the groundwater reservoir [2].

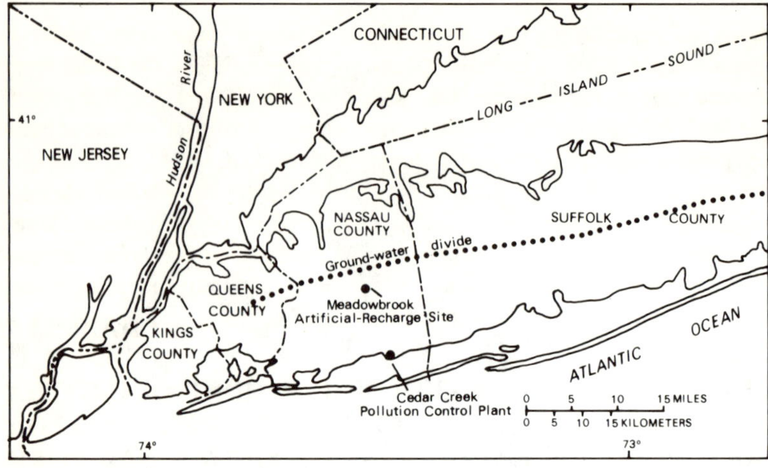

Figure 1. Location of groundwater divide, Meadowbrook artificial recharge site, and Cedar Creek sewage treatment plant.

To prevent further degradation of water quality from domestic wastewaters, a comprehensive sewering program was begun in the 1950s to serve about 85% of Nassau County. The work is expected to be completed in 1985.

Two of the effects of urbanization—sewering and increased groundwater pumpage—have resulted in a reduction in the quantity of ground water; this reduction has, in turn, caused a decline in the water table, a corresponding decrease in freshwater streamflow, increased salinity of the bays and the landward movement of saline ground water in some areas. To offset this decline in the water table and to alleviate the accompanying problems, highly treated sewage that would otherwise be discharged into the sea can be used to replenish the groundwater reservoir.

Comparison of water-quality data collected before recharge with data collected after recharge begins should reveal whether or not artifical recharge with reclaimed water of high quality will be safe and feasible. This report is a result of an ongoing cooperative program with the Nassau County Department of Public Works. Special analyses for selected organic compounds were performed by the New York State Department of Health Laboratory through the cooperation of Nassau County Department of Health.

Purpose and Scope

In an effort to offset the effects of lowered groundwater levels and declining quantity and quality of ground water, reuse of treated waste water and methods of replenishing the groundwater reservoir have been under investigation since 1968. The most recent of these studies is the Cedar Creek water-reclamation recharge project, which will study the feasibility of recharge with 4 Mgal/day of tertiary-treated sewage through a system of basins and injection wells. This report describes the role of the U. S. Geological Survey in monitoring the quality of ground water for selected chemical and microbiological constituents before the introduction of reclaimed wastewater to the system and presents the results obtained to date. Data obtained through this study will provide a basis for future comparisons to determine the chemical and biological effects of artificial recharge and to evaluate its safety.

Previous Recharge Studies in Nassau County

Recharge and reuse of water in Nassau County were proposed as early as 1963. In response to concern that a predicted deficit in the quantity of available ground water could lead to the landward movement of salt water in some areas, the U. S. Geological Survey, in cooperation with the Nassau County Department of Public Works, began an experimental deep-well recharge program in 1968. Between 1968 and 1973, 13 recharge tests with tertiary-treated sewage (reclaimed water) were made through a well screened in the Magothy aquifer between depths of 420 and 480 feet below land surface. However, injection into the fine-grained sediments at that particular site (Bay Park) was found to be unfeasible because the excessive head buildup of injected water required frequent redevelopment of the well [3].

In a study of storm-water basins in Nassau County, Aronson et al. [4] found that large quantities of reclaimed water could be transmitted to the groundwater reservoir through many of the county's stormwater basins, which are excavated in the highly permeable material of the upper glacial aquifer. That study concluded that, under proper conditions, there would be little chance of overflow or excessive water-table buildup during recharge.

HYDROGEOLOGY

The hydrogeology of the major aquifers underlying Long Island is described only in general terms in this chapter. More detailed descriptions are given in reports by McClymonds and Franke [5] and Franke and Cohen [6].

Long Island is underlain by a southward-dipping wedge of unconsolidated deposits (Figure 2). The two aquifers of concern include the upper glacial (water table) aquifer and the underlying Magothy aquifer, which has been

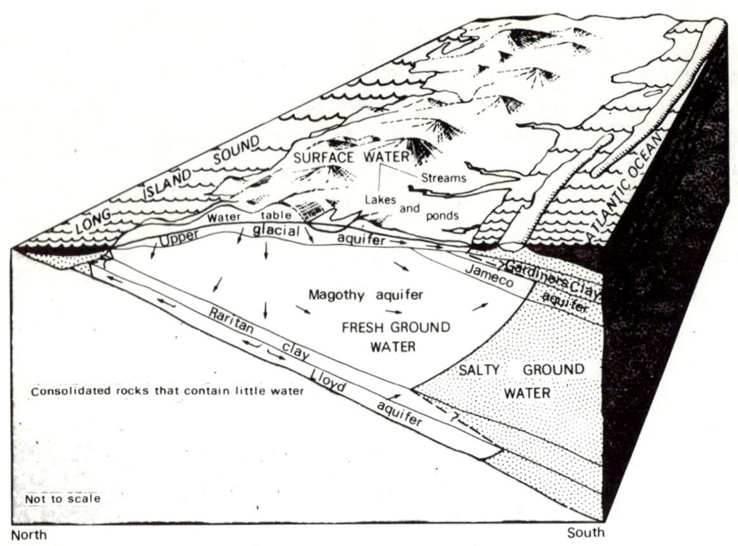

Figure 2. Generalized cross section of Long Island showing major hydrogeologic units and paths of groundwater flow (modified from Cohen et al. [7].

given sole-source designation by the EPA. Within the 1-mi^2 study area these aquifers are hydraulically connected. Here, the upper glacial aquifer is composed of Pleistocene outwash deposits, consisting mainly of highly permeable sand and gravel of glacial origin; depth to the water table generally ranges from 29 to 32 ft.

The Magothy aquifer, which underlies the upper glacial aquifer in the study area, consists of beds or lenses of fine to medium sand, clay, silt and gravel. At present, water from the Magothy aquifer is used for public supply for most of Nassau County because the upper glacial aquifer has become widely contaminated.

Precipitation is the source of all fresh ground water in Nassau County; average annual precipitation is 44 in. Under natural conditions, about half the average annual precipitation percolates to the water table; the rest runs off or evaporates and transpires. Additional recharge to the water table results from infiltration of storm runoff, injection of water used by industry and discharge of domestic and industrial wastewater from cesspools and septic tank systems [6]. The deeper Magothy aquifer is recharged by downward movement of water from the upper glacial aquifer.

In general, ground-water movement north of the regional groundwater divide (Figure 1) is northward toward Long Island Sound; south of this divide, it is generally toward the south shore of Long Island, as indicated by the arrows in Figure 2.

DESCRIPTION OF CEDAR CREEK WATER-RECLAMATION RECHARGE PROJECT

The Cedar Creek water reclamation recharge project was developed in an effort to offset the effects of urbanization, where the quality and quantity of potable ground water have been diminished. The project is a multiagency effort supported by grants from EPA, the New York State Department of Environmental Conservation and the Nassau County Department of Public Works. The U. S. Geological Survey entered into a cooperative agreement with the Nassau County Department of Public Works in 1977 to study the hydrologic effects of water reclamation and artificial recharge. The artificial recharge system consists of three main components—the reclamation plant, the transmission-main system and the recharge facilities.

Reclamation Plant and Transmission-Main System

The Cedar Creek sewage treatment plant is a conventional activated sludge facility designed to treat 45–50 Mgal/day. Approximately 5 Mgal/day of influent sewage will be diverted to the water-reclamation plant after screening and grit removal; the remaining 40–45 Mgal/day will be discharged to the ocean after secondary treatment. The advanced wastewater-treatment process consists of four steps (shown in Figure 3):

1. chemically aided primary treatment to improve the removal of phosphorus, biochemical oxygen demand (BOD), suspended solids and heavy metals;
2. a two-stage biological treatment system consisting of a nitrification-denitrification process to promote biological nitrogen removal in combination with oxidation of remaining carbonaceous material;
3. filtration and carbon adsorption to reduce organic compounds in the effluent to a few parts per billion; and
4. chlorination and storage to disinfect and store this effluent to allow for continuous delivery of 4 Mgal/day to the injection wells and basins at the recharge site. Figure 3 is a diagram of the treatment process.

New York state and federal drinking-water standards must be met before any water will be transmitted from the Cedar Creek sewage-treatment plant (Figure 1) to the Meadowbrook artificial recharge site a distance of 6.25 mi.

Recharge Facilities

At the Meadowbrook artificial recharge site (Figure 1), a system of 5 injection wells and 10 basins will be used to return 4 Mgal/day of reclaimed

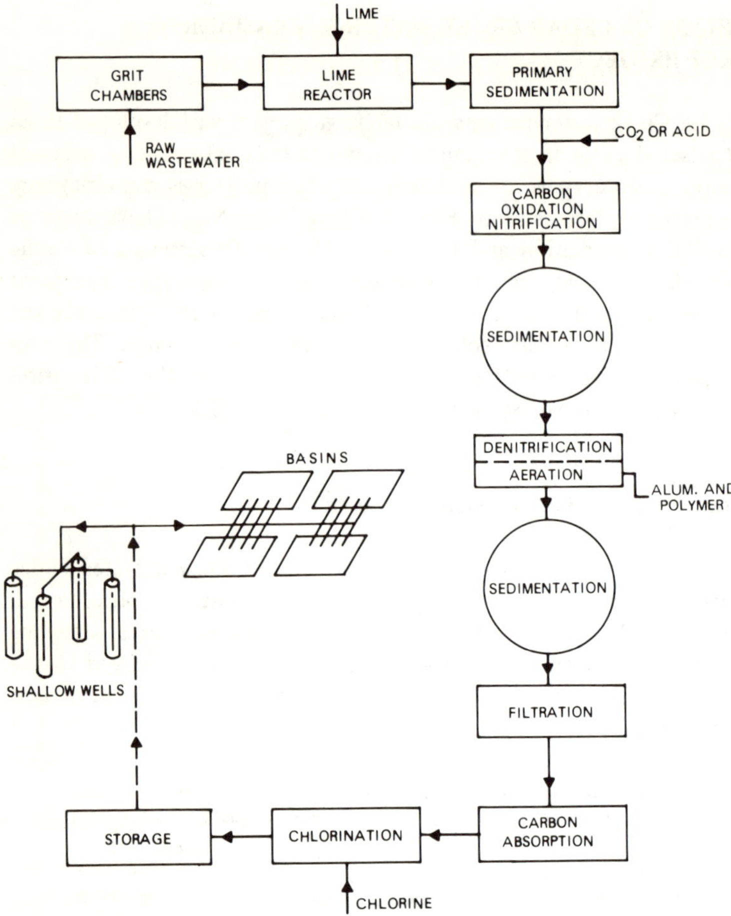

Figure 3. Schematic diagram of water reclamation recharge project facilities [8].

water to the groundwater system. This site, near the geographic center of Nassau County, includes approximately 35 acres. Two major goals of the project [9] are (1) to determine the most efficient methods of artificial recharge; and (2) to monitor and evaluate the chemical, biological and physical effects of artificial recharge on the groundwater reservoir at and downgradient from the recharge site. An observation well network, which will be used to monitor changes in ground-water quality before and after recharge begins, is discussed in the next section.

METHODS OF SAMPLING GROUND WATER FOR SELECTED CHEMICAL AND MICROBIOLOGICAL CONSTITUENTS

A monitoring network of 48 observation wells at 22 sites (Figure 4) was established within the 1-mi^2 area surrounding the recharge site. The well sites contain either 1, 2 or 3 wells screened at depths of 45–50 ft, 95–100 ft, and 195–200 ft below land surface. Hereafter, these screened depths are referred to as 50, 100 and 200 feet.

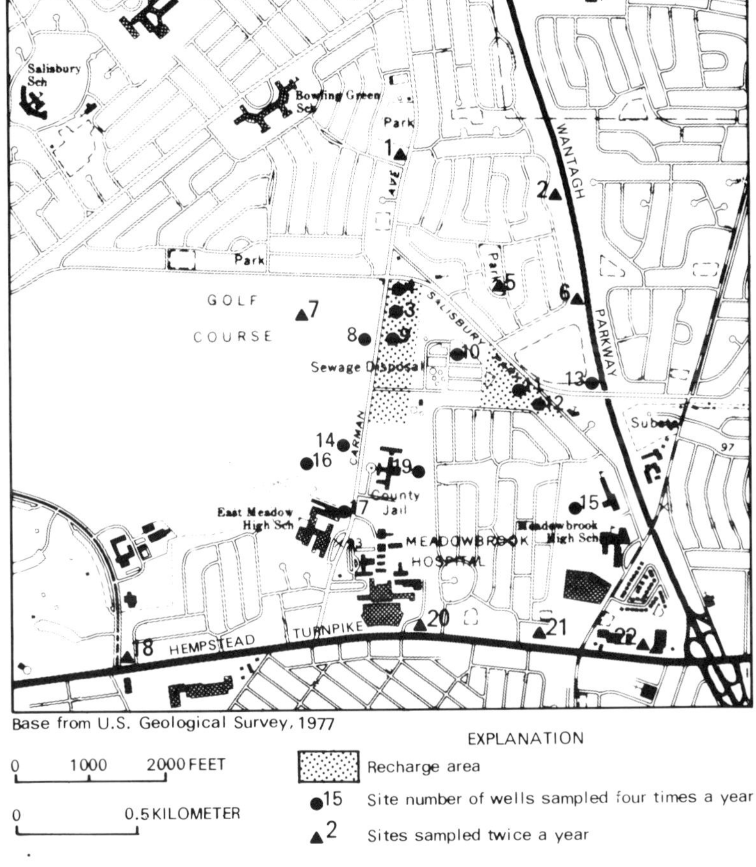

Figure 4. Location of observation wells at Meadowbrook artificial recharge site.

Construction of Observation Wells

All holes for the observation wells were drilled with rotary drilling equipment. Bentonite or fine sand, free of added organic substances, was used as a drilling mud. The wells were cased with 6-in.-diameter noncorrosive fiberglass pipe. Five feet of stainless-steel 6-in.-diameter well screen was attached to the bottom of each casing. The well screen was surrounded by clean, washed gravel-size material. A layer of cement grout was used to fill the annular space above the gravel pack to reduce the possibility of contamination from above; the remainder of the hole was backfilled with the original aquifer material. After construction was completed, the well was developed by surging and pumping until the water was substantially free of suspended material and the turbidity was less than 5 Jackson turbidity units (JTU).

Chemical and Microbiological Constituents

To provide a basis for future comparison, the quality of water from these wells before artificial recharge is being evaluated in terms of major cations and anions, nutrients (nitrogen, phosphorus and carbon), selected low-molecular-weight halogenated hydrocarbons, selected organochlorine and organophosphorus insecticides, polychlorinated biphenyls, selected herbicides, several groups of bacteria including pollution indicator organisms, and organisms important in the nitrogen cycle. A list of all chemical and microbiological constituents is provided in Table I.

Sampling Procedures

Samples of water from observation wells near and downgradient from the basins and injection wells to be used for artificial recharge were collected four times per year; observation wells farther from the recharge area were sampled twice a year.

All wells were sampled by submersible pump lowered down the casing to 10 ft or less below water level. Three times the volume of water in the well casing was cleared before the samples were taken for analysis. The pump was then removed from the well and disinfected for 30 min by recirculating a solution of calcuim hypochlorite containing 25–30 mg/l free chlorine. This is the procedure recommended by *Standard Methods* [10] for sterilizing virus-concentrating equipment that cannot be autoclaved.

After disinfection, the pump was again lowered into the casing, and at least the volume of water in the well casing was removed from the well. In practice, more than the volume of water in the well casing was cleared from

Table I. Selected Chemical and Microbiological Constituents Routinely Analyzed in Samples of Ground Water from Study Area

Major Inorganic Cations and Anions
   Calcium, magnesium, potassium, sodium
   chloride, fluoride, sulfate

Nutrients
   Nitrogen species
   Phosphorus species

Metals
   Aluminum, chromium, iron,
   manganese, zinc

Low-Molecular-Weight Organic Chemicals
   Bromodichloromethane,
   carbon tetrachloride,
   chloroform,
   tetrachloroethylene,
   trichloroethylene,
   1, 1, 1-trichloroethane,
   1, 1, 2-trifluorotrichloroethane

Organochlorine Insecticides
   Aldrin, chlordane, DDD, DDE, DDT,
   dieldrin, endosulfan, endrin, heptachlor,
   heptachlor epoxide, lindane, methoxychlor,
   mirex, perthane, toxaphene

Organophosphorus Insecticides
   Diazinon, ethion, malathion,
   methylparathion, methlytrithion,
   parathion, trithion

Herbicides
   2, 4-D, 2, 4-DP, Silvex, 2, 4, 5-T

Polychlorinated Biphenyls

Bacteria
   Total aerobic heterotrophs,
   total coliforms,
   fecal coliforms,
   fecal streptococci,
   denitrifying bacteria,
   nitrifying bacteria

the shallowest wells sampled. This practice allowed any chlorine solution remaining on the inside or outside of the pump to be removed. Water to be analyzed for bacteria was collected in sterile bottles and processed within two hours.

Analytical Methods

Inorganic substances in samples of ground water were analyzed at a U. S. Geological Survey Laboratory by methods developed by Skougstad et al. [11]. Seven selected low-molecular-weight halogenated hydrocarbons were analyzed by the New York State Department of Health Laboratory with a modified Beller-Lichtenburg technique [12]. The organophosphorus and organochlorine insecticides, along with the polychlorinated biphenyls and selected herbicides were analyzed by the U.S. Geological Survey by gas chromatography/mass spectroscopy (GC/MS) techniques developed and modified by Goerlitz and Brown [13]. Identification and enumeration of bacteria in all water samples were done according to standard procedures of the U. S. Geological Survey [14]. Total aerobic heterotrophs, total coliforms, fecal coliforms and fecal streptococci were analyzed by membrane filter techniques. A multiple-tube most probable number (MPN) method was used to enumerate nitrifying and denitrifying bacteria.

QUALITY OF GROUND WATER BEFORE RECHARGE

The quality of ground water before recharge was determined by analyzing samples from 48 observation wells at 22 sites in the recharge area. Data from the observation well network could be grouped either by depth of the well screen or location within the recharge area. In this chapter, the data are grouped according to screen depth to allow comparison among 50-, 100- and 200-ft levels. It would not be useful at this time to examine areal differences in water quality. Water at any given depth is likely to be uniform within the area because there is no readily identifiable point source for a plume of groundwater contamination.

The alteration of groundwater quality in Nassau County results mainly from nonpoint sources, which include effluent from cesspools, septic tanks, stormwater discharge sumps and sewers; salts for road deicing; and chemicals used on lawns, in agriculture and in industry. Precipitation and stormwater runoff are the principal agents in moving dissolved substances from these sources through the ground to the water table. Ground water in the study area moves southward at approximately 0.5 ft/day [15].

Alteration of Groundwater Quality by Inorganic Chemicals

Chloride, nitrate, sulfate and ammonium were selected for analysis because they are the major inorganic ions that can indicate contamination. In general, the mean concentration of all four ions decreases with depth (Table II), as would be expected, because water from the 100- and 200-ft wells is withdrawn from the deeper regional flow systems, in which water contains lower concentrations of dissolved ions. This decrease in concentration with depth is most evident for ammonium-N; it is due in part to chemical reactions such as ion exchange, sorption and oxidation-reduction.

Alteration of Groundwater Quality by Organic Chemicals

Low-molecular-weight halogenated hydrocarbons are widespread in the ground water beneath Long Island. Trichloroethylene, chloroform, 1, 1, 1-trichloroethane and tetrachloroethylene have all been observed in concentrations greater than 50 μg/l in the upper glacial and Magothy aquifers during a recent islandwide study [2]. In the study area, at least one of these four compounds was detected in water from most wells and from both aquifers (Table III). Concentrations of trichloroethylene and tetrachloroethylene increase with depth below land surface; the highest percentage of samples that exceeded 20 μg/l was in the 100-ft level (Table III). In contrast, both the maximum concentration of 1, 1, 1-trichloroethane and the percentage of samples exceeding 20 μg/l of this compound decrease with increasing depth.

The health risk [16] associated with the presence of these compounds in ground water has prompted the Nassau County Department of Health to undertake a comprehensive survey of industrial sites that use organic solvents and consumer products that contain organic chemicals. The above mentioned low-molecular-weight organic solvents were the most heavily used organic chemicals in 641 industrial plants surveyed. Dry cleaners use large quantities of tetrachloroethylene, and a large aerospace manufacturer uses much 1, 1, 1-trichloroethane and tetrachloroethylene [17].

Results of a consumer-products survey by the Nassau County Department of Health showed that the major source of these low-molecular-weight compounds was cesspool cleaners [17]. In 1977, it was estimated that 18,600 gallons of 1, 1, 1-trichloroethane and 17,600 gallons of other halogenated compounds were sold and presumably used for cleaning cesspools in Nassau County.

In addition to the low-molecular-weight halogenated hydrocarbons, insecticides have been detected in ground water at all three sampling zones

Table II. Mean, Median and Range of Nitrate-N, Chloride, Sulfate and Ammonium-N in Water from Wells in Study Area

Chemical	Depth of Well Screen Below Land Surface (ft)	Mean (mg/l)	Median (mg/l)	Range (mg/l)	No. of Analyses
NO_3-N	50	11	11	0.93–26	45
	100	8.9	8.7	0.26–20	34
	200	9.2	7.9	0.07–25	17
Cl	50	39	33	2.8–220	52
	100	36	34	7.3–67	46
	200	22	21	7.0–46	20
SO_4	50	36	38	6.4–52	53
	100	32	30	2.6–56	46
	200	17	16	0.0–44	20
NH_4-N	50	0.18	0.02	0.0–1.9	50
	100	0.12	0.02	0.0–2.0	41
	200	0.01	0.01	0.0–0.03	20

Table III. Low-Molecular-Weight Halogenated Hydrocarbons Detected in Water from Wells in Study Area

Chemical	Depth of Well Screen Below Land Surface (ft)	Number of Wells Sampled	Number of Wells Where Concentration Equals or Exceeds 20 μg/l	Maximum Concentration in Ground Water (μg/l)
Trichloroethylene	50	21	3	173
	100	15	5	270
	200	9	1	290
Tetrachloroethylene	50	21	5	580
	100	15	5	1500
	200	9	2	1000
Chloroform	50	21	1	21
	100	15	1	20
	200	9	0	17
1, 1, 1-Trichloroethane	50	21	14	208
	100	15	4	87
	200	9	0	16

(Table IV), and polychlorinated biphenyls (PCB) have been detected at the two upper (50- and 100-ft) zones. As depth below land surface increases, the number of wells containing detectable concentrations of dieldrin and PCB decreases (Table IV). However, the maximum concentration of dieldrin was not in the shallowest zone but at the 100-ft depth.

Heavy rates of application of dieldrin, aldrin (which is oxidized to form dieldrin) and heptachlor (which is converted to heptachlor epoxide) are used to control termites in Nassau County. In 1975, it was estimated that more than 150,000 pounds of aldrin and dieldrin was applied for termite control [18]. Apparently these insecticides travel downward through the nonreactive aquifer material without being removed or altered, as indicated by their presence at the 200-ft depth (Table IV).

Microbiological Characteristics

Most public supply wells that obtain water from the deeper parts of the Magothy aquifer are free of coliform bacteria [19]. The shallow water sampled was also relatively free of coliform bacteria; only 5 samples out of 81 analyzed had any fecal coliform bacteria in 100 ml (Table V), and the numbers of fecal coliforms in those samples were low. Similar results were obtained for fecal streptococci.

Bacteria that are important in the geochemical transformation of nitrogen were observed regularly at all three levels sampled. Denitrifying bacteria were observed in all samples; nitrifying bacteria were observed in much lower numbers and in only approximately 50% of the samples at all levels. Substantial variation in numbers was observed among well sites and over time at the same location. No relationship between concentration of nitrogen species and number of nitrogen-metabolizing bacteria in any given water sample was noted.

In general, the total number of bacteria (total aerobic heterotrophs) in the water samples was low, and the mean number of bacteria per milliliter decreased with depth below land surface. A statistically significant difference was noted between the counts obtained at the 50-ft level and those obtained at the 200-ft level, but interpretation is difficult because the numbers within the samples from all three depths ranged over almost two orders of magnitude. This scatter suggests the need for caution in interpretation of data and also points out the importance of collecting several samples over time and at several sites to obtain adequate information.

Table IV. Presence of Dieldrin, Heptachlor Epoxide and PCB at or Above Detection Limits in Water from Wells in Study Area

Chemical	Depth of Well Screen Below Land Surface (ft)	Detection Limit (μg/l)	Percentage of Water Samples from Wells at or Above Detection Limit	Maximum Concentration (μg/l)
Dieldrin	50	0.01	80	0.22
	100		56	1.40
	200		22	0.080
Heptachlor Epoxide	50	0.01	15	0.03
	100		6	0.01
	200		11	0.02
PCB	50	0.1	35	0.6
	100		13	0.6
	200		0	0.0

Table V. Bacterial Indicators of Fecal Contamination in Ground Water from Selected Depths

Bacteria	Depth of Well Screen Below Land Surface (ft)	Number of Samples with One or More Bacteria per 100 ml	Total Number of Samples	Maximum Count per 100 ml
Fecal Coliforms	50	3	34	23
	100	1	32	4
	200	1	15	1
Fecal Streptococci	50	6	33	27
	100	7	31	6
	200	0	13	0

SUMMARY AND CONCLUSIONS

One of the main effects of urbanization in Nassau County has been a reduction in the quantity and quality of potable ground water. The Cedar Creek water reclamation project was designed to determine the feasibility of recharging the groundwater reservoir with reclaimed water to replenish the supply and improve its quality. Starting in late 1980, approximately 4 Mgal/day of tertiary-treated sewage will be used to recharge the groundwater reservoir in central Nassau County through a system of 10 recharge basins and 5 shallow injection wells.

To evaluate the hydrologic and water-quality impact of large-scale recharge with reclaimed water, the U. S. Geological Survey has collected data on groundwater quality in a 1-mi^2 area around the recharge site to provide a basis for comparison after recharge has begun. Water from observation wells screened at depths of 50, 100 and 200 ft below land surface is being analyzed for major cations, major anions, nutrients, selected metals and several groups of bacteria, including pollution-indicator organisms and organisms important in the nitrogen cycle. Samples are also being analyzed for selected low-molecular-weight halogenated hydrocarbons, organochlorine and organophosphorus insecticides, chlorinated phenoxy and acid herbicides and PCB.

Preliminary results indicate that ground water at the recharge site contains elevated concentrations of nitrate, chloride, sulfate and ammonium ions; however, the concentrations of these ions generally decrease with depth below land surface. Water from the upper glacial aquifer and Magothy aquifer contains significant concentrations of low-molecular-weight chlorinated hydrocarbons and detectable concentrations of organochlorine insecticides and PCB. At present no fecal contamination of the aquifers is evident. In the few cases where fecal indicator bacteria were present, numbers were low.

Nonpoint sources provide significant loads of organic and inorganic compounds. Major sources include cesspool and septic tank effluent, cesspool and septic tank cleaners and other over-the-counter domestic organic solvents, fertilizers, insecticides for termite and other pest control and stormwater runoff to recharge basins.

The feasibility of artificial recharge on Long Island depends on whether it will adversely affect the supply of potable water. Any change in the water quality during recharge with reclaimed water should be apparent upon comparison with the background data collected during this study.

REFERENCES

1. "Aquifers Underlying Nassau and Suffolk Counties, Long Island, N. Y.—Determination of Sole-Source Status," *Federal Register*, 43(120) (1978).
2. Nassau-Suffolk Regional Planning Board, "The Long Island Comprehensive Waste Treatment Management Plan, Vols. I and II," Long Island Regional Planning Board (1978).
3. Vecchioli, J., J. A. Oliva, S. E. Ragone and H. F. H. Ku, "Recharge of reclaimed wastewater, Bay Park, New York," *J. Eng. Div., ASCE* 101(EE2):201-214 (1975).
4. Aronson, D. A., T. E. Reilly and A. W. Harbaugh, "Use of Storm-water Basins for Artificial Recharge with Reclaimed Water, Nassau County, Long Island, New York—A Hydraulic Feasibility Study," Long Island Water Resources Bulletin 11, Nassau County Department of Public Works (1979) pp. 53-56.
5. McClymonds, N. E., and O. L. Franke, "Water Transmitting Properties of Aquifers on Long Island, New York," U. S. Geological Survey Prof. Paper 627-E (1972), pp. E1-E24.
6. Franke, O. L., and P. Cohen, "Regional Rates of Ground Water Movement on Long Island, New York," U. S. Geological Survey Prof. Paper 800-C (1972) pp. C271-C277.
7. Cohen, P., O. L. Franke and B. L. Foxworthy, "An Atlas of Long Island's Water Resources," New York State Water Resources Commission Bulletin 62 (1968), 117 p.
8. Beckman, W. J., and R. J., Avendt, "Correlation of Advanced Wastewater Treatment and Ground Water Recharge" U. S. EPA Project R801478 (1973).
9. Aronson, D. A. "The Meadowbrook Artificial-Recharge Project in Nassau County, Long Island, N. Y.," Long Island Water Resources Bulletin 14, (1980).
10. *Standard Methods for the Examination of Water and Wastewater*, 14th ed. (New York: American Public Health Association, 1976).
11. Skougstad, M. W., M. J. Fishman, L. C. Friedman, D. E. Erdman and S. S. Duncan, "Methods for Determination of Inorganic Substances in Water and Fluvial Sediments," in *U. S. Geological Survey Techniques of Water-Resources Investigations* (1979).
12. Smith, E., New York State Department of Health Personal communication (1979).
13. Goerlitz, D. F., and E. Brown, "Methods for Analysis of Organic Substances in Water," in *U. S. Geological Survey Techniques of Water Resources Investigations* (1972).
14. Greeson, P. E., T. A. Ehlke, G. A. Irwin, B. W. Lium and K. V. Slack, "Methods for Collection and Analysis of Aquatic Biological and Microbiological Samples," in *U. S. Geological Survey Techniques of Water Resources Investigations* (1977).
15. Aronson, D. A., U. S. Geological Survey, Syosset, N. Y. Unpublished results (1980).

16. *Drinking Water and Health* (Washington, DC: National Academy of Sciences, 1977), pp. 775-856.
17. Nassau County Department of Health, Mineola, N. Y. Unpublished results (1977).
18. Porter, K. S., Cornell University Cooperative Extension Service, Unpublished results (1979).
19. Nassau County Department of Health, Mineola, N. Y. Unpublished results (1980).

SECTION 2

QUALITY ASSURANCE

QUALITY ASSURANCE FOR WATER AND WASTE WATER ANALYSES

John A. Winter

Environmental Protection Agency
Cincinnati, Ohio

As in other technical agencies, Environmental Protection Agency (EPA) scientists have always incorporated quality control into their routine analyses; however, such quality control practices often tended to be casual, performed when time was available, variable among workers and seldom documented sufficiently. However, EPA is a regulatory agency, where valid analytical data are essential to establish and regulate the standards and limits which are mandated under the water-related laws and their amendments: the Federal Water Pollution Control Act (FWPCA); the Marine Protection, Research and Sanctuaries Act; and the Safe Drinking Water Act. EPA had to have some assurance that its analytical data were of a quality and quantity sufficient for these decisions. This assurance was only possible through the development of a formal quality assurance program. Consequently, in January 1972 EPA established a formal program of Quality Assurance (QA) in its Office of Research and Development (ORD) with responsibilities assigned as follows:

1. program direction—Office of Monitoring and Technical Support, ORD;
2. technical guidance—air programs: Environmental Monitoring Systems Laboratory Research Triangle Park, NC; water programs: Environmental Montitoring and Support Laboratory Cincinnati, OH; radiochemistry—Environmental Monitoring Systems Laboratory Las Vegas, NV; and
3. regional liaison—Regional Quality Assurance Coordinators in the Ten Federal Regions.

SCOPE OF THE EPA QA PROGRAM

Quality assurance, as conceived by EPA, is much more than a simple program of quality control charts, standards and checks. It encompasses all aspects of field and laboratory operations that affect data validity: study and survey design; sampling methodology, equipment and sample preservation; analytical methodology; personnel; facilities; instrumentation; equipment; supplies; analytical quality control practices; and data handling, reporting, storage and retrieval.

This broad quality assurance program for water and wastewater monitoring is addressed through the major functions conducted in the program of the Environmental Monitoring and Support Laboratory (EMSL), Cincinnati, OH.

- develop/select methodology
- conduct an equivalency program for approval of alternative test procedures
- develop manuals and guidelines
- maintain a quality control check sample program
- conduct method validation studies
- conduct performance evaluation studies
- maintain the EPA repository for toxic and hazardous materials

Methodology

EMSL is responsible for guiding the EPA in development of analytical methodology and quality control for all parameters in water-related matrixes: ground, surface, drinking and waste waters, marine water, leachates, sludges, and sediments.

Initially, the parameters of interest were water quality measurements under the FWPCA amendments and the Ocean Dumping Act [1]. However, parameters were soon added for regulation of the Safe Drinking Water Act [2]. Later, the Consent Decree called for establishment of limits for priority pollutants [3]; EPA responded by developing analytical methodology for the 129 priority pollutants. Manuals of methodology have been published for chemistry, biology and microbiology, and methodology for the 113 organic priority pollutants has been published in a proposed regulation in the *Federal Register*. Specifically, these publications are:

- "Methods for Chemical Analysis of Water and Wastes," EPA-600/4-79-020 (1979);
- "Biological Field and Laboratory Methods for Measuring the Quality of Surface Waters and Effluents," EPA-670/4-73-001 (1973);
- "Microbiological Methods for Monitoring the Environment, Water and Wastes," EPA-600/8-78-017 (1978); and

40 CFR 136, "Guidelines Establishing Test Procedures for Analysis of Pollutants," Proposed Regulations, pp. 69464-69575, December 3, 1979 (priority pollutants).

Alternative Test Procedures

EPA recognized that the methods selected for use in the water regulations would not be applicable to all samples, and that new methodology and new instrumentation are constantly being developed. Consequently, a mechanism and protocol for obtaining approval of alternative test procedures (ATP) was developed and incorporated into the Clean Water Act and Safe Drinking Water Act [1, 2].

There are two types of ATP approval based on the proposed method application: (1) limited use ATP proposed by a discharger, water utility, or private, state or EPA regional laboratory for use within that region or state, and (2) nationwide use ATP proposed by manufacturers of instruments and analytical systems. In either case the proposer is required to provide supporting information to EPA that includes a complete description of the analytical method, published papers supporting the methodology and analytical data substantiating method equivalency and applicability.

The appropriate Regional EPA Administrator approves limiting use proposals. The Assistant Administrator of the Office of Research and Development, EPA, approves nationwide use proposals for National Pollution Discharge Elimination Systems (NPDES) while the Deputy Administrator for the Office of Drinking Water approves nationwide use proposals for the Interim Primary Drinking Water Regulations. Approval or denial of the ATP is based on the regional review and the technical review and recommendation of EMSL. Notification of nationwide approval is made in the *Federal Register*.

Manuals and Guidelines

In addition to specified analytical methodology, EMSL provides other quality assurance supports as needed. It is impossible to address in a single document the details of sampling site location, frequency and number of samples, sampling depths, sampling techniques and equipment, and other related factors which must be considered in field or plant surveys. The person on the spot must use his/her professional judgment and experience to prepare the sampling plan or survey design that best meets the particular needs of a program. The same person should also apply specific quality control measures to meet his/her sampling and analytical needs. However, everyone seems to agree that general guidance on sampling and within-laboratory quality control should be available. Data users should know that uniform planning and review of sampling and quality control techniques have been done at least to minimal standards. Hence, EPA manuals were developed on sampling and quality control. Finally, separate manuals are required and input is provided to meet the special needs of programs such as the certification program under the Safe Drinking Water Act regulations. Example documents are:

- "Handbook for Analytical Quality Control in Water and Wastewater Laboratories." EPA-600/4-79-019 (1979);
- "Handbook for Sampling and Sample Preservation of Water and Wastewater" (in final review); and
- "Manual for the Interim Certification of Laboratories Involved in Analyzing Public Drinking Water Supplies, Criteria and Procedures," EPA-600/8-78-008 (1978).

Sample Design

The same sample form and designs are used by EMSL in its quality control (QC) sample program as are used in its method validation and performance evaluation studies. Samples are prepared as concentrates in sealed all-glass ampules to be diluted to volume by the user laboratory according to instructions. Samples contain related analytical parameters dissolved in water or a water-miscible solvent in a balanced chemical system. Because the sample concentrates hold relatively high levels of solutes, the avoidance of interferences, precipitation, selective solubilities and phase changes are critical concerns in sample design. Use of concentrates in ampules has many advantages:

1. Sample concentrates in ampules reduce storage space requirements by a factor of 16 or more over full-volume samples.
2. Sealed all-glass ampules allow steam-sterilization for preservation of many sample types.
3. There is no possibility of contamination from stoppers, caps and liners.
4. Sealed all-glass ampules do not leak or evaporate contents. A properly-designed and preserved sample sealed in an all-glass ampule is stable indefinitely.
5. Preparation of samples as concentrates permits work with solutions at the milligram/liter to microgram/liter levels that are inherently more stable than the microgram to nanogram levels present in the final diluted samples. This stability is one of the keys to our success in preparing a full spectrum of organic and inorganic sample series to meet all EPA needs.

6. Because the user laboratory makes dilution to volume in laboratory-pure, tap, natural or waste water of choice, it is able to determine performance under its own conditions. When such sample concentrates are used in method validation studies, the number of waters and waste waters tested are equal to the number of laboratories and we obtain the maximum measure of possible interferences.

Quality Control Sample Program

QC samples are furnished without charge to interested governmental, industrial, commercial and private laboratories for use as secondary checks on their within-laboratory quality control programs. The samples are intended as independent measures of technique and performance, not as replacements for the standards, replicates or spike samples run routinely as part of the laboratory's own QC program.

There is no certification or other formal evaluative function resulting from the use of QC samples. No reports are prepared and there is no requirement for use of specific methodology in these QC analyses. The Quality Control Sample Program now covers water quality, drinking water, and the priority pollutant parameters shown in Table I to III.

Repository

EMSL has established the EPA repository for toxic and hazardous materials to provide a continuing source of calibration materials, standards and reference compounds needed by EPA. Initially, the repository will provide the priority pollutants, but will be expanded soon to cover all trace organics of interest to EPA. For example, personnel are being added to support the broader needs of the Hazardous Wastes Program. Also, by late 1980, a gas chromatograph/mass spectrometry (GC/MS) library will exist of approximately 2000 trace organic compounds which have been characterized and stored in a computer data base as part of a contract administered by the Health Effects Research Laboratory (HERL), Cincinnati, OH. Because a number of the toxic or hazardous compounds are increasingly difficult to obtain as manufacturing and commercial uses are restricted, the repository has been given the staff, equipment and facilities needed to purify and upgrade chemicals, and to perform complex chemical syntheses for unattainable compounds.

To establish qualitative identification and determine purity, all repository compounds will be measured qualitatively and quantitatively by at least two independent methodologies by the repository staff and subsequently verified

Table I. Quality Control Samples for Water Quality Analyses

Demand Analyses: BOD, COD and TOC

Mineral/Physical Analyses: sodium, potassium, calcium, magnesium, pH, sulfate, chloride, fluoride, alkalinity/acidity, total hardness, total dissolved solids and specific conductance

Nutrients: nitrate-N, ammonia-N, kjeldahl-N, orthophosphate and total phosphorus

Mercury: organic and inorganic

Trace Metals: aluminum, antimony, arsenic, beryllium, cadmium, chromium, cobalt, copper, iron, lead, manganese, mercury, nickel, selenium and zinc

Trace Metals: antimony, thallium and silver

Cyanide

Total Nonfilterable, Filterable and Total Volatile Residue

Linear Alkylate Sulfonate (LAS): the anionic surfactant standard/MBAS test.

Nitrilotriacetic Acid (NTA)

Chlorophyll: spectrophotometric and fluorometric analyses

Municipal Digested Sludge: 26 inorganic and general organic parameters

Oil and Grease

Petroleum Hydrocarbons: two crude oils, #2 fuel oil, Bunker C API reference oils for characterization analyses

Phenol (4AAP Method)

Pesticides: aldrin, dieldrin, DDT, DDE, DDD, heptachlor and chlordane

Polychlorinated Biphenyls (PCB) Aroclor 1254 and 1016

Volatile Organics: 6-9 compounds, including trihalomethanes (THM)

by at least two referee laboratories. Any significant differences will be resolved before release of a compound.

Three grades of materials will be distributed:

1. quality assurance standards (QAS) $\geqslant$ 99% purity;
2. quality assurance reagents (QAR) 95-98% purity; and
3. quality assurance technical materials (QAT) $<$ 95% purity.

Table II. Quality Control Samples for Drinking Water Analyses

Nitrate/Fluoride	Trace Metals: arsenic, barium, cadmium, chromium, lead, mercury, selenium and silver
Herbicides: 2-4-D, 2, 4, 5-TP	
	Pesticides: endrin, lindane, methoxychlor and toxaphene
Turbidity	
Residual Chlorine	Trihalomethanes

Table III. Quality Control Samples for Priority Pollutants, Categories and Number of Compounds/Category

Available in 1980	Available in 1981
Phthalate Esters (6)	Acrolein, Acrylonitrile
Haloethers (7)	Nitrobenzenes/Isophorone (4)
Chlorinated Hydrocarbons (9)	Nitrosamines (3)
Benzidines (3)	Dioxins (3)
Polynuclear Aromatics (16)	Phenols (specific) (11)
Pesticides/PCB (25)	
Purgeables (26)	
Trace Metals (15)	

The repository will move as many compounds as possible from the QAT and QAR categories into the QAS category by use of purification techniques. Of course, multicomponent materials such as PCB and toxaphene will be categorized as QAR and QAT and will not be purified further. Distribution of the first repository materials began in November 1980.

Method Validation Studies

Method validation studies are formal collaborative studies conducted primarily to establish precision and accuracy of a single method of analysis. An exact protocol is follwed, and results are required within a set period of time.

Preparation for a Method Study

Before a study begins, the laboratory must be familiar with the analytical method. It may be necessary to conduct a preliminary study to familiarize laboratories. If this is not done, the analysts' unfamiliarity will show as increased variability in the data, and a falsely high level of imprecision will be attributed to the method.

As an analytical method is developed, the developer should complete ruggedness testing described by Youden [4]. Ruggedness testing is simply performance of the analytical method in a single laboratory while making ordered alterations of test conditions. This testing will determine what effect, if any, is produced by small changes in pH, volumes, weights, times and temperatures as one might expect to occur among a group of laboratories participating in a collaborative study. Most changes should produce minor effects. However, if small changes in one factor produce large changes in results, the factor must be identified and controlled by careful specification in the write-up. If the factor is uncontrollable, the method should not be considered for formal evaluation.

Written instructions for performance in the study (sample handling, dilutions, sequence of analyses, etc.) and the analytical method write-ups are provided to each laboratory. Method alternatives or options are carefully considered as desirable or undesirable variables. If accepted as an option in the study, then data for the option must be segregated for determination of the effect on the results.

Study Design

EMSL design follows the recommendations of Youden for measuring precision without duplication [4]. As Youden observed, when laboratories have qualified staff, facilities and equipment, and when they exercise reasonable quality control over their operations, the random variability of their measurements (titration, meter readings, weighings, pipetting, etc.) is small and predictable within each laboratory, and, further, the random error is quite uniform among laboratories. When random error is small and predictable, replicate analyses are of limited value. Single analyses can be performed on more samples for the same effort, and a better measure can be obtained of the major source of variability between laboratories, i.e., systematic error. Consequently, EMSL conducts its method validation studies involving single analyses on multiple samples by a large number of laboratories.

Sample concentrates are prepared in pairs of similar yet different concentrations at three levels over the testing range, each concentrate containing one or more of a series of related parameters such as trace metals, nutrients or volatile organics. True values are established for each parameter. A "true value" is defined as that amount calculated, added, then verified by repeated analyses over time. Serious lack of agreement between calculated and analyzed values must be resolved or samples are discarded. True values are not adjusted. Samples are designed for minimum stability of one year.

Sample concentrates are sent to participant laboratories for dilution to volume with a laboratory-pure water and one or more "natural" water and waste water of choice. The accuracy and bias of the method are directly calculable from the true value and mean recoveries in pure waters. By analyzing the natural and waste water with and without the increment, by-difference recoveries are calculable and can be compared with recovery in a pure water, and interferences can be distinguished from bias. The overall difference in recoveries in the pure water and the natural/wastewaters is a measure of overall interference problems in the method.

Reports

The results of the study are presented in a formal EPA report, and the statistics are incorporated into the EPA method with full details on how obtained. Although each laboratory's results are coded for confidentiality, the laboratories can compare their performance with overall performance. Data are presented in a number of ways to meet the primary need for statements of precision and accuracy for EPA and to assist participant laboratories and subsequently user laboratories in understanding the scope and limitations of the method. For each sample concentrate in each test water, the raw data are ranked, outlier routines applied and a series of statistics calculated separately for laboratory-pure and natural waters. Table IV shows an example of this statistical presentation for mercury analyses by the cold vapor technique in EPA's Method Study 8 [5].

Regression equations are developed for accuracy and precision which best reflect performance of the analytical method in laboratory-pure water, in natural and waste waters. Study results in the "natural" waters are used to characterize expected performance of the method on real-world samples as shown in Figures 1 and 2.

Note that separate measures of precision are calculated for all laboratories (S) and single laboratory (S_r). The overall S is significantly larger than S_r because as expected, the between-laboratory deviation is a major source of imprecision in the method.

Currently, method validation studies are being performed on the 113 trace organic priority pollutants described in Table III. Each study is being performed by 20 laboratories analyzing paired samples at three levels to meet the statistical requirement for at least 15 sets of complete data for each method. The 20 x 2 x 3 analyses are being performed on a laboratory-pure water, three relevant industrial waste waters, a surface water and a finished water to obtain good measures of matrix effects.

Table IV. Presentation of Data for Method Validation Studies

Ampules 7 Increment = 8.8 μg/l organic + inorganic mercury, natural water

N, All Data	80	Range	27.47000	Coef. Var.	0.42143
True Val.	8.80	Variance	13.65052	Skewness	2.00048
Mean, All	13.62260	Std. Dev.	3.69466	No. of Cells	8
Mean, Ret.	8.76678	Conf. Lim.	±7.24153 (95 Pct)		
Median	8.50000				
Accuracy	- 0.37750	Pct Relative Error			

Data in Ascending Order			Midpoint	Frequency	Histogram (Retained Data Only)
0.53	8.40	12.00	2.2469	7	XXXXXXX
1.88	8.40	12.00	5.6807	12	XXXXXXXXXXXX
2.45	8.43	13.00	9.1144	46	XXXXXXXXXXXXXXXX
3.00	8.50	13.00	12.5482	10	XXXXXXXXXX
3.22	8.50	13.80	15.9819	1	X
3.42	8.50	15.75	19.4157	0	
3.90	8.50	21.30	22.8494	1	X
4.50	8.60	28.00	26.2832	1	X
4,80	8.60	66.00R[a]			
6.10	8.70	340.00R			
6.30	8.80				
6.36	8.90				
6.50	8.90				
6.60	9.00				
6.80	9.00				
7.20	9.00				
7.21	9.00				
7.30	9.00				

7.34	9.10
7.40	9.10
7.50	9.40
7.76	9.45
7.80	9.50
7.80	9.70
8.00	9.80
8.00	10.05
8.10	10.30
8.10	10.40
8.20	10.60
8.20	10.70
8.30	11.00
8.30	11.00
8.30	11.03
8.35	11.20
8.40	11.98

[a]R = rejected data.

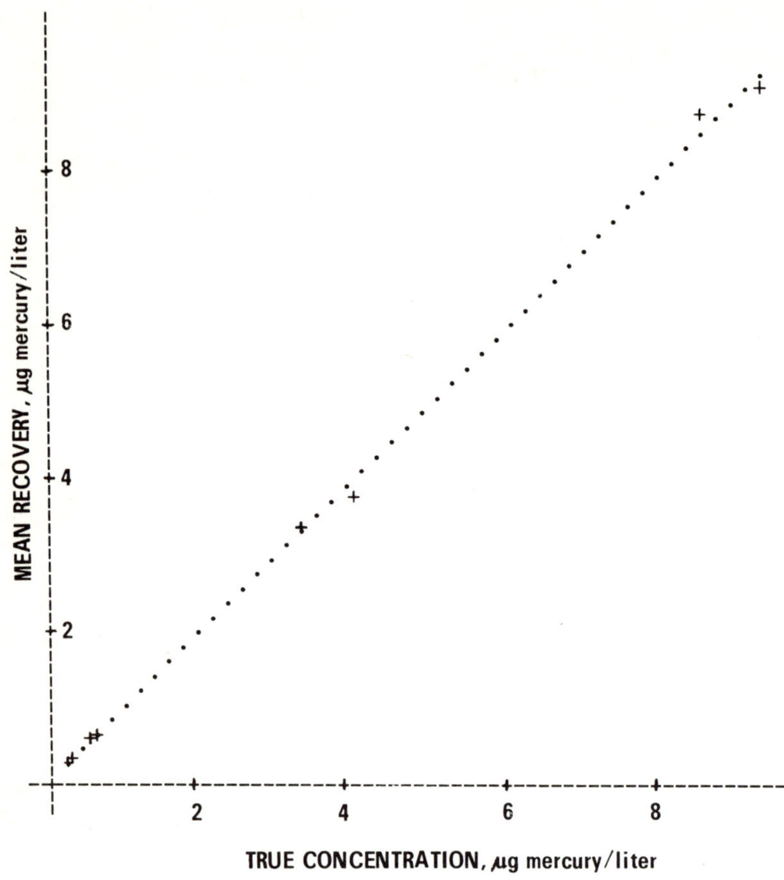

Figure 1. Accuracy of mercury analyses in natural waters [5]. Mean recovery = 0.1373 + 0.9508 (true concentration); R = 0.2 – 10 µg/1.

Performance Evaluation Studies

Performance evaluation studies are modified versions of the method validation studies, with the primary purpose of regularly evaluating performance of laboratories as required by EPA mandates or by program request. One or two sample concentrates for each parameter group of interest are sent to participating laboratories for spiking into a laboratory-pure water and analyses within a set time interval. Performance is judged from analyses in laboratory-pure water rather than waste waters in order to remove the variable of sample interference from the evaluations.

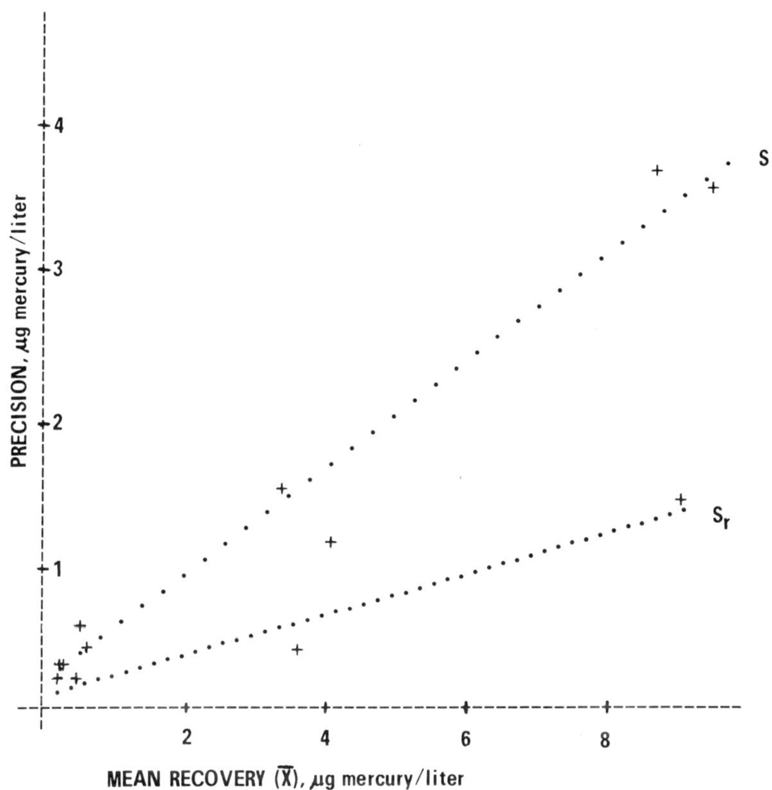

Figure 2. Precision of mercury analyses in natural waters [5]. Overall precision (S) = 0.1661 + 0.3647 $\overline{X}$, R = 0.2 – µg/1; single laboratory (Sr) = 0.0465 + 0.1379 $\overline{X}$, R = 0.2 – 10 µg/1.

Data Handling

Outliner Rejection. In each performance evaluation study, the raw results are compiled and gross outliers are rejected that exceed 0.2–5T (where T = true value). After rejection of gross outliers, the arithmetic mean and standard deviation are calculated and a second outlier test, the student's t test is applied at a 99% confidence interval. Values exceeding this limit are rejected.

Calculation of Limits. After rejection of outliers, which are usually less than 5%, analytical results are compared against performance evaluation limits based on past or current data adjusted for the true values. For a particular sample and parameter with true value equal to T, acceptance and/or warning limits are calculated at the desired level of confidence to fit program needs. For example, limits might be:

$$\text{acceptance limits} = \overline{X} \pm t_{.99} (S)$$
$$\text{warning limits} = \overline{X} \pm t_{.95} (S)$$

where

$\overline{X}$ = the arithmetic mean expected from a sample with a true value of T, based on a regression equation generated from previous study data

S = the standard deviation expected at a true value of T, based on a regression equation generated from previous study data

$t_{.99}$ = the student's t value appropriate for a 99% confidence interval

$t_{.95}$ = the student's t value appropriate for a 95% confidence interval

Reports and Followup

It is very important for an effective QA program that the annual performance evaluation study be incorporated into a continuing communication with the laboratory. The philosophy of the EPA QA program is not to catch the "bad guy" but rather to identify problems, resolve them jointly and improve quality of the data generated.

Values within the specified limits are termed "acceptable" and those exceeding the limits are termed "not acceptable." Values beyond the warning limits are considered questionable and should be checked for possible errors. Each participating laboratory receives a computerized report showing their reported values, the true values, acceptance limits, warning limits and an evaluation statement for each parameter, as shown in Table V.

Unacceptable results in the performance evaluation study must be resolved before approval is given, as part of a five step sequence:

1. Performance evaluation is completed. If results are acceptable, laboratory approval is possible.
2. If some results are unacceptable or beyond warning limits, EPA or the state works with the laboratory to identify and solve the problem. If possible, an onsite visit is made to the laboratory.
3. When the problem is apparently solved, quality control check samples (samples with known values) are provided by EPA to confirm resolution of the problem.
4. If QC results are satisfactory, a followup performance evaluation is conducted on the problem parameters.
5. Satisfactory performance supports approval; unsatisfactory performance does not.

Table V. Performance Evaluation Report Form

WATER POLLUTION STUDY NUMBER___________

Laboratory: DATE:

Parameter	Sample Number	Report Value	True Value	Acceptance Limits	Warning Limits	Performance Evaluation
		Trace Metals in micrograms per liter				
Aluminum	1					
	2					
Arsenic	1					
	2					
Beryllium	1					
	2					
Cadmium	1					
	2					
Cobalt	1					
	2					
Chromium	1					
	2					
Copper	1					
	2					
Iron	1					
	2					
Mercury	1					
	2					
Manganese	1					
	2					
Nickel	1					
	2					
Lead	1					
	2					
Selenium	1					
	2					
Vanadium	1					
	2					
Zinc	1					
	2					

Mandatory Quality Assurance

Both the Clean Water Act and the Safe Drinking Water Act and their amendments include sections which place direct or implied requirements for quality control on persons, systems or agencies covered under the regulations. These requirements have in turn directly impacted the internal Quality Assurance Program of EPA.

Potable Water

With the passage of the Safe Drinking Water Act and the National Interim Primary Drinking Water Regulations (NIPDR—40 CFR Part 141, 12/24/75), Congress set a pattern for incorporating requirements for quality control.

Quality Control and Testing Procedures. Section 1401 (1) of the act defines "primary drinking water regulation" to include "quality control and testing procedures."

> A key element of quality control for public water systems is accurate laboratory analysis . . . [Analyses] . . . for . . . determining compliance with MCLs must be conducted by a laboratory approved by the entity with primary enforcement responsibility. EPA will develop as soon as possible, in cooperation with the States and other interested parties, criteria and procedures for laboratory certification.

EMSL has been conducting a full program of evaluation of EPA and state laboratories involved in the Drinking Water Program for three years. Although only the analytical methodology is mandatory at this time, it is expected that the program will have full mandatory requirements with passage of the Revised National Primary Drinking Water Regulations.

National Pollution Discharge Elimination Systems

Section 308(a) of the FWPCA, as amended, contains statements on data evaluation which have been interpreted to permit assessment of NPDES permittee's ability to analyze, within certain acceptable limits, parameters that are required under the permit:

> Whenever required to carry out the objective of this Act, including but not limited to . . .(2) determining whether any person is in violation of any such effluent limitation, prohibition or effluent standard, pretreatment standard, or standard of performance; . . . (A) the Administrator shall require the owner or operator of any point source to (i) establish and maintain such records, (i i) make such reports, (i i i) install, use, and maintain such monitoring equipment or methods (including,

> where appropriate, biological monitoring methods), (iv) sample such
> effluents (in accordance with such methods, at such locations, at such
> intervals, and in such manner as the Administrator shall prescribe) and
> (v) provide such other information as he may reasonably require . . . [6].

Pilot studies of NPDES dischargers were conducted in two states in 1979. Based on the results of this pilot study, a nationwide evaluation of major dischargers was begun in 1980.

EPA Internal Quality Assurance Program

Since its inception, the EPA internal quality assurance program has been voluntary for EPA laboratories and its program participants. However, D. M. Costle, EPA Administrator, changed the QA program drastically on May 30, 1979 when he stated in a memorandum to the staff:

> . . . .I am making participation in the quality assurance effort mandatory
> for all EPA supported or required monitoring activities. . . .Monitoring
> is defined as *all* environmentally related measurements which are funded
> by the EPA or which generate data mandated by the EPA.

Then on June 14, 1979, the administrator placed a further requirement on extramural efforts:

> In order to assure . . . usable data of known quality, I am making . . .
> "Quality Assurance Requirements" mandatory for all EPA contracts,
> grants, cooperative agreements, and interagency agreements that involve
> environmental measurements.

The central authority for the mandatory program is assigned to the Office of Research and Development but details for implementation of the program are still being developed. However, the basic requirements of all extramural projects will include:

- a written QA Plan that meets EPA criteria;
- an estimate of QA costs;
- acceptable performance on evaluation samples;
- provision for onsite evaluation by EPA; and
- documented QC performance.

Water Reuse

In the future, laboratories involved in water reuse research and development may be required to participate in quality assurance activities which include use of specified analytical methodology, formalization of internal quality control practices, incorporation of quality control samples into their own QC programs, onsite visits, and participation in formal performance evaluation studies.

These QA requirements can occur because of involvement in activities such as: EPA grants, contracts, cooperative agreements, and interagency agreements, the NPDES permit program, and the reuse/recycle studies of industrial wastewaters and drinking water.

In preparation for possible formal involvement, reuse projects for municipal, industrial and agricultual wastewaters should take full advantage of relevant method validation and performance evaluation studies as available, and of the wide range of quality control samples in the EPA QA program for water and wastewater monitoring. Certainly the parameters in the NPDES and priority pollutant series are largely the inorganic and organic parameters of interest to them.

There is no way for a laboratory to monitor the precision and accuracy of its results except by use of reference-type samples from an external independent source. Maintenance of quality control is particularly important in water reuse because the planning requires solid knowledge of the water quality necessary for each application and a good measure of the increased pollution contributed by each succeeding process. Certainly, the complete recycling or reuse of industrial wastewaters which is the 1985 goal of the FWPCA amendments will be attainable only through careful analyses and maintenance of quality control over the processes.

In 1976, the Toxic Substances Control Act and the Resource Conservation and Recovery Act were passed, and toxic and hazardous materials are now being considered for incorporation in further standards and limits in the basic water laws. As these later acts are fully evaluated and implemented, the priority pollutant list will be succeeded by expanded lists of toxic compounds identified in waters and wastewaters for reuse or discharge. President Carter stated in his May 23, 1977 message to Congress:

> I have instructed the Environmental Protection Agency to give its highest priority to developing 1983-best-available-technology industrial effluent standards which will control toxic pollutants under the Federal Water Pollution Control Act, and to incorporate these standards into discharge permits.

With increased interest in water/waste water recirculation and reuse systems, the addition of permit limits for new toxic pollutants will make monitoring ever more critical and complex; and concentrations of toxics could become limiting factors in reuse systems. The EPA quality assurance program intends to conduct the method validation studies and performance evaluation studies, and to provide the quality control and calibration standards necessary to monitor all of these present and future parameters identified under the water laws and their amendments.

REFERENCES

1. Parameters and Test Procedures, P. L. 92-500: 40 CFR 38, Part 126 - Appendix A, pp. 9781-9785, April 19, 1973; 40 CFR Chapter I (H) (Ocean Dumping Criteria) pp. 28610-28621, October 15, 1973; 40 CFR, Part 130. "Test Procedures for the Analysis of Pollutants, Amendments." pp. 17318-17323, June 29, 1973; 40 CFR, Part 136. "Guidelines Establishing Test Procedures for the Analysis of Pollutants." pp. 28758-28760, October 16, 1973; 40 CFR, Part 136. "Guidelines Establishing Test Procedures for the Analysis of Pollutants, Amendments." pp. 52780-52786, December 1, 1976; 40 CFR, Part 136. "Guidelines Establishing Test Procedures for the Analysis of Pollutants; Proposed Regulations." pp. 69464-69575, December 3, 1979. [113 organic toxic pollutants]

2. Paramters and Test Procedures, P. L. 93-923: 40 CFR, Part 141 and 142. National Interim Primary Drinking Water Regulations (NIPDWR), December 24, 1975; 40 CFR, Part 141. National Interim Primary Drinking Water Regulations; Control of Trihalomethanes in Drinking Water, pp. 68674-68707, November 29, 1979.

3. *Natural Resources Council, et al. v. Train*, 8 E. R. C. 2/20 (D. D. C. 1976).

4. Youden, W. J., "Statistical Techniques for Collaborative Tests," AOAC, Washington, DC (1967).

5. Winter, J., P. Britton and H. Clements, "EPA Method Study 8, Total Mercury in Water." U. S. EPA, EMSL-Cincinnati, EPA-600/4-77-012, (1977).

6. Public Law 92-500, 86 Stat. 816, 33 U. S. C., 1251 *et seq.*

CHAPTER 9

PRIORITY POLLUTANT METHODOLOGY QUALITY ASSURANCE REVIEW

Robert D. Kleopfer, Jerry R. Dias and Billy J. Fairless

Environmental Protection Agency
Region VII Laboratory
Kansas City, Kansas

In compliance with a consent decree emanating from civil court actions brought against the U. S. Environmental Protection Agency (EPA) by concerned organizations, EPA has developed a list of 129 priority pollutants with the goal of establishing industrial effluent limitations and guidelines [1]. This chapter will summarize the results of over 10,000 measurements collected by seven laboratories on 104 of the 114 organic priority pollutants, all of the metals (13), and total phenol and cyanide. (Actually there are 130 priority pollutants if one differentiated 1,3-dichloropropene into *cis* and *trans* isomers.) The seven different Aroclors are reported collectively as polychlorinated biphenyls while three base/neutral compounds, two volatile organic compounds, and asbestos are not included in this report. The data assembled here have permitted identification of probable problem areas and provide an initial estimate of the accuracy and precision of the EPA analytical methodology [2] for the priority pollutants.

ANALYTICAL METHODOLOGY

The screening of waste water for 114 organics is accomplished by analyses of five different fractions [2]. The fractions are pesticides (26 compounds), base/neutrals (46 compounds), acids (11 phenols), purgeables (29 volatile organics) and direct aqueous injection (2 compounds).

The base/neutrals (B/N) include compounds such as 1,4-dichlorobenzene, benzidine and benzo [a] pyrene. These compounds are extracted from basic aqueous solution with methylene chloride. The concentrated extracts, are analyzed by gas chromatography/mass spectroscopy (GC/MS) on a 1% SP-2250 column. Columns packed with 3% OV-17 and 3% SP-2250-DB have also been used effectively.

The acid fraction (A) consists of 11 phenols ranging from phenol itself to pentachlorophenol (PCP). These are determined by extraction of acidic solution with methylene chloride and analysis by GC/MS. The preferred column packing for this analysis is 1% SP-1240-DA.

The pesticide fraction (P) is prepared by extraction with 15% methylene chloride in hexane. The pesticides are initially screened by gas chromatography with electron capture detection (GC/EC) [3]. Any compounds which are tentatively identified by that technique must be confirmed by GC/MS.

The purgeable organics, which include compounds ranging in volatility from methyl chloride to ethyl benzene, are measured using a purge and trap analytical technique. In this method, the volatile organics are swept from 5 ml of sample with an inert gas. The purged organics are retained on a trap consisting of Tenax GC activated carbon and silica gel. These organics are then heat-desorbed onto an analytical column consisting of 0.2% Carbowax 1500 on Carbopak C with a 3% Carbowax 1500 percolumn. Again, MS is used for the detection.

Finally, because of their high water solubility, acrolein and acrylonitrile are determined by direct aqueous injection on Chromosorb 101 with MS detection. However, more recent work has shown that these compounds can also be measured by the purge and trap technique [4].

Identification is based on the appearance of mass chromatograms (extracted ion current profiles) for three characteristic ions for each compound. As an example, for the three dichlorobenzene isomers, the molecular ions at 146 and 148 and the fragment ion at 113 are used for identification and quantification. The ions must appear at the correct retention time with the correct relative responses to be considered a positive identification. Quantification is based on external standard methods for the volatile organics and pesticide fractions and internal standard methods for the B/N and A fractions. The internal standard for these fractions is fully deuterated anthracene which has a molecular ion occurring at mass 188.

The metals are analyzed by flameless atomic absorption (AA), and cyanide and total phenols are colorimetrically determined [5]. Cyanide is measured colorimetrically by converting to cyanogen chloride with chloramine-T, which reacts with pyridine to give a product that forms a colored complex with barbituric acid. Total phenol is colorimetrically determined by oxidizing with potassium ferricyanide to give quinones that condense with 4-aminoantipyrine to form colored complexes.

Quality Assurance Requirements

The original priority pollutant methodology document [2] specified several quality assurance (QA) requirements, but additional features were subsequently added [6,7]. These requirements include the following.

Method Blank

The method blank is defined as an appropriate volume of "organic-free" water which has been processed exactly as a sample (same glassware, reagents, solvents, etc.). For the extractable parameters (B/N, A, P) this would require extraction of one liter of water. For the volatile fraction, 5 ml of "organic-free" water should be analyzed by the purge and trap method. One method blank sample should be run with every batch of 20 or fewer samples. Also, a method blank should be run whenever a new source of reagent or solvent is introduced into the analytical scheme. Reagents having background levels that interfere with the compounds to be determined must be purified and shown to be acceptable or replaced with some that are acceptable.

Field Blank

It is the responsibility of the sampling team to provide the appropriate field blanks to the analytical team. For the extractable parameters (B/N, A, P) the minimum requirement is to provide an appropriate volume of blank water which has been processed through the sampling equipment in the same manner as a sample. The field blank is then analyzed in the laboratory as if it were a sample. When interferences occur, the analytical results must be discarded or flagged so as not to result in the reporting of false positives. Field blanks for the volatiles consists of "organic-free" water which has been sent from the laboratory to the sampling site and retained with the samples. The purpose is to check on possible contamination of the sample by permeation of volatiles through the septum seal. Sample blanks should be protected with activated carbon during transit and during storage in the laboratory.

Replicates

To determine the precision of the method, a regular program of analyses of replicate aliquots of environmental samples must be carried out. At least 2 replicate aliquots of a well mixed samples must be analyzed with each set of 20 samples or less analyzed at a given time. For those parameters where a sufficient number of positive results are accumulated over a period of time, precision criteria should be developed. A minimum of 15 replicates at a particular concentration or concentration range where linearity exists is required to start an ongoing program for QA and subsequent estimates of precision.

Standard Addition to Samples

Known amounts of authentic standard are added to the sample. The samples are then processed and analyzed in the same manner as a sample. The percent recovery is then determined as described in Reference 6. At least one spiked sample is analyzed along with each set of 20 samples or less. Spiked data are obtained for each parameter of interest.

Standard Addition to Blank Water (Method Standard)

Known amounts of authentic standards are added to the water blank before extraction. The standard samples are then processed and analyzed in the same manner as a sample. The standard should be approximately equal to the concentration found in routine samples. The percent recovery is determined as described in Reference 6. At least one method standard must be analyzed along with each set of 20 samples or less.

Surrogate Spikes

Standards are added to every sample prior to analysis. The standards chosen should be chemically similar to compounds in the fraction being analyzed. Also, the standards should be compounds which would not likely be found in environmental samples. The purpose of the surrogate spike is to provide quality control on every sample by constantly monitoring for unusual matrix effects, gross sample processing error, etc. The surrogate spike should not be used as an internal standard for quantification purposes.

GC/MS Calibration Check

For the BN, and A fractions, decafluorotriphenylphosphine (DFTPP) is run daily according to EPA procedures. The requirement is for 50 ng of DFTPP to meet relative ion abundance criteria [8]. For the volatile fraction, pentafluorobromobenzene (PFBB) is run daily according to EPA procedures. The requirement is for 100 ng of PFBB to meet the relative ion abundance criteria [7]. More recently, p-fluorobromobenzene has been recommended as a replacement for PFBB [9].

GC Performance Check

For the base/neutral fraction, benzidine is run daily either separately or as a part of a standard mixture. The requirement is to be able to chromatograph the compound at the 100-ng level. For the A fraction, PCP is analyzed daily either separately or as part of a standard mixture. The requirement is to be able to chromatograph the compound at the 100-ng level while giving an acceptable tailing factor [7,10]. For the P fraction, Aldrin is run daily either separately or as part of a standard mixture. An injection of 100 pg should give a recorder response of at least 50% full-scale deflection.

RESULTS AND DISCUSSION

The results have been condensed and summarized in eight tables. In order to maintain readability, information such as the number of data points contributed by each lab for each parameter and the corresponding concentration range have not been included. Also, specific results from each laboratory for each parameter are not included. These data are, however, available from the authors on request.

Interlaboratory Comparison

All the data in this report were collected before June 1979, while the controlling protocol was published in April 1977. There were four primary laboratories contributing almost 10,000 data points to this review. Some of the data were statistically rejected. In Laboratory I, 3% of the B/N and metal analytical data was rejected using the statistical one-sided test at the 2.5% significance level [11]. Similary, 0.3% of the data from Laboratory II was also excluded. A total of 18% of the data from Laboratory III was rejected for two reasons: (1) the spiking levels were not at least two times above the background level (especially at low levels); and (2) application of the one-sided test suggested to a 97.5% confidence that the data removed were likely outliers.

The data are expressed in terms of percent recovery (P) plus or minus ($\pm$) one standard deviation (Sp) where

$$Sp = \sqrt{\frac{\sum\limits_{i=1}^{n} P_i^2 - (\sum\limits_{i=1}^{n} P_i)^2/n}{n-1}}$$

Table I summarized the data for the various classes of priority pollutants. The interlaboratory comparison gives the average recovery of only the compounds analyzed by all the contributing laboratories in the method standard analysis. Once this comparison set had been determined, the same set of compounds was used as the basis for interlaboratory comparison for the matrix spike. This precaution minimized data variability between the laboratories due to properties associated with the individual compounds like volatility, reactivity, etc. Laboratory III had no method standard analytical data and had analyzed fewer compounds. Note that Sp in the interlaboratory comparison is the deviation of the recovery means and not measurement deviation.

Purgeable Organic Compounds

As part of the process of quality assurance, Labs I, II, III and IV analyzed field blanks consisting of organic-free water that was transported to the analytical laboratory along with the samples. Invariably, dichloromethane (methylene chloride), the extraction solvent, was observed in field blanks and laboratory blanks. Typical levels for laboratory IV were 2 ± 1 ppb.

Table I. Interlaboratory Comparison[a]

Priority Pollutant Fraction[b]	Lab I	Lab II	Lab III	Lab IV	Lab V	Lab VII	Average[c]
Volatile (MS)	88 ± 21	95 ± 5		100 ± 8			90 ± 13
Volatile Sample Spike	82 ± 24	101 ± 9	93 ± 13	107 ± 9			92 ± 15
Acid (MS)	90 ± 18	89 ± 5		67 ± 14	82 ± 16		84 ± 13
Acid Sample Blank	92 ± 34	72 ± 10	62 ± 12	60 ± 15	84 ± 17		76 ± 19
B/N-(MS)	95 ± 25	78 ± 41		77 ± 15			84 ± 25
B/N Sample Spike	84 ± 18	61 ± 22	55 ± 24	68 ± 16		63 ± 13	68 ± 21
Pesticide (MS)	73 ± 8	74 ± 19		88 ± 8			78 ± 11
Pesticide Sample Spike	69 ± 7	51 ± 18	33 ± 10	93 ± 5			59 ± 11
Metals (MS)	113 ± 37			103 ± 8			108 ± 22
Metals Sample Spike	100 ± 20	103 ± 14		92 ± 7			96 ± 11
Cyanide (MS)	103 ± 14			103 ± 8			103 ± 7
Cyanide Sample Spike	101 ± 12			93 ± 16			96 ± 14
Phenolics (MS)	10 ± 13	97 ± 6		100 ± 7			101 ± 8
Phenolics Sample Spike	93 ± 15	98 ± 10		97 ± 9			96 ± 11

[a] The values are in units of percent recovery (P) plus or minus (±) one standard deviations (Sp).

[b] MS refers to the method standard or the standard addition to blank water. Sample spike refers to the standard addition to a sample.

[c] P and Sp are weighted averages based on the number of data points contributed by each laboratory.

No data are reported for 2-chloroethyl vinyl ether and *bis*(chloromethyl) ether; the former appears not to have been available and the latter is an alpha-chloromethyl ether which is well known to undergo facile hydrolysis. In a ten-month stability study, Radian Corporation found that the initial concentration of both of these ethers in their methanolic standard was reduced to zero, probably from substitution reactions with methanol [12]. Also, the concentration of acrolein was reduced to zero. Note that acrolein can undergo air oxidation, Michael addition and dimethyl acetal formation in methanol.

Tables I and II summarize the results obtained for the purgeables. The difference in recoveries between the method standard analysis and matrix spiked analysis should indicate sample matrix effects; in this case there appears to be slight increase in recovery in going from the method standard analysis (interlab comparison average = 90 ± 13) to the matrix spiked analysis (interlab comparison average = 92 ± 15).

Volatility is a very important variable in the analysis of the purgeable organic compounds (VOA). Labs I and III have omitted analysis of many of the most volatile compounds, including bromomethane, chloromethane, dichlorodifluoromethane and vinyl chloride. Inspection of the individual laboratory data show significant differences in going from the most volatile group of compounds (bp less than 75°C) to the least volatile compounds (bp greater than 112° C) for Labs I and III only. Recoveries increase and relative standard deviations decrease. Using the two-tailed statistical test for significance of the sample differences between the most volatile group and the intermediate and least volatile groups, assuming there is no difference between the population means from which these samples are drawn [13], it is established to a greater than 99.9% confidence that the observed difference for these groups arises from volatility. Lab II cooled their Tenax sorbent trap with liquid CO_2 and Lab IV used a sorbent trap mix containing Tenax, silica gel, and activated charcoal. In a separate precision study (method standard analysis) by Lab IV involving five measurements at each of four different concentrations, standard deviations of 11.2 ± 3.4 and 10.0 ± 2.4 were obtained for the most volatile group (11 compounds or 220 data points) and the intermediate and least volatile groups (17 compounds or 340 data points), respectively. Thus, to a 95% confidence it appears that even in analytical laboratories where volatility is mainly under control, the more volatile compounds exhibit greater measurement fluctuation.

Comparison of the recovery data of deuterated purgeable organic surrogates in Table VII vs the recoveries of the corresponding nondeuterated priority pollutants in Table II gives, respectively, 93 ± 22 and 93 ± 24 for benzene, 139 ± 96 and 110 ± 21 for chloroform, 146 ± 55 and 88 ± 24 for dichloromethane, 94 ± 18 and 106 ± 28 for ethylbenzene, 92 ± 18 from 97 ± 25 for

Table II. Purgeable Organics[a]

Compound	Method[b] Standard	Standard[c] Spike
Acrolein	77 ± 30	32 ± 30
Acrylonitrile	96 ± 31	102 ± 28
Benzene	89 ± 12	93 ± 24
Bromodichloromethane	97 ± 11	103 ± 31
Bromoform	94 ± 14	88 ± 12
Bromomethane	90 ± 16	78 ± 15
Carbon Tetrachloride	91 ± 23	91 ± 33
Chlorobenzene	94 ± 23	103 ± 24
Chlorodibromomethane	86 ± 12	99 ± 17
Chloroethane	67 ± 22	60 ± 23
Chloroform	90 ± 18	91 ± 26
Chloromethane	91 ± 22	64 ± 28
Dichlorodifluoromethane	108 ± 11	114 ± 8
1,1-Dichloroethane	83 ± 10	87 ± 21
1,2-Dichloroethane	102 ± 12	103 ± 27
1,1-Dichloroethylene	74 ± 24	80 ± 32
trans-1,2-Dichloroethylene	90 ± 25	85 ± 35
Dichloromethane	82 ± 46	66 ± 66
1,2-Dichloropropane	94 ± 26	99 ± 30
cis-1,3-Dichloropropene	95 ± 15	98 ± 20
trans-1-3-Dichloropropene	91 ± 13	93 ± 16
Ethylbenzene	109 ± 19	106 ± 28
1,1,2,2-Tetrachloroethane	81 ± 31	78 ± 31
Tetrachloroethylene	97 ± 13	99 ± 26
Toluene	96 ± 22	97 ± 25
1,1,1-Trichloroethane	92 ± 21	94 ± 36
1,1,2-Trichloroethane	102 ± 14	103 ± 19
Trichloroethylene	106 ± 14	110 ± 22
Trichlorofluormethane	59 ± 23	67 ± 48
Vinyl Chloride	103 ± 30	79 ± 22

[a] The values are in terms of P ± Sp. Data from 2–4 laboratories have been averaged except where noted. In general the concentration added ranged from 10 to 1000 ppb.
[b] Standard addition to blank water.
[c] Standard addition to sample.
[d] Data from only one lab were available.

toluene, 119 ± 29 and 103 ± 27 for 1,2-dichloroethane, and 117 ± 41 and 94 ± 36 for 1, 1, 1-trichloroethane. The large scatter for the data would suggest that d-chloroform and d_2-dichloromethane are poor choices for surrogates.

Phenolic Acids

After the B/N compounds have been extracted from the alkaline sample with methylene chloride (Figure 1), the aqueous phase is made acidic with 6 *N* HC1 and again extracted with methylene chloride. This methylene chloride extract is dried by passing through a column of anhydrous sodium sulfate and the solvent concentrated by distillation in Kuderna-Danish apparatus. Lab III investigated acid carryover in waste samples from the leather tanning industry which tended to form emulsions and therefore were extracted continuously for 24 hours; they noted a pH change from greater than 11 to 8–9. In a separate study, Lab IV demonstrated that, in the absence of emulsions, significant carryover of only the weakest acids, 2-4-dimethyl-phenol (50%) and 4-chloro-3-methylphenol (10%), occurred. Lab V minimized emulsion problems by dilution of the creosote waste from the wood pre-serving industry before analysis.

A strong relationship exits between acid volatility and analytical recovery. In all laboratories contributing to this report, acid recovery was lower for the more volatile phenols (bp less than 211° C: 2-chlorophenol, phenol, 2,4-dichlorophenol and 2,4-dimethylphenol), than for the less volatile ones (bp greater then 211° C: 2-nitrophenol, 4-chloro-3-methylphenol, 2,4,6-trichlorophenol and PCP). In general, the relative standard deviation decreased in going from the most volatile to least volatile group; the *p*-nitrophenols were excluded from consideration in the volatility grouping because of their propensity to thermally undergo intramolecular oxidation to *p*-quinones, which may also explain why these phenols chromatograph poorly. Application of the two-tailed statistical test for significance of the overall difference in recoveries for these two groups of acids results in a greater than 95% confidence that the volatility hypothesis is valid.

Comparison of the recoveries of phenol (Table III, 54 ± 24) and d_5-phenol (Table IV, 55 ± 20) suggests that d_5-phenol is a good choice for a surrogate.

Base/Neutral Compounds

The simultaneous analysis of the 46 B/N compounds probably represents the most difficult task in the analysis of the organic priority pollutants.

Table III. Acid Fraction[a]

Compound	Method[b] Standard	Sample[c] Spike
4-Chloro-3-methylphenol	96 ± 16	99 ± 19
2-Chlorophenol	80 ± 22	71 ± 23
2,4-Dichlorophenol	86 ± 24	84 ± 23
2,4-Dimethylphenol	71 ± 19	72 ± 16
4,6-Dinitro-*o*-cresol	87 ± 34	102 ± 23
2,4-Dinitrophenol	89 ± 22	92 ± 40
2-Nitrophenol	95 ± 22	87 ± 22
4-Nitrophenol	65 ± 33	59 ± 46
Pentachlorophenol	87 ± 24	84 ± 22
Phenol	61 ± 11	54 ± 24
2,4,6-Trichlorophenol	91 ± 22	80 ± 24

[a] The values are in terms of $P \pm Sp$. Data from 2–5 laboratories have been averaged. In general the concentration added ranged from 20 to 2500 ppb.
[b] Standard addition to blank water.
[c] Standard addition to sample.

This report summarizes the analytical results of 43 B/N compounds. No analytical results for *bis*(2-chloroethoxy)methane, 4-chlorophenyl ether, or N-nitrosodimethylamine are contained in this report. Presumably, 4-chlorophenyl phenyl ether and *bis*(2-chloroethyoxy)methane are not readily acquired. Absence of data for N-nitrosodimethylamine probably results because of its poor chromatographic properties (GC gives a low, broad peak). Also, it is questionable whether *bis*(2-chloroethyoxy)methane is stable enough in the aqueous environment, particularly under basic conditions, to be analyzable per the specified methodology. Dibutyl and *bis*(2-ethylhexyl) phthalates were invariably present in the method and field blanks analyzed by Labs III and IV. Since the waste samples from the leather tanning and timber industries were particularly beset with emulsion problems, some carryover of B/N compounds into the acid fraction was observed. The most significant carryover occurred in the lagoon influent samples taken from

Table IV. Priority Pollutant Surrogates[a]

Compound	Lab III[b]	Lab IV[c]	Lab VIII[d]
Purgeable Organics			
d_6-Benzene		95 ± 10	
Bromochloromethane		98 ± 11	
d-Chloroform	139 ± 96		
1,4-Dichlorobutane		96 ± 10	
d_4-1,2-Dichloroethane	119 ± 29		
d_2-Dichloromethane	146 ± 55		
d_{10}-Ethylbenzene	102 ± 25	97 ± 12	
Fluorobenzene	96 ± 20		
d_8-Toluene		95 ± 10	
d_3-1,1,1,-Trichloroethane	117 ± 41		
1-Chloro-2-bromopropane		93 ± 11	
Acids			
2-Fluorophenol	76 ± 36		
Pentafluorophenol	84 ± 30	50 ± 22	101 ± 39
d_5-Phenol	55 ± 20		
Trifluoro-*m*-cresol	72 ± 42		
Base/Neutral			
d_{12}-Benzo[a]anthracene	68 ± 16		
Decafluorobiphenyl	39 ± 18	41 ± 29	46 ± 13
2-fluoroaniline	74 ± 39		
2-fluorobiphenyl	63 ± 5		
1-Fluoronaphthalene	69 ± 18		
2-Fluoronaphthalene	75 ± 20		
d_8-Naphthalene	76 ± 22		
d_5-nitrobenzene	70 ± 21		

[a] The values are in terms of $P \pm Sp$. The concentration added ranged from 20 to 200 ppb.
[b] The matrix for these surrogates included influent and effluent samples from 12 different industrial categories.
[c] The matrix for these surrogates included publicly owned treatment works, detergent, and chemical disposal industries.
[d] The matrix for these surrogates included POTW samples only.

the timber industry, and these samples invariably had a high background of the priority pollutant being analyzed. Analysis for the basic compounds in the acid fraction would probably be futile since they form hydrochloride salts that would most likely remain in the aqueous phase even if they were carried into the acid fraction. If the B/N compounds are separated into reactive and nonreactive groups, then discernible recovery differences become evident (Table V). The reactive group comprises the alkyl amines and chlorides, the phthalate esters and isphorone (an enone); the nonreactive group consists of all the other B/N compounds. It is well known that alkyl amines and chlorides react with each other to produce alkylated ammonium salts [14]. Furthermore, alkyl chlorides can undergo elimination and substitution reactions, like reaction with hydroxide ion to form alkene and alcohols (aryl chlorides are inert to nucleophilic substitution and are placed in the nonreactive group); perchlorocarbons can also form aldehydes, ketones and carboxylic acids by reaction with hydroxide ion. Esters are saponified by bases, and isophorone can undergo condensation and Michael addition [14]. Thus, greater analytical variability is to be expected in the B/N reactive group. The data in Table V appear to verify this anticipation since, in general, recovery decreases and relative standard deviation increases in going from the nonreactive group to the reactive group. Lab II and Lab IV have noted instability indications associated with the standards for the B/N analysis. Radian Corporation did a ten-month storage stability study of their B/N standard in methylene chloride and found significant changes in the concentrations of the chloroalkyl ethers, N-nitrosamines, hexachlorocyclopentadiene, dibenzo [a, h] anthracene and benzidine [12]. Because of emulsion problems, Lab III used continuous extractors in the analytical processing of the samples from the leather tanning and timber industries. As noted earlier the pH was initially adjusted to above 11, but after 24 hours of extraction the pH dropped to 8-9. Lab III performed analyses on sample matrices from a more complex and diverse industrial cross section and has the lowest interlab comparison recovery value (55 ± 24) in Table I, despite statistical refinement of the data. Other problems in the GC/MS screening of the B/N compounds include coelution of anthracene with phenanthrene, benzo [a] anthracene with chrysene, and benzo [b] fluoranthrene with benzo [k] fluoranthrene coupled with the inability to distinguish between these pairs by MS. Thermal decomposition of 1,2-diphenylhydrazine to azobenzene and N-nitrosodiphenylamine to diphenyl amine and tetraphenylhydrazine has been documented [15].

Comparison of the respective recoveries for d_8-naphthalene (76 ± 22) and d_5- nitrobenzene (70 ± 71) in Table IV vs naphthalene (89 ± 51) and nitrobenzene (77 ± 51) in Table VI is favorable.

Table V. Reactivity Groups of the B/N Priority Pollutants[a]

Laboratory	Method Standard Analysis		Matrix Spiked Analysis	
	Nonreactive Group[b]	Reactive Group[c]	Nonreactive Group	Reactive Group
I	102 ± 26	86 ± 21	87 ± 18	78 ± 17
II	93 ± 31	52 ± 46	60 ± 21	64 ± 25
III			58 ± 26	48 ± 12
IV	82 ± 10	67 ± 18	71 ± 12	63 ± 18
VII			65 ± 11	57 ± 17

[a] Expressed as P + Sp.

[b] Nonreactive B/N compounds: acenaphthene, acenaphthylene, anthracene, benzo[a] anthracene benzo[g,h,i] perylene, benzo[a] pyrene, 2-chloronaphthalene, 1,2-, 1,3- and 1,4-Dichlorobenzene, 2,6-dinitrotoluene, fluoranthene, fluorene, hexachlorobenzene, napthalene, nitrobenzene, pyrene, and 1,2,4-trichlorobenzene.

[c] Reactive B/N compounds: benzidine, bis(2-chloroethyl) ether, bis(2-ethylhexyl), diethyl and dimethyl phthalates, 1,2-diphenylhydrazine, hexachlorobutadine, hexachloroethane, and isophorone.

Table VI. Base/Neutral Fraction[a]

Compound	Method[b] Standard	Sample[c] Spike
Acenaphthene	90 ± 22	78 ± 24
Acenaphthylene	83 ± 22	79 ± 27
Anthracene[d]	98 ± 20	79 ± 26
Benzidine	44 ± 27	40 ± 29
Benzo[a]anthracene[d]	105 ± 33	51 ± 24
Benzo[k]fluoranthene[d]	96 ± 68	47 ± 27
Benzo[b]fluoranthene[d]	96 ± 68[e]	41 ± 21
Benzo[a]pyrene	90 ± 22	43 ± 21
Benzyl Butyl Phthalate	49 ± 39	49 ± 22
bis(2-Chloroethyl) Ether	98 ± 48	80 ± 49
bis(2-Chloroisopropyl) Ether	154 ± 136	96 ± 88
bis(2-Ethylhexyl) Phthalate	70 ± 33	66 ± 50
4-Bromophenyl Ether	80 ± 25	63 ± 25
2-Chloronaphthalene	88 ± 20	79 ± 21
Chrysene[d]	105 ± 33	77 ± 27
Dibenzo[a,h]anthracene	80 ± 42	36 ± 29
Di-*n*-butyl Phthalate	80 ± 32	58 ± 27
1,2-Dichlorobenzene	65 ± 24	65 ± 27
1,3-Dichlorobenzene	67 ± 21	62 ± 20
1,4-Dichlorobenzene	67 ± 22	63 ± 21
3,3'-Dichlorobenzidine	71 ± 85	62 ± 45
Diethyl Phthalate	71 ± 37	65 ± 37
Dimethyl Phthalate	43 ± 37	66 ± 43
2,4-Dinitrotoluene	122 ± 55	94 ± 45
2,6-Dinitrotoluene	115 ± 41	104 ± 35
Di-*n*-octyl Phthalate	84 ± 44	88 ± 32
1,2-Diphenylhydrazine (and/or Azobenzene)	97 ± 26	91 ± 32
Fluoranthene	111 ± 26	63 ± 20
Fluorene	98 ± 24	88 ± 25
Hexachlorobenzene	98 ± 31	76 ± 31
Hexachlorobutadiene	76 ± 26	77 ± 45
Hexachlorocyclopentadiene	38 ± 28	27 ± 10
Hexachloroethane	63 ± 22	58 ± 23

Table VI, continued

Isophorone	66 ± 36	67 ± 22
Indeno[1,2,3-cd]pyrene	109 ± 14	40 ± 21
Naphthalene	83 ± 24	89 ± 51
Nitrobenzene	106 ± 31	77 ± 51
N-Nitrosodiphenylamine (and/or Diphenylamine)	72 ± 22	66 ± 25
N-Nitrosodi-*n*-propylamine	86 ± 34	71 ± 22
Phenanthrene	98 ± 20	79 ± 20
Pyrene	142 ± 41	63 ± 20
1,2,4-Trichlorobenzene	74 ± 22	69 ± 24

[a] The values are in terms of P ± Sp. Data from 2–5 laboratories have been averaged except where noted. In general, the concentration added ranged from 10 to 500 ppb.
[b] Standard addition to blank water.
[c] Standard addition to sample.
[d] These isomers pairs are not separated by packed column GC. Also mass spectral data are not sufficiently unique to allow differentiation.
[e] Data from only one lab were available.

Pesticides

The GC/EC analysis of the pesticides is the most straightforward of the organic priority pollutants [3]. Since chlordane, PCB and toxaphene are indistinguishable due to interferences between the many compounds in the spike, the pesticides are spiked in groups of compounds known to be separable. Endrin aldehyde has a high standard deviation in Table VII probably because of partial air oxidation to a carboxylic acid.

Metals

For the most part, none of the laboratories had difficulty in the metals analysis above concentration of approximately 25 ppb. The high standard deviations present in the data of Lab I is no doubt due to the low spiking level, which is closed to the detection limit of the method used by that laboratory. In the flameless analysis of zinc, environmental contamination problems were apparently experienced by Labs I and II. It is also possible that in the method standard analysis of chromium, copper and nickel, Lab I was having contamination problems from copper and stainless steel plumbing.

Table VII. Pesticide Fraction[a]

Compound	Method[b] Standard	Sample[c] Spike
Aldrin	72 ± 13	55 ± 12
alpha-BHC	78 ± 13	55 ± 12
beta-BHC	79 ± 21	57 ± 22
gamma-BHC	78 ± 14	64 ± 11
delta-BHC	82 ± 16	61 ± 16
Chlordane	$81 \pm 17*$	$39 \pm 9*$
4,4 -DDD	82 ± 14	62 ± 16
4,4 -DDE	76 ± 14	57 ± 18
4,4 -DDT	85 ± 17	76 ± 26
Dieldrin	71 ± 14	62 ± 16
Endosulfan I	65 ± 14	61 ± 13
Endosulfan II	67 ± 19	66 ± 14
Endosulfane Sulfate	74 ± 39[d]	84 ± 30[d]
Endrin	82 ± 25	68 ± 18
Endrin Aldehyde	64 ± 76[d]	34 ± 39[d]
Heptachlor	72 ± 12	49 ± 12
Heptachlor Epoxide	82 ± 14	65 ± 11
PCB	83 ± 11[d]	42 ± 13
Toxaphene	89 ± 12[d]	

[a] The values are in terms of P ± Sp. Data from 2–4 laboratories have been averaged except where noted. In general the concentration added ranged from 0.1 to 100 parts per billion.

[b] Standard addition to blank water.

[c] Standard addition to sample.

[d] Data from only one lab were available.

Cyanide and Total Phenols

Total phenols were colorimetrically determined by Labs II and IV via a totally automated procedure [5]. Recovery data for the cyanide and total phenols are presented in Table VIII.

Table VIII. Metals, Cyanide and Phenolics[a]

Parameter	Method[b] Standard	Sample[c] Spike
Antimony	61 ± 47[d]	103 ± 24
Arsenic	120 ± 20	97 ± 25
Beryllium	89 ± 16	94 ± 20
Cadmium	91 ± 18	98 ± 23
Chromium	99 ± 30	106 ± 25
Copper	136 ± 70	99 ± 24
Lead	116 ± 32	93 ± 25
Mercury[e]	83 ± 24	79 ± 38
Nickel	84 ± 62	101 ± 26
Selenium	112 ± 15	93 ± 20
Silver	110 ± 25	80 ± 25
Thallium	99 ± 33[d]	95 ± 23
Zinc	122 ± 44	106 ± 37
Cyanide	103 ± 7	96 ± 14
Total Phenols	101 ± 8	96 ± 11

[a] The values are in terms of P ± Sp. Data from 2–3 laboratories have been averaged except where noted. In general the concentration added ranged from 10 to 1000 ppb.
[b] Standard addition to blank water.
[c] Standard addition to sample.
[d] Data from only one lab were available.
[e] Analyzed by the cold vapor technique.

SUMMARY AND CONCLUSIONS

A summary of the overall recoveries of the priority pollutants is given in Table I. In every case, except for the purgeable organic compounds, the overall recovery decreases and the relative standard deviation increases in going from the method standard to the matrix spiked analysis. Of the organic priority pollutants, the analysis of the purgeable and acid organic compounds results in the highest recoveries with the smallest matrix effect. The increase in recovery for the phenolic acids in going from unadulterated water to a matrix aqueous medium may be a salting-out and salting-in effect, respectively, and needs to be studied. Two major variables for these groups is volatility and carryover of phenolic acids into the B/N fraction, the latter arising mainly from emulsion problems since the data from Lab III suggest that no pKa dependence was associated with their observed acid carryover. The process of continuous extraction itself may be responsible for carry over which needs to be experimentally determined. Volatility can be controlled by close monitoring of the analytical process. Ascertaining that the pH is greater than 11 and, in the case of emulsions, performing an additional base wash of the B/N methylene chloride extract will rectify carryover of the acids into the B/N fraction. The question concerning the stability of methylene chloride during long-term (24-hour) contact with hydroxide solution under emulsion conditions needs to be investigated.

In part, the analysis of the B/N and P compounds gives the lowest recoveries because of their greater propensity to react among each other or with hydroxide or water [14]. Several laboratories have noted that when the four different B/N methanolic standards are mixed in methylene chloride, this solution becomes turbid and eventually changes to a yellow color. Thus, the stability of the priority pollutants, benzidine and 1,2-diphenylhydrazine in contact with methylene chloride needs to be fully determined. Saponification of the phthalate esters has been observed during the B/N analysis by several laboratories. Another major variable in the analysis of the B/N compounds is carryover into the acid fraction in the presence of emulsions; this appears to be particularly acute at high background concentrations of the B/N component.

VOA analysis per the EPA protocol has proved to be the most successful. The acid and pesticide analyses have been moderately successful, and the B/N analysis has been less successful on samples with severe matrix problems, as from the leather tanning and timber industries. Nevertheless, the present methodology is adequate and appears capable of improvement. One can be confident that false positive analyses are considerably less likely than false negative analyses so that when a priority pollutant is detected in the environment we know the measured quantity is probably smaller than the true value. Table IX summarizes problems associated with the current priority pollutant methodology.

Initial recommended performance control limits can be computed by adding (upper control limits, UCL) or substracting (lower control limit LCL) three times the standard deviation from the average percent recovery values presented in Tables I to VIII [16]. Some LCL will no doubt increase as the analytical methodology improves. Also, one should note that these control limits are somewhat dependent upon the level of standard addition, the degree of which has not been investigated fully.

Table IX. Problem Priority Pollutants

Compound	Problem
Dichloromethane	Frequently found in blanks and samples because of in lab contamination
bis(Chloromethyl) ether	Readily hydrolyzed in water
N-Nitrosodimethylamine	Poor chromatographic properties
Di-*n*-butylphthalate	Frequently found in blanks
Bis(2-ethylhexyl) phthalate	Frequently found in blanks
1,2-Diphenylhydrazine	Thermally decomposes to diphenyl amine and tetraphenylhydrazine
Benzidine	Poor chromatographic properties
Hexachlorocyclopentadine	Subject to thermal decomposition
Endrin Aldehyde	Is readily oxidized
Anthracene and Phenanthrene	Coelute on packed columns
Chrysene and Benzo[a] anthracene	Coelute on packed columns
Benzo[b] fluoranthrene and Benzo[k] fluoranthrene	Coelute on packed columns

ACKNOWLEDGMENTS

The major portion of the data in this summary came from EPA-Region VII and EPA Quality Assurance Contract Reports from A. D. Little. Inc., Acorn Park, Cambridge, MA 02140 (EPA contract #68-01-3857); Carborundum Co., 3401 LaGrande Blvd., Sacramento CA 95823 (EPA contract #68-01-4689); and Versar, Inc., 6621 Electronics Dr., Springfield, VA 22151 (EPA contract #68-01-3852).

REFERENCES

1. *Environ., Sci., Technol.* (February 1978) p. 154.
2. "Sampling and Analysis Procedures for Screening of Industrial Effluents for Priority Pollutants," U. S. EPA, Environmental Monitoring and Support Laboratory, Cincinnati, OH (March 1976, revised April 1977).
3. *Federal Register* 38(125):17318 (1973).
4. "Midwest Research Institute Final Letter Report on Evaluation of Test Procedure for Acrolein and Acrylonitrile," MRI Project No. 4719-A, U. S. EPA Environmental Monitoring and Support Laboratory, Cincinnati, OH (June 1979).
5. "Methods for Chemical Analysis of Water and Wastes," U. S. EPA, Enviromental Monitoring and Support Laboratory, Cincinnati, OH (March 1979).
6. "Procedure for Preliminary Evaluation of Analytical Methods to be Used in the Verification Phase of the Effluent Guidelines Division BAT Review," U. S. EPA, Environmental Monitoring and Support Laboratory, Cincinnati, OH (March 1978).
7. "Addendum for Sampling and Analysis Procedures for Screening of Industrial Effluents for Priority Pollutants," U. S. EPA, Environmental Monitoring and Support Laboratory, Cincinnati, OH (April 1979).
8. Eichelberger, J. W., L. E. Harris and W. L. Budde. *Anal. Chem.* 47:995 (1975).
9. "Guidelines Establishing Test Procedures for the Analysis of Pollutants, Proposed Regulations," *Federal Register* 40 (233) (1979).
10. McNair, H. M., and E. J. Bonelli. *Basic Gas Chromatography*, (Berkeley, CA: Consolidated Printing, 1969), p. 52.
11. American Society of Testing and Materials, D2777.
12. Keith, L. H., Radian Corporation., Austin, TX. Personal communication.
13. Spiegel, M. R. *Statistics, Schaum's Outline Series* (San Francisco, CA: McGraw-Hill Book Co., 1961), p. 170.
14. Morrison, R., and R. Boyd. *Organic Chemistry*, 2nd ed. (Boston: Allyn and Bacon Inc., 1969).
15. Seminar on Analytical Methods For Priority Pollutants, U. S. EPA, Denver, Colorado, November, 1977.
16. "Handbook for Analytical Control in Water and Wastewater Laboratories," U. S. EPA, Environmental Monitoring and Support Laboratory, Cincinnati, OH (March 1979).

CHAPTER 10

QUALITY ASSURANCE IN THE EVALUATION OF POLLUTION CONTROL TECHNOLOGY: EXPERIENCES AT THE INDUSTRIAL ENVIRONMENTAL RESEARCH LABORATORY, U.S. ENVIRONMENTAL PROTECTION AGENCY, CINCINNATI, OHIO

Paul E. Mills

Industrial Environmental Research Laboratory
Cincinnati, Ohio

The Industrial Environmental Research Laboratory (IERL), Cincinnati, OH, is responsible to the Deputy Assistant Administrator for Environmental Engineering and Technology for the management and implementation, within the Environmental Protection Agency (EPA) Office of Research and Development (ORD) policies, guidelines and allocated resources, of programs to develop and demonstrate cost-effective technologies and methods to (1) prevent, control or abate pollution from operations associated with industrial processing and manufacturing, and from the extraction, processing, conversion and utilization of energy and mineral resources, and (2) to destroy, detoxify and treat solid and hazardous wastes. IERL also identifies and evaluates environmental quality control alternatives, including conservation measures, and assesses environmental and socioeconomic impacts of alternative control strategies. Work is carried out either through its own facilities and field stations or under contract, grant, cooperative agreement or interagency agreement with other organizations. Additionally, IERL assists in the development of broad research policy, program guidelines and long-range research plans; recommends specific projects and programs, including needed resources; provides technical assistance and technical support to EPA components as requested; assures that the results of its work are disseminated according to ORD guidelines; and provides the administrative and financial framework to assure that the activities of the IERL meet EPA and federal requirements.

IERL conducts environmental assessment and control technology development studies in an effort to provide a multimedia approach to pollution control problems. The multimedia program is concerned with air, water, solid waste and energy-consuming aspects of environmental pollution. Water recycle/reuse projects are an important part of these studies. Water recycle/reuse can reduce pollution entering U. S. water resources, allow for extraction of valuable compounds from the water and provide energy savings.

The first part of this chapter describes the Quality Assurance Program of the IERL. The second part of the paper discusses a number of quality assurance problems involved with chemical analysis in recycle/reuse projects in several industries.

Quality Assurance

Quality assurance (QA) is defined here to be the total program for assuring the reliability of data collected in research activities. Quality control (QC) is the routine application of procedures for controlling the measurement process. QA encompasses all actions taken by the organization to achieve accurate and reliable results for programs undertaken. An established QA program is essential to produce valid sampling and analytical data to support research, demonstration and monitoring efforts. QA applied to appropriate samples, tested by trained personnel, using reliable methods and instrumentation, will yield timely data with the required accuracy, precision, completeness and comparability.

THE IERL QUALITY ASSURANCE PROGRAM

The primary objectives of the IERL QA program are to improve, assure, assess and document the quality of measurements made and reported in connection with IERL research. Management, administrative, statistical, investigative, preventive and corrective techniques are employed to increase the data quality. Since the pollution control technology research conducted by IERL is often used to demonstrate state-of-the-art control technology, the results can be incorporated into national pollution control laws on effluent limitations, with a consequent national economic impact. It is critical that the research results be valid and the process by which they were produced not be subject to legal or scientific attack. The procedures for sample collection, handling and analysis, and data interpretation must be trustworthy and provide accurate, precise and reliable information.

Control of pollution requires measurements to define and describe the pollutants; measurements of the extent of pollution; and measurements of the effectiveness of controls in reducing pollution.

The IERL QA program is focused on the extramural (contractor or grantee) laboratory performing research on the behalf of IERL. The difficulties are twofold: procuring a capable research team, with the appropriate equipment, training and experience to perform the research; and monitoring the work as it is conducted, so that data interpretation proceeds with valid results.

Benefits expected as a result of the QA program include:

1. more complete and quantitive assessment and documentation of the validity of data generated by IERL from environmental assessment and control technology development projects;
2. identification of potential sources of measurement error to correct deficiencies in data quality in a timely and economical manner;
3. increased credibility and wider acceptance and use by the scientific community of the IERL research results;
4. improved quality of the data through increased awareness of common data quality problem areas and the use of effective quality control and quality assurance techniques;
5. a focal point within IERL for documentation of measurement methods and the quality of data resulting from the methods;
6. improved utilization of planned experimental programs in the development and testing of control systems;
7. more effective use of performance evaluation techniques such as the use of standard reference materials, split and blind samples, and inter-laboratory testing programs; and
8. improved decision-making on future allocations of resources based on more complete and better quality data.

Some specific objectives of the QA program are to document that:

* qualified personnel are used;
* appropriate instruments are used;
* representative samples are collected;
* preservation, shipping and handling procedures are appropriate;
* analysis is by EPA-approved methods;
* quality control procedures are followed; and
* analytical results are correlated with plant process data.

The tools used to meet these objectives come from several places: (1) EPA's Mandatory Quality Assurance Policy; (2) the IERL Quality Assurance Policy; and (3) the IERL Quality Assurance Office.

EPA's Mandatory Quality Assurance Policy was issued by the Administrator on June 14, 1979. The policy requires anyone who submits environmental monitoring data to EPA to provide:

1. a written QA plan for the work;
2. an estimate of QA cost;
3. acceptable performance on audit samples;
4. access for field and lab site visits; and
5. documentation for QC performance.

These requirements (Table I) when followed, help ensure that those who perform work for EPA fully understand commitment to QA, and make QA an integral part of the work from beginning to end.

The IERL Quality Assurance Policy requires that the performing researchers must submit a Sampling and Analytical Plan for approval prior to beginning work, and EPA methods must be used unless alternative procedures are approved by the Quality Assurance Office within the IERL. These requirements ensure that there is complete understanding of what is to be done, how, when and where, before the collection of environmental samples begins. The Sampling and Analytical Plan, revised July 6, 1979, is reproduced in Table II. It is part of each project involving collection and analysis of environmental samples.

The IERL Quality Assurance Office is responsible for ensuring that the EPA and IERL QA policies are followed. Some of the functions of the QA office are discussed below.

Technical Review

The IERL QA office provides a review of Sampling and Analytical Plans submitted for approval after the services of an extramural research organization have been procured. The QA officer is available to assist in the technical evaluation of proposals for specific projects. After work is complete, report reviews are provided. The reviews examine in detail the statistical design of the project, site selections, sampling and analytical procedures used, quality control and quality assurance applied, and overall compliance of the work with appropriate policies.

**Table I. Quality Assurance Requirements for All EPA Extramural Projects
Involving Environmental Measurements[a]**

Potential contractors/grantees shall provide the following quality assurance (QA) information and related documentation for EPA funding of any projects involving the collection and analysis of samples. These criteria shall be used in determining the overall acceptability of any contractor or grantee.

1. A written QA plan shall be developed that includes the oversight role of management, identification of personnel responsible for the QA program, proper sample collection, use of approved measurement techniques, calibration standards and their verification, internal quality control practices, and appropriate data management controls. Preaward information can be limited to that required to evaluate capabilities. The specific QA plan may be developed and subsequently approved by EPA after award of funds.

2. An estimate of costs associated with the QA program in terms of percentage of overall project cost shall be provided. Normally, a minimum of 10–30% of the estimated sample collection and analysis costs will be necessary for adequate quality control.

3. As a condition of award, acceptable performance must be demonstrated on audit samples, when available and where applicable, and throughout duration of contract/grant.

4. Provision shall be made for onsite laboratory and/or field sites evaluation at option of the Project Officer for a performance evaluation and for documentation that all equipment and supplies necessary for successful completion of project are available.

5. Documented quality control performance will be submitted with the final report and otherwise as required by the Project Officer.

Criteria and guidelines providing further details and steps to implement these general QA requirements will be forthcoming from the Agency's QA Implementation Work Group.

[a] This includes, but is not limited to, grants, contracts, cooperative agreements and interagency agreements.

Approve QA Review Forms

Before a request for research services leaves IERL, it is examined for policy compliance, and to determine that the principal investigator for EPA (the Project Officer) had made appropriate arrangements for QA throughout the project. Any corrections must be made before work begins.

Table II. Sampling and Analytical Plan

The Grantee or Contractor shall prepare and submit sampling and analytical plans to the Project Officer for review and approval prior to the collection or analysis of any samples obtained under this contract/grant.

Such plans shall include, but not be limited to: (1) summary; (2) purpose of sampling and analysis efforts and use to which results will be put; (3) rationale for sampling; (4) factors which may affect sample reliability and integrity; (5) selection of sampling approach, locations, procedures and alternatives considered as appropriate; (6) flow measurement methods and procedures where applicable; (7) materials to be analyzed for and methods to be used, including references when appropriate; (8) data-handling procedures, including statistical methods to be used if appropriate and format in which sampling and analytical results will be presented; (9) quality assurance/quality control plan; (10) proposed schedule and level of effort and costs; (11) appendices if any; (12) references.

EPA-approved sampling, analytical and quality assurance/quality control procedures will be used when they are available. This requirement may be waived by the Project Officer: (1) if it can be demonstrated that the sample matrix in question produces unacceptable interferences in the standard method; (2) when standard procedures are not available; (3) when other valid reasons exist. Project Officer approval is required prior to using non-EPA–approved methods. The performing organization is responsible for keeping itself informed of the status of standard methods and quality assurance/ quality control procedures published by the Agency.

Where required by the Project Officer, the sampling and analytical plan will include provisions for evidence of satisfactory performance using synthetic samples or appropriate standards provided by EPA. The Project Officer may require the procurement of preliminary samples to test the validity of the sampling and analytical plan. Site inspections to evaluate quality assurance/quality control measures may be required.

Provide Audit Samples

At the request of a Project Officer, the QA office arranges with the appropriate EPA Environmental Monitoring Systems Laboratory (EMSL) to provide audit samples for the performing research organizations. These samples are either air, water or sludge, with specific concentrations of parameters of concern in the project. The researchers receive these samples, with instructions for their analysis, and results are compared to EPA values. The procedure serves to identify analytical difficulties and allow for their prompt remedy. (Winter [1] describes the EMSL program for water at Cincinnati, OH more fully.)

Approve Alternative Methods

If researchers request a variance from an EPA-approved method, they must submit comparability data showing an alternative procedure is equal to or better than that of the EPA. In concert with experts at the EMSL, the IERL QA office either approves or denies the request, based on a review of the comparability data. Since a great deal of effort has been expended in methods development and validation by EPA, these requests are examined closely.

Methods Development

IERL occasionally is involved in development of a new instrument, method or technique. This is done only after contacting other researchers in the EPA whose primary responsibility it is to do such work, and verifying that they are unable to develop and validate the method in the time required with the resources available. These contracts ensure that the most capable researchers available oversee the work, and that the research is coordinated with other interested parties. Duplication of effort is minimized by this policy.

Audit

The IERL QA Office selects projects for detailed audit and review. The format of such review is described by Stratton and Bonds [2]. These audits are independent checks by the QA officer or the IERL QA contractor on the technical aspects of a project's sampling, analysis and quality assurance. Findings are reported to the Project Officer and the research organization for correction of any problems identified. The audits can entail a visit to a field site or to the analytical laboratory, examination of data or sending an audit sample for analysis.

Quality Assurance Training

QA materials are made available in formal courses for the staff professionals at IERL. These courses enable them to learn the latest techniques and QA requirements, and how to apply them in their work. The philosophy is that

QA is not something to be added on, then checked by someone outside the program; but that each professional should build quality assurance into each project from its start.

Quality Assurance in Recycle/Reuse Projects

Water recycle/reuse projects have been conducted for IERL by contractor and grantee laboratories in the following industries: fruit, vegetable and meat processing; pulp and paper production; metal finishing and electroplating; and leather tanning operations. These projects have generally been conducted to evaluate the effectiveness of effluent treatments or process changes in restoring water quality for recycle or reuse.

Eller et al. [3] distinguish "reuse" as the utilization of water that has been used previously for another purpose, and "recycling" as the use of the same water one or more times for the same purposes. An example of water reuse is the direct use of municipal waste waters as cooling waters by steel plants. Recycled water could be used in washing vegetables to chill them for storage. Recycle systems offer more attractions as final effluent requirements become more stringent. Successive use of waters, however, needs to meet certain water quality standards, and often these waters have to be subjected to some form of treatment. A broad spectrum of quality criteria exists for industrial water supplies because of the highly variable nature of industrial water use [4].

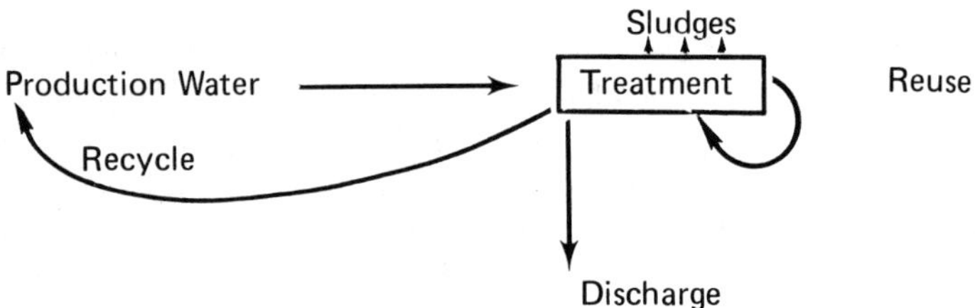

Figure 1. Simplified diagram of water use/reuse system and testing recommendations.

Figure 1 shows a simplified diagram of water use, treatment and recycle or reuse. Some of the testing recommendations are shown, and discussed below, which are subject to QA requirements.

Recycle and Reuse Monitoring of Water and Equipment

The quality of the water which is recycled or reused after treatment needs to be determined. The appropriate parameters for such determinations in past IERL projects have included, for most projects, the gross parameters listed in Table III. Specialized tests were also conducted for certain industry segments (e.g., coliform testing of waters used for food processing). However, these parameters did not indicate the effects of recycle or reuse on the treatment equipment; consequently, it is difficult to determine if some of the contaminant buildup came from equipment corrosion, changes in the character of the water's constituents as an artifact of treatment or other sources. Most sampling was composite grab samples taken over 24 hours. Continuous monitoring instrumentation for pH, temperature, flow and turbidity was used as by Esvelt [6].

Treatment Effectiveness Testing

The degree to which control technology, whether it was reverse osmosis, filtration, carbon adsorbtion, etc., or some combination, removed pollutants and allowed the process water to be reused/recycled was measured in these projects and reported as percent removal of those pollutants. The parameters were taken from the permit discharge effluent limitations. Specific organics were identified in only one project [5].

Sludge Testing for Hazardous Materials

These projects were completed before the Resource Conservation and Recovery Act (RCRA) hazardous materials characterization methods were proposed in the *Federal Register* (December 18, 1978). Sludge composition was not characterized for these processes or control technologies, although future wastes will have to be tested for degree of hazard according to the RCRA protocols.

Discharge Testing for Permit Parameters

Effluents from the treatment processes examined in these reports were tested for the permit parameters listed in the Federal Water Pollution Con-

Table III. Gross Parameters Used in Various U. S. EPA Reuse Projects

Parameters	Number of Projects Used in (Out of 17)	Number Using EPA Standard Method
Suspended Solids	15	14
Dissolved Solids	10	8
Chlorides	13	9
Oil to Grease	8	5
Total Plate Count	7	5
Heavy Metals	7	5
BOD	11	9
COD	12	9
N (NH_4, NO_2, etc)	8	5
pH	13	10
Alkalinity	7	6
Total Solids	8	7

trol Act (FWPCA) Amendments as required for the respective plants; these tests were generally limited to the gross parameters shown in Table III. The organic compounds in the 1976 Consent Decree (priority pollutants) were not specifically quantified; only those projects measuring pesticides looked at some of the listed compounds [6-8].

Some of the problems observed in the IERL water recycle/reuse projects are discussed below. Generally, the most significant flaw, from a quality assurance standpoint, is a failure to document the treatment of samples from collection to analysis. Sample preservation, handling and shipping procedures are not specified; where it is stated that EPA or *Standard Methods* procedures are followed, the reader assumes that preservation was also in accordance with approved practice. Since the projects were not conducted for enforcement actions, chain-of-custody procedures were not followed, and the time from sample collection to receipt and storage in the laboratory, plus the conditions of shipping, are not known. Quality control procedures, such as the use of replicate analyses, spikes, blanks, recoveries and control charts for precision and analysis, again are not supplied for the projects; it is undetermined if the relevant procedures in the EPA QC handbook were followed[9]. No record exists of researchers' performance on audit samples or participation in EPA interlaboratory studies.

Methods used for analysis in the projects were not validated for the recycle/reuse matrixes encountered. As stated in Appendix III of Reference 10,

> initially, the methodology must be validated for each industrial sub-category being measured by the laboratory. The requirement for validation of each subcategory is based on the assumed unique nature of the wastewater associated with most of the subcategories.

This assumption can also apply to other than organic priority pollutant compounds and methods. Information on method validation can be obtained from the EMSL in Cincinnati. An example project [4] indicates that "when water reuse is practiced, the process water chemical concentration increases from 2 to 6 times the concentrations under conventional operating conditions." Depending on the treatments the water receives, there could be a buildup of pollutants, such as chlorides, solids, organics or metals, which can significantly alter the efficiency of methods for these parameters. Therefore, a method validation step is recommended for water recycle/reuse characterizations. A necessary part of the quality assurance for this step is the provision of standard reference materials, substances with a known concentration of pollutant for calibration of instruments, determination of recoveries, etc. Such samples, which would be typical of recycle/reuse waters, would represent for example, high solids, high chlorides and/or high metals matrixes; these are not widely available.

In some instances, inappropriate procedures were used. For example, chemical test kits were used to measure alkalinity, chlorides and phosphates in one project [11]. EMSL has interpreted Section 304(g) of PL 92-500 to mean that the test kits are not equivalent to the procedures promulgated in the *Federal Register* and therefore are subject to the requirements governing alternative procedures. Therefore, data demonstrating the comparability with EPA-approved procedures for these parameters of the test kits should have been presented, prior to use in the project. In another example, composite sampling for oil and grease [12, 13] is impractical, because these pollutants will tend to aggregate and become lost on sampling equipment. Due to process variations, the composite sample may be difficult to relate to actual pollutant levels.

Implementation of the IERL and EPA Quality Assurance Policies should improve the quality of future recycle/reuse project data. The policies require the use of EPA-approved, or alternative validated methods; documentation at every phase of the project; chain-of-custody procedures as part of the documentation where appropriate; and the use of audit samples and site visits to verify satisfactory performance.

Pollution control technology evaluation requires the testing of a number of water quality parameters; in addition, the effects of recycle/reuse on the equipment should be measured. Some of the parameters which are appropriate include:

- corrosion
- scale
- foam
- product quality
- slime
- treatment artifacts

Due to the buildup of salts, metals, hardness, etc. in the recycled/reused waters, the equipment may not last as long as with once-through process water systems. Additional pollution may come from leaching of equipment, or from treatment artifacts. Each industry usually sets standards for its product quality, and methods to measure that quality. Corrosion standards have been devised by ASTM. Surrogate parameters, such as sulfate content, can be used to estimate scaling tendencies of treated waters. Foam and slime are difficult to sample, but chemical treatment can reduce their occurrence. A surrogate for foam could be an anionic surfactant standard such as linear alkylate sulfonate; for slimes, some are organic, some inorganic, and case-by-case methods would have to be devised.

In the projects reviewed in this chapter, such parameters were not quantified; rather, when the equipment fouled, it was cleaned. Continuous monitoring for some of these parameters would ensure that recycle/reuse water was of a known quality, that the operating costs and equipment downtime would be minimized, and the discharged effluent was within permitted limits.

Suggestions for future research in methods development, validation and quality assurance are summarized below.

Specific Compound Identification

Since the 1976 EPA Consent Decree, 129 priority pollutants in industrial effluents have been sought by EPA contractors. A total of 113 specific organic compounds, including pesticides and plasticizers, are looked for in the samples. One can expect additional chemicals to be added to the listing. Also, some of the treatment processes (e.g., chlorination) involved in water recycle/reuse may produce artifacts which are toxic, such as chlorinated organic compounds. Gas chromatography/mass spectrometry (GC/MS) can help identify such compounds.

Continuous Monitoring Instrumentation

The ability to monitor water quality on-line in real time allows plant operators greater flexibility for treatment and greater control. It is certainly preferable to waiting several hours or days to determine what was in the water discharged, recycled to product use or reused for other purposes. Techniques such as high-pressure liquid chromatography (HPLC) for organics and ion chromatography for ions could fulfill some of these requirements for quick analytical response.

Selection of Representative Monitoring Parameters for Pollution Control Technology Evaluation

Some of the parameters or surrogate parameters mentioned above, such as sulfates, should also be measured in an evaluation of the effectiveness of control technology for recycle/reuse. These representative parameters will obviously differ among industries, and even among plants, depending on water quality influent and treatment processes.

Standard Reference Materials Representative of Recycle/Reuse Matrixes

Standard samples, containing known amounts of certain pollutants, determined by validated methods, should be made available for researchers to calibrate their instruments, and to compare alternative procedures. From these standards, audit samples can be prepared for evaluating the analytical performance of researchers. Such parameters as dissolved oxygen, dissolved ions, flow velocity and temperature all affect the rate of corrosion.

Interlaboratory Comparisons Studies

For new methods and reference materials, these should be subjected to round-robin testing by qualified researchers. These studies would help point out weaknesses in the procedures, possible improvements, operator error sources and indicate how much variation in analytical results might be expected.

Sludge Testing Procedures

The sludges which accumulated from the water treatment for recycle/ reuse were tested in only one project [14]. They should be tested, for control purposes, in all; indeed, the RCRA will require they be tested to determine if they are hazardous wastes; the testing protocol includes tests for:

- ignitability
- corrosivity
- reactivity
- toxicity
- radioactivity
- infectiousness
- phytotoxicity
- teratogenicity and mutagenicity

Industries will have to determine if the benefits derived from reduced water consumption make up for the possible production of hazardous wastes which need to be disposed of in an environmentally acceptable fashion.

Biological Testing Procedures

The utility of biological test systems for determination of control effectiveness, toxicity, etc. should be evaluated. Flow-through bioassays, mutagenicity tests, etc. could be examined for their applicability.

CONCLUSIONS

The experience of IERL with water reuse and recycle projects in several industries has shown that the quality assurance aspects of pollution control technology evaluation can be improved. Steps are being taken to accomplish this, but there are a number of suggestions which should aid in more complete characterizations of the qualities of the waters involved and the effectiveness of treatments.

REFERENCES

1. Winter, J. A. "Quality Assurance for Water and Waste Water Analyses," Chapter 7, this volume.
2. Stratton, C., and J. Bonds. "Quality Assurance Guidelines for IERL-Ci Project Officers," Industrial Environmental Research Laboratory—Cincinnati, U. S. EPA Report-600/9-79-046 (December 1979).
3. Eller, J., D. Ford and E. Gloyna. "Water Reuse and Recycling in Industry," *J. Amer. Water Works Assoc.* 62:149 (1970).
4. Streebin, L. E., G. W. Reid and P. Law. "Water Reuse in a Paper Reprocessing Plant," Industrial Environmental Research Laboratory—Cincinnati, U.S. EPA Report-600/2-76-232 (October 1976).
5. Lang, E. W., W. G. Timpe and R. L. Miller. "Activated Carbon Treatment of Unbleached Kraft Effluent For Reuse," National Environmental Research Center, U. S. EPA Report-660/2-75-004 (April 1975).
6. Esvelt, L. A. "Reuse of Treated Fruit Processing Wastewater in a Cannery," Industrial Environmental Research Laboratory—Cincinnati, U. S. EPA Report-600/2-78-203 (September 1978).
7. McFeeters, R. F., W. Coon, M. P. Palnitkar, M. Velting and N. Fehringer. "Reuse of Fermentation Brines in the Cucumber Pickling Industry," Industrial Environmental Research Laboratory—Cincinnati, U. S. EPA Report-600/2-78-207 (September 1978).

8.   Rose, W. W. "Tomato Cleaning and Water Recycle," Industrial Environmental Research Laboratory–Cincinnati, U. S. EPA Draft Report.
9.   "Handbook for Analytical Quality Control in Water and Wastewater Laboratories," Environmental Monitoring and Support Laboratory–Cincinnati, U. S. EPA Report-600/4-79-019 (March 1979).
10. "Guidelines Establishing Test Procedures for the Analysis of Pollutants; Proposed Regulations," *Federal Register* 44(233) (1979).
11. Carawan, R. E., W. M. Crosswhite, J. A. Macon and B. K. Howkins. "Water and Waste Management in Poultry Processing," Office of Research and Development, U. S. EPA Report-660/2-74-031 (May 1974).
12. Clise, J. D. "Poultry Processing Wastewater Treatment and Reuse," Industrial Environmental Research Laboratory–Cincinnati, U. S. EPA Report-660/2-74-060 (March 1974).
13. Gardner, F. H., Jr., and A. R. Williamson. "Naval Stores Wastewater Purification and Reuse by Activated Carbon Treatment," Industrial Environmental Research Laboratory–Cincinnati, U. S. EPA Report-600/2-76-227 (October 1976).
14. Polkowski, L. B., W. C. Boyle and B. F. Christensen. "Biological Treatment, Effluent Reuse, and Sludge Handling for the Side Leather Tanning Industry," Industrial Environmental Research Laboratory–Cincinnati, U. S. EPA Report-600/2-78-013 (February 1978).

BIBLIOGRAPHY

Burton, J., and E. Kreusch. "Industrial Water Softner Waste Brine Reclamation," Office of Research and Development, U. S. EPA Report-660/2-74-007 (February 1974).
Coda, R. L. "Water Reuse in a Wet Process Hardborad Manufacturing Plant," Industrial Environmental Research Laboratory-Cincinnati, OH, U. S. EPA Report-600/2-78-150 (July 1978).
Hamza, A., S. Saad and J. Witherow. *J. Food Sci.* 43:1153-1157 (1978).
"Handbook for Analytical Quality Control in Water and Wastewater Laboratories," Environmental Monitoring and Support Laboratory–Cincinnati, U. S. EPA Report-600/4-79-019 (March 1979).
Lang, W. C., J. H. Crozier, F. P. Drace and K. H. Pearson. "Industrial Wastewater Reclamation with a 400,000-Gallon-Per-Day Vertical Tube Evaporator," Industrial Environmental Research Laboratory-Cincinnati, U. S. EPA Report-600/2-76-260 (October 1976).
"Methods for Chemical Analysis of Water and Wastes," Environmental Monitoring and Support Laboratory-Cincinnati, U. S. EPA Report-600/4-79-020 (March 1979).
Renn, E. "Management of Recycled Waste-Process Water Ponds," Office of Research and Monitoring, U. S. EPA Report-R2-73-223 (May 1973).

Robinson, A. K., and D. F. Sekits. "Aircraft Industry Wastewater Recycling," Industrial Environmental Research Laboratory—Cincinnati, U. S. EPA Report-600/2-78-130 (June 1978).

Rogers, C. J. "Recycling of Water in Poultry Processing Plants," Industrial Environmental Research Laboratory-Cincinnati, U. S. EPA Report-600/2-78-039 (March 1978).

Standard Methods for the Examination of Water and Wastewater, 14th ed. (New York: American Public Health Association, 1975).

Wabraven, G. O., W. R. Nelson, P. E. DeRossi and R. L. Wisneski. "Closed Process Water Loop In NSSC Corrugating Medium Manufacture," Industrial Environmental Research Laboratory—Cincinnati, U. S. EPA Report-600/2-77-241 (December 1977).

Wiley, A. J., L. E. Pambruch, P. E. Parker and H. S. Dugal. "Combined Reverse Osmosis and Freeze Concentrations of Bleach Plant Effluents," Industrial Environmental Research Laboratory—Cincinnati, U. S. EPA Report-600/2-78-132 (June 1978).

CHAPTER 11

USE OF LOG-NORMAL STATISTICS
IN ENVIRONMENTAL MONITORING

Robert B. Dean

Copenhagen, Denmark

Almost all collections of measurements of pollutants and trace substances are log-normally distributed, yet most of the statistical analysis applied to this data is based on normal or Gaussian statistics. Too many investigators let the computer calculate the mean, standard deviation and 95% confidence limits of their data according to normal statistics, even though in the latter case the computer may dutifully report a negative concentration for the lower confidence limit.

The normal distribution is symmetrical about the central value, which is at the same time the mean or average $\bar{x}$; the median M or 50 percentile; and the mode or most frequent value. The normal distribution extends to plus-and-minus infinity and 95% of the data lie inside the range of the mean plus or minus two standard deviations ($\bar{x} \pm 2s$). Real sets of data for example on concentrations, are skewed. The average is greater than the mode or the median. There can be no concentrations less than zero and the proportion of data that is more than two standard deviations greater than the mean is more than 2.5%. Distributions of data displaying these characteristics can usually be converted to a normal distribution by taking their logarithms. Such distributions are called log-normal. It does not matter which logarithmic base is used to make the conversion, except that the actual magnitudes of logarithms depend upon the choice of the base. Most applied scientists, analytical chemists and engineers use common logs to base 10. Mathematicians and pocket calculators seem to prefer natural logs to base e. The properties of the log-normal distribution have been described exhaustively by Aitchison and Brown [1], who use natural logarithms to base e in their tables. Chow [2] has also given the mathematical background of the log-probability law. He has shown that Hazen's charts, the index of variability and Gumbels extreme value law are examples of the log-probability law. His examples use common logs.

The theoretical justification for using log-normal statistics, aside from the fact that it fits the data, is similar to the theoretical justification for the Gaussian or normal distribution. Gauss assumed a large number of infinitesimal fluctuations of arbitrary sign but equal in magnitude. The sums of these fluctuations obey the binomial distribution, which in the limit is the proportional to the magnitude being measured, the fluctuations in their logarithms will be equal in magnitude, and the sum of the logarithms will be normally distributed. In other words, in the Gaussian distribution each datum contains the base value plus or minus a large number of very small fluctuations, ϵ. In the log-normal distribution each datum is multiplied by a large number of terms of the form $(1 \pm \epsilon)$. Since the natural logarithm of $(1 + \epsilon)$ is ϵ the effect of taking logarithms is to convert the log distribution to a normal distribution.

The three-parameter log-normal transformation assumes that there is "zero adjustment", θ, such that $\log (X + \theta)$ is normally distributed. This conversion is useful if the correlation line does not pass through zero. If some of the data points are reported as zero, it is convenient to use an arbitrary θ of about one tenth of the lower limit of detection to avoid attempts to calculate the logarithm of zero. If the data are counts, such as the number of apples on a tree, then θ should be 0.5 [3].

The properties of a log-normal distribution are completely described by the normal or Gaussian distribution of their logarithms. However, the logs do not describe the data itself and it is necessary to convert back to the antilogs. The mean or average of a set of logs, $\overline{lnx}$ is $(lnx_1 + lnx_2 + \ldots)/n$. The antilog of $\overline{lnx}$ is $\sqrt[n]{x_1 . x_2 \ldots})$ which is the geometric mean, $\overline{x}_g$. The easiest way to get the geometric mean with a pocket calculator is to sum the logs of the data, divide by the number of points to get $\overline{lnx}$ and raise e or 10 to the power $\overline{lnx}$ or $\overline{\log x}$.

The standard deviation, σ, is calculated in the same way as for normal distributions using the logs of the data. The value of the standard deviation, σ, will be in logarithms and must be converted back by taking the antilog. $S = e^{\sigma}$ or $S = 10^{\sigma}$. There has been much confusion in the literature regarding the proper name for the antilog of the standard deviation. Aitchison and Brown call σ the (logarithmic) standard deviation and they use natural logs to base e. It must be multiplied by 0.4343 to convert it to base 10 logs. Many authors call S the geometric standard deviation. Because of this confusion, S will here be called the spread factor or spread of the data [4].

In a normal distribution 64% of the data lie between $\bar{x}$ + s and $\bar{x}$ - s. The same is true for the logs of a log-normal distribution. But the antilog of $\overline{lnx}$ + σ equals $\bar{x}_g$S and antilog (lnx - σ) = $\bar{x}_g$/S. Therefore, the geometric mean must be multiplied or divided by S to get the 64% limits. The 95% limits are at approximately X ± 2s and in a log-normal distribution they lie at $\bar{x}_g$S^2 and $\bar{X}_g$/S^2. The one sided 0.1% limit lies at about $\bar{x}_g$S^3. Naming S the spread factor calls attention to the fact that it must be multiplied by or divided into the geometric mean, not added to or subtracted from it.

The relationships between the parameters of a log-normal distribution have been worked out theoretically [1,2,5]. It is possible to extimate $\bar{x}_g$ and S for a log-normal population from calculated values of $\bar{x}$ and s. The relationships of course represent limiting values and are approximate for small samples (Table I).

Table I. Log-Normal and Normal Statistics

$\bar{x}$, the arithmetic mean
$$\bar{x} = (x_1 + x_2 + \ldots x_n)/n$$

$\bar{x}_g$, the geometric mean
$$\bar{x}_g = \text{antilog } [lnx_1 + lnx_2 \ldots + lnx_n]/n = [(x_1)(x_2) \ldots (x_n)]^{1/n}$$

s, the standard deviation
$$s = \left[\frac{(x_1-\bar{x})^2 + (x_2-\bar{x})^2 \ldots + (x_2-\bar{x})^2}{(n-1)} \right]^{1/2}$$

σ, the logarithmic standard deviation
$$\sigma = \left[\frac{(lnx_1-ln\bar{x})^2 + (lnx_2-ln\bar{x})^2 \ldots + (lnx_n \cdot \bar{x})^2}{(n-1)} \right]^{1/2}$$

S: the spread factor = antilog σ
$$S = e^{\sigma} \text{ or } 10^{\sigma}$$

M: the median or 50 percentile

Cv: the coefficient of variation
$$Cv = s/\bar{x}$$

In a large population, the following relationships are approached as a limit:
$$Cv = (e^{\sigma^2} - 1)^{1/2}$$
$$\bar{x}/\bar{x}_g = \bar{x}/M = e^{\frac{1}{2}\sigma^2}$$

The difference between log-normal and normal data depends on S and Cv. Table II shows values of the average $\bar{x}$ the standard deviation s and the coefficient of variation Cv for various values of S calculated from the formulas of Table I when $\bar{x}_g$ equals 1.000. Also included is the probability P that $\bar{x} + 2s$ and $\bar{x} + 3s$ calculated on log normal populations will not be exceeded. P falls from 97.72% for a normal population to 95.54% at S = 1.8, then stays fairly steady around 96% as S increases. To maintain P at 97.72% it is necessary to increase the factor to $\bar{x} + 2.7s$. P for $\bar{x} + 3s$ is only a little more than 98% over most of the range.

Table II should not be used as a short-cut method for determining log-normal parameters from original data. Its principal function is to enable one to estimate how far a previously calculated estimate of significance is apt to deviate from the best obtainable value if normal statistics have been applied to log-normal data.

One way to decide whether a set of numbers obeys normal or log-normal statistics is to plot them on .cumulative probability paper. This is available either with a linear or a logarithmic scale, and the plot that gives the best straight line indicates the appropriate statistic to use. In most cases quite a large number of points must be plotted to show significant deviations from the best straight line. Deviations of greater than $88.6/\sqrt{n + 1.5}$ percentage points are significant at the 95% level. A deviation of 10% is therefore only

Table II. Properties of the Log-Normal Distribution

				Probability that x is less than	
S	$\bar{x}/\bar{x}_g$	s	Cv	$\bar{x} + 2s$ (%)	$\bar{x} + 3s$ (%)
1.0	1.0		$(0)^a$	97.7	99.87
1.2	1.017	0.187	0.184	96.4	99.38
1.4	1.058	0.367	0.346	95.8	98.88
1.6	1.117	0.555	0.497	95.6	98.52
1.8	1.189	0.764	0.642	95.5	98.29
2.0	1.272	0.999	0.785	95.5	98.18
2.2	1.365	1.267	0.928	95.8	98.12
2.4	1.467	1.575	1.073	95.9	98.12
2.6	1.578	1.928	1.221	96.2	98.17
2.8	1.699	2.334	1.374	96.4	98.18
3.0	1.828	2.799	1.531	96.6	98.22
4.0	2.164	5.227	2.415	96.6	98.11

[a] As S approaches 1.00, the data become normally distributed and the coefficient of variation approaches zero.

significant when n is greater than 77. This rejection criteria is based on an empirical modification of the Kolmagorov-Smirnov law [4,6].

There have been several erroneous methods recommended for plotting probability curves [7]. When n is greater than 50 all methods produce about the same result. The correct method plots each point twice and produces a step function. For small numbers of observations some of the suggested methods give erroneous estimates of $\bar{x}$ and s. The rejection criteria is rarely satisfied when n is small no matter what population the data belong to. However, if a plot on linear probability paper shows obvious upward curvature together with no points less than zero and several points greater than twice the mean, the data are probably log-normal.

Figure 1 shows a log-normal curve with spread factor S of 2 and a geometric mean of 1. Superimposed is a normal curve which has the same average and standard deviation as the log-normal curve. This is the curve one would get by calculating the average and standard deviation of the log-normal data and constructing the curve. Note that the average is greater than the median. Also note that the normal curve predicts negative values and most importantly it underemphasizes the fraction of the data which lies beyond 1.7 standard deviations from the average. This last error is most important if one wishes to estimate the reliability of a plant or process. When the spread factor S is 2, 2.5% of the data lie above $4\bar{x}_g$ and 0.1% lie above $8\bar{x}_g$. The range between the 95% limits (2.5-97.5%) is 16-fold, a not uncommon range for real data.

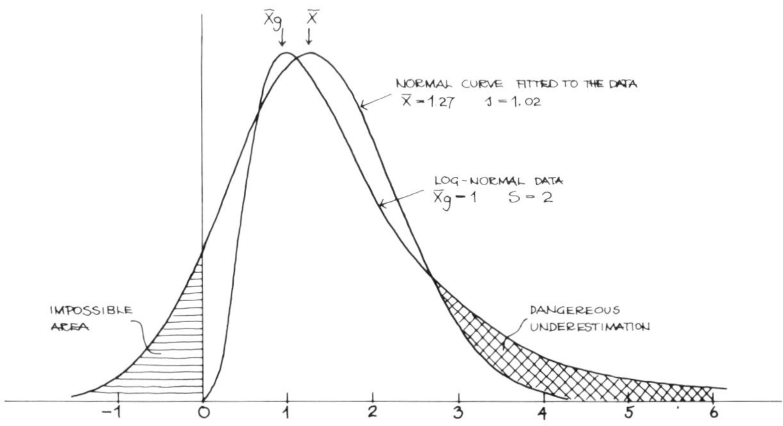

Figure 1. Errors produced by calculating normal statistics for log-normal data.

The cumulative plots corresponding to Figure 1 are shown in Figures 2 and 3. In the first one the shape of the log-normal data is shown as a curved line on an arithmetic scale. The very high probability of extreme values is obvious. Figure 3 shows the same data plotted on log-normal paper. In both of these plots the differences over the range 10-90% are not very great, but the predictive value of the normal curve for extreme values of log-normal data is not acceptable.

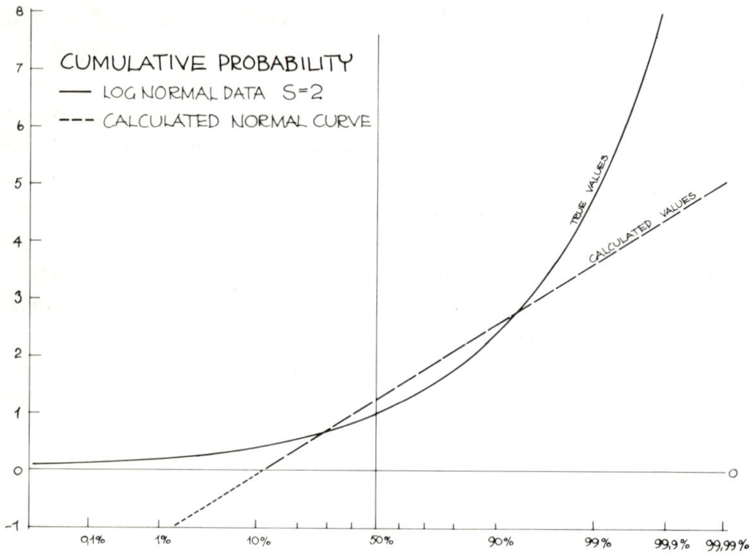

Figure 2. Comparison of log-normal and normal plots on arithmetic probability paper.

The value of S determines how much the data depart from normality. If S is close to 1, the data may be represented equally well by normal or log-normal statistics. However, a large proportion of data related to concentrations of pollutants has S-factors close to 2.0. Figures 4 and 5 show phosphate at South Lake Tahoe [4,8] plotted on normal and log scales. Table III shows spread factors for pollutants in effluents from advanced waste treatment plants.

It is possible to estimate S and decide whether data are likely to be log-normally distributed by looking at the coefficient of variation Cv, which equals $s/\bar{x}$. Figure 6 presents a plot of S as a function of Cv calculated from theory. Over the range of Cv=0.4-1.4, S is given very closely by the equation;

$$S = 1.4\ Cv + 0.9$$

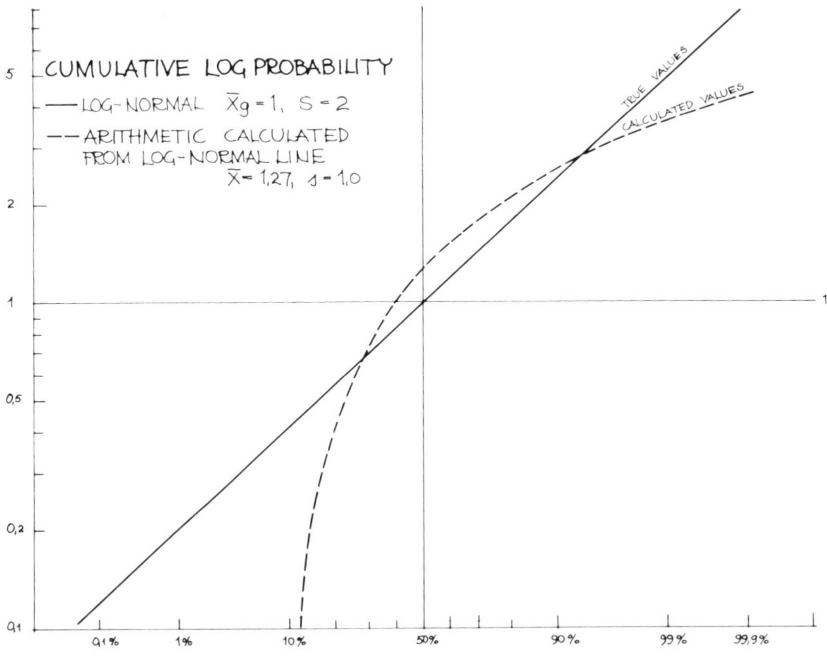

Figure 3. Comparison of log-normal and normal plots on log-probability paper.

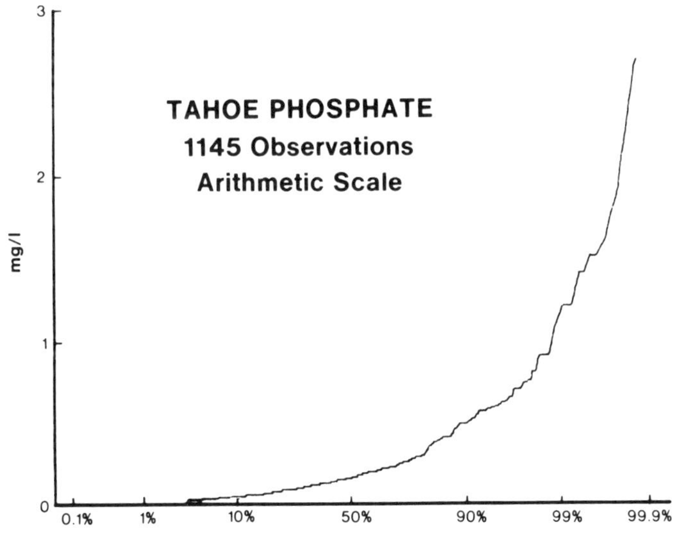

Figure 4. Cumulative probability of phosphate at South Lake Tahoe. Reprinted from *Water and Sewage Works* 123:57-60,81-89 (1976).

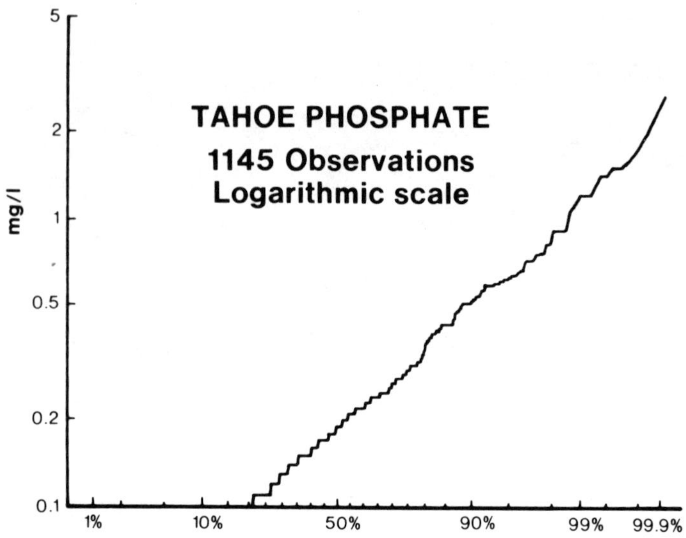

Figure 5. Cumulative probability of phosphate, logarithmic scale, at South Lake Tahoe. Reprinted from *Water and Sewage Works* 123:57-60,81-89 (1976).

Table III. Spread Factors for AWT Effluents [3] (mg/l)

Parameter	Location	$\overline{x}_g \rightarrow$	$S \rightarrow$	$\overline{x} \rightarrow$	$s \rightarrow$
Phosphate P	Tahoe	0.19	2.11	0.20	0.18
	Ely	0.035	1.70	0.043	0.024
	Själevad	0.35	1.85	0.48	0.34
	Leksand	0.30	1.62	0.36	0.20
	Borlange	0.72	1.87	0.91	0.43
	Vikmanshytten	0.53	1.51	0.61	0.27
	Eolshall	0.54	1.64	0.65	0.33
	Grand Haven	0.80	1.54	0.85	0.40
MBAS	Tahoe	0.18	1.72		
BOD		1.3	2.05		
COD		9.6	1.57		
Cl_2		0.90	1.77		

When Cv is less than 0.4, it is difficult to distinguish between a normal and a log-normal distribution unless the data set is very large. As Cv approaches zero, S approaches 1 + Cv. In any actual set of data the relation between Cv and S will only be an approximation. The points shown in Figure 6 are taken

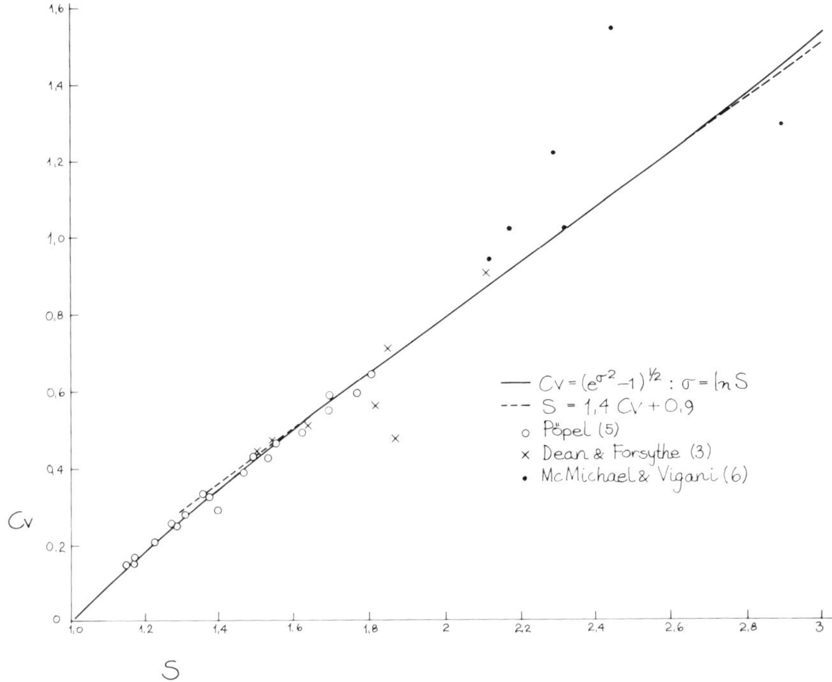

Figure 6. Spread factor vs coefficient of variation.

from data for which both S and s had been calculated. Pöpel [9] examined the effluent data from seven large sewage treatment plants in the United States and Europe. He found that BOD and suspended solids usually followed log-normal statistics. His data sets covered 70-300 observations, and the deviations from the theoretical line when S is calculated from Cv are within 0.02 of S. When Pöpel examined data that he reported as not fitting the simple log-normal distribution, the agreement between Cv and S is, as expected, further from the theoretical line. Dean and Forsythe [4] compared phosphate elimination at a number of plants including South Lake Tahoe, Ely, MN, and several Swedish plants. Here the agreement is not quite so good and the greatest deviation in calculated S is 0.2. McMichael and Vigiana [10] reported six years of phenol concentrations in the Ohio River. Here S lies between 2 and 3, and the deviations are much larger than in the other sets.

When calculating Cv from published data, care must be taken to use the standard deviation of the original data, s, not the standard deviations of the average. This parameter is s divided by the square root of the number of observations and is sometimes called the standard error.

Some data are already logarithmic in nature. pH is the best known, but all chemical potentials are logarithms of concentration ratios. If we admit that chemical potentials are normally distributed then it follows that concentrations obey log-normal statistics.

One type of natural data on concentrations does not follow a log-normal distribution. If an element is essential for nutrition, such as copper or iron, its distribution in an animal or plant will be truncated at an upper and lower limit for physiological reasons, and a normal distribution will fit well except near the truncation limits [11].

Data presented as percent removals or percent remaining do not fit a log-normal distribution unless they are close to zero. This distribution is always limited at the upper end at 100%. However, percentage removal data can be converted to a form that obeys log-normal statistics by converting them to the logarithm of the ratio $p/(1-p)$ or $\log p - \log(1-p)$ where p is the percentage. This conversion is symmetrical and has no upper or lower limit.

If some of the data are reported as "zero," "less than" some lower limit or "greater than" some upper limit, it is not possible to calculate conventional log-normal statistics. In particular, the geometric mean of a set of data containing a "zero" is by definition zero. Since "zero" in an analytical determination actually means "less than the lower detectable limit" it should be so reported. A datum reported as "less than" or "greater than" may be called a deficient datum. Although it is deficient it still carries some useful information. Figure 7 shows coliform data from the Lake Tahoe AWT plant. Here 94% of the values were less than 2 [4].

A very good estimate of the midpoint of a set of data containing one or more deficient (or doubtful) data can be obtained by arranging the data in order of magnitude starting with "less than" proceeding through the numerical values and ending with "greater than." Data are then removed (pruned) symmetrically from both ends until all deficient data are gone. Pruning can, of course, be done before converting to logarithms. The average or geometric average of the pruned residue is very efficient for estimating the central value, but the pruned data give a very poor and badly biased estimate of the standard deviation or spread of the data.

A better estimate of the spread can be made by plotting the data as a cumulative probability. When a "less than" is present, the curve starts with a vertical line from below, terminating at the percentile of the first known value. The area to the left of this line represents the percent of the data that is less than the first known point by an unknown value. Likewise if there is a "greater than" datum, the step plot will terminate in a line going up from the last known point. The best estimate of the distribution is the straight line fitting the steps. The standard deviation, or the spread, can be determined from the slope and is most conveniently measured between the 50th and 84th percentiles.

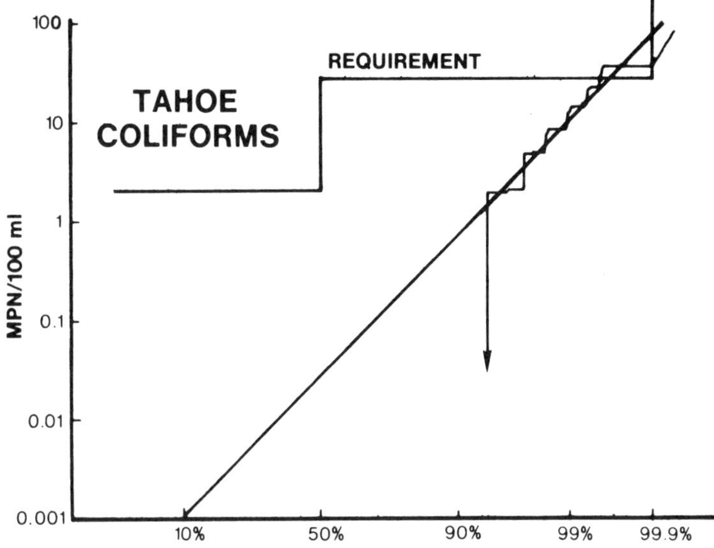

Figure 7. Cumulative probability of coliform bacteria at South Lake Tahoe. Reprinted from *Water and Sewage Works* 123:57-60,81-89 (1976).

Almost all the powerful tools of statistics are based on the normal distribution. If a distribution is not normal, the use of normal statistics can lead to serious errors. One has only two choices: either use less powerful nonparametric statistics or apply a conversion that brings the data close to a normal distribution. In particular one should not attempt to calculate confidence limits or regression coefficients or do analyses of variance on data which is not normal. The Q-test for outliers [12] was also developed for a normal distribution and should be used on the logs of the data if the outlier is on the high side and there is a suspicion that the data is log-normal.

A very important use of log-normal statistics is to equalize variances. One of the most powerful statistical tools for examining data containing several variables is the analysis of variance. The variance, which is the square of the standard deviation, can be divided up into components representing effects and a residual error or background "noise." This separation is only valid if the residual error is essentially independent of the size of the effects. When the data are log-normally distributed this will not be true even if S is close to one because the variance will be proportional to the square of the effect. As early as 1945 Pearce [13] showed that log-normal statistics could be used to "equalize" or homogenize the variances. The practical applications of his suggestion were small because of the computational labor involved at that time.

Youden's analysis of round-robin tests implicitly depends on equality of variances [14,15]. In this test, two samples, A and B, are analyzed by a number of laboratories. The results are plotted as a scatter diagram of A against B with each laboratory represented by a single point. If there are no systematic errors, the scatter diagram will be circular. Bias is shown by a tendency of the points to lie along a diagonal such that high errors in A correspond to high errors in B and vice versa. Youden restricted his method to cases where the magnitudes of A and B were close together. This limitation ensures that the standard deviations will also be close together and the scatter along the two axes will be equal. By plotting logarithms of A and B, it is frequently possible to extend Youden's method to cases where A is not close to B. The actual value of S for any good analytical method is usually very close to one, and the coefficient of variation is correspondingly small. Although normal statistics fit the data for any one sample quite as well as log-normal, much more information can be obtained from the log-normal presentation.

The literature is full of articles pointing out that real data are log-normally distributed [3], yet the text books of statistics rarely give more than a single paragraph to this distribution. Before the days of electronic calculators, the labor of calculation dominated applied statistics, and the log-normal distribution was impractical to apply. Nowadays, conversion to logs calls for a single key entry which can be carried out on any scientific or business pocket calculator and can easily be programmed into any small scientific computer.

One justification that is often advanced for using normal statistics is based on the law of large numbers. This law states that averages of almost any distribution will approach a normal distribution as the number of data becomes large. This law is true but the number may have to be quite large if the data in fact has a log-normal distribution with a large spread.

It has been shown that averages of data drawn from a log-normal population are still log-normal, but S is closer to one. The law that appears to govern averages of log-normal data is similar to that for normal data in the following form. σ_n, the logarithmic standard deviation of the arithmetic average of n values, equals $\sigma/\sqrt{n}$. S_n is therefore $e^{(\sigma/\sqrt{n})}$. When S = 2, S_5, the spread of the averages of groups of 5, is 1.36; S_{25} = 1.15. When n = 100, the averages are fitted equally well by a normal plot or by a log-normal plot with the expected S of 1.07.

The three-tube, three-decimal dilution MPN method for determining bacteria has been shown to be log-normally distributed with S = 2.07 [7]. Actual collections of data from field measurements of bacteria always show a greater spread. S was 13 for the Tahoe coliform data of Figure 7. Bacteriologists have long recognized the log-normal aspects of their data and routinely plot on a log scale. It is well to remember that the standard deviation of the MPN method on a common log scale is not less than 0.3 and attempts to calculate MPN to two significant figures are essentially meaningless.

SUMMARY

Most real data obtained in analytical laboratories follow log-normal statistics. In order to make use of the most powerful statistics that are available the data should be transformed to logs. If log-normal data are treated as if they were normal, quite erroneous estimates of large deviations from the central value will be made. Conversely, if truly normal data are treated as log-normal, the errors will usually be small and conservative. One should justify in each case the use of normal statistics rather than log-normal on all data, such as concentrations, where values less than zero are impossible.

REFERENCES

1. Aitchison, J., and J. A. C. Brown. *The Log-normal Distribution* (London: Cambridge University press, 1957).
2. Chow, V. T. "The Log-probability Law and Its Engineering Applications," *Proc. Am. Soc. Civil Eng.* 80(536) (1954).
3. Gaddum, J. H. "Lognormal Distributions," *Nature* 156:463-466 (1945).
4. Dean, R. B., and S. L. Forsythe. "Estimating the Reliability of Advanced Waste Treatment," *Water Sew. Works* (June 1976) p. 87-89; (July 1976) p. 57-60.
5. Johnson, N. L. "Lognormal Distributions," in *Distributions in Statistics— Continuous Univariate Distributions, Vol. 1* (New York: John Wiley & Sons, 1970), pp. 112-136.
6. Lilliefors, H. W. "The Kolmogorov-Smirnov Test for Normality with Mean and Variance Unknown," *Am. Stat. Assoc. J.* 62:399-402 (1967).
7. Velz, C. J. "Graphical Approach to Statistics," *Water Sew. Works Ref. Data* (1952) pp. R106-R135.
8. Elfving, E., A. Forsberg and C. Forsberg. "Minitest Method for Monitoring Effluent Quality," *J. Water Poll. Control Fed.* 47:720-726 (1975).
9. Pöpel, H. J. "A Concept for Realistic Effluent Standards," *Prog. Water Technol.* 8(1):69-89 (1976).
10. McMichael, F. C., and F. C. Vigiani. "Problems in Phenolics-Modeling Methods in the Ohio River at Wheeling, W.Va." *J. Am. Water Works Assoc.* 55(11):725-731 (1973).
11. Leibscher, K., and H. Smith. "Essential and Non-essential Trace Elements," *Arch. Environ. Health* 17:881-890 (1968).

12. Dean, R. B., and W. J. Dixon. "Simplified Statistics for Small Numbers of Observations," *Anal. Chem.* 23:636-638.
13. Pearce, S. C. "Lognormal Distributions," *Nature* 156:747 (1945).
14. Youden, W. J. *Statistical Techniques for Collaborative Tests* (Washington, DC: The Association of Official Analytical Chemists, 1969).
15. Youden, W. J. "The Role of Statistics in Regulatory Work," *J. Ass. Off. Anal. Chem.* 50:1007-1013 (1967).

SECTION 3
TOXIC CHEMICALS: REMOVAL BY MEMBRANE PROCESSES

REMOVAL OF VARIOUS TOXIC HEAVY METALS AND CYANIDE FROM WATER BY MEMBRANE PROCESSES

Louis J. Kosarek

Trace Metal Data Institute
El Paso, Texas

Subsequent to a suit settled in 1978 against the U.S. Environmental Protection Agency (EPA), the defendants were required to designate a list of toxic compounds that were considered to be aqueous pollutants, prior to specifying technologically based effluent limitations and guidelines. The plaintiffs in this legal action brought suit because it was felt that the EPA was lax in implementing the Federal Water Pollution Control Act (FWPCA) [1,2].

A list of 129 priority pollutants were cited as retaining a nature which justified regulation. Of these constituents, 13 are metals and their associated compounds, and one is cyanide. The metals which were assessed to be toxic pollutants were antimony, arsenic, beryllium, cadmium, chromium, copper, lead, mercury, nickel, selenium, silver, thallium and zinc. When these metals are present at excessive levels in an aqueous discharge or source, this contamination renders the stream unusable because of the adverse impact associated with consumption [3-5].

The undesirable aqueous species which are the subject of this chapter include arsenic, cadmium, chromium, copper, cyanide, lead, mercury, nickel, selenium, silver and zinc. The heavy metal constituents of this group which usually retain a predominant cationic form are cadmium, copper, lead, mercury, nickel, silver and zinc. The compounds which are present within aqueous sources predominantly in the anionic form are: arsenate, chromate, cyanide and selenate [6]. A list of these constituents and their predominant ionized chemical forms is included in Table I.

Table I. Levels, Frequency and Regulations Associated with Undesirable Constituents

Aqueous Constituent	Predominant Ionic Form	Drinking Water Regulations (mg/l)	Frequency of Detection (%)	Average Levels (mg/l)	Problem
Arsenic	$H_2AsO_4^-$	0.05	5.5	0.064	Possibly
Cadmium	Cd^{2+}	0.01	2.5	0.0095	Possibly
Chromium	CrO_4^{2-}	0.05	24.5	0.0097	No
Copper	Cu^{2+}	1.0	74.4	0.015	No
Cyanide	CN^-	0.2	NA[a]	NA	
Lead	Pb^{2+}	0.05	19.3	0.023	No
Mercury	Hg^{2+}	0.002	NA	0.0001	No
Nickel	Ni^{2+}		16.2	0.019	No
Selenium	SeO_4^{2-}	0.01	NA		
Silver	Ag^+	0.05	6.6	0.0025	No
Zinc	Zn^{2+}	5.0	76.5	0.064	No

[a]Not available

HEALTH EFFECTS OF EXCESS LEVELS OF CONSTITUENTS

The human health effects associated with the acute or chronic injestion of these species are diverse and substantiate the basis of regulating these constituents (Table II). The consumption of elevated levels of arsenic results in kidney, liver and bone dysfunction. As a result of these effects, the drinking water regulation was set at 0.05 mg As/l. Chronic ingestion of elevated levels of cadmium causes kidney and bone dysfunction, and this metal also possesses carcinogenic characteristics. Subsequently, the drinking water regulation for cadmium was set at 0.01 mg/l. Chromium taken as the hexavalent chromate ion, is more toxic than the trivalent species, and causes kidney damage, nausea and ulcers. Potable streams containing levels of chromium in excess of 0.05 mg/l are considered to be objectionable and exceed the quality limit. Dissolved copper is usually in the divalent cationic form and imparts an unde-

Table II. Health Effects and Animal Threshold Levels of Undesirable Constituents

Aqueous Constituent	Drinking Water Regulations (mg/l)	Lowest Reported Toxic Concentration (mg/l)	Basis of Regulation (Human)
Arsenic	0.05	1.0	Kidney, liver and bone dysfunction
Cadmium	0.01	0.03	Kidney and bone dysfunction, carcinogen
Chromium	0.05	0.016	Kidney damage, nausea, ulcers
Copper	1.0	0.009	Undesirable taste
Cyanide	0.2		Inhibition of oxygen metabolism, nervous system
Lead	0.05	0.01	Brain damage with accumulation
Mercury	0.002	0.004	Liver damage and circulatory collapse
Nickel		0.8	None
Selenium	0.01	2.0	Similar to arsenic toxicity
Silver	0.05	0.003	Tissue accumulation
Zinc	5.0	0.003	Undesirable taste

sirable taste to the water at a concentration of 1.0 mg/l. Hence, the basis of regulating the level of copper in water is established on esthetic criteria as opposed to health considerations [7-12].

The presence of cyanide in the acidic form results in the formation of hydrocyanic gas, which on inhalation causes inhibition of oxygen metabolism and subsequent nervous-system damage. The drinking water regulation for cyanide is 0.2 mg/l, and the physiological effects of cyanide are definitely pH-related. The accumulation of lead through chronic ingestion results in nervous-system damage and neural impairment; this metal is found to be detrimental at levels exceeding 0.05 mg/l. Mercury, which is especially toxic when ingested as an organometallic complex, causes liver damage and circulatory collapse, and is regulated at a level exceeding 0.002 mg/l. Nickel, which is relatively innocuous as a human health hazard but detrimental to a variety of plants, is not presently regulated with regard to potable water quality [7-12].

The health effects of excess levels of selenium are similar to arsenic toxicity. However, ingestion of selenium at lower levels is considered to be beneficial to human health—resulting in a heated controversy surrounding the potable water regulation of 0.01 mg/l for this trace metal. The precious metal silver accumulates in a variety of tissues, including the skin, resulting in a dermatological condition; this accumulation, with its possible toxic effects, designates the water quality level of 0.05 mg/l. The last constituent included in this discussion is zinc; this metal is regulated at a level of 5.0 mg/l because of the esthetics associated with the undesirable taste of this metal at these concentrations [7-12]. The lowest reported toxic concentration or threshold level for other organisms exposed to these constituents is also listed in Table II [13].

NATURAL BACKGROUND LEVELS

Because of the adverse effects associated with the presence of excess amounts of these species within surface-type drinking water sources, the frequency of their occurrence at elevated levels is a paramount consideration. As designated in Table I, the frequency of the percent of detection of arsenic is very low at 5.5%, and the average levels are present at 0.064 mg/l. The frequency of occurrence of cadmium within most surface waters is very low at 2.5% with concentrations averaging 0.0095 mg/l. Of the objectionable species discussed here and designated in Table I, only arsenic and cadmium are naturally present at average levels which suggest the possibility of a health-related problem. The remainder of these constituents are present at average levels which are significantly below threshold concentrations which are considered safe [14].

INDUSTRIAL SOURCES

The alternative sector which substantiates the focus of the EPA guidelines and effluent limitation is industry. As designated in Table III, industrial use and discharge of the aforementioned constituents establishes a direct need for these facilities either to control or discontinue discharges because of legal criteria. These species (arsenic, cadmium, chromium, copper, cyanide, lead, mercury, nickel, selenium, silver and zinc) are valuable commodities to the manufacturing facilities designated in Table III [3,4,7,10,11,15]. In light of regulations presently being promulgated, these commodities should be conserved by removal prior to discharge and reclaimed. The application of this type of process would result in the conservation of operating capital because of lower input needs, compliance with effluent guidelines because of acceptable discharge levels and overall decreased capital expenditures to meet new legal guidelines.

Because of (1) the detrimental effects of these species at elevated levels, (2) dictated legal compliance, and (3) the economics of conservation, treatment technologies must be implemented so that more diverse water sources can be utilized, and effluent discharge limitations can be met. Water is becoming a more expensive commodity in both the municipal and industrial

Table III. Industrial Sources of Various Undesirable Aqueous Constituents

Constituent	Industrial-Manufacturing Sources
Arsenic	Copper and lead alloy treatments; petroleum industry; glass and ceramic production; cannery operations; dye-paint-pigment manufacture; fireworks production; electrical semiconductors; herbicide production.
Cadmium	Mining and smelting; electroplating plants; pigment works; textile printing and chemical industries; plasticizers; ceramics manufacture; mine drainage
Chromium	Electroplating facilities; manufacture of dyes, ink and paint pigments; chromate water treatment; aluminum anodizing; metal cleaning; tanning
Copper	Electrorefineries; biocide production; mining and smelting; metal cleaning-plating-rinse baths; polymer production; pulp; fertilizers
Cyanide	Mining and extraction facilities; synthetic fiber manufacture; steel treatment; gas scrubbing; electroplating; photographic processing
Lead	Mining and smelting; petroleum additives; paint and ceramic pigments; battery production; manufacture of television tubes
Mercury	Germicidal and fungicidal agents; chloralkali production; explosives manufacture; photography; electrical apparatus and instrumentation; bleach preparations; pharmaceuticals
Nickel	Mining and smelting; metal plating; silver refineries; motor vehicle and aircraft industries; printing; basic steel works
Selenium	Electrolytic refining; electronics manufacture; steel; glass-paint-dye-ceramic-pigments; insecticide production
Silver	Biocide manufacture; photography, procelain manufacture; ink industry; electroplating processes
Zinc	Mining-smelting-refining industries; steel galvanizing; brass metal works; tableware manufacture; rayon fiber production; pulp-paper and news-paper print production; pigment industry

sectors, and the use of treatment technology will be beneficial to both sectors in light of present source limitations and laws governing constituent guidelines.

MEMBRANE PROCESS

Membrane processes have been developed, tested and used to achieve the economical removal of a significant portion of these metals and cyanide from aqueous sources. The purified streams can be acceptable for human consumption as well as recycling, while the valuable commodities are concentrated and reclaimed in reuse. The membrane processes which have been applied are chelation in combination with ultrafiltration, charged ultrafiltration and reverse osmosis. A new membrane process presently in the research phases, which has also demonstrated metal recovery capability, is coupled membrane transport.

The concept of "chelation in combination with ultrafiltration" is based on reacting various types of ligands with the endogenous cationic aqueous metallic constituents to form a metal-containing complex (chelate) and removing these metal containing coordination complexes by ultrafiltration [16]. The opposite charges of the ionized ligand and metal attract each other and form a stable chelate complex. The properties which facilitate ultrafiltration membrane rejection of the metal-containing complex while not rejecting the free metal are thought to be (1) the significantly increased size of the metal chelate comples, (2) a dramatic alteration in the shape of the metal, (3) modified solubility and (4) reversal of charge from cationic metal to a functionally anionic chelate or electroneutral chelate species [17].

The basis of a purification system using chelating compounds in conjunction with low-pressure ultrafiltration is to modify chemically the constituent of interest. This complex makes the constituent more amenable to removal by ultrafiltration; hence, this mode of pretreatment optimizes the purification efficiency and constituent concentration factors ascertained by ultrafiltration for these specific species [18,19]. To facilitate the recovery of valuable commodities, the concentrated metal solution containing the chelate complex (retentate) is chemically treated to elute or decouple the metal from the chelate. On separation of the ligand and metal, this solution is further processed by ultrafiltration to yield a metal-rich solution [20]. The metal concentrates retain endogenous value and can be reused in the process, treated to remove the metal values or marketed to recover costs.

The ligands which have been used successfully are ethylenediamine tetraacetic acid (EDTA), ethylene *bis*(oxyethylene nitrilo) tetraacetic acid (EGTA), 1,2-diaminocyclohexane-N,N'-tetraacetic acid (CDTA), diethylenetriaminopentacetic acid (DPTA), N,N'-ethylene-*bis*-2-(*o*-hydroxyphenyl) glycine (EHPG) proprietary polymers and naturally occurring polyelectrolytes [16-20]. These ligands have been used to chelate (and facilitate removal via ultrafiltration of) chromium, copper, mercury, zinc, calcium and magnesium [16-20]. The binding of the constituent to the ligand is a reversible attachment whose chemical characteristics are governed by a conditional stability constant. The chelate is in equilibrium between the coordination complex, the free constituent and the unbound ligand. In most cases, this equilibrium is governed by the pH of the solution and the concentrations of the various reactants. Because metal recovery is possible in this system, it can be used as either a purification process or as a resource recovery system.

The performance of the chelation-plus-ultrafiltration system is significantly affected by pH, applied operating pressure, type of ultrafiltration membrane, stoichiometric quantities of ligand (the concentration of reactants) and the type of ligand used to chelate the metal. The effect of pH on the chelation-plus-ultrafiltration process is significant because pH directly governs the quantity of chelate complex formed with respect to the reactants; also, over a wide range, pH affects heavy metal solubility [16-19]. The applied pressure which promotes the flow of treated water across the membrane, even though it is normally 40-60 psi, does affect membrane performance. At a very low applied pressure (20-60 psi) the rejection of chelated metals increases directly with increasing pressure. Over a range of 100-300 psi applied pressure, the rejection of complexes plateaus, and additional rejection is not obtained with increasing pressures [17]. At pressures which exceed 300 psi, the threat of membrane plugging and possible metal breakthrough become imminent.

As the type of membrane used varies between applications and consumers, performance will vary between applications and processes. The stoichiometric quantities of ligand which are implemented designate a critical parameter associated with system performance. To optimize performance, the pH of the feed solution should be set at a level which maximizes chelation, and the concentration of ligand be such that all the free metal can be complexed. If deficient amounts of ligand are used, performance will deteriorate [18,19]. Different ligands retain differing chelation affinities for different metals; therefore, the metal which requires removal will determine the best ligand. With different ligands performance will vary [16-19].

The performance of the charged membrane ultrafiltration system is based on fixed functional groups which are an inherent part of the membrane polymer. When dissociated, these functional groups retain a negative charge. The rejection of positively charged aqueous constituents is suggested to be based partially on Donnan ion exclusion. In addition to the removal of metal species by charged ultrafiltration, this system also retains endogenous ultrafiltration capabilities [21].

Charged membrane ultrafiltration incorporates a noncellulosic high-flux membrane which is negatively charged because of dissociated subgroups within the membrane structure. This low-pressure system (40-100 psi) is effective where complete water purification is not required and is appropriate where certain toxic metals are required to be concentrated or removed. Charged membrane ultrafiltration removes metals not only by Donnan ion exclusion but by size and suspended particle characteristics. Because the membrane is negatively charged, it will repel negative species. Concurrently, the membrane retains a nominal pore size such that molecules larger than the pore size will not permeate the membrane. A beneficial aspect of the charged ultrafiltration membrane is that the negative polarization minimizes membrane fouling [21-25].

The charged membrane ultrafiltration process has been applied to remove arsenic, cadmium, copper, lead, mercury, nickel, selenium and zinc. The efficiency of this membrane treatment process to remove toxic metal constituents is based primarily on pH. The pH of the solution controls (1) the magnitude of the charge maintained by the functional groups on the membrane; (2) the extent of the charge on the metal or its compound; (3) the extent of charge retained by the compounds which ionize by dissociation; and (4) the solubility of the metal with reference to the formation of alkaline earth precipitates, hydroxides and carbonate forms [22-25].

The extent to which pH has an effect on the charged membrane ultrafiltration process is shown in Table IV. As can be seen in the equation:

$$pH = pK + \log_{10} \frac{[A^-]}{[HA]} \tag{1}$$

where   pK = negative log of the dissociation constant
$[A^-]$ = molar concentration of dissociated salt
[HA] = molar concentration of undissociated acid

by means of the dissociation constant, pH governs the ratio of the molar concentration of the dissociated-ionized species to that of the undissociated compound. In this membrane process, pH affects the magnitude of the ionization of both the membrane and several of the species which require removal [26,27].

In addition to the profound effect which pH has on the performance of the charged membrane ultrafiltration process, the matrix of the influent water may affect system performance. In solutions containing both monovalent and divalent anions, the divalent anions will decrease the membrane's efficiency to reject monovalent anions [21]. As it is necessary to retain a conservation of charge on both sides of the membrane, the presence of an

Table IV. Effect of pH on Dissociation of Weak Acids, Metals and Complexes

Constituent	Dissociation Constant[a]	pK
H_3AsO_4	5.62×10^{-3}	2.25
$H_2AsO_4^-$	1.70×10^{-7}	6.77
$HAsO_4^{2-}$	3.95×10^{-12}	11.60
H_2CrO_4	1.8×10^{-1}	0.74
$HCrO_4^-$	3.2×10^{-7}	6.49
HCN	4.93×10^{-10}	9.31
H_2SeO_4	10^2	< 0.00
$H SeO_4^-$	1.2×10^{-2}	1.92

elevated level of divalent anions will facilitate the permeation of monovalent species across the membrane and subsequently reduce the effectiveness of the system to remove undesirable monovalent anions.

Reverse osmosis is a membrane-mediated physical separation technique whereby an applied pressure in excess of the solution inherent osmotic pressure forces water to permeate a semipermeable membrane while rejecting the bulk of the dissolved and suspended constituents [28]. A conventional rule of thumb is that 1000 mg/l of dissolved ionized species retain an average osmotic pressure of 10 psi. The pressures applied to the membrane in the reverse osmosis process range from 200 to 800 psi. The means by which a dissolved constituent is removed by reverse osmosis is related to the constituent's charge, size and the applied pressure on the membrane [29]. Reverse osmosis has been used extensively as a sea- and brackish water desalination process, a water purification device and a water reclamation system.

Reverse osmosis can be used to remove a variety of dissolved species [28-30], and is probably the most advanced type of full-scale commercial membrane technology. This diversity in development is reflected in both the types of membrane designs and the membrane materials used within these designs. Reverse osmosis modules are available in the spiral-wound wrap, the hollow fine fiber or the tubular configurations. The membrane materials which are conventionally used include the cellulosic derivatives cellulose acetate, cellulose diacetate, cellulose triacetate and cellulosic blends; and the noncellulosics, of which only the polyamide type has successfully been applied on a commercial basis [28,31].

The operation of a reverse osmosis membrane system is significantly affected by fouling, scaling, pH-temperature-pressure-related hydrolysis and chemical deterioration of the membranes. The reverse osmosis process has very strict feed water requirements related to the concentration of suspended solids. Suspended solids will foul the membrane and significantly decrease the quantity of permeate produced and deteriorate the quality of purified water produced [32]. The manufacturers of reverse osmosis membranes specify operational warrantees which limit the quantity of suspended solids which enter the module. These warranty specifications are: a feed quality of less than 1.0 nephelometric turbidity units (NTU) or a silting density index (SDI) below 4.0 [32].

The measurement of feed-water turbidity is the most generally applicable and easily acquired quantitative measure of suspended solids water quality. Turbidity can be measured by a nephelometer. The nephelometer measures turbidity as NTU, by obtaining the relative level of light dispersion caused by suspended solids in the water [33]. Figure 1 shows the light path of a nephelometer. The nephelometers (or turbidimeters) that have shown numerous advantages when applied in membrane applications are the types as

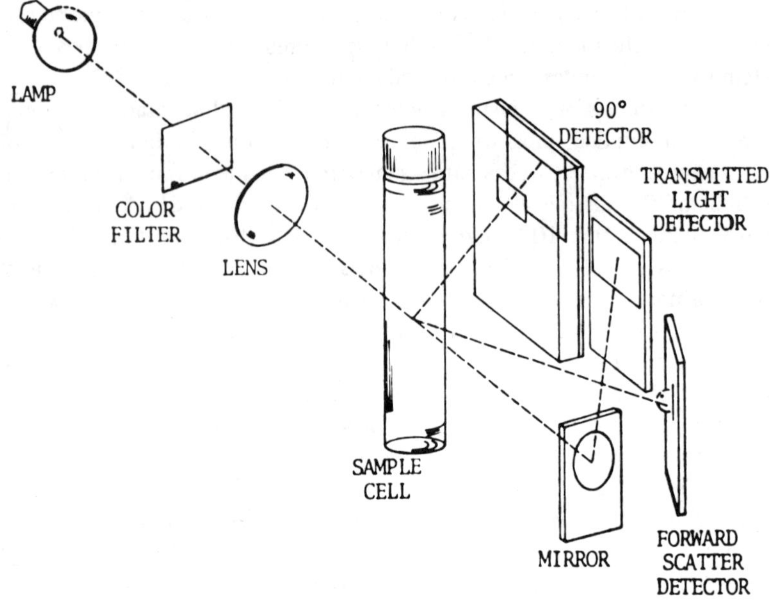

Figure 1. Ratio optics schematic of a Hach Model 18900 turbidimeter. (Courtesy of the Hach Chemical Company, patent pending.)

manufactured by Hach Chemical Company, Loveland, CO. These turbidimeters are used to monitor the feed water quality. These instruments retain analytical ranges of 0.01-1000 NTU and are available in models which are portable, on-line and continuously operating, or portable and designed to compensate for interferences such as color.

A silting density index (SDI) is also a measure of the suspended solids content of influent feed water, and has been used extensively in reverse osmosis applications. An SDI is procured by measuring (1) the length of time for the first quantity of 500 ml of water to permeate a 47-mm-diameter micronic filter with a nominal 0.45-μ rating, and (2) the amount of time required for a final 500-ml sample to permeate the 0.45-μ filter. Dividing this chronological ratio by the interim time period between the measurements results in the SDI [32]. Various types of micronic membranes have been used to measure SDI, but the Gelman GN-6® type (Gelman Sciences, Inc., Ann Arbor, MI) has demonstrated excellent performance [34].

The formation of scale on a reverse osmosis membrane is a specific category of fouling, and occurs when the concentrations of various dissolved salts in the brine stream exceed their solubility and result in the formation of suspended solid nucleation sites. As scaling continues, the permeation of water across the membrane is prohibited by these newly formed suspended solids. The major constituents which usually cause scaling of a reverse osmosis system include calcium carbonate, calcium sulfate, ferric hydroxide and heavy metals [28]. Scaling actually occurs because the levels of these salts increase in the brine and form precipitates at the water-membrane interface.

Hydrolysis of a reverse osmosis membrane occurs when a polymeric linkage within the inherent membrane structure is broken and water is absorbed at this site because of the temporary charges which exist at the severed point of the polymeric structure. Membrane hydrolysis can be caused by extremes in pH [35] (below 3.0 and above 9.0), at temperatures which exceed 90°F [36] or because of applied pressures above 700 psi. Chemical deterioration is also another cause of membrane hydrolysis and can be caused by materials which oxidize or solubilize the membrane polymer. Hydrolysis of membrane materials significantly affects the longevity of reverse osmosis membrane modules.

A relatively new membrane-mediated metal purification process which has been developed and tested over the last five years is coupled membrane transport. This membrane process utilizes the varying solubilities and partitioning coefficients of solutes within organic complexing agents which are incorporated into the pores of hydrophobic microporous membranes to remove and concentrate metals [37]. The solublized metals are extracted from the influent stream by the water-immiscible organic chelating agent and the diffusion of the metal to the product side of the membrane is coupled to the counter-current flow of such species as mainly acidic hydrogen or, to a less extent, chlorides.

Coupled membrane transport is presently not a commercially available membrane separation system, but bench-scale results show significant potential where high concentrations of leached metals are required to be extracted or reprocessed. The coupled transport membrane has been produced from sheets or hollow fibers, and the present status of development is that further work on hollow fiber module development and performance studies, process design studies and field tests are needed [37]. The coupled membrane transport process has been applied on a prototype basis to the removal and acquisition of aluminum, cobalt, copper, chromium, iron, nickel and uranium [37-39].

DEFINITION OF MEMBRANE ENGINEERING PARAMETERS

Because of the diverse backgrounds and technical basis which led to the development of chelation plus ultrafiltration, charged ultrafiltration and reverse osmosis, basic engineering parameters have been used interchangeably. To minimize such confusion, specific operational definitions will be designated here and used to discuss operational performance. These definitions are:

$$\text{Material balance} = C_u F_u = C_p F_p + C_b F_b \tag{2}$$

$$\text{Ion rejection (\%)} = 1 - \frac{2C_p}{C_u + C_b} \times 100 \tag{3}$$

$$\text{Ion reduction (\%)} = 1 - \frac{C_p}{C_u} \times 100 \tag{4}$$

$$\text{Water recovery (\%)} = \frac{F_p}{F_u} \times 100 \tag{5}$$

where
C_u = concentration of constituent(s) within untreated influent stream
F_u = flow of influent stream solution
C_p = concentration of constituent(s) within purified or permeate stream
F_p = flow of purified or permeate solution
C_b = concentration of constituent(s) within brine or retentate stream
F_b = flow of brine or retentate solution

The material balance which must exist within all membrane systems is designated as the sum of the products of both the concentrations and flows of the purified or permeate stream and the concentrated brine or retentate must equal the input or the product of the influent concentrations and flows [30].

Ion rejection is unity minus the concentration of the permeate divided by the average concentration of constituents on the feed side of the membrane. The basis of this definition is the fact that this parameter is constant for a membrane under constant conditions. If water recovery is varied, the level of constituents in the brine will increase and subsequently an equal percent of rejection at a higher concentration will result in an increased level of salt permeation. Therefore, this definition for ion rejection retains this parameter as a constant and compensates for variables such as water recovery, variable influent concentrations, differing levels of brine or retentate quality and the associated increased salt permeation.

Ion reduction is a measure of the membrane system efficiency to remove specific or groups of various constituents. The parameter of ion reduction only incorporates the influent quality and the permeate quality, hence, this measurement directly reflects removal efficiency. The parameter of ion rejection can be utilized to predict the quality of the permeate stream under the specific condition associated with the application; the projected permeate quality value can be used to predict the operational efficiency of the membrane system.

PERFORMANCE OF MEMBRANE SYSTEMS

The performance of chelation plus ultrafiltration, charged ultrafiltration and reverse osmosis regarding arsenic, cadmium, chromium, copper, cyanide, lead, mercury, nickel, selenium, silver and zinc has in most cases been evaluated. These evaluations include laboratory studies, prototype tests and monitoring of commercial-scale facilities. Percent ion rejection will be used here to compare the performances and process amenability of the removal metals and cyanide.

The performance of each membrane system in the removal of metals and cyanide appears in Table V. After the performance of the various membrane systems in terms of range of percent ion rejection are the references which define the conditions under which the data were collected, how the membrane system was operated, what type of membranes were used, the operational parameters of the treatment system and the matrix of the water which retained the constituents.

Table V shows that reverse osmosis is the type of system which has been documented most frequently with regard to all of the listed constituents. Percent ion rejection varied for the three membrane processes, with reverse osmosis retaining the highest level of removal efficiency. Charged ultrafiltration and chelation-plus-ultrafiltration retained somewhat lower removal efficiencies. As based on the previous discussion of these membrane processes, the most applicable system will be ascertained by evaluating the required rejection necessary to meet process specification or water quality criteria, economics associated with process procurement-operation-value of materials conserved and relative reliability and longevity of the process within the application.

CAUSES OF VARIABLE OPERATION IN
MEMBRANE SYSTEM PERFORMANCE

The performance data listed in Table V reflect the operational efficiency of the membrane systems to remove constituents to form a purified stream and to concentrate the metals for subsequent reclamation. In the case of cyanide, which can be concentrated by reverse osmosis (as opposed to the conventional method of cyanide destruction by oxidation [53]), this substan-

Table V. Performance of Various Membrane Systems

	Chelation Plus Ultrafiltration[a]	References	Charged Ultrafiltration[a]	References	Reverse Osmosis[a]	References
Arsenic			88-95	23,24	90-99+	28,40-44
Cadmium			75-92	23,24	95-99+	28,40-48
Chromium	82-93	18,19			94-99	28,29,40-50
Copper	72-93	16,18,19	80-90	22,24	94-99+	28,29,40-48,50-52
Cyanides					80-96	28,40,43,47,51,53
Lead			95-98	24	93-99+	28,29,40,43,44,46,48,50
Mercury	Demonstrated	17	91-93	24	95-97	42-44
Nickel			81-89	22	95-99	28,29,40-42,45,46,48,50
Selenium			85-95	23,24	90-98	28,40-44
Silver					94-96	29,43,45
Zinc	65	18,19	77-90	22-24	95-99	28,40-44,46,48,50-52

[a] Ion rejection as percent

tiates a basis of conservation of materials with cost savings associated with a lower input, because less cyanide is required and no chemical oxidants are necessary.

The performance data specified reflect operation of the membrane systems under certain conditions which are usually optimized to maximize the efficiency of the system. There are, however, operational parameters that will cause variations in system performance, the most obvious of which are the specific membrane used and the membrane preparation procedure. In all three of the membrane processes, different membrane polymers have been applied to ascertain the best performance. The operational data show that different membranes have varying operational characteristics which will cause variation in the rejection of dissolved species. Also, in most cases the membranes are treated or heat-annealed as a final preparation before the membrane is used. Different final preparation procedures will alter the membrane characteristics such that the performance of the membranes will vary.

A key parameter in the operation of a membrane system which affects performance is the applied pressure on the membrane. The applied pressure on the membrane will directly affect performance within a given range, after which breakthrough or compaction may occur. At applied pressures below the optimum level, the rejection of specific constituents will increase in direct proportion with an increase in applied pressure. A point will be reached with increasing applied pressure where rejection will increase only slowly. This plateau is considered to be the optimum level for applied pressure. At applied pressures well in excess of the optimum level, membrane damage, such as deterioration followed by ion breakthrough or compaction followed by densification of the membrane material, causing decreased permeate flows will become prominent. Utilizing the correct level of applied pressure will optimize membrane performance, longevity and economics of operation.

The inherent matrix of the raw water will affect performance because rejection may vary through the necessity to balance charges. As stated earlier, the presence of both monovalent and multivalent species within a solution will result in a lower level of rejection of monovalent species because of a conservation of charge on both sides of the membranes. Hence, the specific matrix of the water must be evaluated prior to implementing a specific process, and this evaluation must consider membrane capabilities in obtaining the required removal efficiencies. Variations in the inherent dissolved solids matrix will alter the performance of all three membrane processes.

The major physicochemical parameter which affects membrane systems is pH. The pH level will affect the dissociation of aqueous, chelated and membrane groups (Table IV) and the solubility of various metals. In all three membrane processes, pH affects the ionization of the dissolved constituents and subsequently the rejection of these species. The greater charge a constituent retains, the higher the level of rejection. The effect of pH can also be seen where ligands are used. The pH will significantly affect the affinity of the ligand for the metal and govern the dissociation of the metal from the chelating complex.

The effect of pH is dramatic where charged membrane functional groups are used. As the pH of the influent stream is decreased, the pH-controlled physicochemical charge of the functional groups becomes less and their effectiveness diminishes. Hence, to optimize the pH within a charged membrane ultrafiltration process one must evaluate, by using the dissociation constants, at what pH the membrane functional groups and the constituents which require removal are sufficiently ionized. The solubility of a variety of these metallic constituents is governed by pH. As the pH increases, the formation of hydroxides is induced, and although these suspended solids will increase the rejection of the metals, they will also scale the membrane. As this scale will impair the production of purified water by the membrane, the formation of scale must be prevented and the scale present on the membranes must be removed.

The membrane system performance as designated by ion rejection is affected by both the overall quantity of water recovered by the process and the quantity of water recovered per pass. As the material balance in Equation 2 demonstrates, as water recovery increases, the concentration of the brine or retentate also increases. The quantity of water recovered per pass in the membrane system determines the concentration of ions at the water-membrane interface, the level of concentration polarization and the concentration of constituents in the brine or retentate. The level of ions which permeate the membrane is in direct proportion with the concentration of ions at the water-membrane interface. As water recovery is increased, the quality of purified water deteriorates.

The overall quantity of water recovered by the membrane process reflects a situation where the original feed water is concentrated many times to optimize the partitioning of the influent stream into the purified portion and the stream containing the metals and cyanide. The number of times the feed water is reprocessed is directly related to the quality of purified water which is produced. Because the level of water recovery per pass affects product water quality and continuous reprocessing further deteriorates the product water quality, the level of overall water recovery is limited by the specification of product water quality that is desired.

Finally, a concept which has predominated this review of membrane processes to remove and conserve metals and cyanides is to retain a physico-chemical environment which minimizes membrane fouling, scaling and excessive deterioration. The membrane fouling can be minimized or alleviated by proper pretreatment to remove suspended solids and microbiological organisms. Membrane scaling occurs at the membrane after the feed water quality has increased in concentration because of the generation of purified water. This fouling can best be evaluated and controlled by reviewing the chemistry and solubility of the constituents. The deterioration of membrane materials will always occur concurrent with system operation, but excessive chemical deterioration can be controlled. When the membrane becomes deteriorated, the quality of purified water decreases and as the deterioration becomes more advanced the membrane process becomes inoperative because there is no longer a partitioning of the feed stream into a purified and concentrated stream.

CONCLUSIONS

The levels, frequency and regulations associated with the metal and cyanide priority pollutants designated by the EPA have been reviewed. The basis for this regulation, health effects and lowest reported toxic concentrations, and the industrial and manufacturing sources which require the removal and reclamation of these constituents from process streams have been presented. In addition, data have been presented which substantiate the amenability of these membrane processes for the removal and conservation of these priority constituents.

REFERENCES

1. Barrett, B. R. "Controlling the Entrance of Toxic Pollutants into U.S. Waters," *Environ. Sci. Technol.* 12(2):154 (1978).
2. Keith, L. H., and W. A. Telliard. "Priority Pollutants," *Environ. Sci. Technol.* 13(4):416 (1979).
3. National Academy of Sciences. "Drinking Water and Health," Printing and Publishing Office of NAS, Washington, DC (1977).
4. "Quality Criteria for Water," U.S. EPA, U.S. Government Printing Office (1976).
5. Underwood, E. J. *Trace Elements in Human and Animal Nutrition* (New York: Academic Press, Inc., 1977).
6. Tate, C. H., and R. R. Trussell. "Developing Drinking Water Standards," *J. Am. Water Works Assoc.* 69(9):486 (1977).
7. U.S. Public Health Service. "Drinking Water Standards," *Federal Register* (March 6, 1962) p. 2152.
8. U.S. Environmental Protection Agency. "National Interim Drinking Water Regulations," *Federal Register* (December 24, 1975) p. 59566.
9. Kimm, V. J. "Report on EPA Regulations," *J. Am. Water Works Assoc.* 69(9):482 (1977).

10. National Academy of Sciences. "Geochemistry and the Environment, Vol. 1," Printing and Publishing Office of NAS, Washington, DC (1974).

11. National Academy of Sciences. "Geochemistry and the Environment, Vol. 2," Printing and Publishing Office of NAS, Washington, DC (1977).

12. Zemansky, G. M. "Removal of Trace Metals During Conventional Water Treatment," *J. Am. Water Works Assoc.* 66(10):606 (1974).

13. Ward, C. H., and E. M. Davis. "Biological Bases for Water Quality Standards," in *Proceedings of the 27th Industrial Waste Conference at Purdue University* (Ann Arbor, MI: Ann Arbor Science Publishers, Inc., 1977).

14. Kroner, R. C. "The Occurrence of Trace Metals in Surface Waters," in *Traces of Heavy Metals in Water Removal Processes and Monitoring*, J. E. Sabadell, ed. EPA-902/9-74-001 (1974).

15. Patterson, J. W. *Wastewater Treatment Technology* (Ann Arbor, MI: Ann Arbor Science Publishers, 1977).

16. Ryan, M. J. "Copper Removal from Water by Chelation and Ultrafiltration," *Diss. Abstr. Int.* B-36:2983 (1976).

17. Hopfenberg, H. B., K. L. Lee and C. P. Wen. "Improved Membrane Separation by Selective Chelation of Metal Ions in Aqueous Feeds," *Desalination* 24:175 (1978).

18. O'Neill, T. L., G. R. Fisette and E. E. Lindsey. "Separation of Metal Ions from Water by Chelation and Ultrafiltration," *Am. Inst. Chem. Eng. Symp. Ser.* 71(151):100 (1976).

19. Bulbula, A., E. E. Lindsey, R. D. Archer and R. W. Lenz. "Separation of Metals from Water by Chelation and Ultrafiltration: Part B," *Am. Inst. Chem. Eng. Symp. Ser.* 71(151):105 (1976).

20. "Ultrafiltration Poised for New Uses," *Chem. Eng. News* 57(39):40 (1979).

21. Bhattacharyya, D., J. M. McCarthy and R. B. Grieves. "Charged Membrane Ultrafiltration of Inorganic Ions in Single and Multi-salt Systems," *Am. Inst. Chem. Eng. J.* 20(6):1206 (1974).

22. Bhattacharyya, D., D. P. Schaff and R. B. Grieves. "Charged Membrane Ultrafiltration of Heavy Metal Salts: Application to Metal Recovery and Water Reuse," *Can. J. Chem. Eng.* 54:185 (1976).

23. Bhattacharyya, D., M. Moffitt and R. B. Grieves. "Charged Membrane Ultrafiltration of Total Metal Oxyanions and Cations from Single- and Multisalt Aqueous Solutions," *Separation Sci. Technol.* 13(5):449 (1978).

24. Bhattacharyya, D., A. B. Jumawan and R. B. Grieves. "Charged Membrane Ultrafiltration of Heavy Metals from Nonferrous Metal," *J. Water Poll. Control Fed.* 51(1):176 (1979).

25. Bhattacharyya, D., S. Shelton and R. B. Grieves. "The Effective Treatment of Acid Mine Wastes by a Low-Pressure Charged Ultrafiltration Process," in *Proceedings of the 33rd Industrial Waste Conference at Purdue University* (Ann Arbor, MI: Ann Arbor Science Publishers, Inc., 1980).

26. Dickerson, R. E., H. B. Gray and G. P. Haight. *Chemical Principles* (New York: W. A. Benjamin, Inc., 1970).

27. *Handbook of Chemistry and Physics*, 54th ed. (Cleveland, OH: CRC Press, 1973).

28. "Industrial Application of Reverse Osmosis Technology: Optimizing Performance," Bulletin 608, Trace Metal Data Institute (1978).
29. "Control of Radiological Contamination Using Reverse Osmosis Technology," Bulletin 604, Trace Metal Data Institute (1979).
30. "Restoration of In Situ Uranium Leaching Sites Incorporating Reverse Osmosis Technology," Bulletin 605, Trace Metal Data Institute (1978).
31. Quinn, R., and W. K. Hendershaw. "A Comparison of Current Membranes and Systems," *Ind. Water Eng.* 13:12 (1976).
32. "An Analysis of Circumventing Reverse Osmosis Fouling," Bulletin 609, Trace Metal Data Institute (1979).
33. *Standard Methods for the Examination of Water and Wastewater*, 14th ed. (Washington, DC: American Public Health Association, 1975) p. 132.
34. Kosarek, L. J. "Water Reclamation and Reuse in the Power, Petrochemical Processing and Mining Industries," in *Proceedings of the International Water Reuse Symposium, Vol. 1* (1979), p. 421.
35. Spatz, D. D., and R. H. Friedlander. "Rating the Chemical Stability of U.C. RO/UF Membrane Materials," *Water Sew. Works* 125(2):36 (1978).
36. Envirogenics Systems Co., Inc., El Monte, CA. Unpublished observation.
37. Babcock, W. C., R. W. Baker, D. J. Kelley and H. K. Lonsdale. "Coupled Transport Membranes for Metal Separations," U.S. Bureau of Mines, Department of the Interior (1978).
38. Babcock, W. C., R. W. Baker, D. J. Kelley and H. K. Lonsdale, R. J. Ray and M. E. Tuttle. "Coupled Transport Membranes for Metal Separation, Final Report, Phases I and II," U.S. Bureau of Mines Open File Report 114-79 (1977).
39. Babcock, W. C., R. W. Baker, M. G. Conrod, H. K. Lonsdale and K. L. Smith. "Coupled Transport Membranes for Removal of Chromium from Electroplating Rinse Solutions," Chapter 13, this volume.
40. Johnston, H. K., and H. S. Lim. "Removal of Persistent Contaminants from Municipal Effluents by Reverse Osmosis," Research Report 85, Ontario Ministry of the Environment, Ontario Canada (1978).
41. Mixon, F. O. "The Removal of Toxic Metals from Water by Reverse Osmosis," Office of Saline Water R&D Report #889, Department of the Interior (1973).
42. Wilmoth, R. C., and J. L. Kennedy. "Removal of Trace Elements from Acid Mine Drainage," EPA-600/7-79-101 (1979).
43. "Drinking Water Standards and RO Performance," Polymetrics, Santa Clara, CA (1972).
44. Argo, D. G., and J. G. Moutes. "Wastewater Reclamation by Reverse Osmosis," *J. Water Poll. Control Fed.* 51(3):590 (1979).
45. Furukawa, D. H. "Removal of Heavy Metals from Water by Reverse Osmosis," in *Traces of Heavy Metals in Water*, EPA-902/9-74-001 (1973).
46. Johnston, H. K. "Reverse Osmosis Rejection of Heavy Metal Cations *Desalination* 16:205 (1975).
47. Hindin, E., G. H. Dunstan and P. J. Bennett. "Water Reclamation by Reverse Osmosis," Washington State University College of Engineering, Bulletin 310 (1968).
48. Rangarajan, R., T. Matsuura, E. C. Goodhue and S. Sourirajan. "Free Energy Parameters for Reverse Osmosis Separation of Some Inorganic

Ions and Ion Pairs in Aqueous Solutions," *Ind. Eng. Chem. Proc. Des. Dev.* 15(4):529 (1976).

49. Cruver, J. E. "Reverse Osmosis—Where It Stands Today," *Water Sew. Works* 120(10):74 (1973).

50. Kremen, S. S., C. Hayes and M. Dubbs. "Large-Scale Reverse Osmosis Processing of Metal Finishing Rinse Waters," *Desalination* 20:71 (1977).

51. McNulty, K. J., D. C. Grant, A. Z. Gollan and R. L. Goldsmith. "Treatment of Metal Finishing Wastes by Reverse Osmosis," *Am. Inst. Chem. Symp. Ser.* 73:(166):176 (1976).

52. Allegrezza, A. E., J. M. Charpentier, R. B. Davis and M. J. Coplan. "Hollow Fiber Reverse Osmosis Membranes," *Am. Inst. Chem. Eng. Symp. Ser.* 73(166):162 (1977).

53. Weiner, R. "Total Recovery: The Final Solution of Waste Problems in the Metal Finishing Industry," *Pure Appl. Chem.* 45:171 (1976).

CHAPTER 13

DEMONSTRATION OF A CROSS-FLOW MICROFILTRATION SYSTEM FOR THE REMOVAL OF TOXIC HEAVY METALS FROM BATTERY MANUFACTURING WASTEWATER EFFLUENTS

Norman I. Shapira, Han-Lieh Liu,
John E. Santo and Daniel Fruman

HYDRONAUTICS, Incorporated
Laurel, Maryland

Charles Darvin

U. S. Environmental Protection Agency
Industrial Environmental Research Laboratory
Cincinnati, Ohio

John Baranski and Kenneth Mihalik

General Battery Corporation
Reading, Pennsylvania

This chapter describes a novel method of suspended solids (SS) removal utilizing cross-flow microfiltration with thickwalled, porous plastic tubes, called Hydroperm™. These tubes, whose wall thickness, inside diameter, porosity and pore-size distribution can all be controlled closely, can be designed for any effluent to obtain optimum performance. The filtration characteristics of the tubes combine both the "in-depth" filtration aspects of multimedia filters and the "surface" filtration aspects of membrane filters. For example, while the removal of micron- and submicron-size particles is often difficult or impossible with conventional through-flow filters, Hydroperm tubes are capable of virtually complete removal of such particles.

In the treatment of domestic and industrial wastewaters, removal of suspended solids is a required operation which arises under a variety of circumstances. For example, this unit operation is often essential as a pretreatment step prior to undertaking further advanced treatment steps. Fine-solids removal is also often necessary as a "polishing" step after chemical or biological treatment. There are several existing physical treatment processes available for this purpose, such as gravity clarification and multi-media filtration. However, most of these processes have inherent disadvantages in dealing with fine and difficult-to-settle solids.

For removal of SS, two types of filtration processes have been used widely: (1) the through-flow (e.g., multimedia) filter; and (2) cross-flow [1,2] filtration (e.g., ultrafiltration and reversed osmosis). Although they can remove some of the SS, conventional through-flow filters have the disadvantage of requiring frequent backwashing, since the filtered particles continuously accumulate on or actually enter the filtration barrier. In other words, through-flow filtration, by its very nature, is a batch process, with the flux declining relatively rapidly when the driving pressure differential across the filtration barrier is held constant (Figure 1).

On the other hand, in cross-flow filtration, the direction of the feed flow is parallel to the filter surface, so that accumulation of the filtered solids on the filter medium can be minimized by the shearing action of the flow. Thus, cross-flow filtration affords the possibility of a quasisteady-state operation with a nearly-constant flux when driving pressure differential is held constant (Figure 1).

It is relevant to point out that cross-flow microfiltration, which is intended primarily for the removal of suspended solids, is significantly different from membrane ultrafiltration (UF) or hyperfiltration (RO), which remove substances on the molecular level in addition to suspended solids. In UF, high filtration pressures ($\sim$50–150 psi) are used, with even higher pressures (600–1200 psi) for RO, compared with only $\sim$15 psi for Hydroperm microfiltration. Furthermore, UF and RO employ relatively thin membranes, compared with the relatively thick-walled ($\sim$1 mm) Hydroperm microfilters. As a result, power requirements and capital and operating costs are higher for membrane systems than for Hydroperm microfiltration.

This chapter describes the results of a two-phase program to demonstrate the applicability of the Hydroperm microfiltration system for the removal of toxic heavy metals from lead-acid battery manufacturing wastewaters after the metals have been precipitated chemically. This program has been conducted under the sponsorship of the U. S. Environmental Protection Agency (EPA) Industrial Environmental Research Laboratory (IERL) in Cincinnati, OH. The results reported here include the Phase I laboratory test program and the Phase II onsite demonstration of a full-scale, 50-gal/min

I. THROUGH FLOW FILTRATION

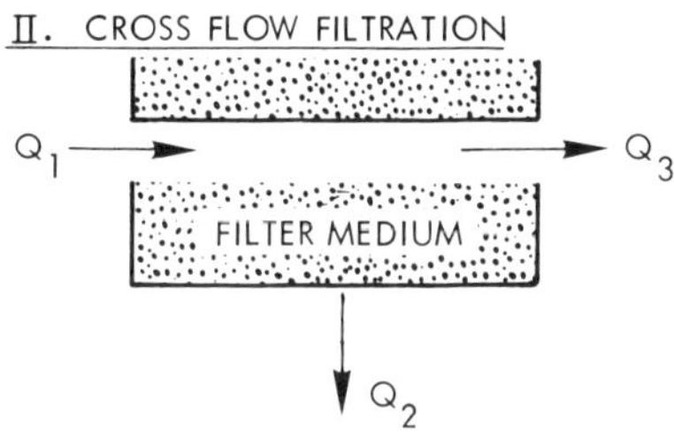

FEATURES:

(i) PERMEATE AND FEED FLOW DIRECTIONS ARE THE SAME

(ii) INHERENTLY UNSTEADY OPERATION

(iii) REQUIRES FREQUENT BACKWASHING

II. CROSS FLOW FILTRATION

FEATURES:

(i) PERMEATE FLOW DIRECTION IS PERPENDICULAR TO THAT OF THE FEED FLOW

(ii) PARTICLE POLARIZATION IS PREVENTED BY SHEAR INDUCED BY THE FEED FLOW

Figure 1. Types of physical filtration.

Hydroperm microfiltration system at the General Battery Corporation plant at Hamburg, PA.

Although the filtrate will be discharged either to the municipal sewer or directly to a local stream, the quality of the filtrate in terms of ss and toxic heavy metals is such that reuse is a feasible option which is under study.

Although the present program is directed toward treating wastes from one specific industry category, it is realized that many other industries including the electroplating and metal refining industries also have toxic heavy metals removal problems. It is expected that the Hydroperm micro-filtration system will find wide application throughout these industries.

In this chapter, Hydroperm microfiltration is discussed in some detail, to provide for a better understanding of the test results. A brief description follows of the typical lead-acid battery wastewater characteristics. A basic discussion of toxic heavy metals removal processes is presented. The Hydroperm laboratory test procedures are then outlined, and the Phase I laboratory test program and results are given. The results of the field demonstration program are presented, followed by a brief discussion of economic factors.

CHARACTERISTIC FEATURES OF HYDROPERM CROSS-FLOW MICROFILTRATION

The method of advanced treatment of lead-acid battery manufacture wastewaters described here is based on cross-flow microfiltration. The principal element of the system is a thick-walled hollow tubular filter ($\sim$6–9 mm i.d.) made of thermoplastic material and containing micron-size pores. Even at relatively low filtration pressures ($\sim$15 psi) these tubes (known by the proprietary name Hydroperm) are capable of virtually total removal of SS while maintaining reasonably high permeation rates.

These tubes can be made from a variety of extrudable thermoplastics, such as nylon and polyethylene, by a proprietary manufacturing process. A brief summary will be given here in order to help set in proper perspective the tests described in later sections.

The filtration characteristics of Hydroperm tubes combine both the "in-depth" filtration aspects of multimedia filters and the "surface" filtration aspects of membrane ultrafilters. For example, in a manner similar to multimedia filters, the Hydroperm tubes may allow some of the smaller particles in the waste streams to penetrate into their wall matrix. However, it should be noted that the pore structure of the Hydroperm tubes differs from those of membrane ultrafilters in that the pore sizes of the former are of the order of several microns, with the "length" of the pores being many times their diameters. It is relevant to emphasize at the outset that because of this unique combination of "in-depth" and "surface" filtration aspects, Hydroperm tubes display excellent performance characteristics, Thus, while the tubes (with micron-size pores) are primarily designed for the separation of suspended solids from the feeds, they frequently achieve substantial levels of dissolved solids separation, because of the formation of a dynamic membrane on the inside walls of the tubes. The characteristics of this membrane vary with the characteristics of the solids content of the feed. The thickness of the membrane is controlled by the shear exerted by the circulation flow. The concentrated solids are not deposited inside the tubes, but settle out in a settling cone (Figure 2) as permeate is removed from the feed.

A schematic view of cross-flow filtration through the tubes is shown in Figure 3. The feed flow is through the inside of the tubes at relatively low pressure ($\sim$15 psi) and the filtrate permeation occurs through the relatively thick ($\sim$1 mm) tube walls. Pore-size distributions of two typical Hydroperm tubes are shown in Figure 4. Tube I has a rather "flat" distribution with the pores ranging in size from 2 to 10 μ. Tube II, however, has a "peaked" distribution, with most of the pores in the 2-μ range. It should be noted that the pore-size distributions shown in Figure 4 are initial values, since the pore-size distribution of the tubes undergoes a change during filtration, because of the penetration of small particles into the tube wall.

Two views of the pore structure of a typical Hydroperm tube are shown in Figure 5. These photographs were taken with the aid of a scanning electron microscope and are of a transverse section of the tubes. The view in 5a has a magnification factor of 200, while that in 5b has a magnification factor

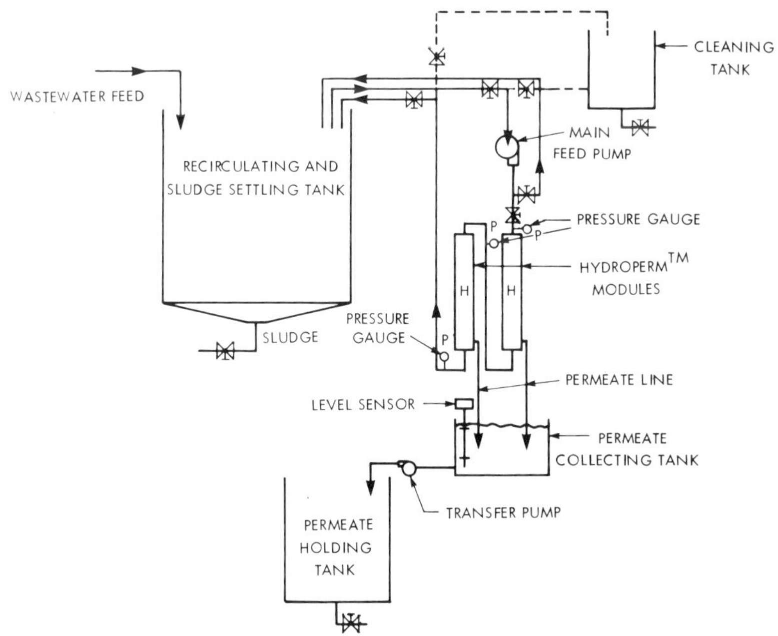

Figure 2. Hydroperm system schematic diagram.

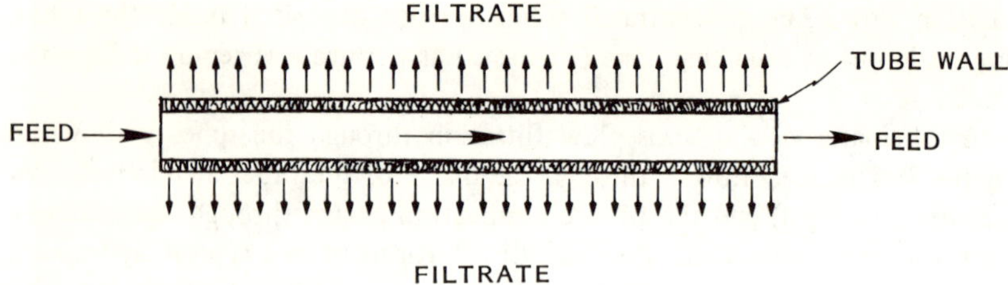

Figure 3. Cross-flow filtration schematic.

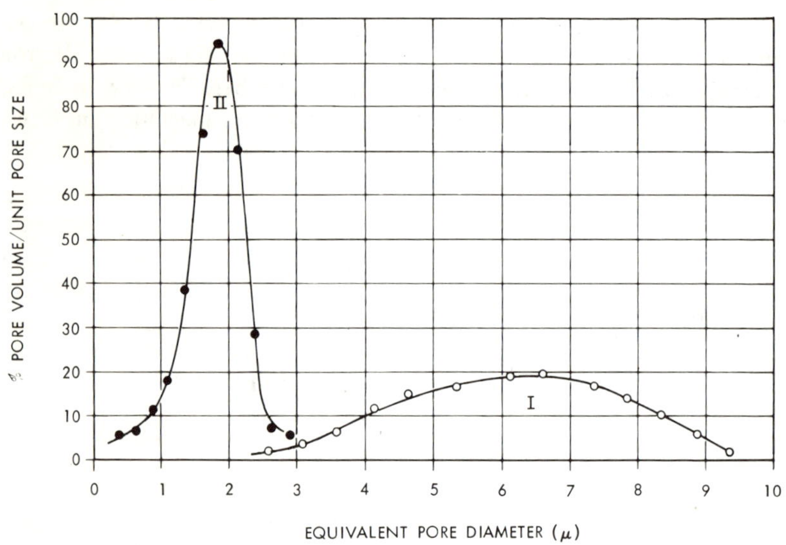

Figure 4. Typical pore-size distribution of Hydroperm tubes.

of 1000. The open-cell, reticulated nature of the pore structure can be appreciated from these photographs. These features (together with operating conditions) are of importance in determining the performance of a given tube when it is used with a specific wastewater, as can be seen by considering a relatively simple model for the filtration process described below.

In general, any effluent for which SS removal is desired will contain a range of particulates, varying in diameter from collodial dimensions to many microns. When such effluents are circulated through the inside of a porous

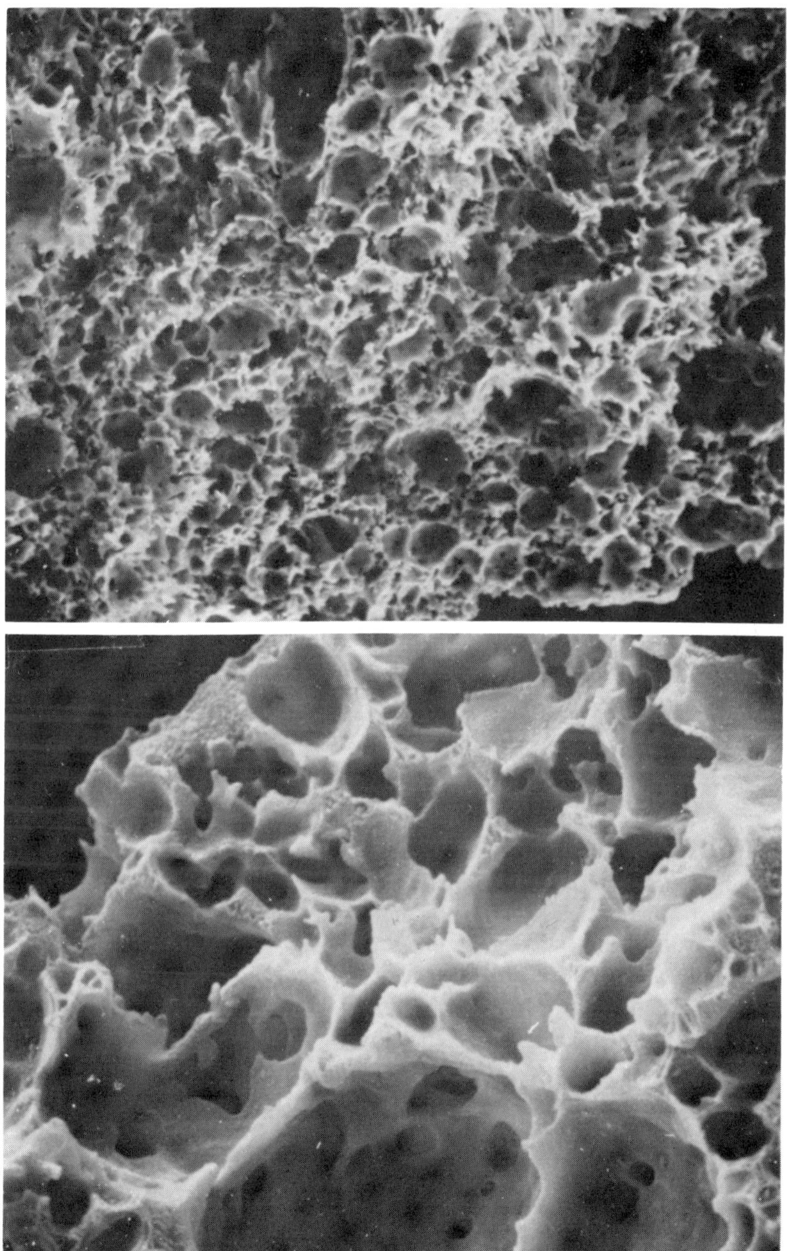

Figure 5. Electron photomicrographs of Hydroperm tube pore structure—transverse section.

tubular filter such as Hydroperm, the solids particles will be driven, with the permeating flow, toward the wall. Thus, the concentration of the particles in regions close to the wall will tend to increase steadily. This tendency will be delimited by the turbulent diffusion of the particles from regions of high concentration to those of lower concentration (i.e., away from the walls and toward the center of the tube).

This turbulent diffusion (which tends to decrease the particle concentration near the wall) depends on the shear stress that is exerted on the walls by the feed circulation, and, hence, its velocity. On the other hand, the permeation rate (which tends to increase the particle concentration near the wall) depends on the pressure differential across the filter surface (Poiseuille's law) as well as the effective pore structure of the tubes (Darcy's law). A quasisteady-state profile of the concentration of the particles will be established near the wall, when the opposing tendencies mentioned above exactly balance each other. The resulting "particle polarization" in this case is analogous to the "concentration polarization" of solutes which occurs close to the walls of ultrafiltration and reverse osmosis membranes.

Because of the in-depth filtration characteristics of the Hydroperm tubes, other factors also come into play. Feed particles which are smaller than the largest pore size of the tubes may actually enter the wall matrix, while particles which are larger than any of the pores in the tubes will be retained at the walls. Some small particles which enter the tube matrix will ultimately become entrapped within the wall of the tube because of the irregular and tortuous nature of the pores, but will not pass through. Thus, as filtration proceeds, the pore structure of the tube as well as its permeability will undergo a gradual change due to the penetration of some of the pores by the intruder particles. However, the tendency of new particles to enter the tube matrix will decrease rapidly as the dynamic membrane forms on the walls due to the particle polarization described earlier.

Even from the relatively simple qualitative discussion given above, it is clear that the filtration performance is influenced not only by such factors as the filtration pressure, circulating flow velocity and temperature (which changes the fluid viscosity and, hence, by Darcy's law, the permeation rate), but also by the pore-size distribution, pore structure of the tubes and the particle-size distribution in the wastes.

THE WASTE CHARACTERISTICS OF LEAD-ACID BATTERY PRODUCTION

The waste waters from lead-acid battery manufacturing are highly acidic (pH $\sim$1), and contain a number of toxic heavy metals, primarily lead, in concentrations varying from 20 to 500 ppm. Other metals in the waste include Cu, Ni, Zn, As and Sb.

In the manufacture of industrial and automobile batteries, waste water is generated from the washdown of sulfuric acid spills which occur during battery filling and the washdown of the stations where lead oxide pasting and forming are done. Contact cooling waste water from the manufacture of lead oxide also contains oil and grease. In addition, used batteries containing spent electrolyte (sulfuric acid, etc.) may also be processed on-site. The tops of batteries are sheared off and contents are dumped. The lead and lead compounds are separated from the electrolyte which is mixed with wash water. This waste water is a major effluent. It is pretreated by settling before being pumped to the treatment plant. Wastewater volumes vary from ∿25,000 to greater than 400,000 gal/day.

The character of the sludge from wastewater treatment is largely determined by the type of treatment process. Current lead-acid battery treatment practices include neutralization, precipitation, sedimentation and filtration. As of 1975, 75% of the lead-acid manufacturing plants (150 plants) within the United States neutralized their waste water and discharged directly to waterways or municipal treatment plants. Fourteen plants used lime treatment to produce a calcium sulfate–lead sludge, while the remaining plants used caustic to produce a lead hydroxide sludge. In most cases no attempt is made to recover the lead from the treatment sludges prior to land-fill. The sludge from all the treatment processes must be disposed of in relatively secure landfill sites to reduce the potential for toxic contamination by migrating lead, and positive safeguards must be considered when disposing of the sludge due to its high lead concentration.

Slaked lime is used in the General Battery Hamburg plant. The previous practice in handling this neutralized wastewater was by trucking to the General Battery Corporation wastewater treatment facility in Reading, PA.

Typical wastewater characteristics from the General Battery Hamburg plant and from a survey conducted by the EPA [3] are given in Table I.

The present effluent standards applied in Pennsylvania are shown in Table II. There are, however, no published federal effluent guidelines at this time.

CONVENTIONAL TOXIC HEAVY METALS REMOVAL

As mentioned earlier, the principal pollutants in battery manufacturing waste waters are toxic heavy metals. When these waste waters are discharged either into the publicly owned treatment works (POTW) or natural streams, they have to meet either the pretreatment standards of the POTW or the water quality standards of the receiving water body. In either case, the

Table I. Battery Wastewater Characteristics

Parameter	EPA Survey (Before Treatment)	Hamburg Plant	
		Before Neutralization	After Neutralization and Before Hydroperm Treatment
pH	1.4	1.1	9.8
Suspended Solids (ppm)		217.0	21,500.0
Dissolved Solids (ppm)		1,711.0	3,070.0
Pb	11.70	21.1	382.0
Cu	0.16	0.52	0.86
Ni	0.14	0.23	1.29
Zn	0.58	0.69	0,09
As		0.027	0.50
Sb	0.80	1.56	3.33

Table II. Pennsylvania NPDES Permit

	Monthly Average	Daily Maximum
Suspended Solids (ppm)	20	40
Lead (ppm)	0.14	0.50
pH	6–9	6–9

intention is to reduce the health and environmental risk of pollution caused by discharging hazardous waste waters. As a general practice, manufacturing waste waters must be treated before being discharged. Neutralization and precipitation of heavy metals in waste waters has been used widely, after which a number of processes can be employed for removal of the suspended solids.

In an evaluation of treatment technology for water reuse and recycling, sponsored by OWRT [4], it was suggested that "lime precipitation tends to be more effective than the addition of alum or ferric chloride." Precipitants, such as lime, and coagulants, such as alum, have frequently been used in metals wastewater treatment processes. A brief discussion of this precipitation process follows.

Heavy metals are usually allowed to precipitate as their hydroxides. Since these precipitates are not usually stoichiometric hydroxides, but oxides of rather indefinite composition with various amounts of more-or-less loosely bound water, the term "hydrous oxides" is more precise. Most hydrous oxides have low solubility; therefore, they can be precipitated by adjusting the pH to an appropriate value, which differs according to the particular metal, and is most critical for amphoteric metals, which dissolve at both low and high pH. Alkali for precipitating metal hydroxides may be soda ash (Na_2CO_3), slaked lime $(Ca(OH)_2)$ or caustic soda $(NaOH)$. The hydrous oxides (metal hydroxides) are in general, gelatinous, slow to settle and difficult to filter with conventional filtration. The sludges formed are very watery, often containing 5% solids or less, and their disposal presents problems. Recently, sulfide precipitation has also been investigated. After these metal hydrous oxides have been precipitated, solids can be removed from the waste water through one of several liquid/solids separation techniques, e.g., gravity separation, centrifugation or filtration.

To enhance the settling characteristics of the suspended solids, flocculating agents are usually employed. In the flocculator, the waste water is agitated gently to allow the precipitated solids to coagulate with the flocculant, such as polymers, alum or ferric sulfate. In conventional treatment, the waste water then enters the clarifier where most of the solids settle out. The solids in the underflow can be discharged to a holding tank for thickening and disposal.

The difficulties in this conventional treatment process lie in the relatively inefficient nature of the gravity separation process. The solids removal efficiency depends on the settling rate of the suspended solids in the waste water. Some of these solids settle very slowly because of their small size and density similar to water. Large land areas are required for lagoon clarification. Furthermore, economical design of the clarifier limits the retention time in the settling ponds, and some level of suspended solids will usually appear in the overflow. Also, the gravity settling technique can only separate suspended solids, not the dissolved solids. Since every metal has its own solubility dependence on pH, control of pH at one value in the treatment process is not enough to precipitate all of the dissolved metals, and the effluent in the treated wastewater will usually still show some trace of dissolved heavy metal contamination.

The solids from conventional clarifiers typically contain very low solids concentrations of $\sim$ 0.5–5%; overflow from the sludge tank is recycled to the clarifier. Usually metal hydroxide solids will concentrate only to approximately 3–5% solids in these tanks if given adequate retention time. The tanks also provide storage time and volume for the sludge before shipment to the disposal site, but require large spaces.

Further concentration of the sludge by conventional methods requires the use of mechanical dewatering equipment. Centrifuges, rotary vaccum filters, belt filters and filter presses have been used to dewater metal hydroxide sludges.

As will be shown later in this chapter, the Hydroperm cross-flow microfiltration process achieves virtually complete removal of suspended solids, and is more effective than conventional physical treatment of the waste water by clarification. Hydroperm can also remove some of the dissolved solids by means of the dynamic sludge layer formed inside the Hydroperm wall.

EXPERIMENTAL APPARATUS AND
LABORATORY TEST PROCEDURES

The laboratory tests described in this chapter consisted both of tests with single tubes, and with modules containing a "bundle" of several tubes. A schematic view of a typical single-tube test loop is shown in Figure 6. As indicated in the figure, the loops normally contain a small feed reservoir ($\sim$ 5* gal capacity), a circulating pump, a flow meter, pressure gauges (to measure pressure drops over the length of the tubing being tested) and appropriate valving.

Basically, three different modes of operation are used when carrying out the laboratory tests. The first involves "constant concentration" operation, with the permeate remixed into the feed reservoir, so that (except for evaporation losses) the volume of the circulating feed and concentration

---

*A 55-gal drum reservoir was actually used in this program.

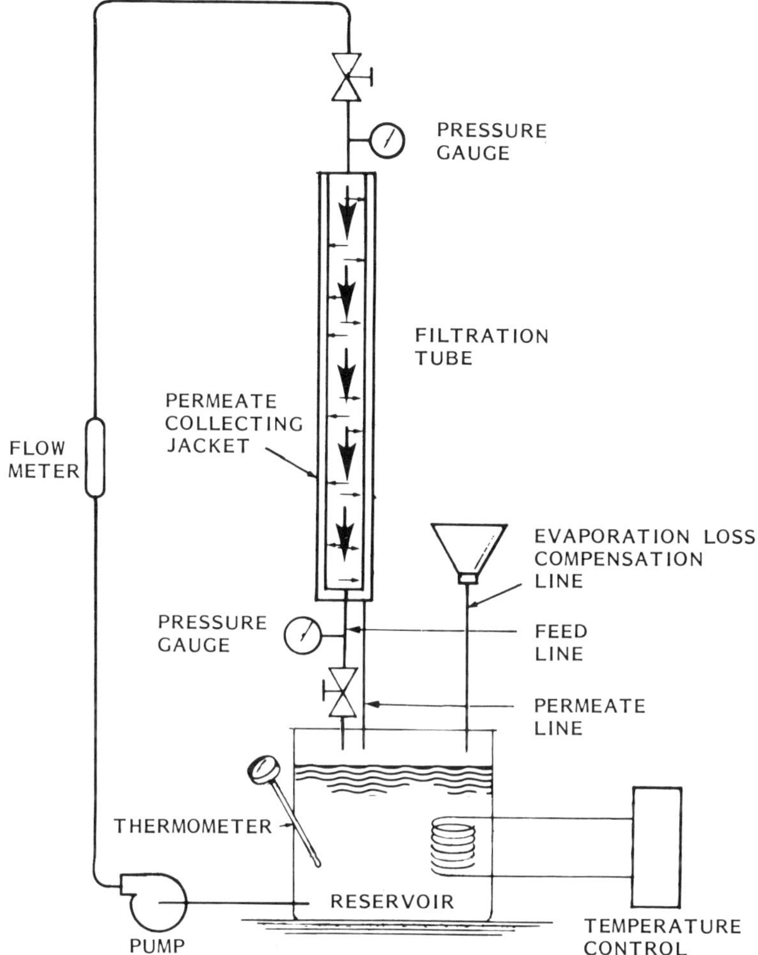

Figure 6. Schematic of a single-tube Hydroperm test loop.

remain constant. The feed in the reservoir is replaced at appropriate intervals to eliminate changes in characteristics due to biological activity and/or constant recirculation.

In the second "concentration" mode of operation, a batchwise process of a field system is simulated. Here the permeate is removed and collected in a separate reservoir, so that the volume of the circulating feed continuously decreases while its suspended solids concentration continuously increases.

The tests are continued until a specified feed concentration is reached or until the volume of the feed becomes so low that adequate pump suction from the reservoir can no longer be maintained.

In the third mode of operation, a "continous" field process is simulated. As permeate is withdrawn by the Hydroperm filtration, comparable amounts of new feed are added. Periodically, sludge is removed.

Essentially, the tube selection test procedures consists of performing several relatively short duration (about 50–100 hr of constant concentration filtration operation) tests with a number of short lengths (about 45 cm long) of Hydroperm tubes of differing materials and pore structures. It has been found, through such a systematic series of tests, that it is usually possible to select optimum tubes and operating conditions for the waste feed under consideration. Optimization is measured in terms of permeate quality and flux, gallons per square foot of filter surface area per day.

Two types of laboratory tests are usually conducted. The first is preliminary determination of effective operating conditions and cleaning procedures. This type of test has the objectives of: (1) investigating various operating velocities to achieve a high filtrate flux and slow initial flux decline; (2) selecting the best tube type; and (3) determining the most promising cleaning method for maintaining high flux for long periods of time. During these tests, the filtrate and the concentrate are usually remixed in the feed reservoir to keep the feed conditions the same throughout the duration of the test.

The second type of test is the long-term module and concentration test. In these tests, an optimum tube is selected for use in a multitube Hydroperm module in both long-term and concentration tests designed to simulate field operation; that is, during the latter stage of this test the filtrate is removed from the feed reservoir. From this test, information is obtained on long-term filtrate flux and permeate quality, the degree of concentration of the feed that can be obtained, and the ability to operate without clogging.

PHASE I–LABORATORY TEST PROGRAM

Phase I of this two-phase program was to select proper Hydroperm tube material, dimensions and operating conditions for later use in the Phase II field demonstration. Phase I of the program was divided into two parts: single-tube and module tests.

Single-Tube Tests

In the single-tube test program the following parameters were investigated: two tube materials (polyethlene and nylon); three tube sizes (4, 6 and 9 mm); one tube length (46 cm); two circulation velocities (4 and 7 ft/sec); one filtration pressure ($\sim$5 psi); a pH range of 7-10; and ambient temperature (about 30°C).

The wastewater used for these tests was obtained from the General Battery Corporation plant at Hamburg, Pennsylvania. The raw wastewaters typically contained $\sim$1500-1900 mg/l of total solids with $\sim$20-200 mg/l of suspended solids and $\sim$ 10-20 mg/l of lead. When received at the Hydroperm Laboratory, the wastewater had a pH of $\sim$1.2. Toxic heavy metals included Pb, Cu, Zn, Ni, Sb and As. The dissolved metals were precipitated at a range of pH by adding hydrated lime. The best lead precipitation results were

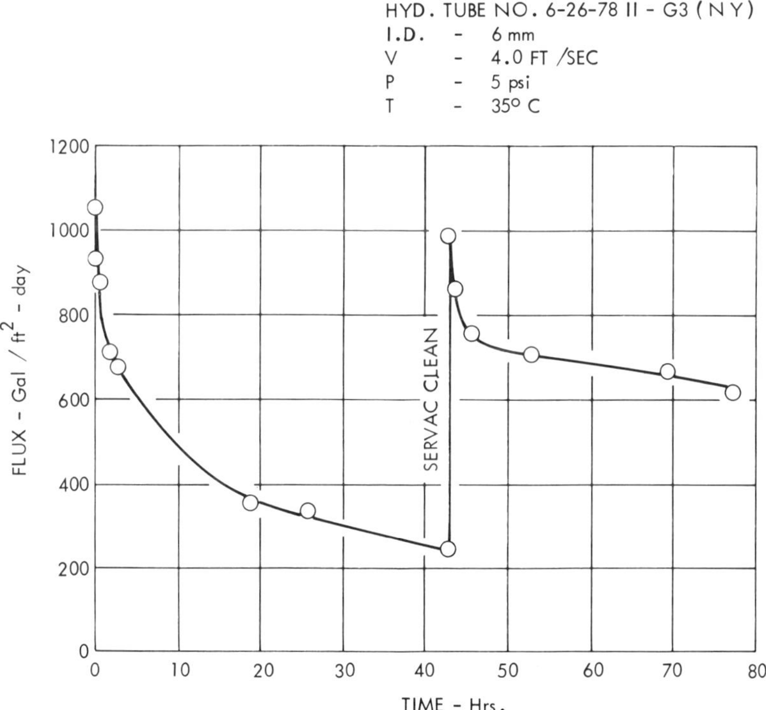

Figure 7. Permeate flux results.

obtained at pH 8.5–9.5. After lime addition, total solids (TS) values increased to ~45,000 mg/l in the feed, most of which were in the form of SS (~40,000 mg/l).

Twelve single-tube tests of up to 160 hours in duration were performed with the lime-precipitated waste described above. In all but one of the tests the filtrate was remixed with the feed, which resulted in a "constant concentration" mode of operation. One test was performed with increasing SS concentration in the feed, with periodic removal of the filtrate from the feed until an 85% reduction in the total volume of the feed had been reached. The purpose of these tests was to provide information of filtrate flux and quality as a function of the type of tube and the operating conditions used. This information was necessary for tube optimization in pilot-plant design.

A typical plot of filtrate flux vs time in hours is shown in Figure 7. The tube used for this test had an internal diameter of 6 mm and the pore structure was that depicted in Curve I of Figure 4. The operating conditions consisted of a feed pressure of 5 psi, a feed velocity of 4 ft/sec and a temperature of 35°C. In this constant-concentration test, the filtrate was remixed with the feed. Note from Figure 7 that the flux begins at ~1000 gal/ft^2/day and typically declines almost immediately, with the rate of decline decreasing with time. From this curve and experience with other wastes it was initially estimated that steady "plateau" fluxes of 150–250 gal/ft^2/day could be maintained for several days without any tube cleaning. However in the present test, the tube was cleaned after 40 hr by pumping through the tube a water solution of Servac (a mild phosphoric acid containing commerical cleaner) in water for a period of 15 minutes. Dilute HCl (~2%) was also used successfully. Note from the figure that when the test was started again, the flux had been restored nearly to its original value (~1000 gal/ft^2/day), after which it began to decline again. The second flux decline was not as rapid as it was in the first part of the test. This behavior occurs frequently, probably because of changes in the feed during pumping, and changes in the pore structure of the tube resulting from penetration of fine suspended solids into the pore matrix.

In studying the influence of tube diameter on the Hydroperm tube performance, several tube diameters were tested. Figure 8 shows the fluxes obtained as a function of the filtration time for four different tests; tests G-3 and G-10 were performed with 6-mm i.d. tubes while G-6 and G-11 were performed with 9-mm i.d. tubes. The circulation velocity and the operating pressure were kept constant during all these tests. As shown in the figure, during the first few hours of operation (<10 hr) the 6-mm tubes seemed to give fluxes higher than the 9-mm tubes. The difference, however, became less significant with increasing filtration time. In either case, a flux over 200 gal/ft^2/day was maintained for 50 hr of continous operation of the tubes. Integrated fluxes over the period of an 8-hr shift would be well over 300 gal/ft²/day.

The results for the constant feed concentration tests in terms of filtrate flux demonstrated the influence of several important parameters. First, the pore structure had a significant influence on flux, since the tube with the wide pore-size distribution had a flux of 250 gal/ft²/day after 40 hr, while the tube with the narrow distribution had a flux of 50 gal/ft²/day after only 2 hr of operation. The effect of feed velocity was also significant, since for the 7-ft/sec tests the filtrate flux after 40 hr was about 100 gal/ft²/day higher than that for the 4-ft/sec tests. The tube diameter, on the other hand, was found to have little or no influence on flux. These findings are consistent with the description of the physics of the filtration process given above. The above results are important for design purposes, since the flux, feed velocity and tube diameters all have an influence on economics in terms of equipment costs and power requirements.

The flux record for the test which was performed with increasing feed concentrations is shown in Figure 9 (∿200 gal/ft²/day 160 hr). The tube type is the same as Tube I shown in Figure 4. However, the internal diameter here is 4 mm. The test conditions are given in Figure 9. Note from the figure that the test was operated in the constant-concentration mode for most of the 160 hours of testing; however, during four 3- to 7-hour intervals, the filtrate was removed from the feed. Thus, the concentration of suspended solids in the feed increased in four steps during the test. During the final part of the test, the total feed volume had been reduced to only 15% of its original value. The flux levels in the figure show that, even with this high feed-solids concentration, fluxes of around 200 gal/ft²/day can be achieved.

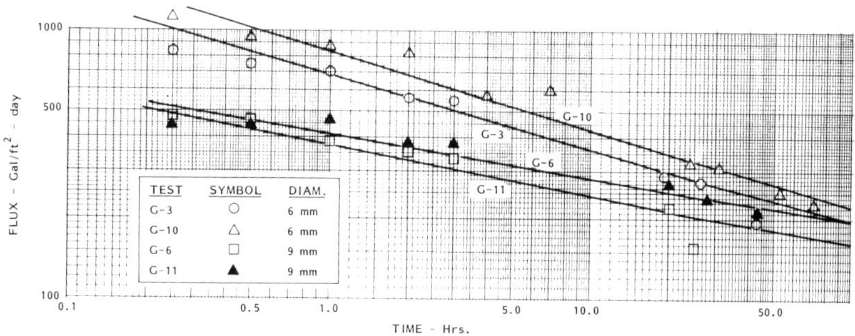

Figure 8. Flux vs time for identical operating conditions in the General Battery wastes.

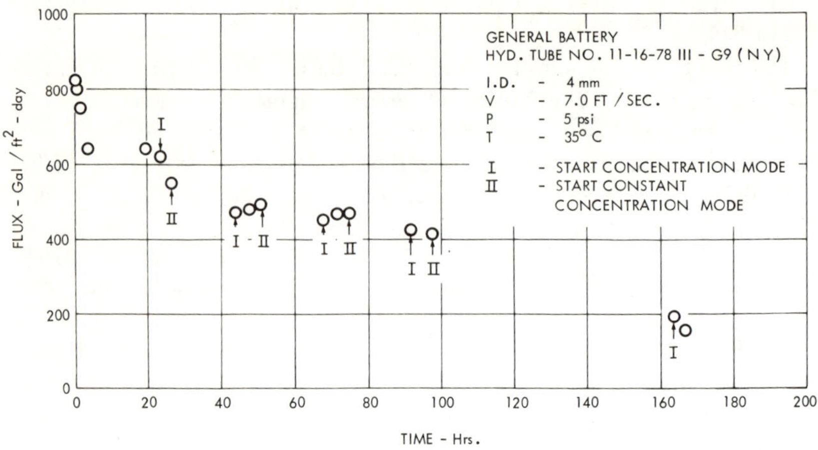

Figure 9. Permeate flux results.

Table III. Neutralized Battery Feed and Filtrate Analysis (mg/1)

	TS	SS	Pb	Cu	Zn	Ni	Sb	As
Feed	50,312	43,762	8.55	1.74	2.56	1.85	0.57	0.027
Filtrate	2,889	5	0.059	0.027	<0.001	0.0028	0.35	<0.002
% Removal	94.26	99.99	99.31	98.45	99.96	99.85	38.6	92.59

Having established typical flux rates and the dependence of these rates on certain operating parameters, one turns now to filtration performance in terms of filtrate quality. During the above tests, samples of the feed and filtrate were analyzed periodically by General Battery Corporation for TS, SS, Pb, Cu, Zn, Ni, Sb and As. Atomic absorption was the analytical method used. From these analyses it was found that the filtrate quality was independent of any of the operating conditions or tube types mentioned above. A typical comparison of the analysis of the feed and filtrate samples is shown in Table III. Note from the table that the suspended solids content of the feed of 43,762 mg/l was reduced to 5 mg/l in the filtrate, and that lead was reduced from 8.55 mg/l to 0.059 mg/l. In some cases, lead was removed to levels as low as <0.01 ppm in the filtrate. These excellent rejection percentages also hold for the other heavy metals tested, as shown in the table. However, it should be pointed out that, while Hydroperm achieves virtually complete removal of suspended solids, the quantity of total metals converted to SS form depends on definition of an optimum pH for precipitation. In this case, the optimum pH for lead removal is ∿8.5–9.5.

In addition to the wastewater treatment studies conducted during these single-tube tests, the sludge dewatering capability of the Hydroperm tubes was also investigated. The conditions for this particular test were: pressure of 15 psi, circulation velocity of 4 ft/sec; and the tube was 6-mm nylon. The initial waste volume in this test was 32 gal. The test proceeded with permeate being continually removed from the test reservoir. A steady solids concentration buildup (or dewatering process) was thus established. Figure 10 shows the flux decline and solids concentration buildup curves in this test. The initially high fluxes in the range of 700–800 gal/ft^2/day decreased to ∿100–200 gal/ft^2/day as the solids concentration increased to the range of 25–35% by weight at the end of this interval. This particular test demonstrated the dewatering capability of the Hydroperm tube.

Module Tests

In these tests, an optimum tube was selected for use in a multitube Hydroperm module in both long-term and concentration tests designed to simulate field operation; that is, during the latter stage of this test the filtrate was

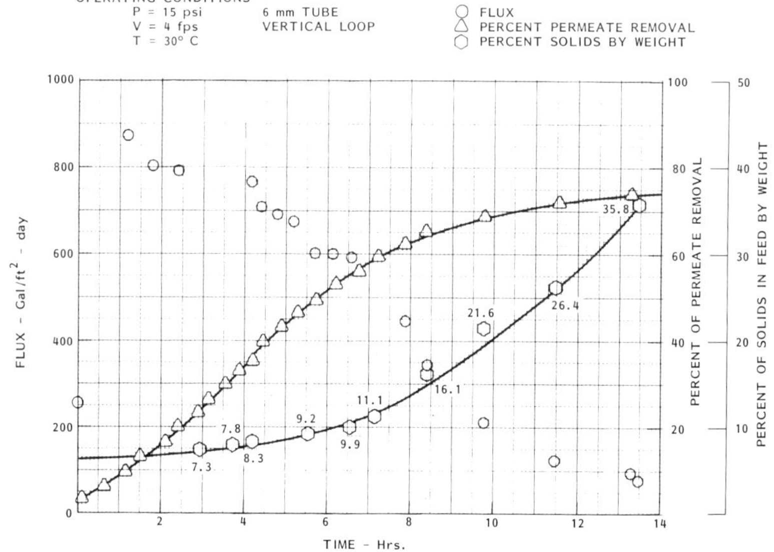

Figure 10. General Battery concentration test.

removed from the feed reservoir. From this test, information was obtained on the long-term filtrate flux and permeate quality of the tube, the degree of concentration of the feed that can be obtained and the ability to operate without clogging.

Two modules were constructed, one with 5 tubes and the other with 13 tubes. The tube selected for these studies was the 6-mm nylon tube shown in Curve I, Figure 4. The modules had an active tube filtration length of 6 ft. The operational conditions were: circulation velocity of 4 ft/sec; pressure of 15 psi; and pH of about 9.5.

The module tests began with the "continuous" mode of testing, where the field treatment process is simulated, i.e., as permeate was withdrawn through the Hydroperm module, comparable amounts of newly neutralized acid wastewater were added. Periodically, sludge was removed from the recirculation tank as would happen in the field operation. The long-term test results are shown in Figure 11.

As shown in the figure, the initial permeate flux was quite high, at about 600 gal/ft^2/day. Each time new feed is added to the recirculation tank, the feed solids concentration in the tank increases. This is reflected in the decline of the flux value. A restoration of the flux is possible only after the sludge in the recirculation tank is removed, and the total solids concentration in the feed is thus reduced. The final plateau value, after nearly

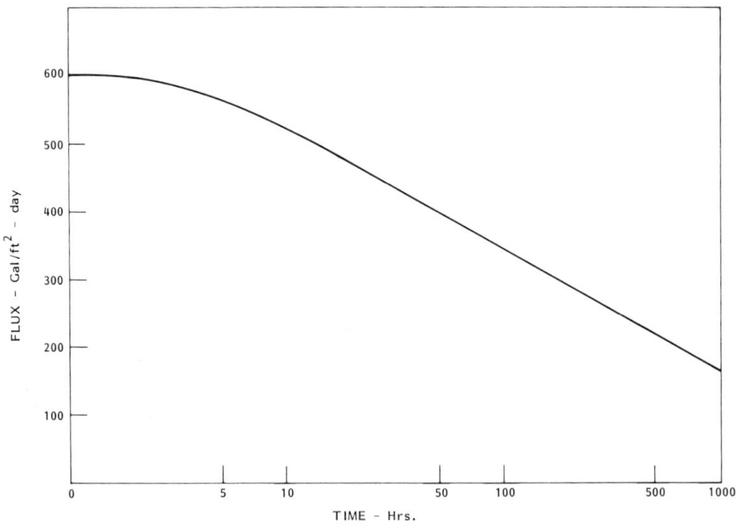

Figure 11. Long-term module test without cleaning.

1000 hours of continuous operation of the module, without cleaning, was still a respectably high 200 gal/ft^2/day.

The permeate quality of these module studies was just as good as in the single-tube tests. Some of the analytical laboratory analyses of the permeate quality are shown in Table IV, at a time when the module was in its 120th hour of operation.

Table IV. Neutralized Battery Feed and Filtrate Analysis (mg/l)

	TS	SS	Pb	Cu	Zn	Ni	Sb	As
Feed	24,058	21,054	3.23	0.25	1.16	0.88	0.818	0.054
Filtrate	2,989	9	<0.01	0.01	0.008	0.058	0.328	<0.002
% Removal	87.58	99.96	99.69	96.00	99.31	93.41	59.90	96.30

PHASE II–FIELD DEMONSTRATION PROGRAM

The successful Phase I laboratory study using Hydroperm tubes and modules in treating the battery manufacturing waste water was followed by the design, construction and final field testing and demonstration of the Hydroperm system with on-line waste water at the Hamburg plant site. A brief description of the system is presented first, and the field test results follow.

The basic element of the Hydroperm cross-flow microfiltration system is, of course, the tubular filter element. The design of the total system first involves assembling large numbers of these tubes into modular form in a manner which is economical of capital and operating costs. The modules make up the basic building blocks of any system (Figure 12).

A number of steps are required in total system design, starting with tube optimization and proceeding to system conponent selection and sizing. The Hydroperm tubes are assembled into modules of optimal length, number of tubes contained, type of end fitting used, etc. The arrangement of the modules can also be arranged in series, in parallel or in combination. The criteria used for the last two optimization steps are ease of handling, ease of maintenance, power requirements and capital and operating costs. The standard Hydroperm module contains ∿250 tubes of 6-mm diameter, with a total filtration surface area of 100 ft^2 (Figure 12).

The modules themselves make up only one part of the "total system," which includes pumps, valves, gauges, feed holding tank, sludge-thickening tank, permeate holding tank and the module cleaning system. The standard, full-size Hydroperm system includes two modules in series with a 250-

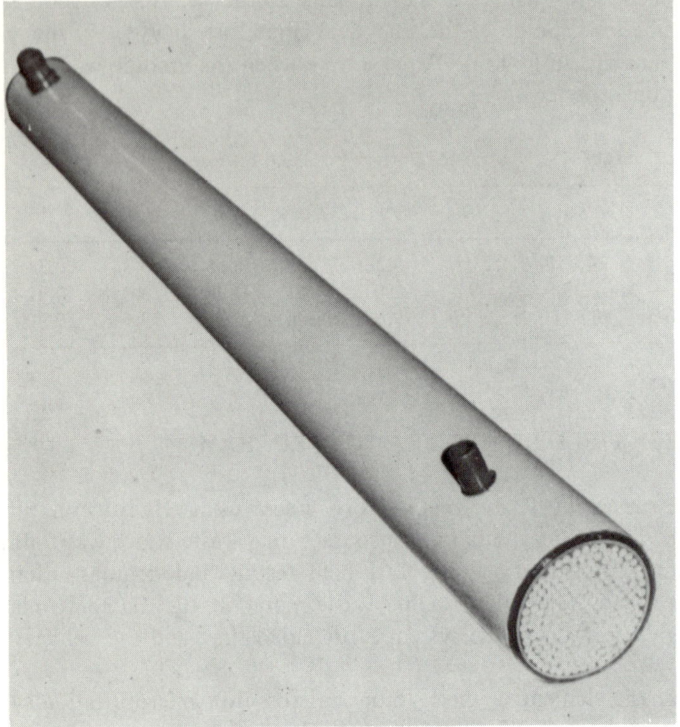

Figure 12. Hydroperm module.

gal/min, 7.5 hp pump, as shown in Figure 13. The total system, as installed in the General Battery Hamburg plant is shown in Figure 14. The entire system includes:

1. a 16,000-gal reaction tank, in which lime is stirred into the acidic waste water to bring the pH up to a range of $\sim$8–9;
2. a 10,000-gal cone, which acts both as a sludge-thickener and a recirculation tank for the Hydroperm modules;
3. a 2000-gal permeate holding tank, from which the permeate can be discharged to the sewer or to a local stream (reuse of the permeate is presently undergoing evaluation); and
4. the two Hydroperm units, each including two modules in series.

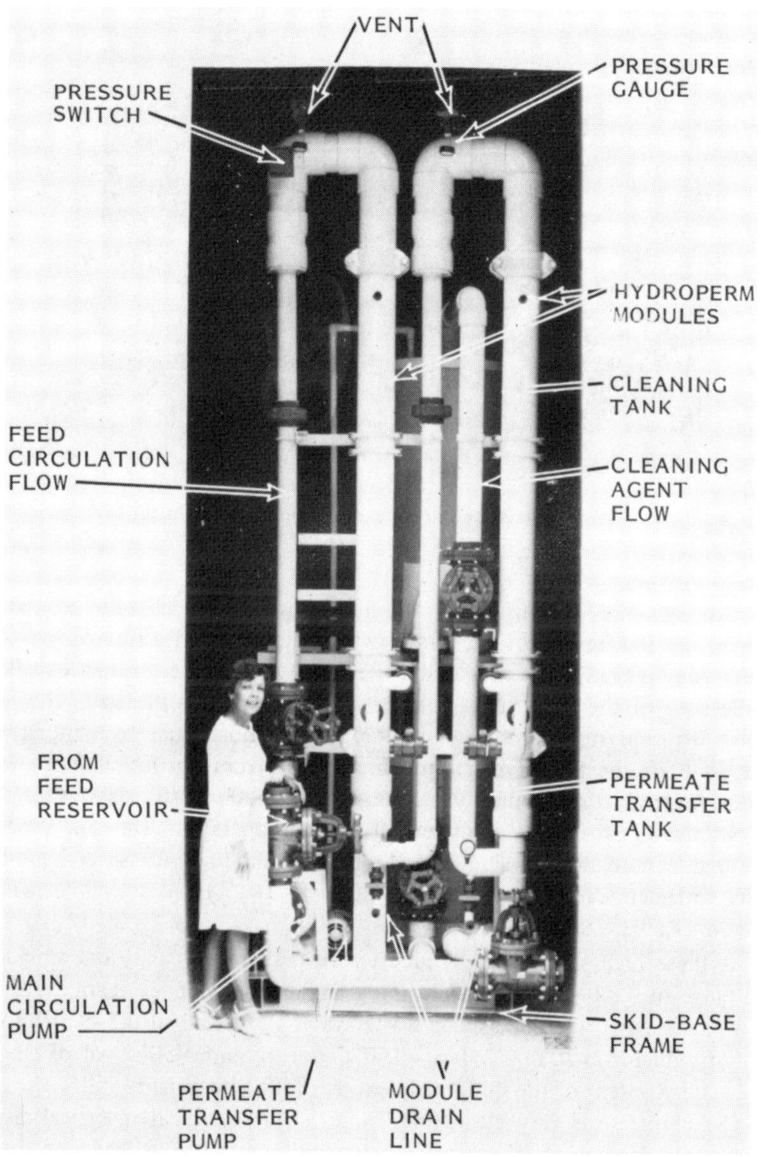

Figure 13. Hydroperm two-module wastewater treatment unit.

Figure 14. Hydroperm system at General Battery plant.

Although the demonstration program now underway will last for 12 weeks, the test results of the first 3 weeks indicate that a flux of over 400 gal/ft^2/day is readily attainable and the quality of the permeate is excellent.

Because of the high fluxes and consequent rapid dewatering of the feed, daily filtration runs have been limited to no more than five hours with only one of the two Hydroperm units, before running out of input feed. Initial flux values are typically in the range of >800 gal/ft^2/day leveling off to ∿400–500 gal/ft^2/day after several hours of operation. The flux is readily restored to the initial value by cleaning the modules for several minutes daily with permeate. Cleaning with dilute HCl (∿2%) will probably be practiced only on a weekly basis.

At these high flux levels, production of permeate by only one unit is in the range of 3500–5000 gal/hr during a filtration run of several hours. Typically, a load of ∿25,000 gal of waste water is filtered in 4.5–5 hr by one unit, producing ∿20,000 gal of clear permeate and ∿5000 gal of sludge. The solids content of the sludge is in excess of 30% by weight.

The permeate is very clear, and Table V shows typical analytical results from an independent laboratory on the pollutants in the feed and in the filtrate. In general, lead contents of the feed are at levels up to 600 ppm,

Table V. Neutralized Feed and Filtrate Analyses, March 4, 1980 (mg/l)

	TS	SS	Pb	Cu	Zn	Ni	Sb	As
Feed	27,260	24,600	54.2	0.330	0.880	0.400	1.350	0.172
Filtrate	2980.6	0.6	<0.04	0.015	0.007	0.032	0.292	<0.002
% Removal	89.10	99.99	99.93	95.45	99.20	92.00	78.37	98.84

and are reduced to a range of $\sim$0.02-0.09 ppm in the filtrate, while the suspended solids are reduced to less than 3 ppm from feed levels ranging up to 180,000 ppm. A summary plot of 40 analytical results during the three weeks of testing is shown in Figure 15.

Study of Figure 15 shows that in 25 of the 40 tests, Pb values were below 0.1 ppm in the permeate; and in 34 of the 40 cases SS were at 3 ppm or less. In all cases, SS values were less than 15 ppm. Maximization of the precipitation of lead depends highly on achievement of an optimum pH. In those cases where lead values were higher than the median range of $\sim$0.04-0.09 ppm, the results can be attributed to pH values greater than 9.5 or less than 8.0.

ECONOMIC FACTORS

Hydroperm represents new, advanced wastewater treatment technology, and no final economic analysis has been made of the cost of a Hydroperm system specifically for application in general wastewater treatment. Such an analysis would depend on such factors as feed characteristics, permeate quality requirements, module configuration, actual flux rates, operating conditions and cleaning requirements.

However, a general preliminary economic study indicates the magnitude of capital equipment costs (Table VI). These costs do not include the land, building foundation or special holding tanks. These costs also do not include any consulting-engineering services. Costs do include pumps, motors, piping, pressure regulators, valving, level controls, steel work, labor and the Hydroperm modules themselves. The costs presented in Tables VI and VII are based on the first-generation, "Model-100" Hydroperm system, shown in Figure 11. New improved design studies currently being conducted should result in a substantial reduction in capital costs. The Model "100" Hydroperm system is a 25- to 50- gal/min system (depending on flux). Costs have not yet been calculated for smaller systems.

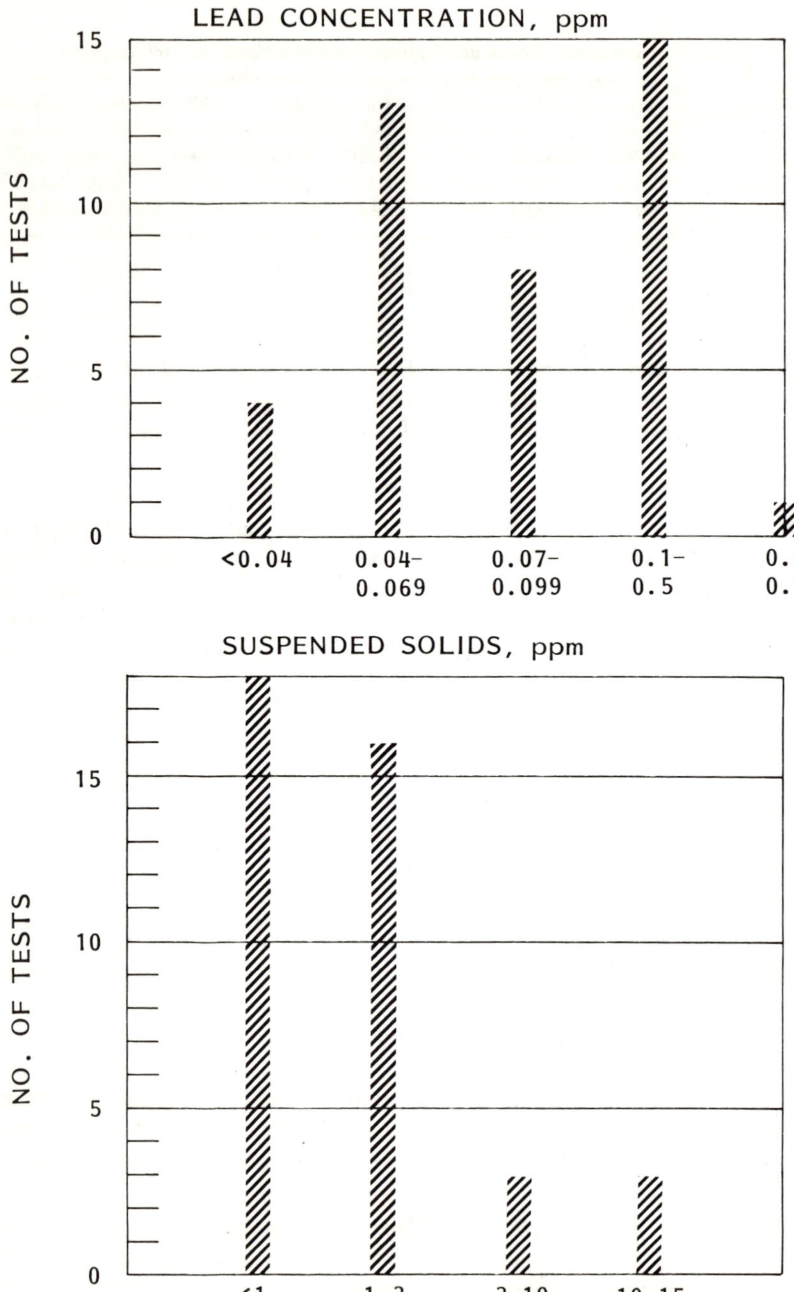

Figure 15. Analytical test results: Hydroperm wastewater treatment system, General Battery demonstration program.

Table VI. Hydroperm Model-100 Capital Cost (1979 $/gal) as a Function of
Flux and Daily Wastewater Treatment Volume

Flux (gal/ft^2/day)	Cost		
	A 50,000-gal/day	B 100,000-gal/day	C 500,000-gal/day
150	1.32	1.25	1.00
200	1.32	0.96	0.69
250	0.70	0.66	0.59

Table VII. Hydroperm Operating Costs ($/1000 gal permeate)

Permeate Production (gal/day)	Operating Costs for a Flux (gal/ft^2/day) of:		
	150	200	250
50,000	1.23	1.23	0.64
100,000	1.19	0.91	0.62
500,000	0.98	0.68	0.58

CONCLUSIONS

This chapter has presented results from laboratory tests and a full-scale field demonstration of the Hydroperm cross-flow microfiltration system with a battery manufacturing waste water.

The results of the tests showed that the Hydroperm system achieves virtually complete removal of suspended solids (the filtrate water had a suspended solids concentration of <5 ppm) regardless of the feed concentration.

The toxic heavy metals, specifically lead, were reduced to a range of $\sim$0.02–0.09 ppm, well below the effluent standards set forth by the state of Pennsylvania for discharge to natural waters (i.e. less than 0.15 ppm). Flux levels of >400 gal/ft^2/day can easily be achieved.

Both from a flux and a filtrate quality standpoint, the test and demonstration program indicates that the Hydroperm system is eminently effective for treating battery manufacturing wastewater for discharge. Reuse of the permeate is now under study.

When compared with other systems, Hydroperm offers several unique advantages:

1. Compactness: Hydroperm filtration systems, unlike biological and chemical systems, do not require large areas. Indeed, they can be engineered to fit available space. Because of its modular construction, the Hydroperm system can be designed to accommodate a full range of wastewater treatment requirements, from small to large.

2. Ruggedness: Since they are made from inert thermoplastics, the performance of Hydroperm tubes does not depend, in general, on changes in influent pH. Moreover, due to their rugged structure and low operating pressure, Hydroperm modules are not subject to the fouling and leaking problems which have plagued some membrane systems.

3. Ease of Maintenance: Because of their ruggedness and modular construction, Hydroperm systems are easy to maintain. They can be engineered in such a way that a failure in a single module causes only a small part of the total system to be shut down.

4. Performance: High flux levels are readily achieved, along with excellent permeate quality, with suspended solids reduced to <5 ppm. Toxic heavy metals in suspended form are removed to levels <0.1 ppm. Sludges in excess of 30% by weight are achieved.

5. Water Reuse: The treated effluent has such high quality that reuse of water will probably be feasible in some of the processes in the daily plant operations.

REFERENCES

1. Henry, J. D., Jr. "Cross Flow Filtration," in *Recent Developments in Separation Science, Vol. 2* (Cleveland, OH: CRC Press, 1972), pp. 205-225.
2. Duncan, J. D., T. R. Sundaram, D. H. Fruman and J. E. Santo. "A Unique Microfiltration System for Treating Industrial Effluents," *Prog. Water Technol.* 8(2/3):181-189 (1976).
3. Mezey, E. H. "Characterization of Priority Pollutants from a Secondary Lead and Battery Manufacturing Facility," U. S. EPA Report 600/2-79-039 (January 1979).
4. Culp, Wesner & Culp. "Water Reuse and Recycling—Evaluation of Treatment Technology," U. S. Department of the Interior, Office of Water Research and Technology, OWRT/RU-79/2 (April 1979).

BIBLIOGRAPHY

Patterson, J. W. *Wastewater Treatment Technology* (Ann Arbor, MI: Ann Arbor Science Publishers, Inc., 1978).

Santo, J. E., J. H. Duncan, H.-L. Liu and N. I. Shapira. "Regeneration of Wastewaters for Reuse Through Hydroperm^{TM} Microfiltration," paper presented at the American Water Works Association Water Reuse Symposium, Washington, DC, March 25-30, 1979.

Santo, J. E., J. H. Duncan, N. I. Shapira, C. Darvin and J. Baranski. "Removal of Heavy Metals from Battery Manufacturing Wastewaters by Hydroperm^{TM} Cross-Flow Microfiltration," paper presented at the AES/EPA Second Conference on Advanced Pollution Control for the Metal Finishing Industry, Kissimmee, FL, February 1979 (EPA 600-8-79-014, June 1979).

Sundaram, T. R., and J. E. Santo. "Microfiltration of Military Waste Effluents," in *Seventh Annual Symposium on Environmental Research—Meeting Report*, J. A. Brown, Ed. (Washington, DC: American Defense Preparedness Association, 1976).

Sundaram, T. R., and J. E. Santo. "The Development of a Hydroperm^{TM} Microfiltration System for the Treatment of Domestic Wastewater Effluents," Technical Report 7658-1, HYDRONAUTICS, Inc., Laurel, MD (1977).

Sundaram, T. R., and J. E. Santo. "Removal of Turbidity from Natural Streams by the Use of Microfiltration," Technical Report 7662-1, HYDRONAUTICS, Inc., Laurel, MD (1978).

Sundaram, T. R., J. E. Santo and N. I. Shapira. "An In-Depth, Cross-Flow Separation Technique for the Removal of Suspended Solids from Wastewaters," *Ind. Water Eng.* (January/February 1978).

CHAPTER 14

COUPLED TRANSPORT MEMBRANES FOR REMOVAL OF CHROMIUM FROM ELECTROPLATING RINSE SOLUTIONS

K. L. Smith, W. C. Babcock,
R. W. Baker and M. G. Conrod
Bend Research, Inc.
Bend, Oregon

In recent years, increasing attention has been placed on removal of toxic materials from industrial waste waters, not only for pollution control, but also to enable reuse of the water and, in some cases, the toxic materials themselves. Most of the waste waters result from cooling and rinsing operations, and many of these waste waters are contaminated with heavy metals. Conventional treatment of these waters is generally expensive, and the metals removed are usually not recovered. Electroplating rinse waters present particularly difficult problems, because of the effect of toxic heavy metal ions on sewage treatment plant operation and because these waters are produced from many thousands of plating shops throughout the country.

These waste streams constitute an important pollution problem that requires expensive treatment methods. They also represent a major reuse opportunity. The discharged water has an inherent value of $0.25–1.25/ 1000 gal, and the metal values involved are typically even greater. Recovery of these values awaits the development of cost-effective, low-energy-requirement technology. Coupled transport is a new, liquid-membrane process that is well suited to this application. The process can be used to separate a waste stream into a purified and reusable stream, while at the same time producing a second stream containing the concentrated metal ions for subsequent recovery. This chapter describes basic studies on the coupled transport of chromium.

DESCRIPTION OF COUPLED TRANSPORT

The coupled transport process relies on a liquid, water-immiscible organic complexing agent held by capillary forces within the pores of a microporous membrane, producing a supported liquid membrane. Such a membrane can transport metal ions from one aqueous solution against a steep concentration gradient to a second solution by coupling the transport of the metal ion to the transport of a second ionic species down its concentration gradient, maintaining electrical neutrality. Recent studies with coupled transport membranes have been focused on hydrometallurgical solutions of copper [1-4], uranium [5-7] and chromium [4,8].

In coupled transport of chromium, a tertiary amine (R_3N) can be used as the complexing agent to extract dichromate ions and hydrogen ions from an acidic solution:

$$2H^+_{(aq)} + Cr_2O_7^{2-}_{(aq)} + 2R_3N_{(org)} \rightleftharpoons (R_3NH)_2Cr_2O_7_{(org)} \qquad (1)$$

where (aq) and (org) refer to species soluble only in the aqueous and organic phases, respectively. In a basic solution, there will be insufficient hydrogen ions to allow the forward reaction in Equation 1 to proceed, the chromium will dissociate from the amine, according to the reverse reaction in Equation 1, and the chromium will be stripped from the liquid membrane. The dependence of this reaction on pH can be used to transport chromium across a liquid membrane. This process is shown in Figure 1, where the coupled transport membrane separates an acidic feed solution, which represents a chromium plating rinse solution, from a basic product solution into which the chromium is concentrated. On the feed side of the membrane, the forward reaction in Equation 1 predominates. The resulting chromium-amine complex diffuses across the membrane to the product side, where the reverse reaction predominates, releasing the dichromate and hydrogen ions and reforming the free amine. The amine then diffuses back to the feed side of the membrane, and the cycle is repeated. The transport of dichromate ions across the membrane is thus coupled to the transport of hydrogen ions in the same direction. By maintaining a sufficient gradient of hydrogen ion concentration, dichromate ions can be transported against a steep concentration gradient, from a dilute feed solution to a concentrated product solution. Thus, "uphill" transport results in a rinse solution suffi-

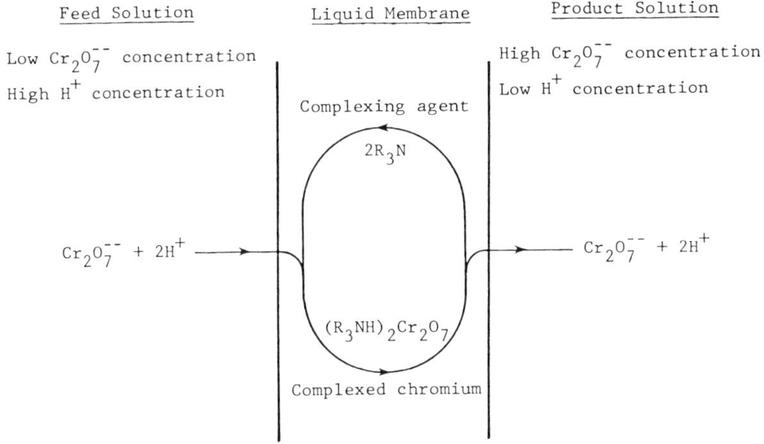

Figure 1. Coupled transport of dichromate ion across a liquid membrane containing a tertiary amine.

ciently free of chromium that it is suitable for reuse, and a concentrated product solution of Na_2CrO_4. In the basic product solution, dichromate ions are converted to sodium chromate via the reaction:

$$2H^+ + Cr_2O_7^{2-} + 4NaOH \rightleftharpoons 2Na_2CrO_4 + 3H_2O \qquad (2)$$

APPLICATION OF COUPLED TRANSPORT TO PLATING RINSE SOLUTIONS

Chromium plating has two major applications: decorative and industrial (hard) plating. Both types of chromium plating baths contain two essential constituents: chromic acid and a catalyst anion, typically sulfate. A common plating bath contains about 150 g/l chromium as CrO_3 and 3 g/l H_2SO_4. The rinse solution waste stream can vary widely, depending on the number and type of rinse tanks used. In acidic solutions, the chromium is primarily hexavalent, in the form of $H_2Cr_2O_7$ [9]. Chromium is extremely toxic: typical allowable sewer entry concentrations in the United States are 0.1 ppm or less [10]. Thus, wastewater disposal poses a substantial problem.

In a typical chromium plating operation, the metal article is dipped into the plating bath, and then successively dipped into several rinse baths prior to drying. The rinse baths are necessary to remove excess plating solution that can cause staining of the surface; this occurs at chromium concentrations in excess of about 40 ppm [11]. A typical plating operation is shown sche-

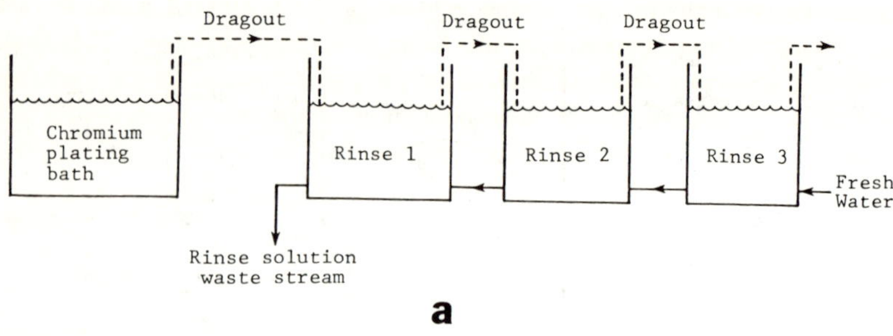

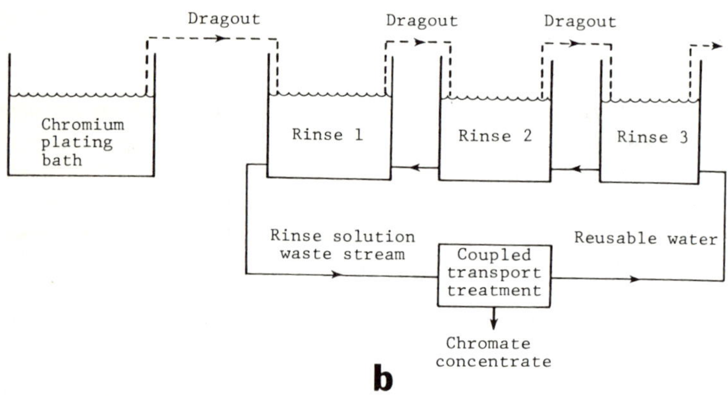

Figure 2. Typical chromium plating operation without coupled transport treatment (a) and with coupled transport treatment (b).

matically in Figure 2a. As the metal article is transferred from one bath to the next, it carries with it some of the solution from the previous bath. This is referred to as dragout. Due to dragout, the chromium concentration in the rinse baths increases. To counteract this increasing concentration, fresh water is introduced into the last (most dilute) rinse bath, the overflow is routed to the previous rinse bath, and this counterflow of dilute solution continues until it exits from the first rinse bath. At this point, the solution, which contains considerable chromium, is disposed of directly or treated to reduce the chromium concentration in the waste stream. In most cases, the metal values are not recovered, and the volume of waste water is large.

Coupled transport membranes would be used to process these waste streams as shown schematically in Figure 2b. The primary functions of this treatment are to conserve water by removing the chromium from the waste stream, allowing water reuse, and to concentrate the chromium into a small, manageable volume, relative to the volume of the untreated waste stream. The chromium concentrate is in the form of sodium chromate. This could be sold or converted to chromic acid by passing it through an ion exchange column or by treatment with a second coupled transport process in which sodium ions are exchanged for hydrogen ions.

EXPERIMENTAL

Two types of experiments were performed in this study: liquid-liquid extractions (LLE) and membrane permeation experiments. Details of the experimental procedures are discussed below.

The tertiary amine complexing agent used in these studies was Alamine 336®, a commercial solvent extraction agent (Henkel Corp., Minneapolis, MN). This agent has the approximate formula $(C_{8-10}H_{17-21})_3N$, and is referred to as tricaprylyl amine. All aqueous solutions were prepared from reagent grade chemicals.

LLE experiments were performed to characterize the equilibrium properties of the complexing agent. The distribution of metal ions between aqueous and organic phases was determined as a function of pH by measuring the concentration of chromium in the aqueous phase at equilibrium. Operating conditions for coupled transport membranes could be selected based on the results of LLE.

The LLE were performed by shaking equal volumes of organic and aqueous solutions in a separatory funnel for 20 min at room temperature. The organic solution consisted of 10 vol % agent in Kermac 470B®, an inert kerosene diluent (Kerr McGee Petroleum Refining Co., Oklahoma City, OK). The aqueous phase initially contained 1.0 g/1 chromium as CrO_3. The pH was adjusted with H_2SO_4 and NaOH.

Membrane permeation studies were performed to determine the performance of coupled transport membranes in removing chromium from synthetic chromium plating rinse solutions. These membranes were formed by immersing a microporous polypropylene membrane, Celgard 2400® (Celanese Plastics Co., Greer, SC), in the organic solution. This membrane is approximately 25 μm thick and has a porosity of about 38% and a pore diameter of about 0.02 μm. The pores are immediately filled with the organic phase by capillary foces. The organic phase used in membrane experiments contained 30 vol % Alamine 336 in Aromatic 150®, an aromatic hydrocarbon diluent (Exxon Corp., Houston, TX).

For permeation measurements, the coupled transport membranes were placed between the two compartments of standard glass permeation cells described elsewhere [1]. One compartment was filled with 100 ml of a synthetic rinse (feed) solution, and the other compartment was filled with

100 ml of an appropriate product solution. The membrane surface area was 20 cm^2. All permeation experiments were performed at 30°C, and the aqueous solutions were stirred at 180 rpm. Other experimental conditions are listed in the figures. By measuring the change in metal concentration in either compartment, the metal flux across the membrane can be calculated:

$$\text{Flux} = \frac{(\text{change in concentration}) \times (\text{volume})}{(\text{time}) \times (\text{membrane area})} \tag{3}$$

Metal flux is expressed in $\mu g/cm^2$-min.

RESULTS AND DISCUSSION

The results of LLE experiments with Alamine 336 are presented in Figure 3, showing the percentage of chromium extracted by the organic phase as a function of the aqueous pH. Chromium is extracted in essentially the manner predicted by Equation 1. No chromium is extracted at pH higher than 7, and the extraction increases with decreasing pH until nearly complete extraction occurs at pH between 1 and 3.5. There is some evidence that at still lower pH, bisulfate ions can displace dichromate ions from the amine. The present studies were conducted at pH 1.3 or above, so this was not a problem. These extraction data indicate that nearly all of the chromium should be extracted from a feed solution at pH 1.3, and that all of the chromium should be stripped from the membrane into a basic product solution.

During LLE studies, a precipitate formed when Kermac 470B was used as the diluent. Therefore, Aromatic 150 was used in membrane studies, since aromatic liquids are reported to have better solvation properties for amines than do other diluents [12].

The chromium flux through the membrane depends on the concentration of agent in the liquid membrane, as shown in Figure 4. There are apparently two completing factors that cause the observed dependence on agent concentration. As the concentration of agent in the membrane is increased, the concentration gradient of the chromium-agent complex increases as well. At low agent concentrations, this results in a higher flux. However, the viscosity of the organic solution also increases with increasing agent concentration. Since the diffusion coefficient is inversely proportional to the solution viscosity, the increased viscosity results in a lower flux. It is postulated that the viscosity of the liquid membrane increases nonlinearly with the agent concentration. Apparently, at an agent concentration of 30 vol %, the flux increase due to the increasing concentration gradient is no longer sufficient to overcome the flux decrease due to the increasing viscosity. At this agent concentration, the flux reaches its maximum value

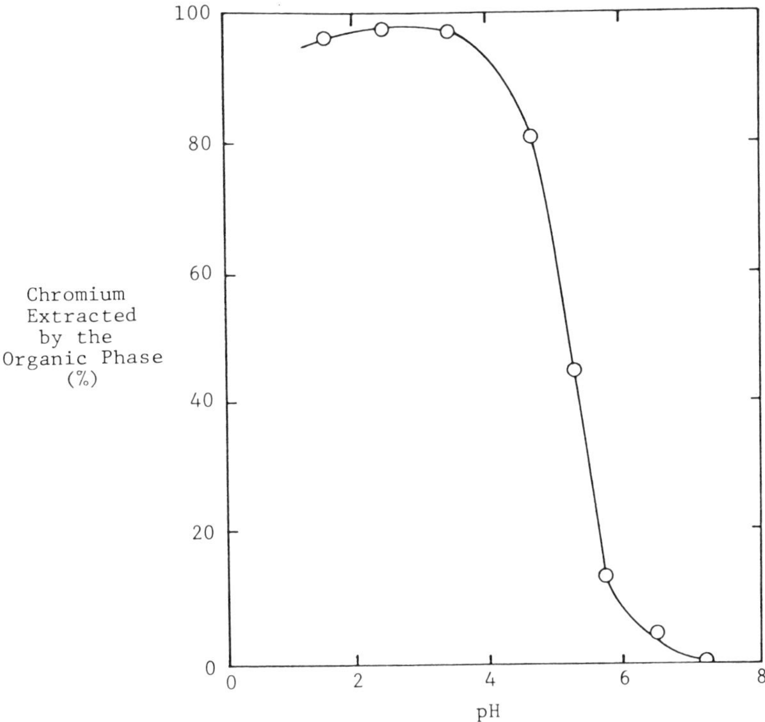

Figure 3. Chromium extracted by the organic phase as a function of pH. Organic phase: 10 vol % Alamine 336 in Kermac 470B. Aqueous phase: 1.0 g/l chromium as CrO_3, pH adjusted with H_2SO_4 and NaOH.

and subsequently decreases with further increases in agent concentration. All the experiments reported below were therefore carried out with a liquid membrane of this optimum composition.

Coupled transport of chromium against a concentration gradient was demonstrated in a membrane permeation experiment in which the feed solution contained 5 g/l chromium at pH 1.3 and the product solution contained 150 g/l chromium in 40 g/l NaOH. The concentration of chromium in the feed solution as a function of time is shown in Figure 5. Chromium was transported from the relatively dilute feed solution to the concentrated product solution until nearly all of the chromium had been removed from the feed solution. At the completion of the experiment, only 16 ppm Cr remained in the feed solution, with 155,000 ppm in the product solution. This represents a concentration factor of about 9700. The coupled transport flux of chromium across the membrane can also be calculated from the data in Figure 5. The initial chromium flux was 55 μg/cm^2-min, which

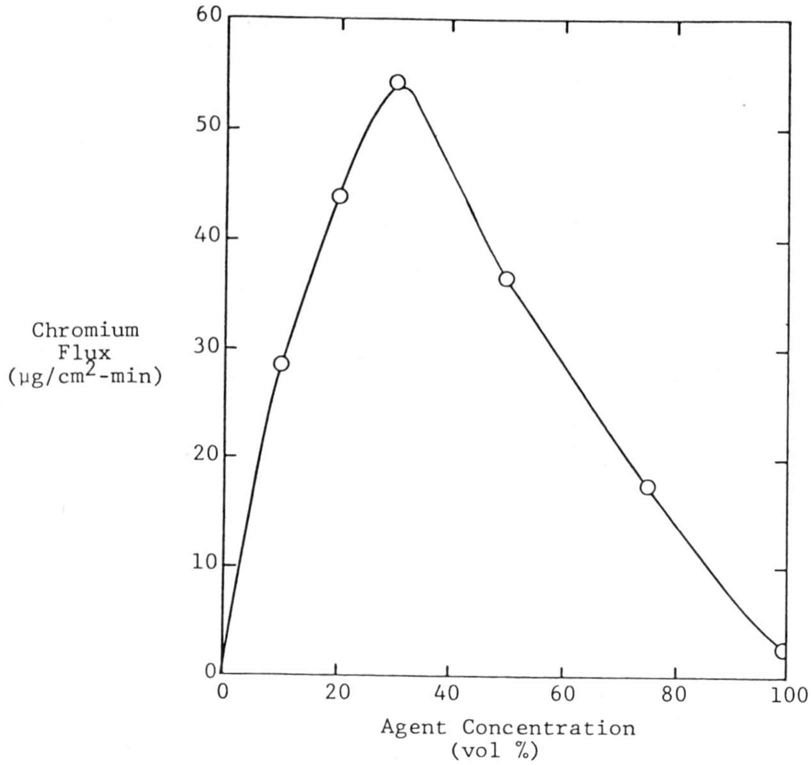

Figure 4. Chromium flux as a function of agent concentration in the membrane. Agent: Alamine 336 in Aromatic 150; feed solution: 2.0 g/l Cr (as CrO_3); product solution: 40 g/l NaOH.

is quite high compared with the coupled transport fluxes of metals such as copper, nickel, iron and cobalt (1–10 $\mu g/cm^2$-min) from acidic solutions [1,13].

Clearly, chromium can be transported against large concentration gradients with these membranes. In fact, it has been found that the chromium flux is nearly independent of the concentration of chromium in the product solution. Figure 6 shows the results of an experiment to demonstrate this. The chromium flux decreased only slightly, from nearly 60 to about 45 $\mu g/cm^2$-min, as the chromium concentration in the product solution was increased from 0 to nearly 200 g/l, the latter being a saturated solution of Na_2CrO_4. Because chromium is not extracted by the amine at basic pH (see Figure 3), the membrane transport of chromium was essentially unaffected by even the highest chromium concentration in the product solution. Thus, these coupled transport membranes demonstrate the re-

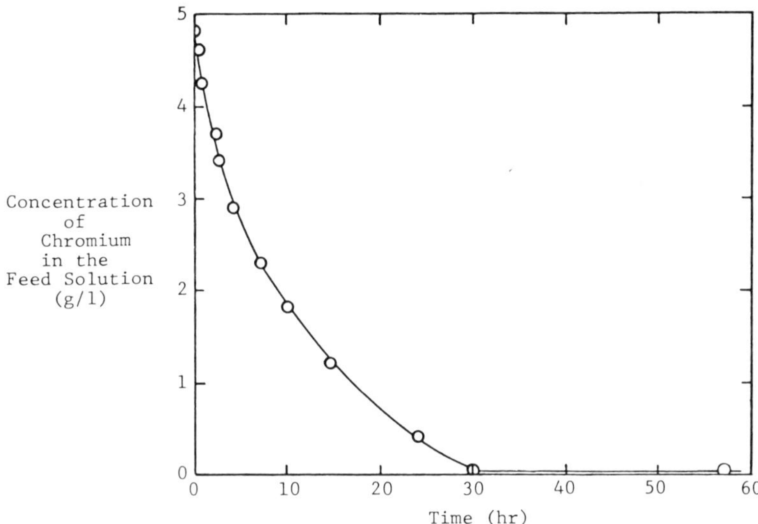

Figure 5. Concentration of chromium in the feed solution as a function of time. Agent: 30 vol % Alamine 336 in Aromatic 150; feed solution: 5 g/l Cr (as CrO_3), 0.1 g/l H_2SO_4; product solution: 150 g/l Cr (as Na_2CrO_4), 40 g/l NaOH.

quired features for rinsewater treatment: (1) the ability to remove chromium from the rinse solution, and (2) the ability to efficiently concentrate the chromium by many times.

One of the criteria for successful application of coupled transport membranes to wastewater treatment is the long-term performance of the membranes. There are many factors that could cause deterioration of performance such as agent degradation, polymer support degradation, macroscopic loss of agent and agent solubility loss. Although the final measure of long-term flux stability must be made on a system treating actual waste waters, some indication of performance can be obtained from preliminary laboratory permeation experiments.

The chromium flux through an Alamine 336 membrane was monitored for 110 days. Feed and product solutions were replaced periodically so that the membrane was always transporting chromium from the feed solution to the product solution. Each flux measurement was made using fresh feed and product solutions. Figure 7 shows the results of the long-term flux stability experiment. The flux rapidly declined from an initially high value of 54 to 12 μg/cm^2-min in 14 days, and then increased to a steady flux of about 30 μg/cm^2-min after 50 days. The reason for this type of behavior is not known; however, based on these results, Alamine 336 coupled transport membranes appear to maintain high chromium fluxes for longer than 100 days.

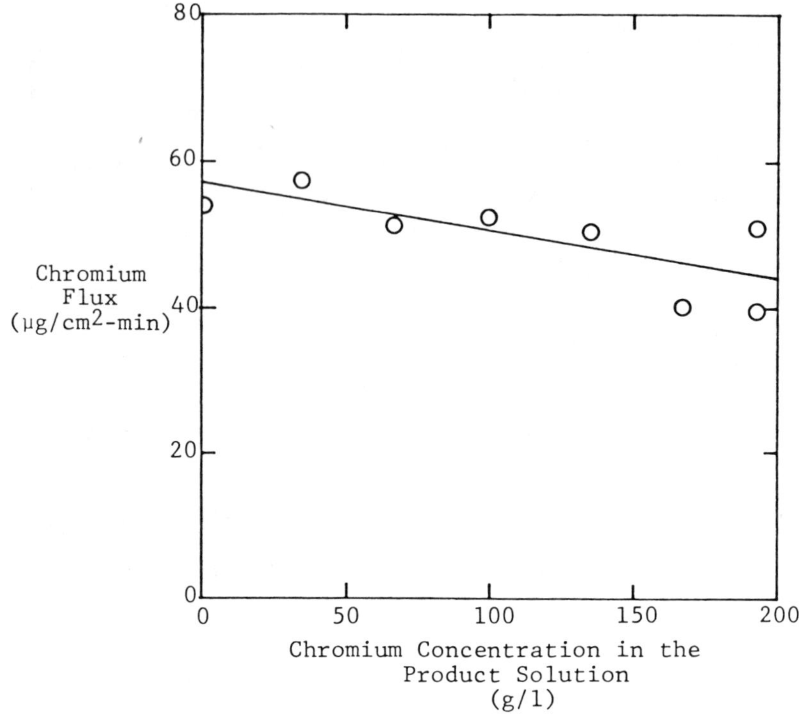

Figure 6. Chromium flux as a function of the chromium concentration in the product solution. Agent: 30 vol % Alamine 336 in Aromatic 150; feed solution: 2.0 g/l Cr (as CrO_3); product solution: Na_2CrO_4, 40 g/l NaOH.

ECONOMICS OF COUPLED TRANSPORT PROCESSING

An accurate evaluation of the economics of coupled transport processing will require a long-term test with a suitably sized pilot plant, operating under plating shop conditions. However, an approximate analysis is possible based on these laboratory results and on similarities between coupled transport operation and operation of reverse osmosis and ion-exchange units in current use for treatment of plating rinse solutions. The analysis assumes 5000 hours of operation per year, and treatment of 4000 lb of chromium per year. The analysis is analogous to and borrows from economic analyses reported for ion-exchange and evaporation treatments of chromium plating rinse solutions and reverse osmosis treatment of nickel plating rinse solutions [14].

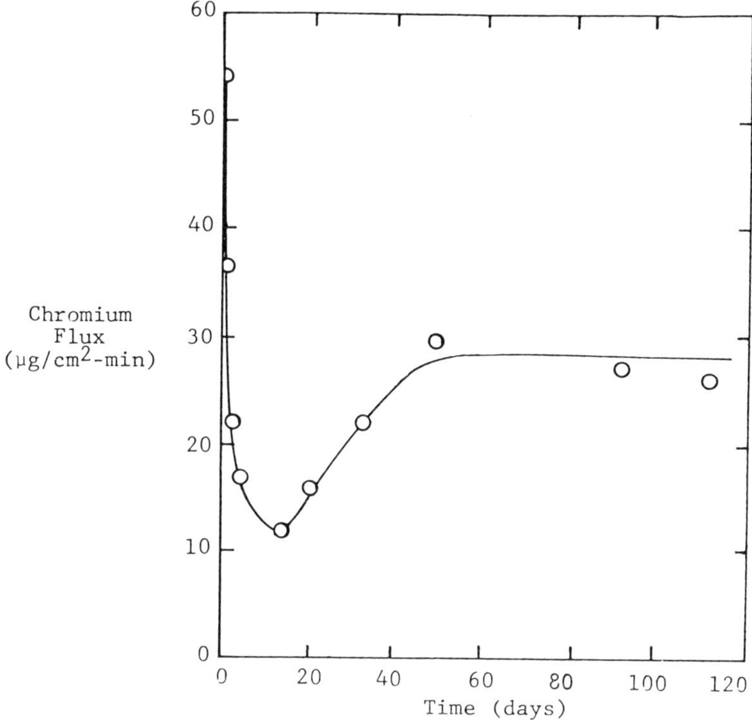

Figure 7. Chromium flux as a function of time. Agent: 30 vol % Alamine 336 in Aromatic 150; feed solution: 2.0 g/1 Cr (as CrO_3); product solution: 40 g/1 NaOH.

The economics of coupled transport processing are summarized in Table I. The coupled transport process operates at low pressures and should not be subject to the membrane fouling that is a problem with reverse osmosis. The capital equipment costs should therefore be less than in reverse osmosis, but a conservative estimate is $30/ft^2 of membrane as the cost of the unit, which is comparable to reverse osmosis costs [14]. A conservative chromium flux of 20 μg/cm^2-min has been assumed, although fluxes more than twice as high were demonstrated in laboratory studies.

The support equipment includes a cation-exchange unit for recovery of chromic acid from the sodium chromate product from coupled transport. Annual operating costs were borrowed from the reverse osmosis analysis [14], including module replacement based on a two-year lifetime. The annual savings were identical to those estimated for ion-exchange processing, resulting

Table I. Economics of Chromic Acid Recovery by Coupled Transport Processing (5,000 hr/yr, 4,000 lb Cr/yr)

Capital Costs	
Coupled Transport Unit (400 ft^2 at \$30/ft^2)	\$ 12,000
Support Equipment	9,200
Installation, Labor and Material	1,500
Total Capital Costs	22,700
Annual Operating Costs	
Labor, Maintenance and Utilities	3,500
Module Replacement, 2-yr life (\$2,000/2)	1,000
Total Operating Costs	4,500
Annual Fixed Costs	
Depreciation, 10% of Capital Costs	2,270
Taxes and Insurance, 1% of Capital Costs	230
Total Fixed Costs	2,500
Total Cost of Operation	7,000
Annual Savings	
Plating Chemicals, 2 lb/hr H_2CrO_4	7,020
Waste Treatment and Disposal	7,200
Water Use, 18 gal/hr at \$1.10/1000 gal	100
Total Annual Savings	14,320
Net Savings (14,320 - 7,000)	7,320
Net Savings after Taxes, 48% tax rate	3,510
Cash Flow From Investment (3,510 + 2,270)	5,780
Payback Period (22,700/5, 780)	3.9 years

mostly from savings in chromic acid purchases and in waste treatment and disposal costs. The payback period for recovery of capital costs is 3.9 years. Payback periods for ion exchange and evaporation treatments are 5.2 years and 7.9 years, respectively. Reverse osmosis treatment of nickel plating rinse solutions has a payback period of 4.3 years, but current reverse osmosis membranes cannot be used with chromic acid solutions.

CONCLUSIONS

It has been demonstrated that coupled transport membranes can be used to concentrate chromium against steep concentration gradients, from very dilute acidic solutions to saturated solutions of sodium chromate. The flux of chromium through these membranes is high, and can be maintained at a high level for longer than 100 days. The economics of coupled transport appear to be competitive with other treatment processes.

ACKNOWLEDGMENTS

This work was supported by the Office of Water Research and Technology, Department of the Interior, under OWRT Contract No. 14-34-0001-8802.

REFERENCES

1. Baker, R. W., M. E. Tuttle, D. J. Kelly and H. K. Lonsdale. "Coupled Transport Membranes I: Copper Separations," *J. Membrane Sci.* 2:213 (1977).
2. Largman, T., and S. Sifniades. "Recovery of Copper (II) from Aqueous Solutions by Means of Supported Liquid Membranes," *Hydrometallurgy* 3:153 (1978).
3. Lee, K. H., D. F. Evans and E. L. Cussler. "Selective Copper Recovery with Two Types of Liquid Membranes," *Am. Inst. Chem. Eng. J.* 24:860 (1978).
4. Frankenfeld, J. W., and N. N. Li. "Waste Water Treatment by Liquid Ion Exchange Liquid Membrane Systems," in *Recent Developments in Separation Science, Vol. III (B)*, N. N. Li, Ed. (Cleveland, OH: CRC Press, Inc., 1977), p. 285.
5. Bloch, R. "Hydrometallurgical Separations by Solvent Membranes," in *Membrane Science and Technology*, J. E. Flinn, Ed. (New York: Plenum Press, 1970), p. 171.
6. Babcock, W. C., R. W. Baker, E. D. LaChapelle and K. L. Smith. "Coupled Transport Membranes II: The Mechanism of Uranium Transport with a Tertiary Amine," *J. Membrane Sci.* 7:71 (1980).
7. Babcock, W. C., R. W. Baker, E. D. LaChapelle and K. L. Smith. "Coupled Transport Membranes III: The Rate-Limiting Step in Uranium Transport with a Tertiary Amine," *J. Membrane Sci.* 7:89 (1980).
8. Hochhauser, A. M., and E. L. Cussler. "Concentrating Chromium with Liquid Surfactant Membranes," *Am. Inst. Chem. Eng. Ser.* 71(152):136 (1975).
9. Dubpernell, G. "Chromium," in *Modern Electroplating*, 3rd ed., F. A. Lowenheim, Ed. (New York: John Wiley & Sons, Inc., 1974), p. 94.
10. "Development Document for Proposed Existing Source Pretreatment Standards for the Electroplating Point Source Category," U.S. Environmental Protection Agency Report EPA 440/1-78/085 (1978).
11. Kushner, J. B. *Water and Waste Control for the Plating Shop* (Cincinnati, OH: Gardner Publications, Inc., 1976), p. 38.

12.   "Diluents for Metal Extraction," Shell Chemical Co. Technical Bulletin INT:77:M5 (1977).
13.   Babcock, W. C., M. G. Conrod, D. J. Kelly and E. D. LaChapelle, Bend Research, Inc., Bend, OR. Unpublished results, (1976-1979).
14.   "Environmental Pollution Control Alternatives: Economics of Wastewater Treatment Alternatives for the Electroplating Industry," U.S. Environmental Protection Agency Report EPA/625/5-79/016 (1979), pp. 47-62.

CHAPTER 15

HYPERFILTRATION OF TEXTILE PROCESS WATER FOR REUSE

H. Garth Spencer

Clemson University
Clemson, South Carolina

Craig A. Brandon

CARRE, Inc.
Seneca, South Carolina

Grant Goodman and John J. Porter

Texidyne, Inc.
Clemson, South Carolina

Max Samfield

U.S. Environmental Protection Agency
Research Triangle Park, North Carolina

Many processes in industry require large quantities of water at elevated temperatures [1]. In some processes the amount of auxiliary nonconsumable materials exceeds that of the consumable components. Thus, recycle of the water and process materials is motivated by the potentially significant reduction in freshwater demand and waste treatment requirement and the conservation of energy and materials.

In textile dyeing the water is principally a carrier of the process materials used in their application to and removal from the fabric. Since application and removal are essentially controlled by concentration (activity) differences, at least a partial separation of process effluent into solute components and water is required in most reuse applications.

High-temperature hyperfiltration provides a method for renovating the hot process effluents for reuse. Experience using membranes to separate textile process effluents (feed to the membrane unit) into a diluted stream (product) and a concentrated stream (concentrate) and in the evaluation of both streams has been obtained in laboratory and pilot studies through a series of government-sponsored grants with industrial and university researchers. Reuse has also been evaluated in a limited plant-scale beck dyeing demonstration [1-4]. Although recycle evaluation requires a plant-scale separation and reuse demonstration, the studies have indicated high potential for hyperfiltration as a suitable separation procedure in the recycle of industrial process effluents.

A project is in progress to demonstrate the closed-cycle operation of a production dye range. The results will determine the practicality of the hyperfiltration recycle procedure. The contract with La France involves the U.S. Environmental Protection Agency (EPA), the Department of Energy (DOE) and the Department of the Interior. The objective of the project is to design, install and operate a full-scale commercially available hyperfiltration system at La France and to demonstrate its practicality. Phase I, the design phase, has been completed [1]. The second phase of the project, equipment installation, is in progress. Associated with the installation is the continuation of the pilot- and laboratory-scale investigation to optimize membrane performance and evaluate product and concentrate reuse and/or disposal procedures. The technical aspects of this investigation are reported here.

SUMMARY OF PREVIOUS STUDIES

Three previous studies provide experience leading to the work reported in this chapter. The first study involved the pilot-scale separation of composite waste water from a beck dyeing process and full-scale reuse of the hyperfiltration products and concentrates. Polyamide (hollow-fine fibers), cellulose acetate (spiral and tubular) and hydrous Zr(IV) oxide-polyacrylate membranes were used. A total of 18 production dyeings involving 1348 m of cloth were carried out in a two-piece dye beck.

The purified product water was a satisfactory substitute for normal tap water in all production dyeings for water recoveries ranging from 75 to 90%. Membranes used in the renovation of the waste water had conductivity rejections of 65–95% and color rejections from 86 to greater than 99%.

It is also technically feasible to reuse all the concentrates. In 11 production dyeings over 700 m of cotton velour fabric was produced, graded as first quality and sold commerically. In 10 tests standard shades were produced with an average dyestuff savings of 16% [2].

In the second study [3] composite wastewater obtained from the several processes occurring in a dyeing and finishing plant were separated by hyperfiltration and the cumulative product and concentrate of a 90% recovery run tested for reuse as process water in laboratory dyeing. Precast and dynamically formed membranes were used in eight dyeing and finishing plants. The processes encountered were: dyeing of nylon using premetallized dyes, dyeing of acrylic fabric using basic dyes, and the scouring, desizing and dyeing of cotton and polyester. In all cases the product water was acceptable for replacement of fresh water as determined in a laboratory dyeing using standard production evaluations. Analyses indicated higher chemical oxygen demand (COD) and dissolved solids (DS), and lower concentrations of metals in the product water than in the fresh plant-process water. The concentrate from the premetallized dye process was suitable for dyeing very deep shades on addition of appropriate dyes. Laboratory dyeing tests using concentrates from the other processes were unsuccessful. Perhaps this result is not surprising because the concentrates were obtained from a feed containing a composite of effluents from the plant processes and not from a single process.

The first two studies dealt with renovation of the composite waste water from dyeing and finishing plants. Because of the obvious advantages for reclamation and energy conservation, a third study [4] evaluated hyperfiltration for direct recycle of unit process effluents. The five major water- and energy-consuming preparation and dyeing processes were studied with high-temperature hyperfiltration membranes.

The product water produced by the membranes was again found to be universally useable. In some cases the reuse of the concentrate from the individual process effluent streams was estimated to be practical.

In Phase I of the current study [1], three hyperfiltration membranes were evaluated in a pilot unit under test conditions consistent with the membrane manufacturer recommendations and the anticipated operating conditions at La France Industries. The technical feasibility for using hyperfiltration in closed-cycle operation of a dye range involves the performance of the membrane equipment and the reuse of the product (95% of the range effluent volume) and the concentrate (the remaining 5% of the effluent volume). The products were judged suitable for recycle as process water. The reuse and/or disposal of the concentrates from a multiple-process range requires detailed studies of each type of process and the compatibility of the concentrate with subsequent processes. These studies will be concluded in later phases of the project.

RANGE OPERATION

Most of the production at La France Industries is done on the dye range. It consists of a dye pad, spiral atmospheric steamer and the washing section shown schematically in Figure 1. The washing section contains in sequence a jet washer, a dip box and two rotojet washers; followed by a final wash/fix box. Fixing chemicals are not included in the counterflow washing system. The range can operate at speeds of 9–36 m/min selected as needed for the fabric and process. The dyeing or bleaching is done on the range as a steady-state process. Cotton, acrylic, nylon, rayon and polyester fabrics and their blends are processed. Several classes of dyes are used; direct, disperse, acid, basic premetallized and reactive. The wash water effluent is variable. However, there are some common elements in all the wash water effluent from dyeing operations. The dye bath contains dyes, a thickener and, in some cases, dye solvents. While about 85% of the dyes are exhausted on the fabric, most of the auxiliary components are removed by the wash water.

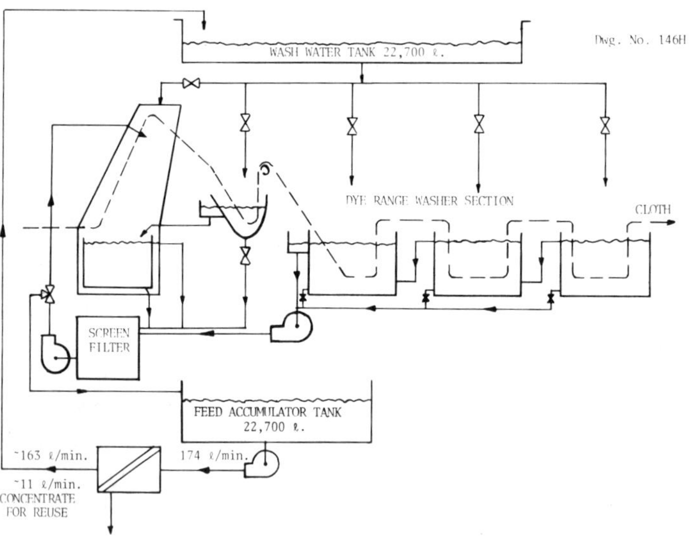

Figure 1. Continuous dye range modified arrangement.

RECOVERY CONCEPT

The recovery system is presented schematically in Figure 2. Discharge from the range occurs in two forms; the wash water discharged continuously during the dyeing run and the dye drop from the dye pad following each run. There is a period between each run needed for cleaning and reloading the dye pad and/or rinsing out the washing train. This permits each dye run to be treated in the hyperfiltration system as a batch containing the components washed from a single dye run. The dye drop is of known composition and its reuse in a subsequent dye pad preparation or its disposal can be planned. The wash water effluent from the run constitutes the feed for the pilot unit and eventually the full-scale hyperfiltration unit. Most feeds are highly colored, and the dyes must be removed to avoid possible staining during the reuse of the product as wash water. The auxiliary components must also be removed to provide wash water with the concentration differences suitable for effective removal of these components from the fabric during the washing step.

The concentrate contains the dyes at concentrations much lower than those in the dye pad solution, but concentrations of the auxiliary components are comparable. Reuse of the concentrates as a dye pad solution depends on the ability to add dyes and auxiliary components to develop a needed production shade that possesses the necessary shade, hue and crocking characteristics. Generally this reconstitution must be accomplished by trial and error in the dye laboratory. Even beginning with pure materials each dye bath requires an adjustment based on the routine tests accomplished in the dye laboratory. Eventually, optimum reuse of concentrate may involve judicious scheduling of dye runs and the use of guidelines based on the properties of dye classes. Some sequences are expected to be favorable for reuse, e.g., those using the same dye class and those of increasingly darker shades in the same general hue. Other sequences are expected to be unfavorable for reuse resulting in disposal of the concentrate. Although, some crossovers between dye classes are feasible, it could be difficult for example, to cross from a basic dye to an acid dye. Crossovers between dye classes will generally require addition of auxiliary components not in the original concentrate.

MEMBRANE PERFORMANCE

Figure 3 is a schematic diagram of the high pressure test unit. The primary mechanical components of this skid-mounted hyperfiltration unit are:

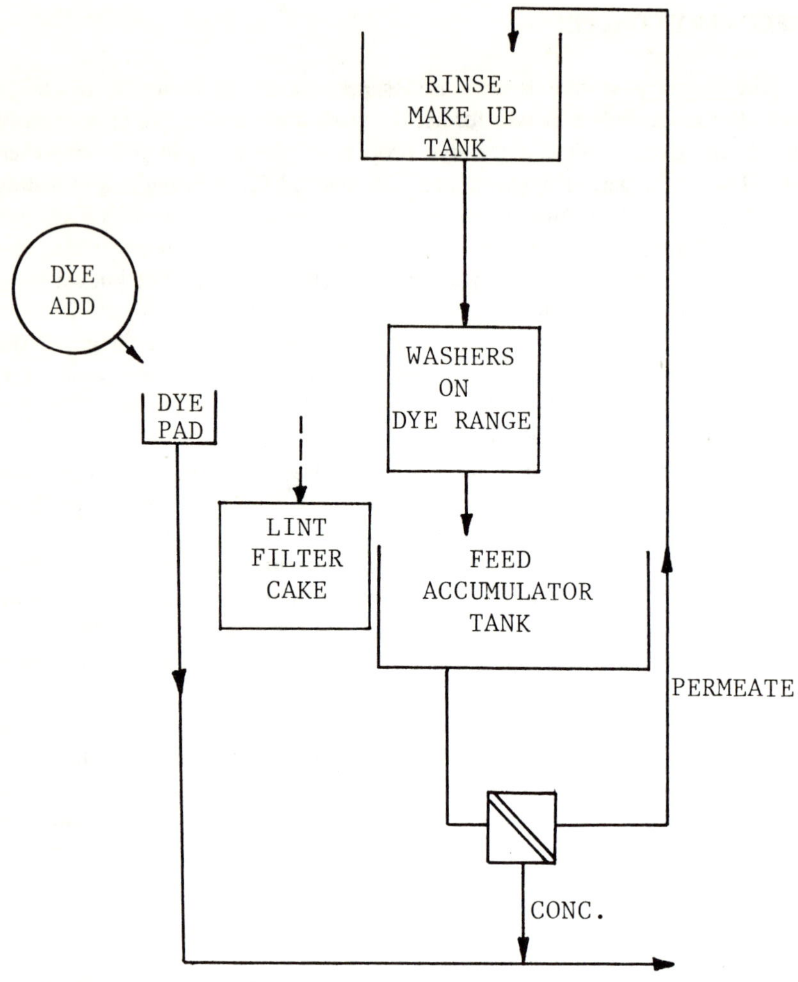

Figure 2. Conceptual diagram of the recovery system.

1. feed reservoir,
2. prepressurizing pump,
3. high-pressure pump
4. heat exchanger,
5. hyperfiltration module and
6. pressure control valve.

The feed reservoir is a 100-liter stainless steel tank. Wastewater flow into the tank is regulated by a liquid level control valve. A centrifugal pump

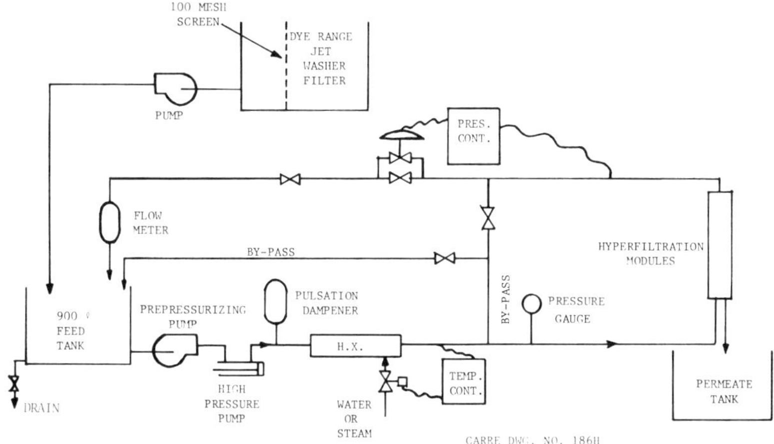

Figure 3. High-pressure skid-mounted unit.

maintains a supply pressure to the suction of the positive displacement pump. The heat exchanger is for control of the fluid temperature in this recirculating system.

Modules of various sizes can be tested in this unit. Since the positive displacement pump provides a constant volume of flow, a manual bypass valve is used to control the flow to a module. An air operated control valve maintains a set pressure in the system.

Conductivity measurements were made with a conductivity bridge with a dip cell, color measurements were made with a Bausch and Lomb Spectronic 20, and later a Spectronic 70, spectrophotometer at a single wavelength, 460 nm. Flux measurements were made by measuring volumes obtained in a specific time.

Figure 4 provides information describing the performance of a dynamically formed zirconium oxide poly(acrylic acid) (ZOPA) membrane module, a Mott-Brandon Corporation module. The feed is range wash water effluent from single dye runs. Flux and rejections at initial (0% recovery) and final (95% recovery) are provided. Generally, the initial flux is high, 50–120 gal/ft^2/day, decreasing during the concentration to values usually in the 40- to 70-gal/ft^2/day range, at 800 psi and 85°C. The flux obtained using a 2-g/l salt solution after washing was normally higher than the initial flux obtained using wash water feed. The color rejection is of great concern. It is expected that product suitable for washing will be obtained if the color rejection is 0.97 for the cumulative product of a 95% recovery hyperfiltration. The color rejections are generally 0.99 at 95% recovery, with low values occurring at 0% recovery only for some basic/direct dye mixtures used on acrylic fabric. Conductivity rejections normally fall in the 0.70–0.85 range.

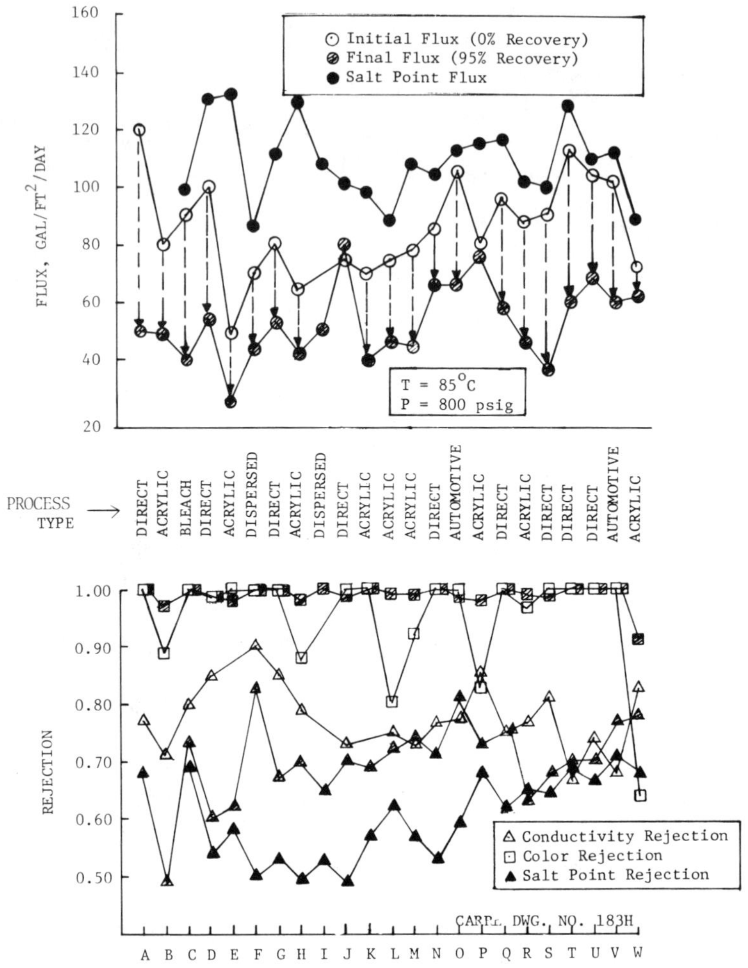

Figure 4. Membrane performance summary data for Module 453.

PRODUCT REUSE

Product samples were generated by hyperfiltration of range wash water effluents to 95% recovery using the pilot test with ZOPA membrane modules. The samples were evaluated in the laboratory for reuse as wash water by: (1) simulated rinsing using product water on La France greige fabric samples, (2) chemical analyses, and (3) foam tests.

The suitability of the product samples for reuse was determined primarily from the results of rinsing tests. This test consisted of stirring the four 3- x 3-in. fabric samples in 400 ml of product water for 10 min at 60°C. The fabric patterns used were: (F-1) 100% cotton, (F-2) 65% cotton, 35% rayon, (F-3) 41% cotton, 59% acrylic and (F-4) 68% cotton, 15% rayon, 17% polyester. The dried test fabric samples were evaluated for staining on an AATCC chromatic transference scale against untreated fabric samples. The scale assigns 5.0 to practically unstained fabric and 1.0 to the fabric most heavily stained. A stain index (SI) will be used in the paper, calculated by subtracting AATCC scale reading from 5.0 to provide an index which increases from zero as the extent of staining increases.

Chemical analyses were conducted on the initial feed, final concentrate and cumulative product obtained from a 95% recovery hyperfiltration. Analytical parameters included COD, total solids (TS), suspended solids (SS), volatile solids (VS) and pH. The absorbance at 460 nm and the conductivity of the feed and the concentrate and product at the final recovery were also determined. Color rejections refer to rejections at 95% recovery and not the cumulative samples.

Selected samples were evaluated for foaming. A 50-ml sample was added to a 100-ml graduated cylinder, and the cylinder was stoppered and agitated vigorously. The foam volume five minutes after agitation was recorded and interpreted as a qualitative indication of the presence of foaming agent or surfactant. Generally, the higher the concentration of surfactant the greater the foam volume. However, the relationship is not linear and cannot be interpreted as a quantitative estimate of surfactant concentration.

Some results from these tests are presented in Table I. Only one sample exhibits a color rejection below the expected acceptable value of 0.97. However, if a stain index of 1.0 is arbitrarily set as an acceptable level six samples fail to meet this criterion: mostly very dark shades of basic/direct formulation. Generally, the color rejection of the basic/direct formulation rinse water is less than those of the direct formulation rinse water, and the staining index is less than those of the direct formulation rinse water, and the staining index is greater when compared at the same product absorbance. The staining index, of course, varies with the fabric composition. The direct formulation wastewater products stain fabrics high in cotton content more readily than the samples with a high acrylic content. The relative staining of these two classes of fabrics by the basic/direct formulation wastewater products depends on the weight ratio of the basic-to-direct dye content in the dye bath (Figures 5 and 6).

Staining is not the only possible criterion for qualifying product water for reuse in washing. The concentration of the auxiliary components should also be low. The ratio of COD, DS and VS in the cumulative 95% recovery

Table I. Reuse Sample Characteristics

Product Sample Number	Type of Dye Formulation	Absorbance, Feed	Absorbance, Product[a]	COD	DS	VS	Stain Index[b]	Color Rejection[a]
3196	Bleach	0.06	0.00	354	920	467	0.0	1.00
3127	Bleach	0.20	0.00	596	1380	744	0.0	1.00
3175	Direct	0.03	0.00	71	104	50	0.0	1.00
3178	Direct	0.06	0.00	122	154	75	0.0	1.00
3157	Direct	0.23	0.00	141	270	114	0.0	1.00
3166	Direct	0.24	0.00	158	273	201	0.0	1.00
3139	Direct	0.79	0.02	157	373	169	0.0	0.99
3148	Direct	0.79	0.04	29	101	40	0.0	0.99
3205	Direct	0.92	0.08	88	500	105	0.4	0.99
3130	Direct	1.50	0.22	140	454	195	0.6	0.99
3172	Direct	4.00	0.22[c]	94	448	169	0.4	0.95[c]
3211	Direct	7.70	0.84	404	1340	357	2.7	0.99
3181	Acid/Direct	0.15	0.01	193	148	113	0.0	0.99
3163	Acid/Direct	1.90	0.22	167	116	65	0.0	0.98
3160	Basic/Direct	0.07	0.01	1030	291	123	1.1	0.99
3193	Basic/Direct	0.16	0.01	1020	307	207	0.0	1.00
3169	Basic/Direct	0.38	0.03	1430	363	220	1.2	0.99
3133	Basic/Direct	0.39	0.10	834	376	239	0.0	0.98
3154	Basic/Direct	0.74	0.08	730	249	148	0.6	0.99
3151	Basic/Direct	0.85	0.13	807	310	175	1.1	0.99
3184	Basic/Direct	1.10	0.34	981	547	264	1.6	0.96
3142	Basic/Direct	2.00	0.35	845	243	136	2.2	0.98
3145	Disperse	1.10	0.00	148	42	26	0.0	1.00
3136	Disperse	1.40	0.27	128	172	111	0.6	0.98
	(Plant Water)			9	57	18		

[a] At 95% recovery.
[b] Average for the four fabrics.
[c] At 89.4% recovery.

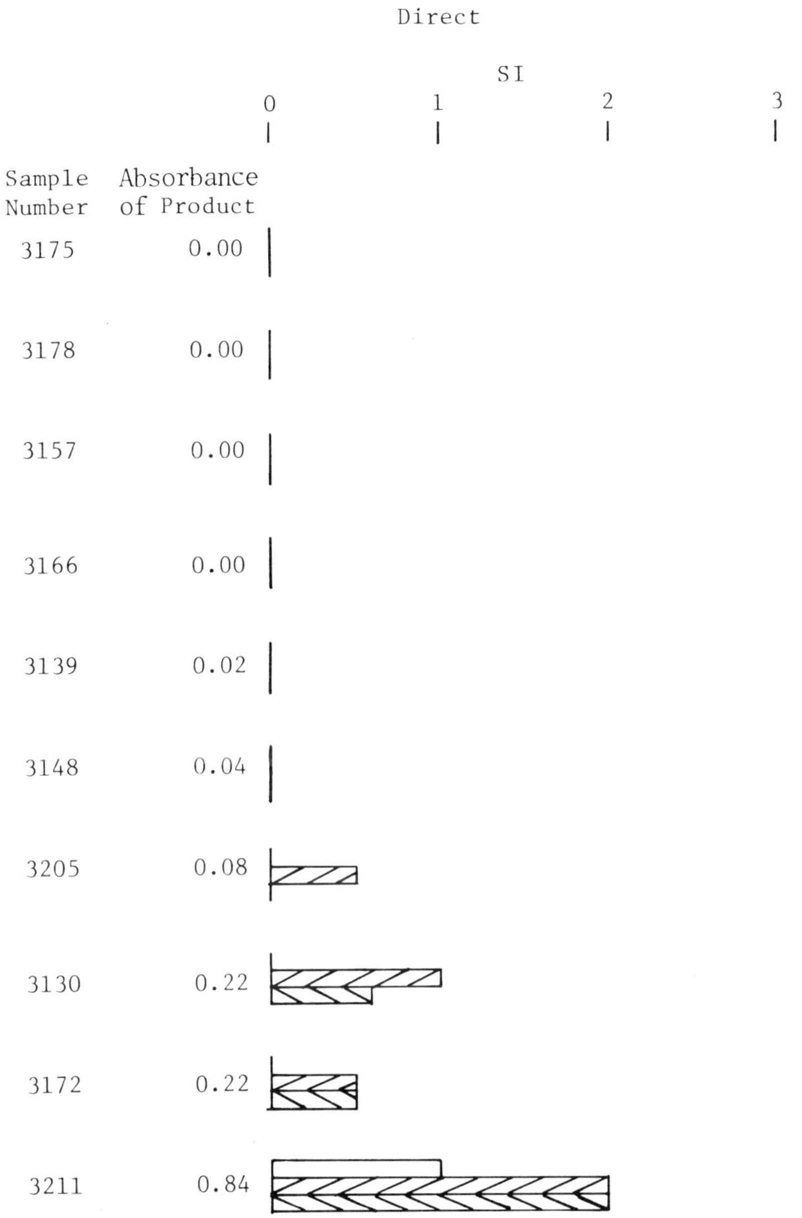

Figure 5. Staining index on fabrics by hyperfiltration products from direct dye range effluents: ☐ = F–3; ▨ = F–1 & F–2; ◪ = F–4.

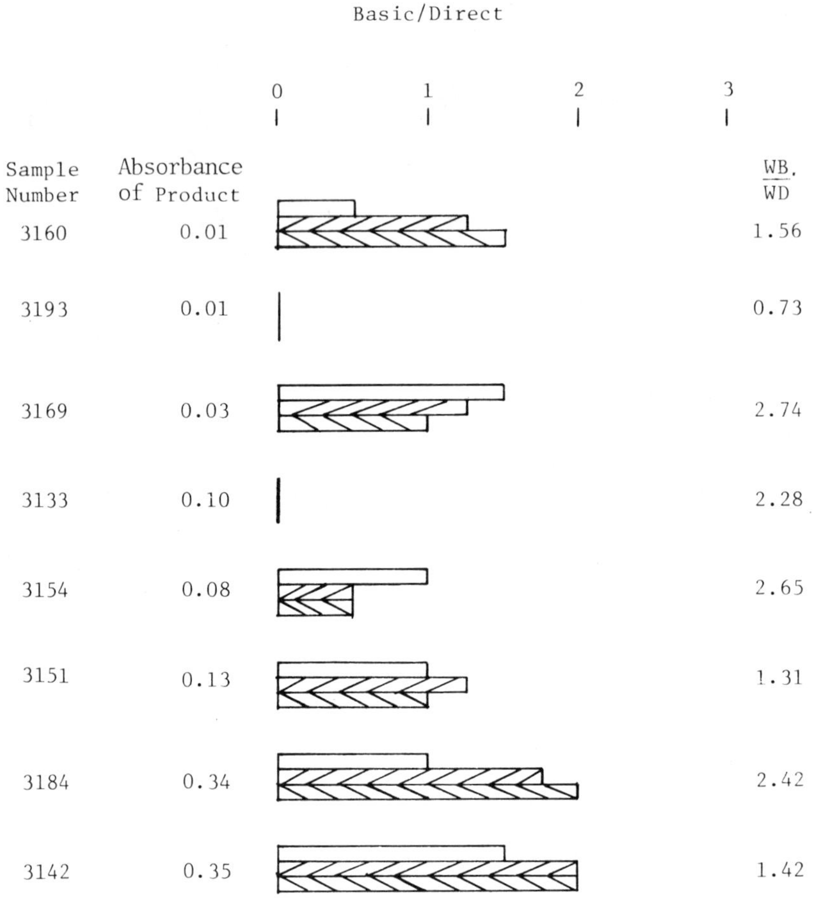

Basic/Direct

Sample Number	Absorbance of Product		$\dfrac{\text{WB.}}{\text{WD}}$
3160	0.01		1.56
3193	0.01		0.73
3169	0.03		2.74
3133	0.10		2.28
3154	0.08		2.65
3151	0.13		1.31
3184	0.34		2.42
3142	0.35		1.42

Figure 6. Staining index on fabrics by hyperfiltration products from basic/direct dye range effluents: ☐ , F-3; ▨ , F-1 and F-2 ▧ , F-4. WB/WD is ratio of weight of basic and direct dyes in range dye pad.

product to their values in the corresponding feed is given in Table II. The $\overline{P}/F$ ratio for COD, DS, and VS is less than 0.3 for direct formulations. It is greater for COD than from DS and VS for basic/direct and acid/direct where a low-molecular-weight (volatile) dye solvent is used in the formulation, which would contribute to COD but not to DS or VS. The ratio is less than 0.1 for the disperse dye formulations.

Table II. Ratio of Chemical Oxygen Demand, Dissolved Solids, Volatile Solids
and Foam Volume for the Cumulative Product to Feed

Product Sample Number	Type Due Formulation	$\overline{P}/F$			
		COD	DS	VS	Foam Volume
3175	Direct	0.23	0.33	0.23	
3178	Direct	0.26	0.36	0.26	
3157	Direct	0.21	0.27	0.21	
3166	Direct	0.20	0.23	0.32	
3139	Direct	0.22	0.31	0.25	0.71
3148	Direct	0.15	0.37	0.17	0.2
3199	Direct	0.08	0.38	0.22	
3130	Direct	0.15	0.29	0.28	0.2
3172	Direct	0.09	0.34	0.23	
3211	Direct	0.40	0.87	0.50	
3181	Acid/Direct	0.66	0.42	0.48	
3163	Acid/Direct	0.46	0.35	0.29	
3160	Basic/Direct	0.47	0.24	0.21	
3193	Basic/Direct	0.52	0.38	0.33	
3169	Basic/Direct	0.33	0.18	0.14	
3133	Basic/Direct	0.33	0.27	0.22	0.45
3154	Basic/Direct	0.49	0.25	0.20	0.3
3151	Basic/Direct	0.49	0.31	0.25	0.3
3184	Basic/Direct	0.46	0.29	0.30	
3142	Basic/Direct	0.33	0.16	0.13	0.6
3145	Disperse	0.07	0.07	0.04	0.7
3136	Disperse	0.09	0.09	0.10	0.0

CONCENTRATE REUSE

Laboratory tests are in progress to determine the reuse potential of the cumulative 95% recovery concentrates. The concentrate was provided as the starting solution to prepare a production dye bath, dye a test fabric on a pilot range and compare the product with production standards for hue, shade and crock. Three types of tests are involved:

1. reformulation of the original dye bath;
2. preparation of the dye bath of the same dye class but different hue and deeper shade; and
3. preparation of the dye bath in a different designated dye class with different hue and deeper shade.

Table III. Type-1 Reformulation Results

Dye Class	Weight of Dyes in Production Formula (g/1)	Weight of Dyes in Text Formulas; Designated Percent of Auxiliary Components			
		0%	25%	50%	75%
Direct	16.35	15.7	15.5	15.2	14.9
Basic/ Direct	0.82	0.87	0.85	0.83	0.84
Direct	5.26	5.72	4.58	4.35	4.58

The quantities of dyes and auxiliary components required to prepare the production dye bath are then compared to the standard production formulation and reported as reduced or excess quantities.

Some results are available for type-1 reformulation tests. A series of tests were run using direct, basic/direct and disperse formulations. In the initial tests 100% of the quantities of the auxiliary components prescribed in the production pad were first added to the concentrate, a test dyeing was performed, additional dyes were added to prepare the production hue and shade and the test fabric was compared to the production standard. The average dye reduction was 23% for direct, 10% for basic/direct and 0% for disperse dye classes. Based on the production levels 67% direct, 24% basic/direct and 1% disperse, and weighted dye reduction (savings) is approximately 19%.

This procedure involves no savings in auxiliary components.

A second set of type-1 reformulation tests are in progress involving the preparation of the test dye formulation using 0, 25, 50 and 75% of the specified auxiliary components. The results of a direct bath and a basic/direct bath are provided in Table III. The dye requirement was reduced by 5-9% for the direct dye, with about a 4% increase for the basic/direct dye. The production tests indicated a reduction of 50-75% of the auxiliary components gave satisfactory dyeings.

These preliminary tests indicate a modest savings in dye and at least a 50% savings in auxiliary components may be possible. Extensive research at this pilot/laboratory scale and full-scale reuse is required to determine the practicality of concentrate reuse.

BLEACHING PROCESS

Fabric bleaching in preparation for dyeing is also carried out in the range. This process occurs for several continuous hours each week, providing an opportunity for recycle for the same process. The bleach pad contains sodium hydroxide, a surfactant wetting agent and hydrogen peroxide. The wash water effluent contains dilute sodium hydroxide and wetting agent; the peroxide is barely detectable. Hyperfiltration to 95% recovery, after reducing the pH from approximately 11 to 9–10, reduced the COD and VS to about 30% of the feed, indicating good rejection of the wetting agent. The pH values of the product and concentrate remain high and the reduction of DS is about 40%. Although some sodium hydroxide is rejected, the product (the new rinse water) is now at a pH about two units higher than fresh plant water. Reuse of the concentrate was carried out using product water with only hydrogen peroxide additions of 0, 25, 50, 75 and 100% of the production formula. The addition of 100% of the production formula produced excellent bleaching. The 50 and 75% additions were also satisfactory.

CONCLUSIONS

The pilot study of renovation of waste water from a production dye range by high temperature hyperfiltration using dynamic zirconium oxide-poly(acrylic acid) membranes (Mott-Brandon) continues to indicate high potential for closed cycle operation by reuse of water and chemicals. The membrane stability, flux and rejection are satisfactory. Simple laboratory reuse tests indicate only a small percentage of the hyperfiltration products may be unsuitable for general reuse as wash water. Judicious planning of the sequence of dyeing processes should permit reconstitution of many hyperfiltration concentrates for reuse in the dye pad. Significant savings (50–70%) of auxiliary components and minor savings of dyes in the concentrate reuse are indicated by the laboratory tests. Reuse of the hyperfiltration products and concentrate obtained with bleach waste water appear suitable for reuse with savings of wetting agent in the pad formulation.

The payback period, calculated by the Riegel Textile Corporation's standard procedure, is 3.8 years with chemical recovery and 5.2 years without chemical recovery. The annual energy savings are about 2×10^{10} Btu/yr. The operating costs are estimated based on February 1979 prices for energy and materials.

ACKNOWLEDGMENTS

This study was conducted by a team, and major contributions were made by a number of people. The cooperation and assistance of the La France staff members is particularly acknowledged: Perry Lockridge and Charles Smith and several machine operators and laboratory technicians.

This demonstration is an interagency program and thus benefited from the guidance of Dr. Sarah Allen, Environmental Protection Agency, as Principal Project Officer; John Rossmeissl, Department of Energy, and Frank Coley, Department of the Interior, as Project Officers; and Mr. Robert Mournighan, Environmental Protection Agency, as Technical Advisor.

REFERENCES

1. Brandon, C. A. "Closed Cycle Textile Dyeing Hyperfiltration Demonstration Phase I Final Report," U.S. Environmental Protection Agency.
2. Brandon, C. A., and J. J. Porter. "Hyperfiltration for Renovation of Textile Finishing Plant Wastewater," U.S. Environmental Protection Agency, EPA-600/2-76-060 (March 1976).
3. Brandon, C. A., J. J. Porter and D. K. Todd. "Hyperfiltration for Renovation of Composite Wastewater at Eight Textile Finishing Plants," U.S. Environmental Protection Agency, EPA-600/2-78-047 (March 1978).
4. Gaddis, J. L., C. A. Brandon and J. J. Porter. "Energy Conservation Through Point Source Recycle with High Temperature Hyperfiltration," U.S. Environmental Protection Agency, EPA-600/7-79-131, (June 1979).

CHAPTER 16

REMOVAL OF CHEMICAL CARCINOGENS FROM
WATER/WASTEWATER BY REVERSE OSMOSIS

William G. Light

Technology Development Department
Walden Division of Abcor, Inc.
Wilmington, Massachusetts

The effectiveness of reverse osmosis (RO) for removing organic chemicals
from water primarily depends on successfully exploiting physicochemical
differences between the organic chemicals to be removed and water [1].
Mechanisms for RO membrane separations, such as the solution-diffusion
mechanism [2] and the preferential sorption-capillary flow mechanism [3],
have been derived which complement experimental observations [4-6] that,
in general, large nonpolar organics are readily removed by RO, while small
polar organics are poorly removed. As many carcinogens are high-molecular-
weight and/or nonpolar organic chemicals, it was reasoned that RO would be
an effective process for removing chemical carcinogens from water and waste
water.

The objective of the work presented here was to determine the feasibility
of using RO to remove suspect carcinogens from actual carcinogenesis laboratory and synthetic waste waters [7]. The technical approach for treating the
actual waste waters was based on the premise that a two-stage system, consisting of suspended solids (SS) removal followed by dissolved solutes removal
by RO, would be required for reliable performance of a wastewater treatment
system. To be an effective process for removing the many possible chemical

carcinogens from water and waste water, it was necessary to demonstrate the applicability of RO for treating a wide variety of chemicals rather than a small number of particular chemicals or groups of chemicals. Accordingly, RO tests were also performed using synthetic wastewaters which contained individual chemicals representative of ten groups of organic chemical carcinogens.

WASTE WATER SAMPLING AND CHARACTERIZATION

Three carcinogenesis laboratories were selected to provide representative wastewater samples. Contaminated waste streams from normal operations were identified and sampled, and the two most highly contaminated waste waters from each laboratory were processed in pilot-scale tests. Flow measurements of waste streams were made to calculate the hydraulic loading for the waste waters. Contaminant loading was determined from analysis for SS, total dissolved solids (TDS), total organic carbon (TOC), pH, conductivity and turbidity. To identify specific chemicals and their concentrations, gas chromatography/mass spectrometry (GC/MS) analyses were performed. Chemicals tested at the laboratories included those in the following groups: plasticizers, aromatic amines, isocyanates, amino acids, organophosphate pesticides, chlorinated hydrocarbons and gums.

On assessing the actual waste waters, it was found that carcinogenesis laboratories generate significant quantities of suspect carcinogen-contaminated waste water (at a flowrate on the order of 1100 liter/hr). Current practices in the laboratories result in approximately 5% of the total amount of suspect carcinogens used becoming a component of waste water. For the three laboratories assessed in this study, it was found that waste waters from the cage washing and dosage preparation areas were the most highly contaminated, representing limiting cases for high hydraulic load and high contaminant load, respectively. The approximate average daily volume of waste water discharged from the cage washing area was 7500 liters and from the dosage preparation area was 230 liters. The approximate concentration of suspect carcinogens from the cage washing discharge was 0.5 mg/l and from the dosage preparation area was 200 mg/l. For combined cage washing and dosage preparation waste waters, 1% of the TOC consisted of suspect carcinogen test chemicals.

The pilot-scale tests were performed to determine the effectiveness of RO for removing dissolved solutes from carcinogenesis laboratory waste waters with minimal emphasis on the particular chemicals contained in the waste waters. To relate the effluent TOC concentrations and removal efficiencies to specific suspect carcinogens, bench-scale tests were performed to determine how effective RO is for removing specific chemicals and groups of chemicals from waste water. These tests were carried out with 20 individual chemicals which are representative of ten groups of organic chemical carcinogens, but which themselves are not known carcinogens [8]. Chemicals were selected which display chemical (i.e., functional groups) and physical (i.e., molecular size) properties representative of the carcinogens in each group. This indicates that the selected chemicals would exhibit RO removal efficiencies similar to carcinogens in each group. In addition to the carcinogenlike chemicals, six widely used solvents and other laboratory chemicals also were tested. The 26 chemicals which were tested are listed in Table I. Chemicals which display suitable water solubilities were tested at a concentration of 100 mg/l. Other chemicals having lower water solubilities were tested at concentrations corresponding to their solubilities, as indicated in Table I.

PILOT-SCALE TESTS WITH ACTUAL WASTE WATERS

Materials and Methods

Representative 380-liter samples of actual cage-washing and dosage-preparation waste waters were collected at the three selected carcinogenesis laboratories and processed in the pilot-scale system shown schematically in Figure 1. Because the high loading of suspended solids in the raw actual waste waters (an average of 300 mg/l for the six waste waters) made direct treatment by RO unsuitable, a two-step purification sequence was employed, with a filtration step preceding removal of dissolved solutes by RO. The filtration of suspended solids was performed using an ultrafiltration (UF) system consisting of four 2.54-cm diameter, 3.05-m long tubular modules (two HFM and two HFD, Abcor, Inc.). The UF system was operated at 345 kPa feed pressure, actual waste water temperature, and 6800 liter/hr feed circulation rate. The characteristics of the UF permeates are listed in Table II. For all waste waters tested, the UF system reduced the concentration of SS to less than 5 mg/l, which satisfied the pretreatment requirements for RO. The UF system also reduced the concentration of total organic carbon in the raw waste waters by an average of approximately 50%. The same UF modules were used for processing all six waste water samples with intermittent cleaning to restore to the original flux values.

The UF permeate used as feed in the RO tests was pumped at 5.5 MPa feed pressure, 32°C feed temperature, and 1200 liter/hr feed flowrate to the first module through two 10-cm diameter, 1-m long RO modules in series: a spiral-wound PA-300 module (Fluid Systems Division, UOP, Inc.) followed by a hollow-fiber B-10 module (Permasep Products Division, E. I. du Pont de

Table I. Chemicals in Synthetic Wastewater

Chemical Group	Chemical	Test Concentration (mg/l)
Aromatic Amines	N,N,N′,N′-Tetramethyl-benzidine,	2
	p-aminodiphenylamine,	100
	aniline,	100
	o-toluidine	100
Alkyl Halides	Chloroacetic acid,	100
	trichloroethylene	100
Polycyclic Aromatics	Pyrene,	1
	anthracene,	4
	phenanthrene,	3
	fluorene,	2
	naphthalene	6
Esters, Epoxides Carbamates	Vinyl acetate,	100
	methyl carbamate	100
Pesticides	Malathion	100
Nitro Aromatics	p-Nitrophenol	100
Alkyl Amines and Amides	Di-n-propylamine,	100
	urea	100
Nitrosamines	Diphenylnitrosamine	12
Fungal Toxins and Antibiotics	Gibberellic acid	100
Azo Dyes and Diazo Compounds	Methyl orange	100
Organic Acids	Acetic acid	100
Solvents	Formaldehyde,	100
	phenol,	100
	corn oil,	3
	ethanol	100
Detergent	Dodecyl sodium sulfate	100

Nemours & Company). Batch-concentration tests were used to determine the maximum water recovery (conversion), and total-recycle tests were performed to determine the potential for membrane fouling. For both membrane modules, analyses were made for suspended solids, total dissolved solids, pH, conductivity and TOC in the feed, concentrate and effluent.

The entire pilot-scale system was contained within an isolated test chamber which had its own exhaust ventilation system (6 air changes/hr).

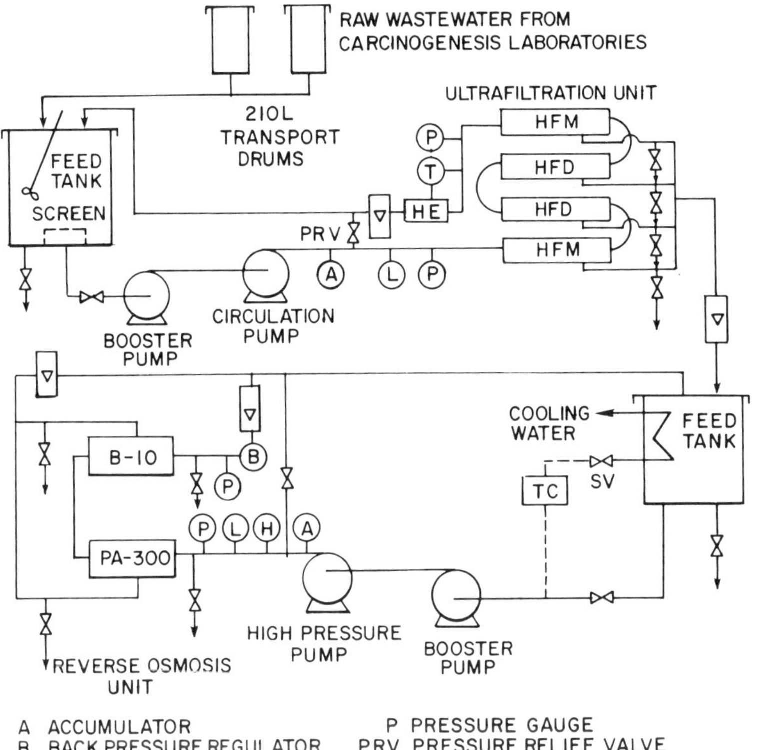

Figure 1. Flow schematic of ultrafiltration/reverse osmosis pilot-scale systems.

Table II. Ultrafiltration Permeate Characterization

Wastewater	Laboratory	TDS (mg/l)	TOC (mg/l)	pH	Conductivity (μS/cm)
Cagewash	A	259	20	8.7	270
	B	750	15	8.2	1100
	C	1060	49	9.5	1850
Dosage	A	821	45	7.2	1200
Preparation	B	2890	218	10.0	5000
	C	909	293	9.1	1200

All personnel wore disposable protective clothing and respirators (fitted with activated charcoal prefilters and dust and mist particulate filters) when performing experiments in the chamber with the carcinogen-contaminated waste waters. Contaminated wastes were disposed of by a U.S. Environmental Protection Agency (EPA) licensed disposal company.

Results

During the course of the RO tests, the performance characteristics (productivity and rejection) of the membrane modules were determined using a 1500-mg/l sodium chloride solution at standard operating conditions. Before each sodium chloride performance test, the RO system was flushed with chlorine-free water. The system was then operated in a total-recycle mode at standard conditions until steady state was achieved. At steady state, productivity and sodium chloride rejection were determined for each module. By conducting sodium chloride performance tests before and after the treatment of actual waste waters, any degradation of membrane performance during waste water treatment could readily be identified. For the PA-300 membrane both productivity (273 liter/hr) and rejection (98%) remained reasonably constant throughout the test program (620 hr), indicating good membrane stability for the waste waters tested. For the B-10 membrane, however, although productivity held constant (454 liter/hr), some declines in rejection were noted. The overall loss in B-10 rejection (from 99.7 to 95%) can be partially attributed to the loss of PT-B, a "tightening" agent used to pretreat the membrane for improved rejection, which is gradually extracted from the membrane, especially at pH 10 and above [9].

Batch concentration tests were performed to determine the flux and contaminant rejections of the membrane modules as a function of water recovery or volumetric concentration ratio (i.e., initial feed volume divided by volume of concentrate remaining at any time). The system was run by recirculating the concentrate to the RO feed tank while collecting the permeate in a separate auxiliary tank. The experiments were performed with ultrafiltration permeate under standard operating conditions and were continued until the volume of the concentrate decreased to a minimum, approximately equal to the system hold-up volume (11 liters). In these tests, the feed flowrate to the PA-300 module was 1200 liter/hr, and the average membrane productivity was 200 liter/hr. This corresponds to a conversion per pass of 17% for the PA-300 module. For a feed flowrate of 1000 liter/hr to the B-10 module, the average membrane productivity was 400 liter/hr, which corresponds to a conversion per pass of 41%. The membrane module productivity for the PA-300 and B-10 modules is plotted in Figure 2 as a function of volumetric concentration ratio (VCR) for the most highly contaminated

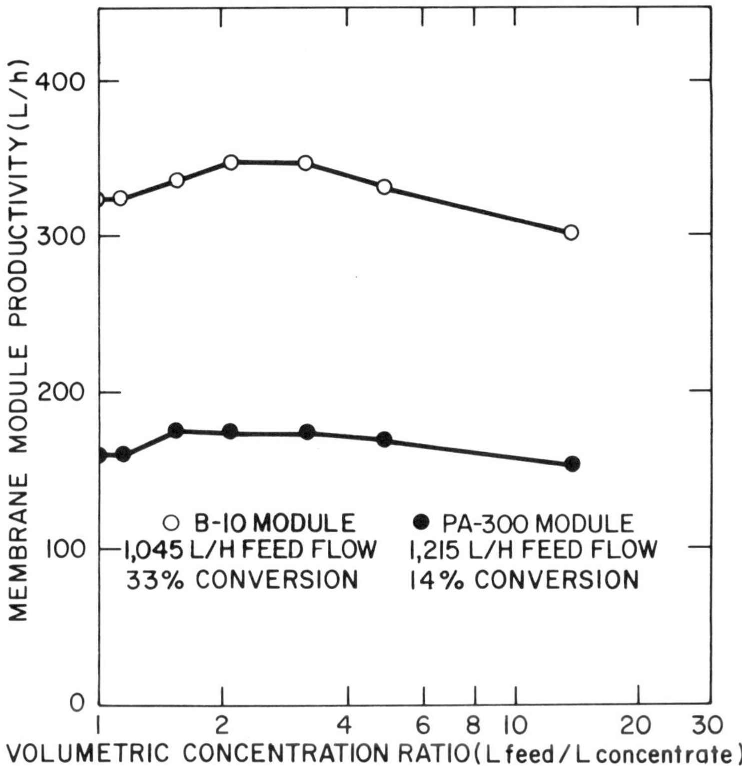

Figure 2. Membrane module productivity as a function of volumetric concentration ratio for batch-concentration tests with dosage-preparation ultrafiltrate from Laboratory B.

ultrafiltrate of the six tested, laboratory B dosage preparation ultrafiltrate. For this worst-case feed, it can be seen that the two membrane modules both gave acceptable performance, with the productivities remaining high throughout the test. In all six tests, high and stable productivity levels for feed concentrations up to 40 times the initial feed concentration were obtained, which indicates that this is a favorable application for RO and that even higher volumetric concentrations could be achieved.

Total-recycle tests were performed to determine the potential for membrane fouling. These experiments were carried out continuously for up to six days using feed having a concentration analogous to approximately 80% water recovery (i.e., five times the original ultrafiltration permeate concentration). Measurements of membrane solute rejection and productivity were employed

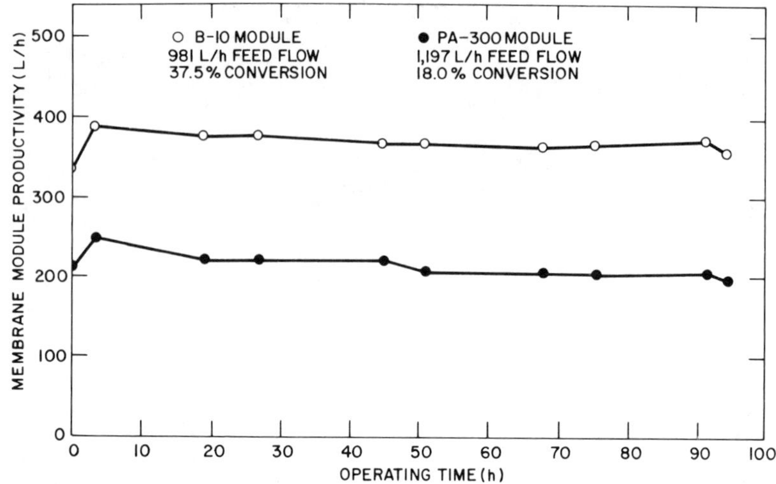

Figure 3. Membrane module productivity as a function of operating time for total-recycle tests with dosage-preparation ultrafiltrate from Laboratory B at 5X concentration.

to indicate membrane fouling. Figure 3 is a plot of membrane productivity as a function of operating time for the most highly contaminated ultrafiltrate, that from the dosage preparation area of laboratory B. It can be seen that the productivities for both the PA-300 and B-10 modules remained high and virtually constant for the duration of the tests. The fact that there was no evidence of fouling during any of the reverse osmosis tests indicates that ultrafiltration is an effective pretreatment method.

A summary of analytical results for the RO tests is given in Table III. The TOC permeate concentrations for each waste water are average values for four grab samples of RO permeate: two collected during the batch concentration tests and two collected during the total-recycle tests. The VCR values listed are the average values corresponding to the times the grab samples were collected. A TOC concentration comparable to that for tap water (3–4 mg/l in the three laboratories) was obtained for all RO permeates from both modules. These TOC concentrations in the permeates correspond to average TOC removal efficiencies for the PA-300 and B-10 modules of about 92%. These removal efficiencies calculated relative to concentrations in the UF permeate would be lower in practice, however, if a usual mode of operation were employed, such as semibatch mode in which the concentrate is recycled back to the feed tank and VCR values on the order of 100 are obtained. For the data reported here, the average VCR was approximately 5.

Table III. Summary of Results for Reverse Osmosis Tests Using Actual Wastewaters

Wastewater	Average VCR for Four Grab Samples	Ultrafiltrate TOC Concentration (mg/l)	Permeate TOC Concentration (mg/l)		TOC Removal Efficiency (%)	
			PA-300 Module	B-10 Module	PA-300 Module	B-10 Module
Cage Washing	5.2	20	2.2	2.5	89.0	87.5
	3.8	15	1.4	0.5	90.7	96.7
	6.0	49	5.0	7.5	89.7	84.5
Dosage	3.3	45	1.3	1.4	97.1	96.8
Preparation	3.5	218	8.0	8.3	96.3	96.1
Average	4.4	70	3.6	4.0	92.6	92.3

BENCH-SCALE TESTS WITH SYNTHETIC WASTE WATERS

Materials and Methods

Bench-scale experiments were performed to determine how effective RO is for removing specific chemicals and groups of chemicals from water. A flow schematic for the RO test system in a total-recycle mode is shown in Figure 4. The system was operated at approximately 25°C, 4.1 MPa feed pressure, and 60 liter/hr feed flow at essentially zero conversion. The stainless steel, flow-through, RO test cells employed in the system were designed to accommodate flat-sheet membranes having a functional surface area of approximately 20 cm^2. The two membranes tested (PA-300 and B-9) were supplied by the Fluid Systems Division, UOP, Inc., and the Permasep Products Division, E. I. du Pont de Nemours & Co., respectively.

Membrane performances (solute rejection and permeate flux) were determined for synthetic waste waters containing individual test chemicals at a concentration of 100 mg/l in deionized water or their solubility limit if lower. Solute removal efficiencies were calculated from measurements made using a gas chromatograph (GC) or a TOC analyzer. In addition, membrane solute rejection and permeate flux for a 1500-mg/l sodium chloride solution were determined prior to each experiment with synthetic waste waters in order to characterize the membranes and determine whether any degradation or fouling of the membrane had occurred during the previous test. All work with test chemicals, except the GC and TOC analyses, was performed under laboratory-type open face hoods by personnel who wore suitable protective clothing.

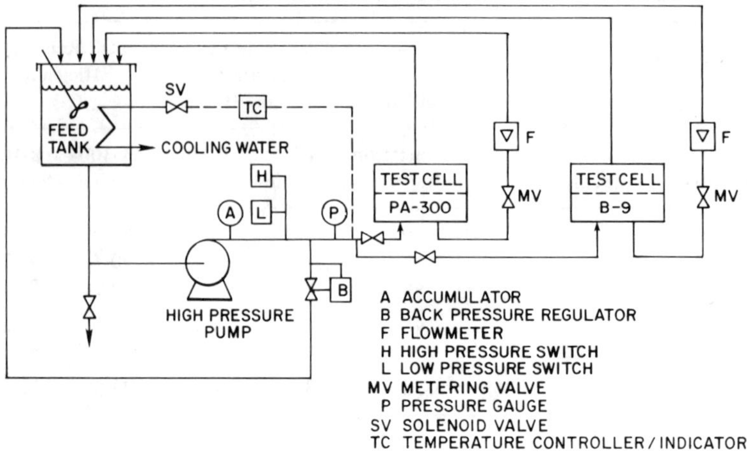

Figure 4. Flow schematic of bench-scale reverse osmosis system.

Results

Reverse osmosis solute rejections for the PA-300 and B-9 membranes were calculated for synthetic waste waters containing individual water soluble test chemicals and are listed in Table IV. In commercial membrane modules, the PA-300 is in a spiral-wound configuration and the B-9 is in a hollow-fiber configuration. The flux values for both membranes remained stable throughout the tests, with the average flux values for the PA-300 (41 liter/m^2-hr) being consistently higher than those for the B-9 (25 liter/m^2-hr). In general, the PA-300 membrane also displayed slightly higher rejections than the B-9 membrane. For the water-soluble chemicals tested, the average solute rejection for the PA-300 was 72%, while that for the B-9 was 66%. The results show that removal efficiencies for both membranes were high for high-molecular-weight chemicals such as methyl orange and malathion, and low for small polar chemicals such as formaldehyde and ethanol. Accordingly, the removal efficiency for the carcinogenlike chemicals only (not solvents and detergent) is slightly greater than that for all chemicals: 75% for PA-300 and 71% for B-9. Removal efficiencies for eight chemicals having water solubilities less than 100 mg/l were all greater than 99%. These chemicals included polycyclic aromatic hydrocarbons (pyrene, anthracene, phenanthrene, fluorene, naphthalene), aromatic amines (tetramethylbenzidine), nitrosamines (diphenylnitrosamine) and corn oil. If a removal efficiency of 99% is assumed for the water-insoluble chemicals, the average removal efficiencies for all chemicals

Table IV. Reverse Osmosis Removal Efficiencies for Water-Soluble Test Chemicals[a]

Chemical Group	Test Chemical	Removal Efficiencies (%)	
		PA-300 Membrane	B-9 Membrane
Aromatic Amines	*p*-Aminodiphenylamine,	84.4	90.6
	aniline,	75.8	76.9
	o-toluidine	75.0	71.0
Alkyl Halides	Trichloroethylene,	57.3	74.2
	chloroacetic acid	55.8	61.1
Esters, Epoxides, Carbamates	Vinyl acetate,	75.8	50.3
	methyl carbamate	45.0	46.0
Pesticides	Malathion	96.1	80.6
Nitro Aromatics	*p*-Nitrophenol	68.2	47.7
Alkyl Amines and Amides	Di-*n*-propyl amine,	92.3	94.0
	urea	69.5	55.2
Organic Acids	Acetic acid	64.5	53.2
Fungal Toxins and Antibiotics	Gibberellic acid	85.1	88.9
Azo Dyes and Diazo Compounds	Methyl orange	100.0	100.0
Solvents	Ethanol,	67.4	45.7
	formaldehyde,	23.6	8.8
	phenol	69.5	56.9
Detergent	Dodecyl sodium sulfate	96.0	87.7
Overall average		72	66
Carcinogen-like chemical average		75	71

[a]Conditions: 100 mg/l feed concentration, 25°C, and 4.1 MPa feed pressure.

by the PA-300 and B-9 membranes are 82 and 78%, respectively; for the carcinogenlike chemicals only, the removal efficiencies are 85 and 82%, respectively.

The results of tests with a 1500-mg/l sodium chloride solution parallel those with the test chemicals. The flux values for both membranes remained stable, indicating that there was no significant fouling. The rejections (an average of 98%) for the PA-300 were stable except for measurements made after testing methyl orange. On the other hand, a gradual decrease in rejection (from 99 to 96%) for the B-9 suggests a possible tendency toward membrane degradation. This tendency can be effectively kept to a minimum by periodically retreating the B-9 membrane with appropriate membrane solutions [9].

DISCUSSION

Effluent discharge requirements have not been established for waste waters containing suspect carcinogens. The results of the tests reported here, however, show that RO is a feasible method for removing suspect organic chemical carcinogens from waste waters. Because of the many types of chemical carcinogens, it is probable that a general measure of chemical content, such as TOC, would be used to indicate the concentrations of the organic chemicals contained in water and waste waters. A TOC concentration comparable to that in tap water (3-4 mg/l) was obtained for all RO permeates from both RO modules in the pilot-scale tests, which corresponds to an average TOC removal efficiency of 92% relative to the TOC concentration in UF permeate. For the individual carcinogen-representative chemicals in bench-scale tests, the average RO removal efficiency was 84%. Assuming an average TOC removal by UF of 50% as was obtained in pilot-scale tests, overall removal efficiencies for a two-stage system consisting of UF followed by RO would be 96% for TOC and 92% for individual suspect carcinogens.

The data obtained in this study were used to demonstrate the effectiveness of RO for removing suspect carcinogenic compounds from laboratory waste waters. For this purpose, an approximate average TOC concentration of 100 mg/l was employed, which has been shown by GC/MS to contain approximately 1% test chemicals. On this basis, the removal efficiencies shown would predict an effluent TOC concentration of 4 mg/l and a suspect carcinogen level of 80 μg/l following combination of UF and RO treatment. It is important to note, however, that the overall removal efficiency for RO in actual concentrate-recycle operation would be lower than that predicted.

Comparing RO with other separation processes is complicated by the fact that most other effective processes, such as activated carbon adsorption (ACA), are similar to RO in being effective for removing large, nonpolar chemicals from water. For processes such as RO and ACA which are feasible methods for removing suspect carcinogens from water and waste waters, selection of the preferred method can be based on economic factors. Economic comparisons between RO and ACA at the same level of performance can be made in principle by calculating costs for systems in which the frequency of carbon regeneration or replacement is selected to give an effluent TOC removal efficiency equal to that obtained using RO, approximately 90%. Such an economic analysis was performed for a two-stage wastewater treatment system having a capacity of 1100 liter/hr [7]. The results indicate that the capital costs for an ACA system ($15,000) would be only two-thirds of that for a RO system ($22,000). However, primarily because of the costs of carbon regeneration or replacement, the total operating cost for ACA ($13,000 per year) would be 1.3 times greater than that for RO ($10,000 per year).

It is not known on what basis future effluent discharge requirements will be established for water and waste waters containing suspect carcinogens. Of the possible alternatives, it seems likely that discharge requirements will be based either on removing a certain high percentage of the suspect carcinogens, on removing them to some low absolute concentration, or on setting a "zero discharge" limit. Whichever option is chosen, a system including an RO unit could be designed to meet the effluent requirements. For the test chemicals employed in this study, an overall removal efficiency for two-stage systems containing UF and RO was greater than 90%. Under the test conditions, these removal efficiencies can produce effluent concentrations below 100 μg/l. If lower concentrations are required, they could be obtained by using several RO modules in series. A "zero discharge" requirement could be met by recycling the RO permeate.

REFERENCES

1. Sourirajan, S., and T. Matsuura. "Physicochemical Criteria for Reverse Osmosis Separation," in *Reverse Osmosis and Synthetic Membranes*, S. Sourirajan, Ed. (Ottawa: National Research Council Canada Publications, 1977) pp. 5-43.
2. Lonsdale, H. K., U. Merten and M. Tagami. "Phenol Transport in Cellulose Acetate Membranes," *J. Appl. Polymer Sci.* 11:1807-1820 (1967).
3. Sourirajan, S. "Reverse Osmosis—A General Separation Technique," in *Reverse Osmosis and Synthetic Membranes*, S. Sourirajan, Ed. (Ottawa: National Research Council Canada Publications, 1977) pp. 1-4.
4. Fang, H. H. P., and E. S. K. Chian. "Reverse Osmosis Separation of Polar Organic Compounds in Aqueous Solution," *Environ. Sci. Technol.* 10:364-369 (1976).
5. Riley, R. L., R. L. Fox, C. R. Lyons, C. E. Milstead, M. W. Seroy and M. Tagami. "Spiral-Wound Poly (ether/amide) Thin-Film Composite Membrane Systems," *Desalination* 19:113-126 (1976).
6. Chian, E. S. K., W. N. Bruce and H. H. P. Fang. "Removal of Pesticides by Reverse Osmosis," *Environ. Sci. Technol.* 9:52-59 (1975).
7. Light, W. G., K. J. McNulty and R. L. Goldsmith. "Development of Decontamination Procedures for Aqueous Chemical Carcinogens," report to National Cancer Institute, Contract N01-CP-43350 (August 1978).
8. McCann, J.,E. Choi, E. Yamasaki and B. N. Ames. "Detection of Carcinogens as Mutagens in the *Salmonella*/Microsome Test: Assay of 300 Chemicals," *Proc. Nat. Acad. Sci. US.* 72:4135-5139 (1975).
9. "Permasep Permeator Engineering Design Manual" (Wilmington, DE: E. I. duPont de Nemours & Co., 1976).

CHAPTER 17

POTENTIAL OF REVERSE OSMOSIS FOR RECYCLING AND REUSE OF SEWAGE EFFLUENTS.

Brian A. Winfield

Department of Civil Engineering
Portsmouth Polytechnic
England

The conventional approach to sewage treatment has been to reduce its organic load to an extent where it can be discharged to the environment without causing undue nuisance.

Increasing pressures to improve the quality of the environment, coupled with an ever-increasing demand for water, necessitate the review of traditional treatment procedures.

This chapter summarizes certain aspects of a program which is investigating the recycling of liquid domestic wastes. The studies have attempted to develop an engineering system which would be complementary to existing treatment procedures and have the following economic and environmental benefits:

1. to minimize pollutants in the environment;
2. to optimize food production from waste product nutrients; and
3. to create an additional industrial or potable water resource.

REVERSE OSMOSIS TREATMENT OF SEWAGE–THE FUTURE

Estimates indicate that the demand for water in the United Kingdom is likely to double in the next 30 years [1]. Similar increases seem inevitable in other Western societies and the developing world, as higher standards of living make demands on limited conventional water resources. Israel already uses 95% of its water resources, and it is acknowledged that in the future the use of unconventional water supplies is inevitable [2]. Shelef's studies [3] of membrane processes show how Israel is trying to overcome this problem.

Against this backcloth, the last ten years have seen the development of expertise in the use of low-quality waters for industrial purposes [4], and also of investigations where these waters have been renovated for potable use.

In many instances the poor-quality waters which have been studied have originated from brackish supplies and even from the sea. However, several research experiences, notably those at Windhoek, South West Africa [5] and Pomona, CA, [6] have succeeded in reclaiming potable water from treated sewage effluents while still retaining realistic treatment procedures.

There are many esthetic grounds on which these sewage reclamation practices seem unacceptable. However, there are many reasons why this technique is likely to gain favor in the future as prejudices are gradually overcome.

For instance, if sewage effluent and sea water are compared as potential sources of potable water, it will be seen that the former is a far more economical proposition. Sea water has an approximately 35 times higher level of dissolved contaminants and can only be reclaimed by energy-intensive processes such as distillation or reverse osmosis. If a comparison of treatment by reverse osmosis is made, it will be seen that potable water reclamation from sea water is approximately three times as expensive as that from sewage due to its greater osmotic pressure and need for multistage treatment. In addition, the use of sewage effluent in potable water supply already has several precedents. Probably the most controversial example occurred in Chanute, KS [7]. Sewage effluent was pumped directly into a water supply reservoir during a period of drought. The water was treated by a conventional system and despite poor color and taste characteristics, no adverse effects were noted in the local population receiving this supply.

The use of low-quality water as a potable resource inevitably means that the cost of the water will be high when compared with that provided by local borehole and river sources. However, once a situation is reached in which the conventional local water supplies are fully committed, reclamation of sewage effluents is likely to become a cost-effective source of water.

Many reasons may be cited why sewage effluents should be considered as a source of potable water. One major argument in favor of this type of recycling is the proximity of effluent production to the location of demand. Thus, in circumstances where alternative supplies of water are distant from the demand situation, recycling of sewage may obviate the need for the expensive practices of river recharging, long-distance pipelines and reservoir construction. Where the sea is the only alternative water supply for a region, sewage effluent is a viable alternative because of its considerably lower level of saline contamination and, hence, lower treatment cost.

Having considered the likely benefits of utilizing sewage effluents it is necessary to explore the various methods of treatment for this exacting requirement.

During the process of contamination of domestic water, approximately 300 mg/l of dissolved salts are added to the solution. In typical situations where this liquid is discharged to a river before being abstracted for reuse, dilution of these salts means that they do not usually present a problem. Alternatively, in circumstances where the sewage is used in industry [1], the increased salt concentration is often of little consequence. However, multiple reuse of sewage in an intensive situation would quickly lead to an unacceptable buildup of dissolved salts in the water being recycled. In this type of situation membrane processes, notably reverse osmosis, could be effective in reducing these salt concentrations. In addition, the presence of a membrane provides a physical barrier for the exclusion of many potentially toxic chemicals and organic materials as well as pathogenic organisms which may be present in the sewage effluent. Indeed, in many situations the reverse osmosis treatment would be capable of producing an extremely soft water which might find additional demand in local industry as a high-quality process water [8]. In this application the water would inevitably command a high price. The potential of reverse osmosis in this type of application has been noted by the U.K. National Standing Committee on Water Treatment [9], which states that "there is increasing interest worldwide in the use of reverse osmosis for treatment of both sewage and industrial effluents." The report also states that the process should be the subject of further research and development.

The above treatise has considered the sewage effluent solely as a source of high-quality water. In most investigations the concentrate which results from the reverse osmosis treatment is an embarassment and has had to be disposed of by incineration [10], activated sludge treatment [11] flash distillation or solar evaporation ponds. The potential which resides in the use of this concentrate has still to be realized, and yet it is the use of concentrate which can make the reverse osmosis system an even more economically attractive proposition.

During the production of potable water from sewage effluent by reverse osmosis, the nutrient components of the sewage are concentrated in a small proportion of the total water flow. This solution may be used as the feed in a range of biological growth chains producing either a food or other useful commodities.

Biological systems which have been investigated involve the growth of hydroponic tomatoes and the marine aquaculture of phytoplankton, oysters and macroalgae [12]. In each case a high-value cash crop can be obtained which could further offset the initial cost of water production.

Current trends indicate the inevitability of severe water shortages in various areas of the world over the next few decades. Indeed, in some regions of the world this shortage has already arrived. Given this situation, reverse osmosis has a great deal to offer in terms of water recovery and crop production. The implementation of this system is something which needs to be evaluated in the context of each specific location. At present the energy-rich areas of the world are those where reverse osmosis could be used to produce water and food since both are in short supply. Thus, it is here that the potential of reverse osmosis should not be underestimated.

The economic validity of these systems is something which will continually vary as the importance placed on environmental quality changes and as the cost of labor, materials and energy fluctuates. However, it is felt that in time the extreme demands made by man on the hydrobiological cycle will necessitate the development of these intensive systems.

In the quest to recover potable water from a brackish or sewage supply it must be realized that processes other than reverse osmosis are applicable. The contaminants in these low-quality waters are both organic and inorganic, and so could, in principle, be removed by activated carbon followed by ion exchange. However, the use of these processes would require that the water was treated extensively to remove colloidal and suspended material in advance of this type of process. Such pretreatment might well prove very costly. In addition, reverse osmosis has the already established advantage of presenting a physical barrier to toxic materials which may be present in the feed water and also enables the recovery of waste nutrient impurities for use in biological food chains.

The research conducted at Portsmouth Polytechnic [13] (Figure 1) has used the reverse osmosis sustem to fulfill two roles outlined above, namely to recover a high quality water and a nutrient concentrate. The water has been investigated fully to determine its potability, and the nutrient solution has been used in a number of biological growth systems. The performance of the reverse osmosis process has also been evaluated.

The use of the concentrate may be considered from several perspectives. It may be used for the growth of tomatoes, cabbage and marrow by the nutrient film hydroponic technique under intensive horticultural conditions and also in admixtures with seawater to promote the marine aquacultural growth of phytoplankton or valuable carageenan and alginate-producing macroalgae. The phytoplankton may be harvested directly or fed to *Doonaliella* or various types of mollusc (Figure 1).

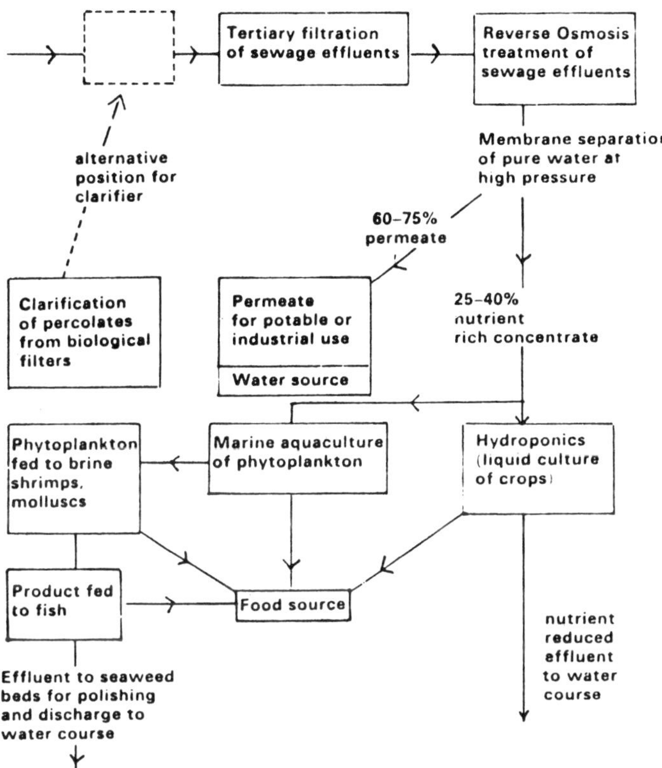

Figure 1. Liquid waste cycle.

EXPERIENCES OF TREATING SEWAGE EFFLUENTS BY REVERSE OSMOSIS

Many research workers have investigated the treatment of sewage effluent by reverse osmosis. These studies started with the work undertaken by U.S.

Department of Health in 1965 [14]. Since then, various types of sewage effluents have been used as feed liquors to the reverse osmosis units, from raw sewage effluents containing very high concentrations of dissolved and suspended material to polished activated carbon treated effluents [15-17]. These liquids have been used in all types of reverse osmosis units from the flat plate to the hollow fiber system. Cellulose acetate has been the most commonly used membrane material.

Detailed comparisons of the reverse osmosis studies are difficult because of an inherent inability to characterize adequately the feed liquids and also because of the fundamental differences between the membrane types and operating conditions of the various trials. The results of the above experiments are well-documented [6], and all tend to indicate that the rate of membrane fouling and hence the flow of product water decreases as the degree of sewage pretreatment increases.

Fouling of the reverse osmosis membranes occurs as material in the solution is transported up to the membrane with the flow of water through the membrane surface. Once adjacent to the membrane surface, fouling components may be precipitated through concentration polarization, adsorbed or trapped by existing fouling material. When this has occurred it is likely that the fouling which is deposited will present a restriction to continued water flow. Thus, when considering the treatment of sewage effluents by reverse osmosis, fouling is quickly identified as a major problem. Research work in this field has been concentrated in the following areas: (1) minimizing the rate of membrane fouling, and (2) developing techniques of rapidly removing fouling and restoring the membrane to its original condition.

Despite the importance of fouling, few research workers have attempted to correlate feed water quality to membrane fouling. Many parameters have been used, including the feed water chemical oxygen demand (COD) and turbidity.

Inevitable problems occur in attempting to compare studies made at various locations with different equipment and operating conditions. Several studies have looked at the possibility of developing high-velocity, turbulent flows against the membrane surface [18] and these ideas have led to concepts of a threshold velocity below which the rate of membrane fouling is high for given sewage feeds and membrane types. Turbulence generators have also been used [19].

Once formed on the membrane, the fouling deposited during sewage treatment has proven consistently to be refractory given the limitations of the cleaning procedures to which the membrane may be exposed. A range of flushing techniques has been developed. These techniques have been investigated by many researchers who have used water, air and detergent solutions [6,17,19,20]. In addition, depressurization of the membrane system has been investigated as a cleaning technique. Many of the techniques have proven partially successful, but few appear to be consistently effective. In many cases the effectiveness of the technique has reduced when used repeatedly. The most acclaimed technique has been flushing with an enzymatic detergent [17], although no attempt has been made to understand this in terms of the chemical interactions which may be involved. To summarize, the studies which have considered a reverse osmosis plant operating on sewage effluents tend to have been semiempirical, primarily as a result of the unpredictable nature of the feed liquid, and as a result have been inconclusive.

The studies described below have attempted to look at a spectrum of membrane fouling characteristics. The prime concern has been to probe the features of the fouling layer by integrating information obtained from studies of its deposition, physicochemical characteristics and removal. The objectives have been:

1. to study the effect of operating conditions on the rate of membrane fouling;
2. to study the physicochemical characteristics of the fouling layer; and
3. to investigate techniques of removing the fouling layer from the membrane.

Although studies have been conducted on one type of reverse osmosis unit (Appendix) many aspects of the work are widely applicable, especially to systems employing a cellulose acetate membrane.

The effects that feed water quality and pH had on the fouling of the membrane were quite marked. Studies in which the feed to the reverse osmosis unit was filtered to remove particles of specific sizes showed clearly that those with a diameter greater than 5 μ played a minor role in the fouling of the reverse osmosis membranes as measured by the flux decline index (Table I). This was despite the fact that the feed liquid was shown to contain a substantial proportion of particles above this size (Figure 2). In contrast, the dissolved organic content of the reverse osmosis feed water measured after GF/C filtration by its absorption at 275 nm was a major factor. Increasing the concentration of this component of the feed water around the membranes by variations in plant operation or the quality of the sewage effluent produced a good correlation with the rate of membrane fouling measured by the flux decline index [21] (Tables II and III). No correlation was found between feed water turbidity and flux decline.

The above results have obvious repercussions when considering the design of a pretreatment plant for this type of reverse osmosis installation. The data indicate that it is unlikely that tertiary filtration through sand or anthracite will have worthwhile benefits in reducing the rate of flux decline.

Table I. The Relationship Between Sewage Filtration and the Reverse Osmosis Membrane Flux Decline Index

Filter in Line to Unit	None	50-μ	25-μ	10-μ	5-μ
Flux Decline Index	−0.125	−0.133	−0.118	−0.121	−0.109

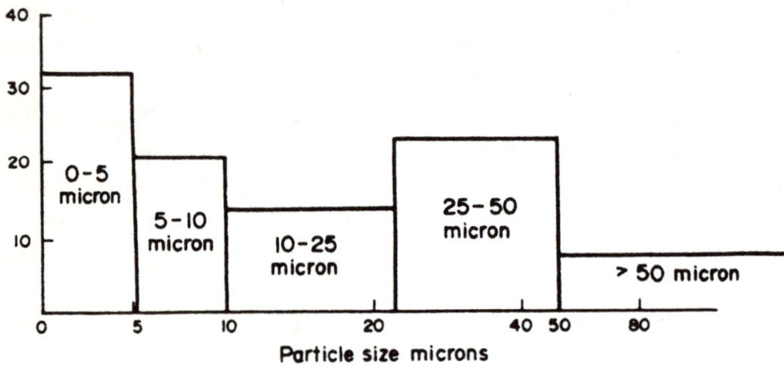

Figure 2. Diagrammatic representation of percentage of total weight of particles with given particle size ranges.

Table II. Relationship Between the Ultraviolet (UV) Absorbance and Turbidity of the Feed Water and the Flux Decline Index for the Reverse Osmosis Unit Operated at Various Water Recovery Ratios[a]

Permeate Water Recovery (%)	(A) Flux Decline Index	(B) Mean (UV) Absorbance of Feed	(C) Mean Turbidity of Feed (FTU)	A/B	A/C x 10^{-3}
31	0.107	0.54	23	0.198	4.65
43	0.111	0.58	16	0.191	6.94
56	0.129	0.70	18	0.184	7.17
64	0.150	0.80	27	0.179	5.56

[a] The pH of the feed water during these trials was maintained at 6.0.
The correlation coefficients are: A & B = 0.989; A & C = 0.585.

In contrast, a process involving coagulation is likely to be beneficial.

The pH of the feed water presented to the reverse osmosis unit has also been found to be critical in determining the productivity of the membrane

Table III. Relationship Between the UV Absorbance and Turbidity of the Feed
Water and the Flux Decline Index for the Reverse Osmosis Unit Operated
at Various Sewage Effluent Feed Water Qualities[a]

Experiment Number	(A) Flux Decline Index	(B) Mean UV Absorbance of Feed	(C) Mean Turbidity of Feed (FTU)	Ratios	
				A/B	A/C x 10^{-3}
1	0.106	0.58	12	0.183	8.8
2	0.135	0.74	18	0.182	7.5
3	0.124	0.63	24	0.197	5.2
4	0.151	0.80	16	0.189	9.4
5	0.098	0.48	10	0.204	9.8
6	0.120	0.61	26	0.197	4.6
7	0.115	0.60	28	0.192	4.1
8	0.110	0.62	13	0.177	8.5

[a] The reverse osmosis unit was operated at a permeate water recovery ratio of 50%
during these trials, and a feed water pH of 6.0. The correlation coefficients are:
A & B = 0.913; C = 0.065.

[22]. Studies have shown that an operating pH of 6.0 produces the lowest
rate of flux decline (Figure 3). Operation at pH 5.0 or 7.0 produces a more
severe flux deterioration. Studies showed clearly that fouling removed from
the membrane after operation at pH 7.0 contained a significantly elevated
level of calcium when compared with samples taken at pH 5.0 and 6.0.
When the pH of operation was lowered the calcium level gradually fell and
the permeability of the membrane/fouling composite rose. The process was
seen to be reversible. This phenomenon is consistent with observations of
other workers who have correlated high calcium levels in the fouling layer
with a rapid rate of flux decline [23].

To understand the reason for the high rate of flux decline under acidic
conditions it is necessary to probe the physicochemical characteristics of
the fouling layer. It was felt that the difference in the rate of flux decline
was a function of the permeability of the fouling since there was no evidence
to suggest a differential rate of fouling layer deposition.

Zeta potential measurements of the fouling material dispersed in solutions of varying pH but equal ionic strengths showed a marked trend (Figure
4). The fouling material became less charged as the pH was lowered, thus
becoming potentially hydrophobic. Solubility trials also showed clearly
that the fouling was less soluble in acidic than alkaline solutions, adding
further weight to the thesis of an increasing hydrophobicity at low pH.

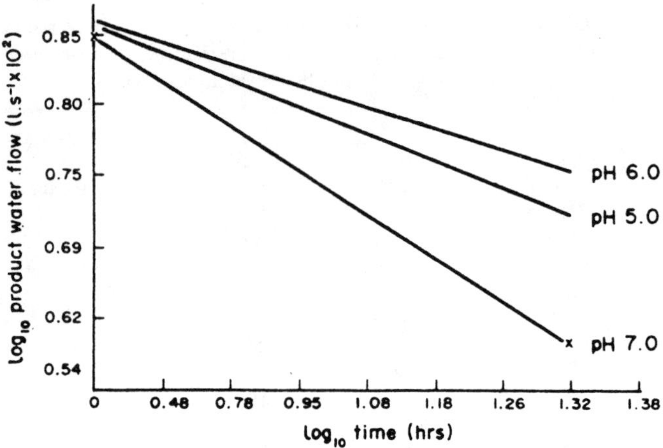

Figure 3. Graph of log flow vs log time for the reverse osmosis membranes operated at a range of feed water pH conditions.

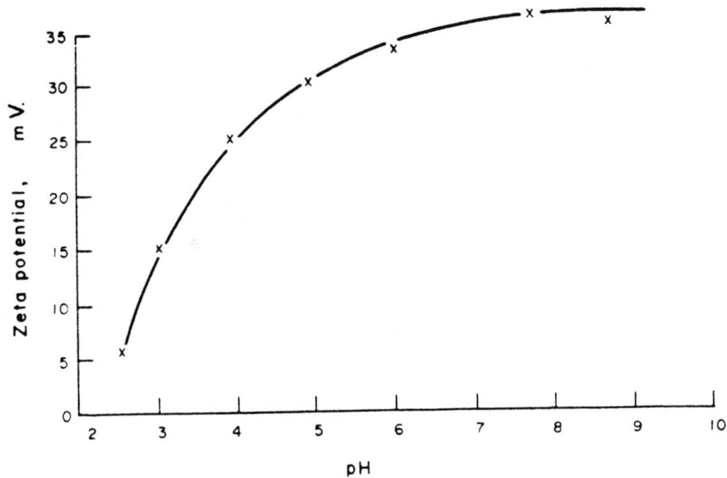

Figure 4. Zeta potential vs pH.

Thus, it appears that reducing the pH of the sewage feed makes the fouling layer less permeable to the flow of water, effectively producing a high rate of flux decline. This may be due to interactions between the fouling matrix and the water and also because as the negative charges on the fouling are neutralized, its collodial structure may collapse, thereby increasing its density. None of these effects are observed with clean membranes. For fouled membranes the phenomenon is reversible and independent of salt effects.

It is reasonable to assume that if the proposals outlined above are correct then the change in charge on the fouling might also affect the flow of salts across this boundary. The results (Table IV) show clearly the trend with decreasing pH of a decreasing sodium ion permeability but of increasing sulfate ion permeability.

The observed changes in salt permeability with pH maybe the result of several contributory mechanisms. As the pH is reduced, the hydrogen ion concentration increases, and in response there is a greater diffusion of these mobile ions across the membrane barrier. This behavior can produce a "streaming potential" whose charge would enhance the flow of negative ions across the membrane; that is, increase the sulfate and reduce the sodium ion flow. However, fouled membranes produce far more pronounced effects. The changes in fouling charge with pH would be expected to produce this.

The obvious conclusion from these studies is that the fouling layer is a negatively charged colloidal layer whose deposition is governed primarily by the level of dissolved organic material in the feed water. As described this has repercussions for pretreatment design. However it is evident that feed water pH is also a crucial aspect of operation if membrane productivity is to be maintained. This also has a significance for the salt rejection performance of the membrane.

Research involving cleaning of sewage-fouled membranes has entailed investigation of physical and chemical techniques. However, almost without exception the methods which have been studied have failed to produce consistently successful results. The procedures can be summarized either as those which attempt to solubilize the fouling once formed or those which remove the fouling by physical scouring or pressure shock. In the work described below a knowledge of the physicochemical nature of the fouling has been used to anticipate techniques by which it may be loosened or dissolved. The study may be divided into two parts: pH studies which have

Table IV. The Effect of pH on Salt Flow Through Fouled Reverse Osmosis Membranes

	Solute Permeability at pH:				
Salt	3	4	5	6	7
Sulfate	9.4	6.8	5.3	5.4	5.15
Sodium	0.38	0.54	0.70	0.71	0.71

investigated the desorption of fouling from the membrane and studies of detergents which are used to assist in its dissolution and dispersion.

The fouling layer has been seen to have a charge which reduces with pH. Thus the effect of pH on membrane cleaning was investigated. In a colloidal matrix like the fouling layer, changes in its natural charge are likely to have a marked effect on its structural integrity and the bond between fouling and membrane.

It was found that when the pH of the solution around the fouling/membrane was reduced below 3 under static conditions, there was a dramatic effect. At these pH values there was noticeable lifting and curling of the fouling from the membrane surface as large sheets. Slight turbulence or agitation completely displaced the fouling. The effect was evident when using mineral and organic acids.

The observed phenomenon can be explained if the nature of the fouling is considered. It is proposed that under acidic conditions the colloidal matrix of the fouling contracts or coagulates as its charge is neutralized. The shrinking tends to break the bond between the fouling layer and the membrane with the effect that the fouling is readily displaced. It is interesting to note that since the acid-induced changes are physical, the cleaning solution can be reused extensively.

As a complementary study the effect of detergents on the removal of fouling was investigated. The group of detergents which was of particular interest were nonionic. These tend to produce clear, pH-stable solutions which have excellent wetting and cleaning properties. They are also produced with various degrees of ethoxylation or propyloxylation and can thus have a range of hydrophobic/hydrophilic characteristics. Thus, it is conceivable that one could be selected which was particularly effective in dispersing the fouling at low pH. A fatty alcohol base material was obtained in a range of stages of ethoxylation and was tested on fouled membranes at a range of pH conditions. At a pH less than 3, a 1% solution of the detergent produced a dramatic effect. The fouling was dispersed more rapidly in the presence of detergent to produce a finely divided suspension. The procedures described above have been consistently capable of displacing sewage fouling from membranes and restoring membrane flux, when accompanied by mild agitation. It can only be inferred that these techniques would permit continuous operation of a plant with a secondary sewage feed water.

The feasibility of operating a reverse osmosis plant with a sewage effluent feed is outlined above. However it remains to establish the value of this process in terms of uses for both the permeate and concentrate liquors.

Subject to certain chemical and microbiological constraints it is easy to envisage uses for the permeate as an industrial or potable water. Uses for the concentrate are more obscure. A system utilizing both liquids is outlined below. The vital concept in proposing uses for the permeate and concentrate is that both are utilized in a common program of reuse; only then can the system be a viable proposition.

THE USE OF SEWAGE EFFLUENT CONCENTRATE

Hydroponics has long been known as a technique by which plants may be grown in an entirely soilless environment. All the nutritional requirements of the plant are met by the liquid which is fed to the root system. Various methods of water culture have been used in recent years, but these have rarely been implemented for commercial scale crop production.

Development of the nutrient film technique (NFT) of hydroponics at the Glasshouse Crops Research Institute, Littlehampton, England in the early 1970s has brought water culture to the forefront of commercial crop production. The system is now used widely; usually within a horticultural glasshouse.

A typical NFT system is shown diagrammatically in Figure 5 where it can be seen that the plants are grown in an enclosed channel set at a gradient of 1 in 50. The nutrient solution is fed to the top of the channel and allowed to percolate to the catchment tank at the base from where it may be recirculated or bled off from the system. The young plant roots are usually allowed to develop in an inert medium such as rock wool or gravel. Once in the NFT system the roots spread extensively throughout the channel.

The NFT system has been seen to have the following benefits:

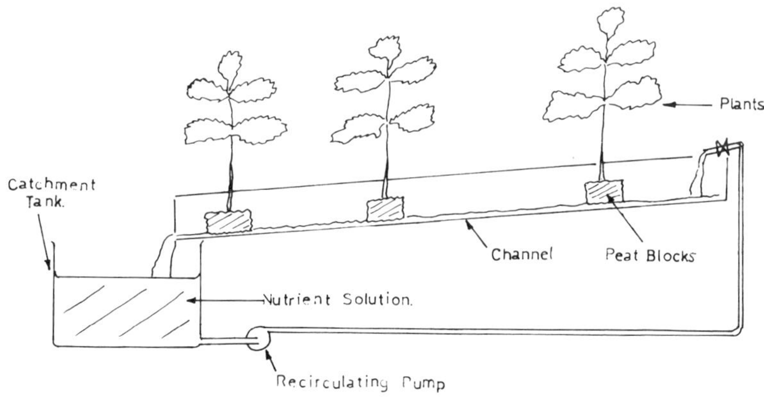

Figure 5. The NFT system.

1. close control of plant nutrition which has resulted in rapid, healthy plant growth with high yields;
2. energy saving without loss of crop yield by heating the nutrient solution and reducing the ambient air temperature;
3. hygienic crop husbandry in a system which does not harbor disease and where rapid crop rotation is possible.

The NFT system is capable of reducing horticultural unit production costs. However one major expense is the cost of the nutrient solution which is continually supplied to the plants. The solution is normally made from completely synthetic chemicals and has a composition similar to that shown in Table V. A comparison of this solution with a typical reverse osmosis sewage concentrate indicates a distinct similarity between the liquids. Many of the essential plant nutrients are present in the sewage effluent, although in most cases at a proportionately lower concentration than in the proprietary solution.

The ability of plants to tolerate a wide range of nutrient levels is well established [23]. The only criterion for healthy growth is that the mass flow of each nutrient is at an adequate level. Thus it seems realistic to assume that sewage effluent could be used as a nutrient solution in an NFT system.

Studies described below have investigated the use of a concentrate obtained from the reverse osmosis treatment of secondary sewage effluent for the growth of tomatoes. Tomato plants were selected because they are a discerning crop which has specific growth requirements. The tomato also exhibits clearly defined deficiency and toxicity symptoms and produces a high-value crop which is remote from the nutrient solution. Where a sewage effluent is used as the nutrient solution the latter fact means that the product is less likely to become contaminated by sewage-borne organisms.

Table V. Comparison of Compositions (mg/l) of NFT Solution and Sewage

Constituent	NFT Solution	Sewage Concentrate
Nitrate (as N)	120	35
Ammonia-N	0.1	7
Phosphate (PO_4^{2-})	160	17
Potassium	270	25
Chloride	35	140
Sodium	60	170
Iron	45	0.4
Manganese	2	0.001
pH	6.3	6.5
Conductivity (μS)	2300	1700

Initial trials showed clearly that plants grown in sewage concentrate exhibited severe interveinal chlorosis which was attributed to an iron and manganese deficiency. Both of these deficiencies could have been predicted from the data in Table V. Thus, the main part of the research has concentrated on techniques of supplementing these unacceptably low levels.

Iron is normally stabilized in NFT solutions by the use of a salt of ethylene diamine tetracetic acid (EDTA) or a similar complexing agent. While this may be acceptable for solutions which are extensively recirculated, it is not feasible for the large volumes of water which would be handled in a sewage-based system. Many techniques exist for introducing iron into solution. However the presence of a well-aerated solution containing a significant level of phosphate places severe constraints on these possibilities.

Experiments which produced encouraging results were those where the tomato plants were established in peat blocks. When an iron supplement was introduced to each peat block in the form of freshly precipitated hydrous iron oxide/hydroxide or as finely divided elemental iron, the plants grew well, showing no signs of iron deficiency. Comparison of the yields of these tomatoes with those of a control crop grown in a commercial-type NFT solution indicated a very comparable performance. Manganese deficiencies were still visible in the above trial but did not appear to significantly affect crop yield. However they may be overcome using a supplement which is introduced in the same way as the iron.

The success which has been found when the plants are grown in supplemented peat blocks has been attributed to the presence of organic acids. These compounds are able to sequester iron from an insoluble state and make it available locally to the plant roots. Using this technique the iron is only introduced to a small proportion of the plant roots, and the interaction with phosphate in the feed solution is minimized. If the iron is applied more liberally so that there is intimate contact between it and the phosphate-rich nutrient solution, iron phosphate salts are produced. This condition can induce both iron and phosphorus deficiencies in the plants.

The above description outlines in principle the major problem in using sewage as a nutrient feed in NFT systems. However it also shows clearly the feasibility of reusing sewage for plant growth. The benefits in operating such a process are manifest in the production of a crop and the purification of an effluent. The development of plant growth systems should therefore be considered as a treatment process in the same way as an activated sludge or biological filter unit. There are constraints imposed by the plants which ultimately control the effectiveness of the system as a treatment procedure. However the ability of plants to obtain their nutrient requirements from dilute solution means that a high quality effluent standard can be attained (Table VI).

Table VI. Influent and Effluent Water Quality Using the Nutrient Film Technique

Constituent	NFT Solution	Sewage Concentrate
Nitrate (as N)	5	35
Ammonia-N	0.1	7
Phosphate (PO_4^{2-})	5	17
Potassium	6	25
Chloride	190	140
Sodium	230	170
Iron	0.1	0.4
Manganese	ND[a]	0.01
pH	7.2	6.5
Conductivity (μS)	2400	1700

[a] Not determined.

The potential of this technique as a means of crop production is outstanding. The system has already been developed for the growth of marrow squash, cabbages and lettuce [24]. However any system of reuse like this shortens the recycling pathways which exist in the natural environment and does not permit dispersion and dilution of toxic or pathogenic agents.

The tomatoes grown in the above system were studied for their heavy metal uptake and for viral penetration from the nutrient solution to the fruit crop. The results of metal uptake studies are entirely consistent with expectations since the metal levels in the secondary sewage effluents studied were low enough to meet the World Health Organization (WHO) drinking water standards. The results presented below (Table VII) compare the observed results with tomatoes from more conventional sources. The studies of viral uptake were made using a model bacteriophage (MS2) for the poliomyelitis virus. This viral unit was injected into the recirculating liquid at high titer but was never detectable in any of the fruit produced. This result is completely in accord with understanding of plant root characteristics and further substantiates claims to the potential of hydroponics for removal of nutrients from sewage and production of a food crop.

Table VII. Typical Analyses of Tomatoes from Various Sources ($\mu g/g$ dry weight)

Tomato Source	Fe	Mn	Zn	Cu	Ni	Cr	Pb
NFT, Commercial	55	19	22	10	7	1	2
NFT, Sewage	34	9	34	12	6	1	3
Soil, Commercial	74	18	34	2	2	1	3

USE OF REVERSE OSMOSIS PERMEATE

The major advantage of using reverse osmosis as a technique for recovering pure water from wastewater is that it presents a physical barrier to the passage of undesirable components in its feed water.

Most chemical or biological contaminants in the waste water which might concentrate within other recycling systems are efficiently rejected by the reverse osmosis membrane. As a consequence it is relatively easy to produce a high-quality water. The chemical quality of the reverse osmosis permeate is shown below (Table VIII) where it is compared with a secondary sewage effluent and the WHO recommended standards for drinking water. The high quality of the permeate means that it could find wide application in industry. In this situation it might reduce the cost of producing a pure "process water" from more traditional main sources.

Some of the chemical components within the water which are potentially troublesome when reuse is considered are described below.

Table VIII. Quality of Reverse Osmosis Permeate

Analysis (mg/l)[a]	WHO Standards (Permissible)	Sewage Effluent	RO Permeate (50% Recovery)
TDS	500	700	50
Color ($^\circ$H)	5	50	ND[b]
Turbidity (FTU)	5	20	ND
Calcium	75	100	2
Magnesium	50	40	1
Sulfate	200	50	0.5
Chloride	200	70	7
Nitrate	11.5	30	6
Ammonia	0.5	9	4
Sodium		80	12
Potassium		19	4
Iron	0.3	0.2	ND
Manganese	0.1	0.01	ND
PAH, ng/1	200	90	9
pH	7.0-8.5	7.4	5.5

[a] Except where otherwise noted.
[b] Not detectable.

Ammonia

The ammonia molecule is poorly rejected by reverse osmosis membranes and as a result can be found at unacceptable levels in the permeate. The rejection of ammonia is pH-dependent because of the ammonia-ammonium ion equilibrium. However at a typical reverse osmosis operating pH the rejection is approximately 70–80%. It is important to be able to control the final ammonia level; this may be done in a number of ways. The most satisfactory technique is to ensure that the sewage effluent feed is fully nitrified. Thus, with a low level of ammonia in its feed the membrane system is able to produce a treated water of acceptable quality. Chemical techniques are also capable of reducing the level of ammonia. Oxidation with chlorine is effective and removes the ammonia as nitrogen gas. However, the technique requires ten times the theoretical dose of chlorine and is thus very costly. Ozonation has been studied but is completely ineffective in oxidizing residual ammonia. Bubbling ozone through the permeate only influences the ammonia at alkaline pH values when ammonia is lost as a gas.

Nitrate

In common with ammonia, the nitrate ion is poorly rejected by normal reverse osmosis membranes. Many potable waters in the industrialized world already have nitrate levels which approach the WHO limit of 11.4 mg/l (as nitrogen), above which cyanosis can occur. This level is increased in a sewage effluent. Therefore it is imperative to ensure that the direct reuse of water does not progressively elevate concentrations by recycling nitrogen.

If the nitrate level of the recovered water is unacceptably high it is possible to reduce it in a number of ways. This may be by ion exchange or denitrification in a fluidized sand filter with an injected carbon source. Both of these techniques tend to be costly and so the problem is best solved by introducing a denitrification stage to the sewage treatment process. Existing technology enables the level of nitrate in a sewage effluent to be reduced by more than 60%.

If the latter course of action is taken the needs of the water recovery program would need to be rationalized with those of any nutrient utilization scheme.

Organic Molecules

The performance of reverse osmosis membranes with respect to their rejection of organic molecules is vitally important when contemplating the reuse of any product water. Notable among the waste water contaminants which would have to be traced in a system contemplating reuse are pesticides, detergents, polynuclear aromatic hydrocarbons (PAH), polychlorinated biphenyls (PCB) and various steroidal compounds. Several studies have investigated the performance of reverse osmosis membranes toward these types of compounds [25,26] and have attempted to predict their rate of transportation across the semipermeable barrier. The results are often inconclusive but tend to indicate that permeation is dependent on polarity and steric configuration for molecules with a molecular weight below 200. Those molecules with a molecular weight above 200 are highly rejected because of their molecular size. The first studies of organic solute rejection in this work considered the behavior of the membrane toward wastewater organics by measuring the total organic carbon content of the feed and product waters. The rejection of organics was always high and was seen to increase as the membranes became fouled (from 88 to 93 % rejection). Characterization of the organic molecules showed that those which penetrated the membrane were small polar compounds. The organic molecules which were well rejected had high molecular weights and frequently had charged sites (for example organic acids). While the effective removal of organic compounds from the waste water by reverse osmosis was approximately 90%, those penetrating the membrane were particularly amenable to conventional treatment by activated carbon. By combining these two processes it was possible to remove more than 98% of dissolved molecules in the sewage feed water. The total organic carbon content of the final treated water was then within the range 0.2–0.5 mg/1.

The scope for studies of the rejection of specific organic solutes is obviously enormous. Many compounds can be cited which should be absent from water reclaimed for drinking purposes and thus must be traced throughout a program of reuse. One group of compounds which have been studied in depth in this research are the PAH. Several compounds within this group have been identified as active carcinoma-inducing agents. In a response to this information WHO has set standards to control the levels of six selected PAH in water abstracted for drinking purposes. The PAH to which the standard applies are ones which can be determined readily and are thought to indicate the presence of other similar compounds. This group of compounds is particularly interesting when considering reuse of effluents because they are known to occur in variable but often significant concentrations in sewage. The compounds also have relatively low molecular weights and may thus penetrate the reverse osmosis membrane and as a result reach the product water.

Analysis showed that the sewage feed to the reverse osmosis unit had significant levels of the PAH compounds (Table IX). The most abundant PAH was fluoranthene. However even in the sewage feed the total levels of PAH did not exceed the WHO standards for a water resource. The reverse osmosis unit was operated at 50% water recovery and produced a permeate with extremely low levels of PAH. Many of the PAH compounds were not detectable in the product water at all. The overall rejection of PAH was approximately 90% for those PAH which could be detected in both feed and product waters. The final water was thus well within the recommended standard set by WHO (Table IX).

Table IX. Comparison of PAH Concentration

PAH Type	WHO Std. (ng/1)	Sewage Feed. (ng/1)	RO Permeate. (ng/1)
Fluoranthene		60	7.0
Benzo[b]fluoranthene		14	1.1
Benzo[k]fluoranthene		10	0.6
Benzo[a]pyrene		11	ND[a]
Indenopyrene		9	ND
Benzo[g,h,i]perylene		14	ND
Total	200	118	8.7

[a] Not determined.

Microorganisms

Microorganisms are known to be present in large numbers in waste waters which derive from domestic sources. The conventional processes of sewage treatment are not specifically designed to remove these organisms and yet in any process that is designed to reclaim water their removal or inactivation is of paramount importance. Viruses are particularly interesting because they are small and may be active at very low numbers. From dimensional considerations, virus particles which range in size up to 250 Å should not be able to penetrate the active surface of a reverse osmosis membrane. In agreement with this, workers [27] have consistently observed extremely high rates of virus exclusion from the reverse osmosis product water.

To monitor virus rejection by the reverse osmosis membranes, an artificial situation was established using the MS2 bacteriophage. This virus, which is a model for the pathogenic poliomyelitis virus because of its similar size, shape and surface charge characteristics was injected into the reverse osmosis sewage effluent feed water at high titer.

The results of the trials showed clearly an extremely high rejection which approached a six logarithm drop in concentration from the feed to the product water. There was a gradual trend of increasing virus penetration with increases in the water recovery ratio and the pressure of operation. However in all cases the rejection of viruses was far higher than could be explained by normal mechanisms. Even in the most efficient reverse osmosis systems the leakage flow may correspond to 0.5% of the total.

Studies of the recovery of the MS2 viruses which had been injected into the system were undertaken in an attempt to explain the observed phenomenon. These trials showed that there was an enormous loss of viral titer somewhere within the system. This loss could be attributed to a number of mechanisms, which included toxic inactivation or adsorption onto particular surfaces. Further investigation showed clearly that the viruses became associated with the solid material deposited on the membrane and were thus effectively "held" in the system. In addition the organic material in the sewage feed tended to inactivate the viruses and hence reduce the recorded titer.

The results indicated that the product water from the reverse osmosis treatment of sewage should only contain very low viral titers if the MS2 virus behaves typically. Viruses are very unlikely to be present in the permeate from reverse osmosis treatment at levels which cannot be removed by conventional sterilization procedures.

Studies of total viable bacteria and coliforms have produced results which concur with the data obtained for viruses. All experiments indicated that there was a very large drop in bacterial numbers across the membrane and that in many cases this was above a six logarithm difference.

CONCLUSIONS

The potential benefits of fully utilizing the intrinsic resources of a sewage effluent are well established and easy to justify. The only questions which may be asked relate to the technological feasibility of implementing recycling systems and also to their economic viability.

This chapter has outlined one system which is proposed as an integrated technique capable of recycling treated sewage. While any economic analysis of the program depends on time and location, the proposals are technologically encouraging. The use of the concentrate has centered on one reuse system selected from a myriad of possibilities which are as numerous as the growth chains of the natural environment. The establishment of a nutrient utilization program appropriate to local conditions must inevitably play a major role in determining the economic validity of the system.

Time will dictate that all man's resources are conserved; waste products in the form of sewage effluents should not, and will not, escape this quest for efficient utilization. This can only be done effectively by integrating man's needs with those of established natural biological systems in which water and nutrients are continuously and efficiently recycled.

APPENDIX

A commercially available "spaghetti"-type reverse osmosis unit supplied by Paterson Candy International Ltd., England, was used for all the experimental work which involved a full-scale unit. The unit was equipped with 15 R.3.37. type membrane bundles producing a nominal product water flow of 0.08 liter/sec when operated at a pressure of 2.5 MPa. The membrane material was made of cellulose acetate which gave a 93% rejection of 0.5% sodium chloride at 4.0 MPa. The reverse osmosis system was pressurized by a 35-stage, 5 kW centrifugal pump.

The pH within the reverse osmosis unit was controlled by sulfuric acid injection, and the permeate pH was monitored with a process pH meter. The normal permeate operating pH was 5.5.

The reverse osmosis unit was fed with secondary sewage from biological filter beds after conventional settlement.

The wetted perimeter Reynolds number of the water flowing past the membranes under normal operation was approximately 3700.

A diagrammatic layout of the reverse osmosis unit is given below:

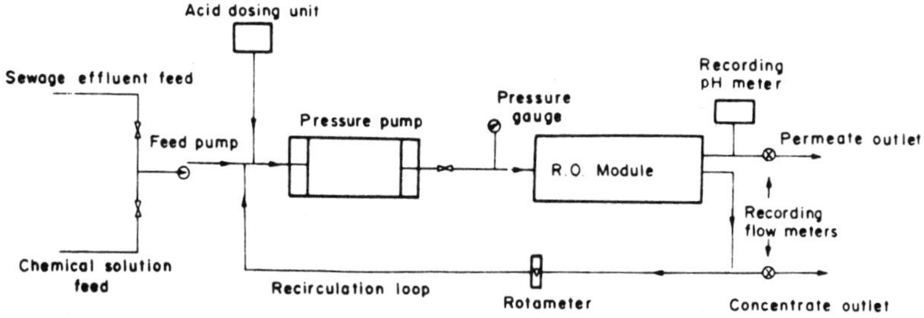

REFERENCES

1. "Annual Report of the Director," Water Research Centre, England (1973), pp. 124-128.
2. Shuval, H. E. *Water Renovation and Reuse* (London: Academic Press, 1977).
3. Shelef, G. "Ultrafiltration and Microfiltration Membrane Processes for Treatment and Reclamation of Pond Effluents in Israel," in *Applications of New Concepts of Physical-Chemical Wastewater Treatment*, W. Eckenfelder and L. K. W. Cecil, Eds. (Oxford: Pergamon Press, 1975), pp. 335-345.
4. Robertson, W. J. "Reverse Osmosis in Industrial Water Treatment," *Water Services* (January 1975).
5. Van Vuuren, L. R. J. "The Reuse of Sewage," in *Proceedings of the International Conference on Water Pollution Research, Vol. 1* (San Francisco: IAWPR, 1970), pp. 1-32.
6. Sammon, D. C., and B. Stringer. "The Application of Membrane Processes in the Treatment of Sewage," *Process Biochem.* (March 1975).
7. Metzler, D. F., and R. L. Culp. "Emergency Use of Reclaimed Water for Potable Supply at Chanute, Kansas," *J. Am. Water Works Assoc.* (June 1976).
8. Doud, D. H. "Field Experience with Five Reverse Osmosis Plants," *Water Sew. Works* (June 1976).
9. "Desalination 1977," Standing Committee on Water Treatment, National Water Council (1977).
10. Besik, F. "Some Aspects of Renovation of Domestic Sewage," *Water Res.* 8:111-117 (1975).
11. Jennett, J. C. "Wastewater Reclamation by Reverse Osmosis—Treatment of Wastes Produced," *J. Water Poll. Control Fed.* 16(3) (1971).
12. Guiry, M. "Potential of Reverse Osmosis Concentrate as a Basic Substrate in the Production of Seafoods," *Water Poll. Control* 77(4): 462-463 (1978).
13. Stead, P. A. "The Potential of Integrated Recycling of Solid and Liquid Wastes," *Water Poll. Control* 77(44):452-453 (1978).
14. "The Advanced Waste Research Program, Jan. 1962–June 1964," U.S. DHEW, PHS, AWTR-14 (April 1965).
15. Smith, J. M. "Hydrodynamic Flux Control for Wastewater Application of Hyperfiltration Systems," U.S. EPA Report R2-73-228 (May 1973).
16. Bailey, D. A. "The Reclamation of Water from Sewage Effluents by Reverse Osmosis," *Water Poll. Control* 73(3) (1974).
17. Cruver, J., and R. Nusbaum. "Application of Reverse Osmosis to Wastewater Treatment," *J. Water Poll. Control Fed.* 46(2) (1974).
18. Smith, J. M. "Hollow Fiber Technology for Advanced Waste Treatment," U.S. EPA Report R2-72-103 (December 1972).
19. Feige, W. A., and J. M. Smith. "Wastewater Applications with a Tubular Reverse Osmosis Unit," *Water–1973, Am. Inst. Chem. Eng.* (April 1974).

20. Beckman, J. E. "Control of Fouling of Reverse Osmosis Membranes When Operated on Polluted Surface Waters," Gulf Environmental Systems Report GA-10232, U.S. Dept. of the Interior (1970).
21. Winfield, B. A. "A Study of Factors Affecting the Rate of Fouling of Reverse Osmosis Membranes Treating Secondary Sewage Effluents," *Water Res.* 13:565-569 (1979).
22. Winfield, B. A. "The Treatment of Sewage Effluents by Reverse Osmosis—pH Based Studies of the Fouling Layer and Its Removal," *Water Res.* 13:561-564 (1979).
23. Cooper, W., Glasshouse Crops Research Institute, Littlehampton, England. Personal communication (1979).
24. Stead, P. A., D. Bone and B. A. Winfield, Portsmouth Polytechnic, England. Unpublished results (1979).
25. Duvel, W. A. "Removal of Wastewater Organics by Reverse Osmosis," *J. Water Poll. Control. Fed.* 47(1):57-65 (1975).
26. Hindin, E., P. J. Bennett and S. S. Narayanari. "Organic Compounds Removed by Reverse Osmosis," *Water Sew. Works* (December 1969).
27. Sorber, C. A. "Virus Rejection by the Reverse Osmosis Ultrafiltration Process," *Water Res.* 6:1377-1388 (1972).

SECTION 4

DISINFECTION

CHAPTER 18

RATIONAL APPROACHES IN THE ANALYSIS
OF CHEMICAL DISINFECTION KINETICS

Charles N. Haas

Department of Chemical and Environmental
   Engineering
Rensselaer Polytechnic Institute
Troy, New York

Disinfection has been regarded as the final and crucial barrier preventing the exposure of human populations to pathogenic organisms via ingestion of potable water or recreational exposure to waste water. Until recently, only economic constraints prevented the operation of the disinfection systems of water or wastewater treatment plants using excessive disinfectant dosages to assure the continual satisfaction of microbial standards. Recent concern for the health effects of organic reaction by-products from wastewater disinfection has forced a reexamination of this approach.

In order to begin attempts at optimal design and operation of disinfection processes, it is necessary to proceed toward development of kinetic models which may readily be interfaced with various transport and reactor models currently being used in environmental engineering. In particular, in the application to wastewater disinfection, where disinfectant decomposition is an integral facet of the process, the examination of detailed kinetics is most critical.

It is the object of this chapter to outline procedures for development of a model for chemical disinfection, focusing on chlorine inactivation of enteric viruses as a case study. Disinfectant decomposition and the effect of mixing rate at the point of application will be considered, and needs for future research will be discussed.

PROCESS CONCEPTUALIZATION

The development of the model will proceed from a division of the overall disinfection process into several distinct subprocesses. For each subprocess, previously developed theory will be employed for quantitative descriptions.

If it is considered that a water or waste stream containing microorganisms is combined with a fluid (liquid or gas) stream containing disinfectant, the overall process which is descriptive of the microbial inactivation may be considered as the result of the following:

1. The disinfectant and water or waste streams are mixed, and the degree of disinfectant heterogeneity is reduced as mixing proceeds. Simultaneously the disinfectant may be decomposed or transformed to less active species, such as chloramines, by contact with the waste stream.
2. The disinfectant must be transported from the bulk solution to the external surface of the microorganism which is to be inactivated.
3. The active species of disinfectant must migrate to the site within the microorganism at which it exerts a lethal effect.
4. Inactivation, per se, occurs at a rate determined by the local disinfectant concentration at the lethal site.

The quantitative discussion that follows is designed to determine when, and if, any of the above steps may limit the disinfection process. By restricting analysis to the inactivation of viruses by chlorine, it may be assumed that the viricidal effect of chloramine species is negligible with respect to free chlorine [1,2]. In addition, if the pH is assumed to be below the pK for hypochlorous acid dissociation, only the cidal effect of the free acid need be considered.

Two extreme cases will be analyzed as an example of rational approaches to analysis. These are (1) where chlorine demand is very low, and the initial rate of mixing sufficiently high so as to make the resistance of step 1 negligible, and (2) where chlorine demand is much greater than the chlorine dose, and mixing may be limiting. These can be recognized as likely situations in potable water treatment, and in disinfection of nonnitrified wastewater effluents, respectively.

CASE I–NEGLIGIBLE DEMAND, HIGH MIXING

In this situation, three potential limiting steps are important: transport from the bulk solution to the organism, intraorganism transport and intrinsic kinetics of the lethal effect per se. In addition to being applicable to potable water disinfection, this case is also pertinent to the analysis of laboratory studies in chlorine demand–free systems.

Under what circumstances may mass transport from the bulk solution to the microbial surface be limiting? From a traditional mass transport view, one can write:

$$R_{\ell m} = k_{\ell m}\, a(C_\ell - C_m) \tag{1}$$

where $R_{\ell m}$ = the rate of mass transport of disinfectant to the microbial exterior
     $k_{\ell m}$ = the mass transfer coefficient
        a = the surface area
C_ℓ and C_m = the liquid and surface concentrations of disinfectant, respectively

The mass transfer coefficient must be estimated from available correlations.

The disinfection process may be viewed as mixing approximately spherical particles in suspension in a liquid. It is suggested [3] that the most applicable mass transfer correlation for this case is that of Brian et al. [4]. In this procedure, the quantity $(\epsilon d_p^4 \rho_\ell^3 / \mu^3)^{1/3}$ is correlated to $N_{sh}/N_{sc}^{1/3}$, where ϵ is the power input per unit mass; d_p is the particle, or microorganism, diameter; ρ_ℓ is the liquid density; μ is the viscosity; N_{sh} is the Sherwood number, incorporating $k_{\ell m}$; and N_{sc} is the fluid Schmidt number.

Using the relation of Camp and Stein [5], the power input per unit mass may be determined as a function of the rms velocity gradient (G):

$$\epsilon = G^2 \mu / \rho_\ell \tag{2}$$

Equation 2 holds, providing that the mass fraction of particles (microorganisms) is negligible. From the use of Equation 2 and the mass transfer correlation of Brian et al. [4], it can be shown that, for typical situations at normal temperatures, particle diameters below 1 μm and G values less than 10^5 sec^{-1}, assuming a bulk liquid chlorine diffusivity (D_ℓ) of 10^{-5} cm^2/sec, the Sherwood number is essentially 2.0. In other words, the primary transport mode is simple diffusion, which would yield such a minimum Sherwood number [3,6]. The value for the mass transfer coefficient may thus be written as:

$$k_{\ell m} = 2.0\, D_\ell / d_p \tag{3}$$

Combining Equations 1 and 3 and assuming spherical particles results in:

$$R_{\ell m} = 2.0\, \pi D_\ell d_p\, (C_\ell - C_m) \tag{4}$$

The analysis of bulk to exterior mass transfer indicates that, for particles less than 1 μm, and at all mixing intensities commonly employed ($G < 10^5$ sec^{-1}), no significant increase in the rate of this process will occur with mixing. If this process were the only significant one limiting disinfectant uptake by a microorganism, a characteristic time could be defined as:

$$t_\varrho = d_p^2/12\,D_\varrho \tag{5}$$

For the case of chlorine and viruses, this characteristic time is on the order of 1 μsec. Because the apparent inactivation rate and rate of chlorine uptake [7-9] of viruses is much slower than this, it would appear that bulk to surface transport is neither a significant limiting process, nor likely to be affected by these mixing levels.

The next factor for consideration then becomes the intraorganism transport of disinfectant. Two questions arise in determination of an appropriate modeling approach for this step. First, is simple intraorganism diffusion from the microbial exterior sufficient to explain the observed behavior, or is it necessary to postulate an internal chemical reaction? Second, if a chemical reaction occurs within the organism, is it significantly limited by internal diffusion to the extent that a gradient of concentration exists within the particle?

To answer these questions it becomes necessary to examine experimental results on chlorine dynamics during disinfection of various microorganisms. In particular, the uptake of chlorine from solutions of free chlorine at pH values below the pK of hypochlorous acid has been determined using ^{36}Cl by Dennis and co-workers [8,9] and Haas and co-workers [10-12] for f2 and for various vegetative microorganisms, respectively.

If simple diffusion were sufficient to explain the uptake of chlorine by microorganisms, one would expect that, at steady-state, the amount of chlorine found associated with the organism would be a linear function of external solution concentration. In other words, a linear adsorption isotherm should result. Furthermore, if simple diffusion were the sole mechanism governing intraparticle dynamics, and it was assumed that the microorganism can be represented as a sphere, an unsteady-state diffusion equation could be solved for conditions of constant surface composition. Using the solution of Dankwerts [13] for such a system, it can be shown that:

$$Q/Q_{max} = 1 - \frac{1}{6\pi^2} \sum_{n=1}^{\infty} \frac{1}{n^2} \exp(-n^2 t/t_d) \tag{6}$$

where $\quad t_d$ = a characteristic diffusion time, defined as $t_d = d_p^2/4\pi^2 D_m$

$\qquad D_m$ = the intraorganism diffusivity

$\qquad Q$ = the mass of disinfectant associated with each organism at time t

$\qquad Q_{max}$ = the ultimate amount of such mass

At sufficient intervals such that t/t_d is greater than 0.5, an error of less than 6% will be made by dropping all but the first term in Equation 6, thus yielding:

$$Q/Q_{max} = 1 - \frac{\exp(-t/t_d)}{6\pi^2} \tag{7}$$

Clearly, by Equation 7, the process is essentially complete when $t = t_d$. Using this equation, if the intracellular transport were solely governed by diffusion, a kinetic measurement of chlorine uptake could be used to determine a t_d from the time when 99% of ultimate uptake occurred. Since the microbial size is known, this implies a diffusivity, and it would be expected that this value be in the same range of diffusivities of similar materials. The data of Haas and co-workers [10-12] show that chlorine uptake by *Escherichia coli* reaches completion in about 300 seconds. The data of Dennis and co-workers [8,9] do not precisely indicate a completion time for uptake of chlorine by f2 virus; however, at least 30 seconds would appear to be required. These values imply a D_m of 4.75×10^{-12} cm^2/sec in *E. coli*, and less than 1.3×10^{-14} cm^2/sec in bacteriophage f2. These apparent diffusivities are 7-9 orders of magnitude less than diffusivities in aqueous solution. It would therefore appear that the diffusion model is inconsistent with experimental observations of the kinetics or chlorine uptake. Furthermore, steady-state values of cell-associated chlorine in *E. coli* and two other vegetative microorganisms [10-12] are inconsistent with a linear model of C_ϱ vs Q, as would be predicted. It, therefore, appears that influences other than diffusion govern intraorganism chlorine dynamics.

Strong support for the argument that chemical reaction, including formation of organochlorine complexes, must be significant arises from calculations of the equilibrium concentration of free chlorine assuming only diffusion. The concentration of free chlorine to be expected within the microorganism at equilibrium should be identical with the external concentration, if the activity coefficients of the species are identical in bulk and within the organism. On this basis, a solution of 1 mg/l chlorine should produce approximately 0.2 and 60,000 molecules of chlorine/organism in f2 and *E. coli*, respectively, at steady-state. Observed values for organism-associated chlorine after long contact times are several orders of magnitude in excess of concentrations predicted on this basis [8-12]. The potential for hypochlorous acid to react with various biological compounds is well known and has been reviewed [10,12].

Since the above argument indicates that some intraorganism reaction is an important factor in determining uptake from solution, it now becomes important to ask whether this reaction is limited by intraorganism diffusion.

This will determine whether spatial concentration variations exist within the organism, and whether the organism can be modeled using a pseudohomogenous kinetic paradigm.

This question is equivalent to a determination of the influence of intraorganism diffusivity on the overall rate of disinfectant uptake from the bulk, since it has been suggested by the above discussion that bulk to organism transport is relatively fast. If a local intrinsic reaction rate for adsorption, reduction and other transformations of chlorine were available or presupposed, local mass balances within the organism could be developed in the form of partial differential equations. For various simple geometries and reaction rates, these mass balance equations have been solved analytically [13, 14]. The degree to which diffusivity influences overall reaction could then be determined by direct calculation. Unfortunately, this method assumes that the actual reaction rate expressions are known a priori.

An alternative approach was proposed by Weisz [15], who developed the concept of an observable reaction modulus in heterogenous systems. In the present case, this modulus may be defined as:

$$\Phi = (d_p^2/4D_m) \, (1/C_\ell) \, (dn/dt) \tag{8}$$

where D_m = the disinfectant diffusivity within the microorganism
     dn/dt = the observed rate of uptake per unit volume of microorganisms

Weisz [15] has shown that, for arbitrary reaction kinetics and particle geometry, intraparticle diffusion provides negligible resistance if Φ is less than 0.3. If diffusional resistances within the particle could be neglected, the concentration of disinfectant within the microorganism could be regarded as being constant at all locations. If this were the case, the process of disinfectant uptake from the cell exterior could be modeled as a set of ordinary differential equations, rather than a computationally more difficult set of partial differential equations.

For the disinfection of viruses and bacteria by free chlorine at low pH, data are available to estimate the reaction modulus. Dennis and co-workers [8,9] and Haas and co-workers [10-12] determined the kinetics of chlorine uptake from HOCl solutions using ^{36}Cl by bacteriophage f2 and *E. coli*, respectively. Using these data, and estimates for D_m, assuming $D_m = 10^{-5}$ cm^2/sec, the reaction modulus may be calculated as in Table I. The results of these calculations indicate that in the case of bacteriophage f2, unless D_m is three orders of magnitude lower than typical aqueous diffusivities, intraorganism diffusion is not a significant barrier to chlorine uptake. In the case of *E. coli*, the modulus is only 1/3 of a value where diffusion might be significant. It would presently appear to be reasonable to assume that, based on these calculations, a homogenous model would be appropriate for f2 virus, while a heterogenous model might be necessary to describe *E. coli* chlorine dynamics.

The above considerations may be used to formulate a mass balance for chlorine within the microorganism. If it is accepted that the interior of the microorganism is uniform in concentration, this can be written as:

$$dC_m/dt = (6k_{\ell m}/d_p)(C_\ell - C_m) - R \qquad (9)$$

where C_m = the internal concentration of chlorine within the microorganism
R = the rate of reaction of chlorine within the microorganism

Equation 9 cannot be evaluated until a functional form for R is assumed.

A rate law of the form $R = kC_m^x$ can be ruled out on the basis of experimental evidence [8-13]. If such a form was appropriate, under constant external conditions the total amount of chlorine found within the microorganism, including that chlorine which had reacted, would increase indefinitely with time. Kinetic studies on chlorine uptake previously discussed [8-13] appear to show the existence of a diminishing reaction rate with time, which is inconsistent with this model. Furthermore, a model of the form $R = kC_m^x S^y$, where S represents the intraorganism concentration of presumed binding or reaction sites which are irreversibly consumed by chlorine, is inconsistent with the observed data, since this would predict that, as long as C_ℓ was constant, the steady-state value of total organism associated chlorine would be constant, and would be determined by the organism concentration of reaction sites.

Therefore, it would seem that a reversible binding model may be descriptive of the data observed. A model of the following form will be postulated, and the fit to experimental data of the final integrated rate law tested:

$$R = k_1 C_m S - k_{-1} [CS] \qquad (10)$$

Table I. Calculated Reaction Moduli for Chlorine Uptake by
Bacteriophage f2 and *E. coli*

Quantity	f2	*E. coli*	Comments
d_p (cm)	2.25×10^{-6}	1.5×10^{-4}	
dn/dt (mol/1-sec)	2.09×10^{-2}	9.53×10^{-3}	Initial rate
C_ℓ (mol/1)	4.29×10^{-6}	5.71×10^{-5}	
Φ	6.2×10^{-4}	9.4×10^{-2}	D_ℓ assumed 10^{-5} cm^2/sec

If it is assumed that each organism possesses β binding sites, and that the decomposition of the complex, [CS], is totally reversible and not restricted by diffusion, then the following three equations, together with initial conditions, specify the overall chlorine uptake process:

$$dC_m/dt = (6k_{\ell m}/d_p)(C_\ell - C_m) - k_1 C_m S + k_{-1}[CS] \tag{11}$$

$$d[CS]/dt = k_1 C_m S - k_{-1}[CS] \tag{12}$$

$$N_o \beta = S + [CS] \tag{13}$$

In Equation 13, N_o represents the total number of organisms per volume of liquid.

The first term in Equation 11 is due to liquid-to-surface mass transfer. As discussed above, this process should be completed in a characteristic time on the order of microseconds or less for viruses. The observed kinetics of uptake indicate a much slower overall process, which suggests that a useful approximation to Equations 11 and 12 may be made assuming $C_\ell = C_m$. Combining this assumption, and the substitution of Equation 13, the following are obtained:

$$dC_m/dt = -k_1 C_\ell (\beta N_o - [CS]) + k_{-1}[CS] \tag{14}$$

$$d[CS]/dt = k_1 C_\ell (\beta N_o - [CS]) - k_{-1}[CS] \tag{15}$$

Since the major aim of this paper is the modeling of inactivation, it is now necessary to relate microbial survival to the concentration parameters in Equations 14 and 15. It is postulated that Chick's law holds, with an inactivation rate proportional to the mass of complexed chlorine per organism. Implicit in this mathematical presentation has been the assumption that inactivation per se does not affect binding capabilities. Therefore one can write:

$$dN/dt = -k_2 [CS] N/N_o \tag{16}$$

It should be noted that the group $[CS]/N_o$ represents the amount of complexed chlorine per organism. If the initial concentration of chlorine is in excess of the maximum binding capacity C_ℓ is nearly constant, and Equations 15 and 16 may be solved analytically to obtain:

$$ln \frac{N}{N_o} = \frac{-k_2 C_\ell \beta}{C_\ell + K_D} \left(t + \frac{e^{-k_i t(C_\ell + K_D)} - 1}{k\,(C_\ell + K_D)} \right) \tag{17}$$

It is to be noted that the model represented by Equations 15 and 16 is formally analogous to Michaelis-Menten enzyme kinetics [16]. Therefore, in Equation 17, K_D is defined as a Michaelis constant, i.e., $K_D = k_{-1}/k_1$.

This model has three salient features important to the current analysis. First, it predicts an apparent lag which is concentration-dependent: $t_{lag} = 1/k_1(C_\ell + K_D)$. Second, it predicts that the rate of inactivation, defined as the slope of the linear portion of a plot of $ln(N/N_o)$ vs t, can be given as a Monod-type function:

$$R_I = k_2 C_\ell \beta/(C_\ell + K_D) \tag{18}$$

Equation 17 predicts an inactivation curve which has a "shoulder" on a classical semilog plot. This type of inactivation curve has usually been modeled using multihit or multitarget models [17]. In such classical models, the apparent hit number is fixed regardless of concentration, whereas in Equation 17 it can be shown that the hit number is a function of concentration.

As previously described [18], the above model has been tested quantitatively against the data of Floyd et al. [7] for poliovirus Type I inactivation by free chlorine at pH 6. For each of the four experimental temperatures used, the inactivation data for all concentrations of chlorine employed were used to determine best fit values of the constants in Equation 17. Values obtained are given in Table II. Correlations between predicted inactivation fractions (log N/N_o) and observed results were highly significant (P < 0.001), with correlation coefficients ranging between 0.969 and 0.981. Regression slopes were not significantly different from unity, and y intercepts were negligible. It would therefore appear that the model as derived is accurate at describing inactivation of viruses, assuming poliovirus Type I is a representative member, by hypochlorous acid under demand-free conditions.

The determination of each of the model parameters at several temperatures permits the determination of activation energies and reaction enthalpies (Table 2). This analysis indicates that the binding reaction is moderately exothermic ($\Delta H^{binding} = -23$ kcal/mol), with an activation energy of 17.8 kcal/mol, suggesting a chemical complex as the chlorine-site species—in accord with model assumptions. The activation energy for the second step inactivation process is relatively low, suggesting that the inherent sensitivity of the virus to chlorine is not as affected by temperature as the binding and transport properties of the disinfectant within the organism.

Table II. Kinetic Coefficients for HOCl Inactivation of Poliovirus Type I (Mahoney) (Calculated from Data of Floyd et al. [7])

Temperature ($^\circ$C)	k_1 (liter/μmol-sec)	$k_2\beta$ (sec^{-1})	K_D (μmol/1)
2	0.00279	1.335	51.8
10	0.00973	0.721	12.3
20	0.0238	1.490	9.7
30	0.0389	3.085	2.9
kcal/mol	ΔE_a=17.8	ΔE_a=5.85	ΔH=+23

In addition to the reasonable fit to previously reported data on HOCl inactivation of poliovirus [7], the model appears capable of explaining certain other phenomenological observations of disinfection kinetics. For example, Engelbrecht et al. [19] observed a plateau of the inactivation rate of poliovirus contacted with free chlorine. Data from the author's laboratory show that *E. coli* is inactivated by free chlorine at pH 7 at a rate approaching a maximum value with increasing chlorine concentrations [20]. Scarpino et al. [21] report a plateau of the inactivation rate in the case of poliovirus contacted with chlorine dioxide. Data by Floyd et al. [22] indicate that the plateau effect and the presence of a shoulder may be real phenomena in the inactivation of poliovirus by various bromine species.

In the early literature, attempts were made to describe the shoulder, or apparent lag phase, by semiempirical models of the type shown in Equation 19 [23-25]:

$$ln(N/N_o) = -kC^n t^m \tag{19}$$

Typically, for free chlorine n is close to one, and m has been reported to be on the order of 2.3-3.2. If the product $k_1 t(C + K_D)$ is sufficiently small, the exponential term in Equation 17 may be replaced by the first three terms of a series expansion to obtain:

$$ln(N/N_o) = -k_1 k_2 \beta C t^2 /2 \tag{20}$$

Therefore the common practice [25] of linearizing survival curves by plotting log survival vs t^2 appears consistent with the derived model.

CASE II–HIGH CHLORINE DEMAND

In the more general, and complex, case where chlorine demand is high, bulk solution disinfectant decomposition and mixing of the disinfectant and waste streams causes C_ϱ to decrease with time, mixing intensity and chlorine demand. Before proceeding to determine survival, therefore, the decay of chlorine in a turbulent mixing field must be modeled.

Presently, the discussion of case II may be greatly simplified by the following assumptions:

1. Chlorine demand is sufficiently high so that it may be regarded as constant, with a rate of exertion first order in both demand and free chlorine.
2. Within the time scale of interest, the biocidal effect of free available chlorine in solution is much greater than the effect of the decomposition products (i.e., combined chlorine).
3. The solution of chlorine and the waste stream are introduced at zero time into a plug flow contactor at steady state. The contactor imparts a uniform and isotropic turbulent mixing field upon the two streams.

Assumption 1 restricts our consideration to systems at high $N:Cl_2$ ratios, and will be violated as the breakpoint is approached. A generalization of this assumption would require detailed modeling of the system chemistry.

Assumption 2 would appear true for viruses at contact times typically employed. For more sensitive microorganisms, such as coliforms, it is necessary to have information on the kinetics of combined chlorine inactivation prior to model formulation.

Assumption 3 restricts consideration to an idealized chemical reactor. Back-mixing and nonideal flow patterns can be treated by subdivision of the reactor into spatial regions in the manner employed by Patterson [26]. Nonuniform application of mixing can be treated similarly. A generalization of this assumption requires the formulation and solution of a system of partial differential equations.

It should be recognized at the outset that the modeling of turbulent reactive flows is currently an active area of research in fluid mechanics. Consequently, the techniques and equations used to describe this aspect of the process must only be seen as best currently available techniques.

Assumptions 1 and 3 can be used along with arguments outlined by Patterson [26] to write:

$$\frac{d\overline{B}}{dt} = -k_3 (\overline{AB} - \frac{b'^2}{\rho_o}) \tag{21}$$

$$\frac{db'^2}{dt} = + r \tag{22}$$

$$r = \frac{-2b'^2}{4.1\left(\frac{L_S^2}{\epsilon}\right)^{1/3} + \left(\frac{\nu}{\epsilon}\right)^{1/2} \ln N_{Sc}} - 2k_3\left[\bar{A}\, b'^2 - \frac{\bar{B}\, b'^2\,(1-\gamma)}{\rho_o\,(1+\gamma)}\right] \tag{23}$$

$$\gamma = \frac{\overline{B^2} - \rho_o b'^2}{\overline{B^2} + \rho_o b'^2} \tag{24}$$

where t = residence time in the steady state plug flow reactor
     $\bar{A},\bar{B}$ = mean concentrations of chlorine demanding material and chlorine, respectively
     b' = root mean square chlorine concentration fluctuation
     k_3 = rate constant for chlorine decomposition
     L_s = scalar macroscale of turbulence
     ρ_o = molar ratio of chlorine to chlorine demand at $t = 0$
     ν = kinematic viscosity

In the above equations, four parameters require evaluation for particular disinfection applications. The apparent rate of chlorine decomposition may be considered to be the product of k_3 and $\bar{A}$, and may be evaluated from measurements of the rate of disappearance of chlorine residual. Alternatively, this product may be evaluated from separate measurements of chlorine demand and utilization of reported kinetic constants [27]. The power input per unit mass (ϵ) may be evaluated from a knowledge of the "G" value, using Equation 2. The scalar macroscale may be approximated as shown by Brodkey [28] as 0.239 R_c, where R_c is the characteristic radius of the reactor in which turbulent mixing is occuring.

The discussion of this case may be simplified if it can be shown that the time constants associated with turbulence and chemical reaction are slow relative to those associated with bulk liquid to microbial surface mass transfer. In particular, the time constant associated with chemical decay of free chlorine appears to be on the order of milliseconds to seconds [27]. The time constant for the turbulent mixing step may be given by the following, which results from the first term in Equation 23 [28]:

$$t_{turb} = \frac{1}{2}\left[1.578\left(\frac{L^2}{\epsilon}\right)^{1/3} + \left(\frac{\nu}{\epsilon}\right)^{1/2} \ln N_{Sc}\right] \tag{25}$$

The most extreme case, at G values approaching 10^6 sec^{-1} and characteristic radii as low as several centimeters, yields characteristic times from Equation 25 approaching 100 μsec. Because this is still several orders of magnitude slower than the characteristic time defined by Equation 5, the process of bulk to microbial surface mass transfer may be regarded as essentially instantaneous, as in the previous case.

If it is assumed that microorganisms randomly sample all parcels of fluid during mixing, then it may be shown that the model describing this situation consists of Equations 14 through 16 and 21 through 24. Any alternative assumptions of nonideal micromixing of microorganisms with the chlorine-waste stream would require experimental characterization of micromixing imperfections. It is simpler to discuss these equations once transformations to dimensionless variables are effected. This results in the following:

$$dy/d\tau = -(y - \sigma^2 y_o)d \tag{26}$$

$$d\sigma^2/d\tau = -2\sigma^2 \left[1/\Omega + d - \frac{y(1-\gamma)}{y_o(1+\gamma)}\right] \tag{27}$$

$$d\theta/d\tau = y - \theta(y + a) \tag{28}$$

$$dx/d\tau = -x\theta \tag{29}$$

$$\gamma = (y^2 - \rho_o\sigma^2 y_o^2)/(y^2 + \rho_o\sigma^2 y_o^2) \tag{30}$$

Dimensionless parameters are defined, along with useful interpretations of their meaning, in Table III.

Table III. Dimensionless Parameters in Turbulent Disinfection Model

Parameter	Defined Value	Interpretation
a	$k_{-1}/k_2\beta$	Normalized desorption rate
d	$k_3\bar{A}/k_2\beta$	Normalized chlorine decay rate
x	N/N_o	Fractional microorganism survival
y	$k_1\bar{B}/k_2\beta$	Normalized bulk concentration
θ	$[CS]/N_o\beta$	Fractional site coverage
σ	$b'/\bar{B}_o$	Normalized concentration fluctuation
τ	$k_2\beta t$	Dimensionless time
Ω	$k_2\beta\left[4.1\left(\dfrac{L_S^2}{\epsilon}\right)^{1/3} + \dfrac{\nu}{\epsilon}^{1/2} \ln N_{Sc}\right]$	Normalized inverse turbulence intensity

Sufficient data exist in the literature for the estimation of most of the parameters in the above model. If it is assumed that hypochlorous acid is the only significant biocidal species, it may be presumed that data obtained under conditions of no chlorine demand, i.e., d = 0, would yield constants of equal validity to the present case. This would allow use of data, such as in Table II, in this more complex situation.

The complexity of Equations 26–30 makes numerical integration of the model mandatory. To permit such integration, initial conditions must be assumed. By virtue of parameter definitions, at zero time, x = 1 and θ = 0. An initial value for y is set by assuming a value for applied disinfectant concentration. It is necessary to consider an appropriate initial condition for the relative rms concentration fluctuation, σ^2.

If it is considered that the initial situation at the point of mixing consists of a fluid stream of disinfectant and a fluid stream of waste water, then a fraction, α, may be defined as the ratio of disinfectant stream flow to total flow. The application of conservation of mass to that point, therefore, shows that the concentration of disinfectant in the applied stream instantaneously prior to mixing is $\overline{B}_0/\alpha$. The rms concentration fluctuation at this point may be calculated from the statistical definition, and it may be shown that the following holds:

$$\sigma_0^2 = (1 - \alpha)/\alpha \tag{31}$$

Given Equation 31, and values of all constants, the model may be solved numerically to yield inactivation curves under a variety of conditions.

As an illustration of the nature of the model simulation, numerical solution of Equations 26–31 under various conditions noted in Figure 1 was carried out. The solution was obtained using a program (SIMBAS) written by Bungay and employing a fourth-order Runge–Kutta algorithm. The program was written in BASIC and executed on a Data General NOVA minicomputer. The parameters chosen were designed to simulate a high level chlorination of a waste water at various mixing intensities and chlorine to total flow ratios. The nature of the solutions obtained under these particular conditions are typical of those which can be obtained under a variety of possible parameters.

The first conclusion that may be made is that these curves possess the declining rate feature often seen in the disinfection of wastewater [24, 29,30]. The declining rate obscures any shoulder to the inactivation curves which would be seen in the absence of decomposition.

A second conclusion which can be reached from this model is that even though the rate of disinfectant decomposition may be very high—in this case, d = 50 (this may be regarded as the rate of disinfectant decomposition divided by the maximum rate of inactivation)—substantial inactivation can occur over time intervals greater than the lifetime of measurable chlorine in solution. Mechanistically, what is shown by the simulations is that as long as free chlorine is present initially, some microbial uptake will occur and inactivation will continue beyond the time at which solution chlorine vanishes due to the slower rate of desorption from chlorine binding sites relative to bulk solution chlorine decomposition. Hence it is possible, within the framework of this model, for HOCl to serve as the proximate lethal agent even though it may have disappeared from solution by the time substantial inactivation develops. It has been reported that inactivation of microorganisms in wastewater can occur under conditions where HOCl has previously been consumed by decomposition [31]. While the authors of that study invoked the presence of unidentified reactive intermediates as the lethal agents, the current model appears to permit the rationalization of such results with known facets of the biocidal activity of free chlorine.

A third conclusion to be reached from the simulation is that the ratio of chlorine feed flow to wastewater flow may have a substantial influence on inactivation efficiency. White [32] restates a postulate that the use of the strongest possible feed concentration, and consequently the lowest disinfectant stream flowrate, is desirable. This is quantitatively shown by the above model.

The major disagreement between the derived model and experimental results concerns the effect of turbulence on disinfection efficiency. Several authors [32], notably Longley [29], have shown that increased turbulence during wastewater chlorination enhances inactivation of viruses. The results of the model (Figure 1) appear to show that turbulence has a negative effect on disinfection, due to an enhancement of disinfectant decomposition with mixing. It is necessary to discard the model or to postulate additional aspects of the wastewater chlorination process not covered in the above analysis.

For the present, the author would choose to speculate that the experimentally observed improvement of disinfection efficiency with increasing turbulence is due to the disruption of microorganisms from protective particulates at high power inputs. Insufficient data exist to incorporate such factors into the current analysis. However the ability of fluid shearing forces similar to those encountered in turbulent mixing to disrupt smaller macromolecules is well known [33]. It has been shown that a number of particulates can protect virus particles from the action of disinfectants [21,34,35]. It is postulated, without proof, that incorporation of some version of this shearing effect into the above model would rectify predictions with experimental studies of Longley [29] and others.

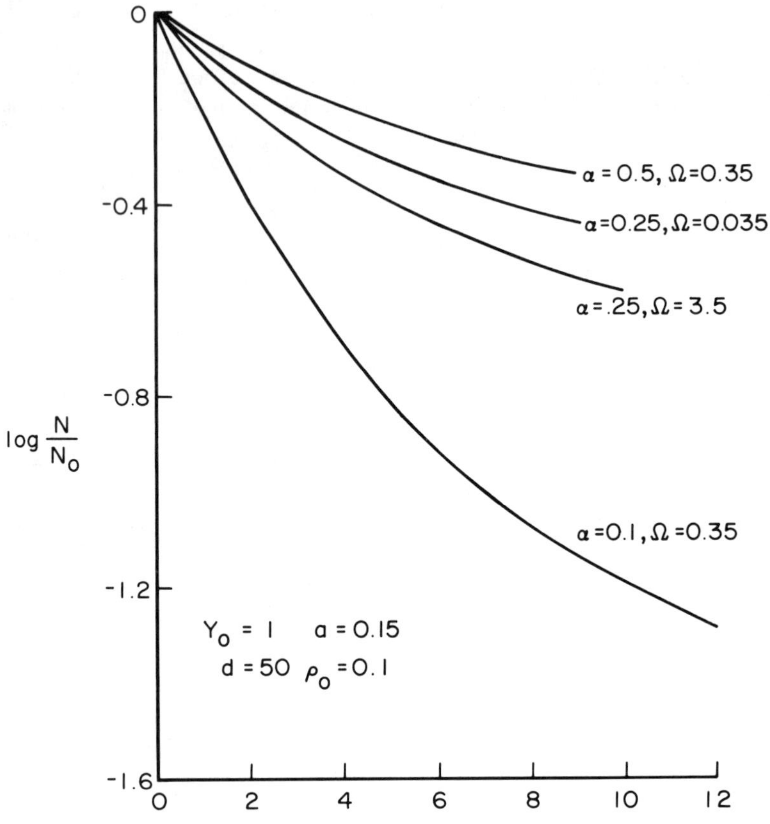

Figure 1. Numerical solution to the turbulent disinfection model.

SUMMARY

A conceptual model for the analysis of the disinfection process has been presented. Application of these ideas to the particular cases of potable water and wastewater chlorination has revealed many areas of agreement between the models and experimental results. Some areas of disagreement remain, notably the possibility that turbulence may influence apparent microbial sensitivity via particular disruption, and it is hoped that further research of both an experimental and analytical nature will enhance the understanding of this ultimate barrier against disease transmission.

ACKNOWLEDGMENTS

The ideas in this paper are entirely the responsibility of the author but have evolved after numerous fruitful conversations with colleagues at Rensselaer Polytechnic Institute and elsewhere. The author particularly acknowledges the constructive comments of D. Dadyburjor, C. Kleinstreuer, R. Rajagopalan, K. E. Longley and V. P. Olivieri.

REFERENCES

1. Chang, S. L. "Modern Concept of Disinfection," *J. San. Eng. Div., ASCE* 97:689-707 (1971).
2. "The Disinfection of Drinking Water," Report of the NAS Subcommittee on Efficacy of Disinfection to the U.S. Environmental Protection Agency (1979).
3. Satterfield, C. N. *Mass Transfer in Heterogenous Catalysis* (Cambridge, MA: MIT Press, 1970).
4. Brian, P. L. T., H. B. Hales and T. K. Sherwood. "Transport of Heat and Mass Between Liquids and Spherical Particles in an Agitated Tank," *Am. Inst. Chem. Eng. J.* 15:727-733 (1969).
5. Camp, T. R., and P. C. Stein. "Velocity Gradients and Internal Work in Fluid Motion," *J. Boston Soc. Civ. Eng.* 30:219 (1943).
6. Geankoplis, C. J. *Mass Transport Phenomena* (New York: Holt, Rinehart and Winston Inc., 1972).
7. Floyd, R., D. G. Sharp and J. D. Johnson. "Inactivation by Chlorine of Single Poliovirus Particles in Water," *Environ. Sci. Technol.* 13(4): 438-442 (1979).
8. Dennis, W. H. "The Mode of Action of Chlorine on f2 Bacterial Virus During Disinfection," PhD Thesis, Johns Hopkins University (1977).
9. Dennis, W. H., V. P. Olivieri and C. W. Kruse. "Mechanism of Disinfection: Incorporation of Cl-36 Into f2 Virus," *Water Res.* 13:363-369 (1979).
10. Haas, C. N. "Mechanism of Inactivation of New Indicators of Disinfection Efficiency by Free Available Chlorine," PhD Thesis, University of Illinois at Urbana-Champaign (1978).
11. Haas, C. N., and R. S. Engelbrecht. "Chlorine Dynamics During Inactivation of Coliforms, Acid-Fast Bacteria and Yeasts," *Water Res.* 14:1749-1757 (1980).
12. Haas, C. N., and R. S. Engelbrecht. "Mode of Microbial Inactivation By Chlorine," *Proceedings of the ASCE Environmental Engineering Division Specialty Conference* (New York: American Society of Civil Engineers, 1979), pp. 646-652.
13. Danckwerts, P. V. "Absorption by Simultaneous Diffusion and Chemical Reaction into Particles of Various Shapes and Into Falling Drops," *Trans. Faraday Soc.* 47:1014-1023 (1951).
14. Roughton, F. J. W. "Diffusion and Chemical Reaction Velocity in

Cylindrical and Spherical Systems of Physiological Interest," *Proc. Roy. Soc. B* 140:203-229 (1952).

15. Weisz, P. B. "Diffusion and Chemical Transformation," *Science* 179: 433-440 (1973).

16. Bailey, J. E., and D. F. Ollis. *Biochemical Engineering Fundamentals* (New York: McGraw Hill Book Co., 1977).

17. Smith, J. M. *Mathematical Ideas in Biology* (Cambridge: Cambridge University Press, 1968).

18. Haas, C. N. "A Mechanistic Kinetic Model for Chlorine Disinfection," *Environ. Sci. Technol.* 14:339-340 (1980).

19. Engelbrecht, R. S., M. J. Weber, C. A. Schmidt and B. L. Salter. "Virus Sensitivity to Chlorine Disinfection of Water Supplies," U.S. EPA Report 600/2-78-123 (1978).

20. Haas, C. N., and E. C. Morrison. "Repeated Exposure of *E. coli* to Free Chlorine: Production of Strains Possessing Altered Sensitivity," paper presented at the 80th Annual Meeting of the American Society for Microbiology, Miami Beach, FL, May 13, 1980.

21. Scarpino, P. V., F. A. O. Brigano, S. Cronier and M. L. Zink. "Effect of Particulates on Disinfection of Enteroviruses in Water by Chlorine Dioxide," U.S. EPA Report 600/2-79-054 (1979).

22. Floyd, R., D. G. Sharp and J. D. Johnson. "Inactivation of Single Poliovirus Particles in Water by Hypobromite Ion, Molecular Bromine, Dibromamine and Tribromamine," *Environ. Sci. Technol.* 12 (9):1031-1035 (1978).

23. Hom, L. W. "Kinetics of Chlorine Disinfection of an Ecosystem," *Proceedings of the National Specialty Conference on Disinfection* (New York: American Society of Civil Engineers, 1970), pp. 515-537.

24. Severin, B. F. "Inactivation of Proposed Indicator Organisms by Inorganic Chloramines," MS Special Problem Report, Department of Civil Engineering, University of Illinois at Urbana-Champaign (1975).

25. Fair, G. M., J. C. Geyer and D. A. Okun. *Water and Waste Engineering* (New York: John Wiley and Sons, 1968).

26. Patterson, G. K. "Simulating Turbulent Field Mixing and Reactors," in *Turbulence in Mixing Operations*, R. S. Brodkey, Ed. (New York: Academic Press, 1975).

27. Lietzke, M. H. "A Kinetic Model for Predicting the Composition of Chlorinated Water Discharged from Power Plant Cooling Systems," in *Water Chlorination: Environmental Impact and Health Effects, Vol. 1*, R. L. Jolley, Ed. (Ann Arbor, MI: Ann Arbor Science Publishers, Inc., 1978), pp. 367-378.

28. Brodkey, R. S. "Mixing in Turbulent Fields," in *Turbulence in Mixing Operations*, R. S. Brodkey, Ed. (New York: Academic Press, 1975).

29. Longley, K. E. "Turbulence Factors in Chlorine Disinfection of Wastewater," *Water Res.* 12:813-822 (1978).

30. Olivieri, V. P., T. K. Donovan and K. Kawata. "Inactivation of Virus in Sewage," *Proceedings of the National Specialty Conference on Disinfection* (New York: American Society of Civil Engineers, 1970), pp. 365-384.

31. Selleck, R. E., B. M. Saunier and H. F. Collins. "Kinetics of Bacterial

Deactivation with Chlorine," *J. Environ. Eng. Div., ASCE* 104:1197-1212 (1979).

32. White, G. C. *Disinfection of Wastewater and Water for Reuse* (New York: Van Nostrand Reinhold Co., 1978).

33. Charm, S. E., and B. L. Wong. "Enzyme Inactivation with Shearing," *Biotechnol. Bioeng.* 12:1103-1109 (1970).

34. Sproul, O. J., C. E. Buck, M. A. Emerson, D. Boyce, D. Walsh and D. Howser. "Effect of Particulates on Ozone Disinfection of Bacteria and Viruses in Water," U.S. EPA Report 600/2-79-089 (1979).

35. Hejkal, T. W., F. M. Wellings, P. A. LaRock and A. L. Lewis. "Survival of Poliovirus Within Organic Solids During Chlorination," *Appl. Environ. Microbiol.* 38:114-118 (1979).

CHAPTER 19

BIOLOGICAL EVALUATION OF METHODS FOR THE DETERMINATION OF FREE AVAILABLE CHLORINE

Michael C. Snead and Vincent P. Olivieri

The John Hopkins University
School of Hygiene and Public Health
Baltimore, Maryland

William H. Dennis

U.S. Army Medical Research and Development Laboratory
Ft. Detrick
Frederick, Maryland

The marked difference in the biocidal activity of the free and combined species of chlorine has repeatedly been noted in the literature [1-6]. The demonstration of a free available chlorine (FAC) residual provides a rapid method to evaluate the disinfection process and assess residual biocidal activity. The National Interim Primary Drinking Water Regulations [7] allow the substitution of FAC residual measurements for up to 75% of the samples for microbiological analysis in water systems. Thus, the FAC residual measurement becomes an indirect measure of microbiological quality, and the reliable differentiation of FAC from combined available chlorine is imperative. False positive measurements of FAC due to the less biocidal chlorine species provides a false assurance of safety.

Standard Methods [8] lists five methods for determination of free chlorine in water: (1) amperometric titration; (2) stabilized neutral orthotolidine

(SNORT); (3) N,N-diethyl-*p*-phenylene diamine (DPD), titrimetric and colorimetric; (4) leuco crystal violet; and (5) syringaldazine (FACTS). The National Interim Primary Drinking Water Regulations sepcify the DPD colorimetric procedure for free chlorine measurements that are to be substituted for microbiological samples. As a result of this there has been a proliferation of manufacturers marketing DPD test kits with little interest in alternative procedures. However, there appears to be some controversy concerning the specificity of this and other methods for free chlorine and the applicability of one test to all situations.

Field test kits are generally limited to colorimetric methods. Amperometric titration, recognized in the United States as reliable procedure for the determination of FAC does not lend itself to field use. Cooper et al. [9] and Sorber et al. [10] evaluated six existing field test procedures for the determination of FAC. The three most promising were FACTS, DPD and SNORT. In a subsequent study, Meier et al. [11] evaluated the specificity of the glycine modification of the DPD method and the FACTS procedure. The DPD method gave false positive readings for FAC in the presence of monochloramine (NH_2Cl), dichloramine ($NHCl_2$), trichloramine (NCl_3) and chlorinated natural water known to contain only combined chlorine. The FACTS procedure did not yield false positive measurements at equivalent levels of NH_2Cl and $NHCl_2$ and in the chlorinated natural water, but did give false positive measurements with NCl_3. DPD was found to be more precise and accurate than FACTS.

Strupler [12], in a study of interferences of free chlorine measurement by the DPD and FACTS procedures, found no interference in the DPD procedure at monochloramine levels up to 4 mg/l and at dichloramine levels up to 10 mg/l. FACTS gave false positives with trichloramine; the response obtained was on the order of 70% of that obtained for equivalent concentrations of free chlorine. The DPD procedure correctly distinguished free chlorine and trichloramine in a mixture of the two compounds.

Palin [13] reported a DPD modification, coined DPD-Steadifac, that was more specific for FAC than were prior DPD methods. DPD-Steadifac has not been evaluated in other laboratories.

Another untested procedure, the free chlorine membrane electrode, has recently become available [14]. The membrane electrode uses a microporous membrane and a positive cathode potential to obtain selectivity for HOCl. Monochloramine and dichloramine were found to produce electrode responses of 1.3-3.0% of that of HOCl, while trichloramine produced a response 7.25 times that of HOCl. No interference from OCl^- was found. The authors suggested that the membrane electrode can be calibrated directly as a function of disinfection efficiency since it measures only the actively germicidal species (HOCl).

Since the intent of the chlorine residual determination is to reflect the biocidal activity of the solution, a biological system would provide a more meaningful procedure to evaluate the performance of existing and proposed methods for the measurement of chlorine residuals. The biological system can serve as the ultimate referee to allow a practical interpretation of the results of chlorine residual measurements by different procedures. This approach was used by Savage and Stratton [15] to determine qualitatively the performance of the orthotolidine chlorine test and syringaldazine test strips in assaying the microbiological quality of swimming pool water. The presence of viable bacteria (*Streptococcus faecalis*, *Escherichia coli* and *Pseudomonas aeruginosa*) was used to determine the validity of free chlorine measurement. Although this procedure was adequate for this short-term qualitative test, the susceptibility of bacteria to combined chlorine makes them a poor choice for a more rigid test, specifically for free chlorine.

The purpose of this study was to develop a biological reference procedure (Biofac) for the qualitative and quantitative determination of free chlorine in solutions containing compounds that may interfere with the colorimetric chemical methods and to use this procedure to compare the specificity of the DPD, FACTS, amperometric and electrode procedures for free chlorine. The bacterial virus f2 was chosen as the test organism for the development of the Biofac procedure, since f2 is resistant to inactivation by combined chlorine [16,17] and sensitive to free chlorine [18].

METHODS

Chlorine

All chlorine, chloramine and buffer solutions were prepared either in water from an acid permanganate distillation (organic-free distilled water) using the method developed by Soper [19] as described in detail by Olivieri [20] or in double-distilled deionized water. Stock hypochlorous acid solutions were prepared by washing and collecting high-purity chlorine gas in organic-free distilled water utilizing the method outlined by Olivieri [18]. Stock solutions contained approximately 4000 mg/l of FAC.

Monochloramine

Stock monochloramine solutions were prepared using the method developed by Granstrom [21] as described by Johnson and Overby [22]. Equal volumes of 0.0100 M ammonium chloride and 0.0033 M hypochlorite at pH 10 were mixed and allowed to react for at least one hour before use.

Amperometric titration and ultraviolet (UV) absorption spectrophotometry showed that stock solutions contained about 100 mg/l of monochloramine. Previous work [16,22] has shown that monochloramine solutions prepared by this procedure do not contain detectable amounts of free chlorine or other inorganic chloramines.

Dichloramine

Stock dichloramine solutions were prepared using a method employed by Chapin [23] and modified by Richfield [17]. Equal volumes of 0.0264 M ammonium chloride and 0.0132 M hypochlorous acid buffered at pH 4.6 with 0.08 M acetate were mixed and allowed to react overnight. Amperometric titration and UV absorption spectrophotometry showed the stock solutions to contain about 400 mg/l of dichloramine. Richfield indicated that this procedure results in a solution containing 95-97% of the available chlorine in the form of dichloramine, with the remaining chlorine as monochloramine. No free chlorine was detected.

Trichloramine

Stock trichloramine solutions were prepared using the method of Saguinsin and Morris [24]. Equal volumes of 0.004 M ammonium chloride and 0.0120 M hypochlorous acid at pH 2.3 were mixed and allowed to react overnight. Amperometric titration and spectrophotometry showed that stock solutions contained about 300 mg/l of trichloramine. Dilution and adjustment of pH to that of the experimental run was done immediately before the run. A 30-fold molar excesses of ammonia was also added at this time to suppress free chlorine formation from the NCl_3 [24].

Chlorine Determination

Chlorine and chloramine concentrations were measured by amperometric titration [8] using a Sargent-Welch model XVI polarograph. UV absorption spectra were determined with a Heath-Schlumberger model 707 spectrophotometer. Concentrations were determined spectrophotometrically using the molar absorptivities determined by Galal-Gorchev and Morris [25]. Measurement of free chlorine concentration with DPD was accomplished by adding 20 ml of test solution to two standard scoops (approximately 0.2 g) of LaMotte DPD powder or two crushed Hellige DPD tablets. The solution was mixed immediately and the absorbance at 515 nm in 1-in. cells was read within 30 sec on a Bausch and Lomb Spectronic 20 spectrophotometer. Two drops of 10% thioacetamide were added within 30 sec of mixing of the test solution and DPD for the DPD-Steadifac tests [13]. The measurement of free chlorine concentration by syringaldazine (FACTS) was accomplished by adding 5 ml of test solution to 0.17 ml of FACTS buffer and 1.7 ml of syringaldazine solution [8] followed by mixing and an immediate absorbance reading at 530 nm in 0.5-in. cells. Measurement of free chlorine concentration by the membrane electrode was accomplished by exposing the electrode to approximately 100 ml of test solution with mixing. Concentration readings were taken after three minutes of exposure.

Biological Preparations and Assays

The f2 bacterial virus (ATCC #15766) was prepared by the method of Loeb and Zinder [26] and purified by polyethylene glycol precipitation [27]. The purified virus at concentrations tested produced no chlorine demand. The f2 bacterial virus was assayed by the agar overlay method described by Adams [28] on tryptone yeast extract (TYE) media, with *E. coli* K-13 (ATCC #15766-B) used as host bacterium.

Experimental Methods

The reaction system shown in Figure 1 was used in the study and has been described in detail [18]. The system was modified by using a 500-ml baffled beaker in place of a trypsinizing flask and by using a pump-driven syringe instead of a manually operated syringe. Before the experiments stock hypochlorous acid solutions were diluted in organic-free distilled water to the needed concentrations and the pH was adjusted and maintained by 0.008 M phosphate buffer. Monochloramine stock solutions were diluted in double-distilled deionized water and the pH was adjusted and maintained with 0.008 M phosphate buffer. Dichloramine stock solutions were also diluted in double-distilled deionized water, and the pH was adjusted and maintained by 0.08 M phosphate buffer. Trichloramine stock solutions were diluted in double-distilled deionized water to the needed concentration, and the pH was adjusted to the experimental level and maintained by 0.17 M phosphate buffer. A 30-fold molar excess of ammonium chloride was added to ensure the absence of free chlorine just prior to experimentation. A baffled beaker was filled with chlorine or chloramine solution at the experimental concentration and pH and was allowed to equilibrate to 20°C in a water bath. Samples were withdrawn for free chlorine residual analysis by amperometric titration, DPD with tablet reagents (DPDT), DPD-Steadifac with tablet reagents (DPDTSF), DPD with powder reagents (DPDP), DPD-Steadifac with powder reagents (DPDPSF) and FACTS immediately before the f2 inactivation test. Since the membrane electrode was not received until the latter part of

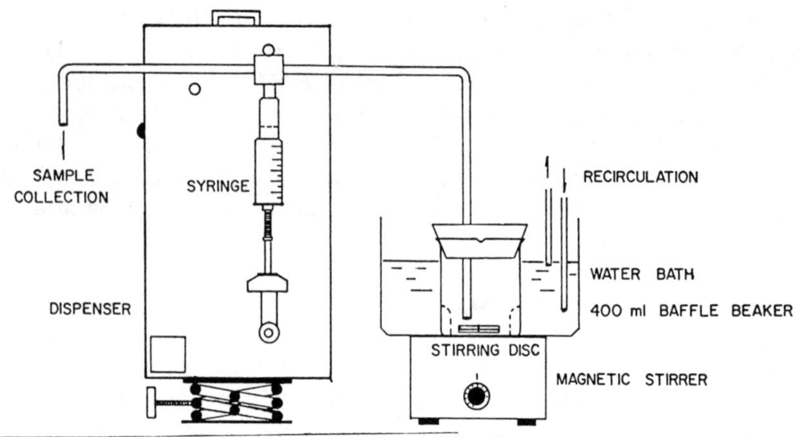

Figure 1. Reaction system used for inactivation experiments.

the study, data for this procedure were obtained only for the later experimental runs. At time zero, f2 virus was added, with mixing, to the chlorine or chloramine solution in the baffled beaker to give an f2 concentration of approximately 10^6 PFU/ml. Samples were withdrawn with time into sterile tubes containing thiosulfate. At the end of the inactivation run, samples were again taken for free chlorine residual measurement by all of the procedures. The survival of the virus was assayed by the procedure given above.

Biofac Calibration

Fair and Geyer [29] give the following relationship for contact time and the inactivation of microorganisms with disinfectant.

$$\frac{N}{N_o} = e^{-kt}$$

where N_o = the number of microorganisms at time zero
N = the number at any time t
k = the coefficient of proportionality or rate constant.

By taking the natural logarithm

$$ln\frac{N}{N_o} = -kt$$

Thus a plot of lnN/N$_0$ v time yields a line with slope -k, the rate constant. Taking the logarithm to base 10

$$\log_{10} \frac{N}{N_0} = -k't$$

A plot of $\log_{10}$ N/N$_0$ v time gives a line with slope -k$'$, the coefficient of proportionality. k$'$ is related to k by

$$k = (k')(2.303)$$

k$'$ was used in the subsequent calculations and was referred to as the rate of inactivation.

Rates of inactivation of f2 virus by 0-1 mg/l of free chlorine at pH 7.0 were determined by the procedure given above. Free chlorine solutions were prepared in double-distilled deionized water and measured by amperometric titration.

Free Chlorine Measured After Exposure to Chloramine Solutions

A series of experiments was run to compare free chlorine residuals measured by DPD, FACTS and Biofac after exposure to monochloramine, dichloramine and trichloramine. These tests run in the same manner as the tests with free chlorine except triple-distilled deionized water was used in place of organic-free distilled water. Additional thioacetimide was added to the DPD-Steadifac test at higher levels of combined chlorine, as called for by Palin [13].

Nitrogen Breakpoint

The addition of chlorine to an ammonia solution to the point at which an irreducible concentration of ammonia remains is referred to as breakpoint chlorination. As chlorine is added to the ammonia solution the species of residual chlorine changes. The accurate measurement of these residuals is essential if a reliable estimate of the biocidal properties of the chlorine residuals are to be obtained.

A series of experiments was set up to enable a comparison of the free chlorine residuals measured by the chlorine probe, DPD, FACTS, Biofac and amperometric titration along several stages of the breakpoint chlorination of 2-mg/l and 0.30-mg/l ammonia nitrogen solutions. The ammonia solution was reacted with increasing doses of free chlorine at pH 7.0 and 20°C for 30 min. Free and combined chlorine residuals were determined by the methods given above.

RESULTS

Chloramine Preparation

Since the formation and stability of the inorganic chloramines is dependent on pH, the effect of the pH values used in the inactivation experiments on chloramine stability was studied.

Figure 2 shows the effect of lowering the pH of a monochloramine solution formed at pH 10.0, to pH 7.0. The data are plotted as log C/C_o v time, where C is the concentration at any given time and C_o is the time zero concentration. The results show little loss of monochloramine over five hours. The effect of raising the pH of a dichloramine solution from the formation pH of 4.6 to the experimental pH value of 7.0 is also shown in Figure 2. The increase to pH 7.0 resulted in substantial loss of dichloramine, with a half life of 210 min. However, inactivation runs were always performed within 60 min of the pH adjustment for dichloramine. The dichloramine concentration remaining at pH 7.0 over this time period was approximately 75% of that at zero time, with 25% converted to monochloramine. No free chlorine was detected. The time for trichloramine concentration to decrease by 50% ($t_{1/2}$) at pH 7.0 was found to be 293 min for a C1:N ratio of 3:1. However, in the inactivation experiments a 30-fold molar excess of ammonia was maintained to suppress the formation of free chlorine. This excess of ammonia decreases the $t_{1/2}$ to 17 min, as shown in Figure 2. This effect of ammonia on trichloramine stability was noted and discussed in greater detail by Sanguinsin and Morris [24].

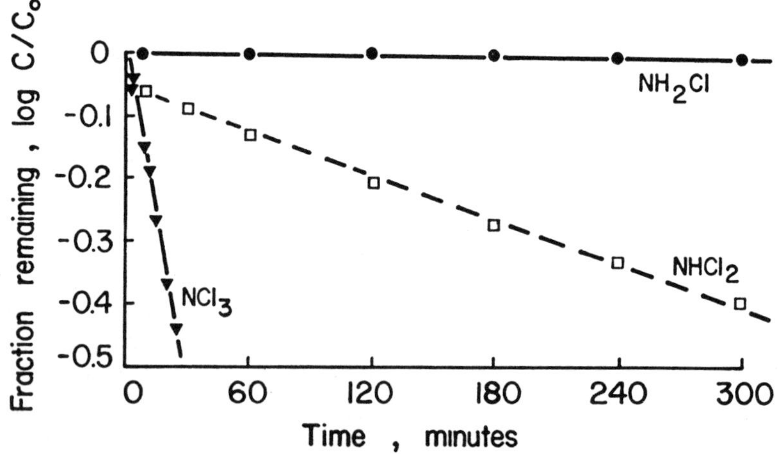

Figure 2. Stability of NH_2Cl, $NHCl_2$ and NCl_3 after adjustment of pH from the formation pH of 10.0, 4.5 and 2.3 respectively, to 7.0. NH_2Cl initial concentration was 87 mg/l with Cl:N = 1:3. $NHCl_2$ initial concentration was 96 mg/l with Cl:N = 1:3. NCl_3 initial concentration was 24 mg/l with Cl:N = 1:30.

Colorimetric Tests

Standard curves were prepared for DPD and FACTS using amperometric titration as the referee procedure. All standard curves were plotted using linear regression procedures to force the line through the origin, since a sample with zero chlorine will have zero absorbance. Linear calibration techniques [30] were used to determine free chlorine concentrations from any absorbance reading and to determine the 95% confidence bands around the standard curve. Of particular interest is the point where the lower 95% confidence band intersects the x-axis. This point is the lowest chlorine concentration with a lower confidence limit that does not include zero. In the part of this study dealing with false positive results in the colorimetric tests, any chlorine residual value determined by the test to be greater than this point of intersection was said to be significantly ($p = 0.05$) greater than zero. Stated another way, this point is the quantitative limit of detection for the test. For the FACTS procedure, this limit was found to average 0.19 mg/l FAC.

The lower quantitative limit of detection for DPD and DPD-Steadifac tablet reagent was 0.17 and 0.15 mg/l FAC respectively. The limit for DPD powder reagent was 0.10 mg/l and for DPD-Steadifac powder reagent the limit was 0.13 mg/l.

Since the DPD-Steadifac procedure was relatively untested, some effort was directed toward chemical evaluation of this procedure, with emphasis on determining the importance of time of addition of the thioacetamide reagent. Figure 3 shows the development of color with time for DPD with monochloramine and dichloramine solutions in the absence of free chlorine. DPD was found to produce substantial color with monochloramine, with the color increasing over the five-minute period observed. A noticeable difference was observed in the different commercially available DPD reagent preparations. The DPD powder reagent produced an absorbance at 515 nm of greater than 1.0 for 10-mg/l NH_2Cl while the DPD tablet yielded only an absorbance at 515 nm of 0.2 for the same NH_2Cl preparation. The DPD reagent, both tablet and powder, was relatively insensitive to dichloramine. In a subsequent experiment, monochloramine color development with DPDT was allowed to proceed for varying times. At the end of the time period, the sample was removed from the spectrophotometer and the appropriate amount of thioacetamide was added. The sample was returned to the spectrophotometer and the absorbance at 515 nm was determined. The results of this experiment are shown in Figure 4, upper panel, for times of 1-5 min. The dashed lines represent the time when the sample was out of the spectrophotometer. The addition of thioacetamide resulted in a decrease in the absorbance in all cases. A similar experiment with free chlorine is shown in Figure 4, lower panel. In this case, thioacetamide addition had no effect on the developed color.

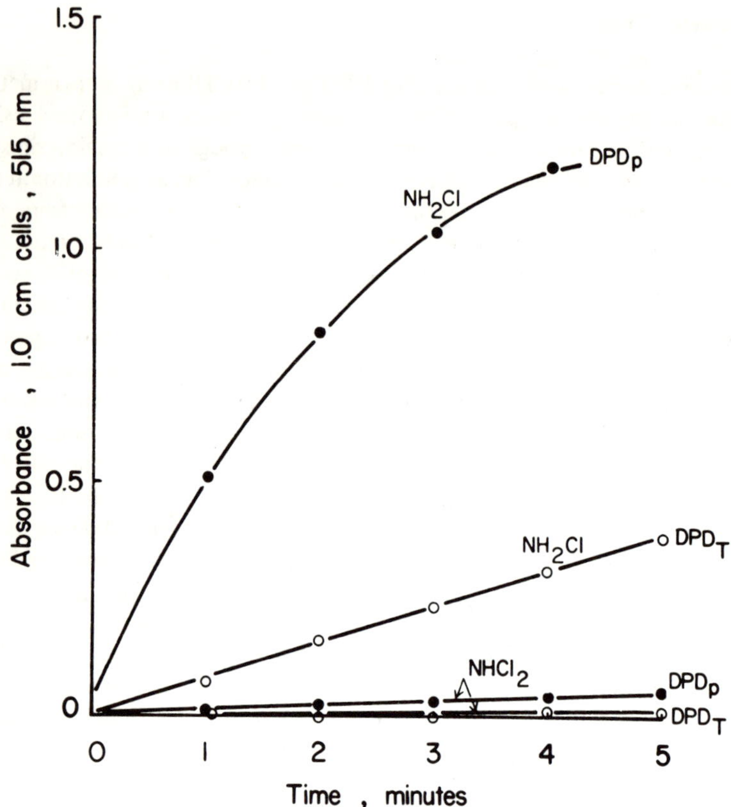

Figure 3. Time course of color development for the DPD reagents with 10.0 mg/l monochloramine and 10.0 mg/l dichloramine at 20°C.

Chlorine Membrane Electrode

The chlorine membrane electrode was calibrated daily or twice daily according to the manual provided [31]. The response of the electrode to a solution containing only free chlorine at pH 7.4 is shown in Figure 5. The response was linear over the range observed with a slope of 0.63.

The response of the electrode, given as apparent HOCl concentration, to solutions of monochloramine and dichloramine containing no free chlorine is also shown in Figure 5. The meter reading was 2% of the actual NH_2Cl concentration over the range 5-25 mg/l. The response to dichloramine was 18% of the dichloramine concentration over the range 2-15 mg/l. The electrode was found to be very sensitive to NCl_3, with a 0.59-mg/l NCl_3 solution,

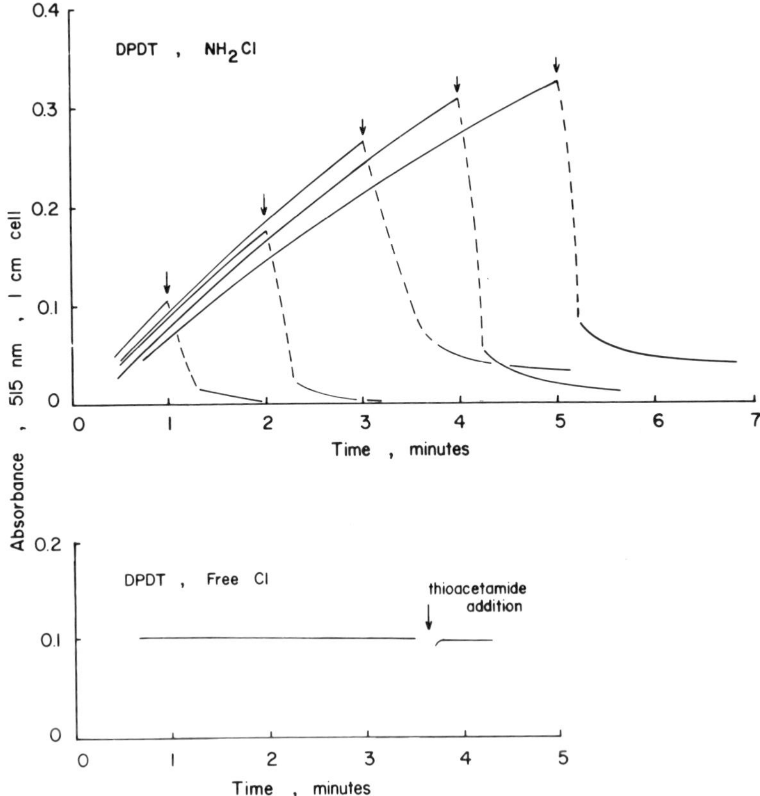

Figure 4. Effect of thioacetamide addition on the color developed by DPD, tablet reagent, with monochloramine (top) and with free chlorine (bottom). Arrows indicate time of additions of thioacetamide.

determined by UV absorption, yielding a meter reading greater than 5.0 mg/l HOCl and a 0.11-mg/l NCl_3 solution giving a meter reading of 2.7 mg/l HOCl. False positive readings increased as the number of chlorine atoms on the inorganic chloramine increased.

Biofac Calibration

Figure 6 shows the inactivation rates, taken from the log N/N_o vs time plots, plotted against the free chlorine concentration at pH 7.0. Although there was some scatter, the relationship appears to be sufficiently linear

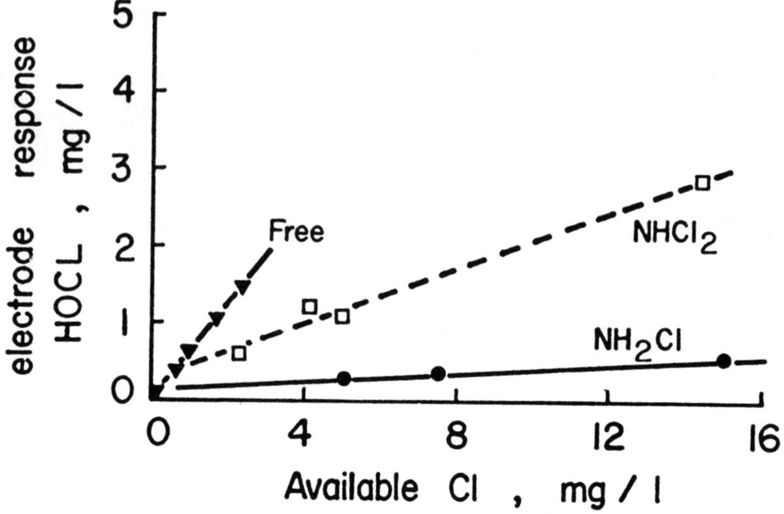

Figure 5. Response of the chlorine membrane electrode to free chlorine, monochloramine and dichloramine at pH 7.0, 20°C.

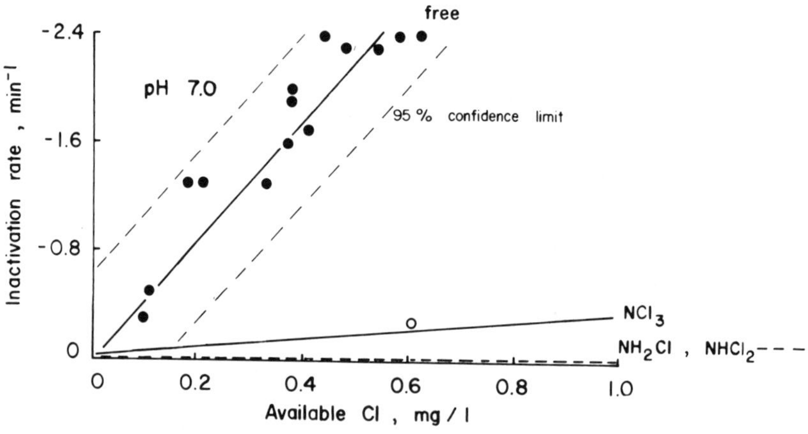

Figure 6. Rate of inactivation (K') of f2 by varying concentrations of free and combined chlorine at pH 7.0, 20°C.

(r=0.932, slope = 3.9) for a quantitative determination of free chlorine concentration to be made from the inactivation rate. This figure also gives the inactivation rate of f2 by NH_2Cl, $NHCl_2$ and NCl_3 at pH 7.0. The great disparity between the rates of inactivation of f2 by the chloramines and free chlorine forms the basis of the Biofac calibration procedure. The Biofac calibration procedure can be used to determine quantitatively free chlorine levels, and to determine qualitatively the presence or absence of free chlorine for unknown solutions.

Specificity of the Colorimetric Tests

The specificity of the colorimetric tests was determined by comparing the values of apparent free chlorine obtained by the test to the value of free chlorine obtained by the Biofac procedure for chemically defined chloramine solutions. In these experiments, chloramine solutions were added to the reaction vessel, f2 was added at time zero, and colorimetric tests for free chlorine and the f2 inactivation rate was determined. Levels of free chlorine were determined from the standard curves for the colorimetric tests and from the appropriate Biofac calibration curve. A false positive for the colorimetric test is defined as a significant (absorbance $\neq$ 0, p = 0.05) indication of free chlorine by the test where the Biofac procedure indicates no free chlorine. Levels of significance for DPD and FACTS were given in a preceding section.

Table I gives the monochloramine concentrations used in this set of experiments, determined amperometrically and spectrophometrically, and the apparent free chlorine level measured by the other tests. The FACTS procedure was the most specific, with no false positives occurring at monochloramine concentration of up to 20.0 mg/l, the highest level tested. The DPD-Steadifac procedure using the tablet reagent gave significant false positive readings at 19.5 mg/l. The DPD test with powder reagent was the least specific test for free chlorine, showing false positive indications of free chlorine at monochloramine levels as low as 1.0 mg/l. The addition of thioacetamide was effective in reducing, but not eliminating, the false positives obtained with DPD.

All of the tests performed well with dichloramine, as shown by the results in Table II. It should be noted that the pH value used was the pH at which the f2 inactivation was determined. The colorimetric procedures all incorporate a buffer to maintain the optimum pH for the test. No false positives were obtained with FACTS, DPDT-Steadifac and DPDT at dichloramine concentrations of up to 17.6 mg/l. False positives were given by the DPD and DPD-Steadifac procedure with the powder reagent at $NHCl_2$ levels of 17.2 mg/l.

Trichloramine was found to produce false positives in all the tests at relatively low levels (Table III). Indications of free chlorine levels of approxi-

Table I. Apparent Free Chlorine Concentration Measured by the DPD Procedures, FACTS and Biofac
at Varying Levels of Monochloramine at pH 7.0, 20°C

Chlorine (mg/l)			Free Chlorine (mg/l)					
Amperometric		Spectrophotometric	Colorimetric					
NH_2Cl	Total	NH_2Cl	FACTS	DPDP	DPDPSF	DPDT	DPDTSF	BioFAC
1.0			0	0.26^a	0.05	0.07	0.03	0.03
1.01			0	0.17^a	0.12	0.02	0	0.02
5.01			0	0.77^a	0.17^a	0.24^a	0.09	0.02
5.04			0	1.20^a	0.10	0.32^a	0.04	0.04
9.10	9.10	9.70	0.03	1.38^a	0.19^a	0.37^a	0	0.03
10.93			0.04	$\geqslant3.44^a$	0.23^a			0.03
11.10	11.10	9.70	0.04	1.46^a	0.12	0.26^a	0.02	0.02
11.24			0.03	1.93^a	0.82^a			0.04
19.46		20.31	0.03	$\geqslant3.44^a$	1.17^a	0.81^a	0.75^a	0.02
20.03		18.75	0.06	$\geqslant3.44^a$	2.26^a	1.04^a	1.12^a	0.02

[a]Significant (p = 0.05) false positive.

Table II. Apparent Free Chlorine Concentration Measured by the DPD Procedures, FACTS and Biofac at Varying Levels of Dichloramine at pH 7.0, $20^{\circ}C$

$NHCl_2$ by Spectrophotometer (mg/l)	Total Cl, Amperometric (mg/l)	Free Chlorine (mg/l)						
		Colorimetric					Electrode	BioFAC
		FACTS	DPDP	DPDPSF	DPDT	DPDTSF		
5.10		0	0.03	0.02	0.03	0.02	1.7	0
5.10		0.03	0.03	0.02	0.02	0.02	1.0	0
8.85	10.40	0.06	0.09	0.09	0.04	0.08		0
9.00	10.63	0.06	0.06	0.08	0.04	0.08		0.01
17.24	18.40	0.06	0.19[a]	0.13[a]	0.07	0.12		0.01
17.58	19.38	0.06	0.14[a]	0.14[a]	0.04	0.12		0.02

[a] Significant (p = 0.05) false positive.

mately 0.2 mg/l were seen for all tests at trichloramine concentration of 0.6 mg/l. The results for DPDT at this level were not significantly greater than zero, although close in magnitude to the values obtained by the other tests, which accounts for the higher level of NCl_3 required to produce a false positive given in Table III.

The Biofac procedure showed significant free chlorine levels at trichloramine concentrations of 3.2 mg/l at pH 7.0. Trichloramine was more effective in inactivating f2 than NH_2Cl and $NHCl_2$, which resulted in this indication of free chlorine.

Nitrogen Breakpoint

Figure 7 (top) shows the breakpoint curve for a 2.0-mg/l NH_3-N solution for a 30-min reaction time. The breakpoint of this solution was 17 mg/l chlorine. Free chlorine residuals were determined at points along this curve by the DPD, FACTS, free chlorine electrode, amperometric and Biofac procedures. The shape of the breakpoint curve indicates that no free chlorine should be present below 17.0 mg/l, and this was found to be the case as shown by the Biofac procedure (Figure 7, middle). The FACTS and DPDTSF procedures were found to be specific for free chlorine under these conditions with no significant levels of free chlorine measured until after the breakpoint. The amperometric procedure and the free chlorine electrode gave significant false positives for samples below the breakpoint at chlorine residuals of approximately 2-8 mg/l.

Figure 8 gives a comparison of the DPD methods for points along the same breakpoint curve. While false positives were observed for DPDP and DPDT, the addition of thioacetamide resulted in a reduction in the magnitude of the false positive and gave results comparable to the FACTS and Biofac procedure.

Figure 9 shows the results of a similar experiment performed at a lower nitrogen concentration (0.3 mg/l). The breakpoint at 2.8 mg/l available chlorine was analogous to that obtained with many surface waters and reflects concentrations that might be encountered at a water treatment plant. No free chlorine was detected by the FACTS, DPD-Steadifac and Biofac procedures below the breakpoint. A slight color was developed by the DPD procedure, but the absorbance was below that required for the sample to be designated as significantly false positive. The amperometric titration procedure was found to be the least specific in this experiment, giving false positive indications of free chlorine of approximately 0.1 mg/l before the breakpoint dose was reached. Figure 10 gives a comparison of the four DPD procedures for the same experiment. Again, the tablet reagent was more specific than the powder, and the addition of thioacetamide resulted in an improvement in specificity.

Table III. Apparent Free Chlorine Concentration Measured by the DPD Procedures, FACTS and Biofac
at Varying Levels of Trichloramine at pH 7.0, 20°C

$NHCl_3$ by Spectrophotometer (mg/l)	Total Cl, Amperometric (mg/l)	Free Chlorine (mg/l)						
		Colorimetric					Electrode	BioFAC
		FACTS	DPDP	DPDPSF	DPDT	DPDTSF		
0.58		0.28[a]	0.19[a]	0.20[a]	0.15	0.16[a]		0.06
0.59		0.31[a]	0.20[a]	0.20[a]	0.16	0.17[a]		0.06
3.16		1.56[a]	0.77[a]	0.64[a]	0.53[a]	0.52[a]		0.25
3.24		1.60[a]	0.76[a]	0.63[a]	0.53[a]	0.54[a]		0.25
3.16	6.69	1.85[a]	0.93[a]	0.75[a]	0.58[a]	0.51[a]		0.57
3.16	7.03	1.91[a]	0.91[a]	0.68[a]	0.60[a]	0.47[a]	>5.0	0.57
8.90	9.10	1.85[a]	0.96[a]	0.77[a]	0.70[a]	0.53[a]		0.48
9.10	16.60	>2.84[a]	1.14[a]	1.03[a]	0.81[a]	0.82[a]		0.86
16.54	16.70	>2.84[a]	1.72[a]	1.15[a]	1.23[a]	1.09[a]		1.29

[a] Significant (p = 0.05) false positive.

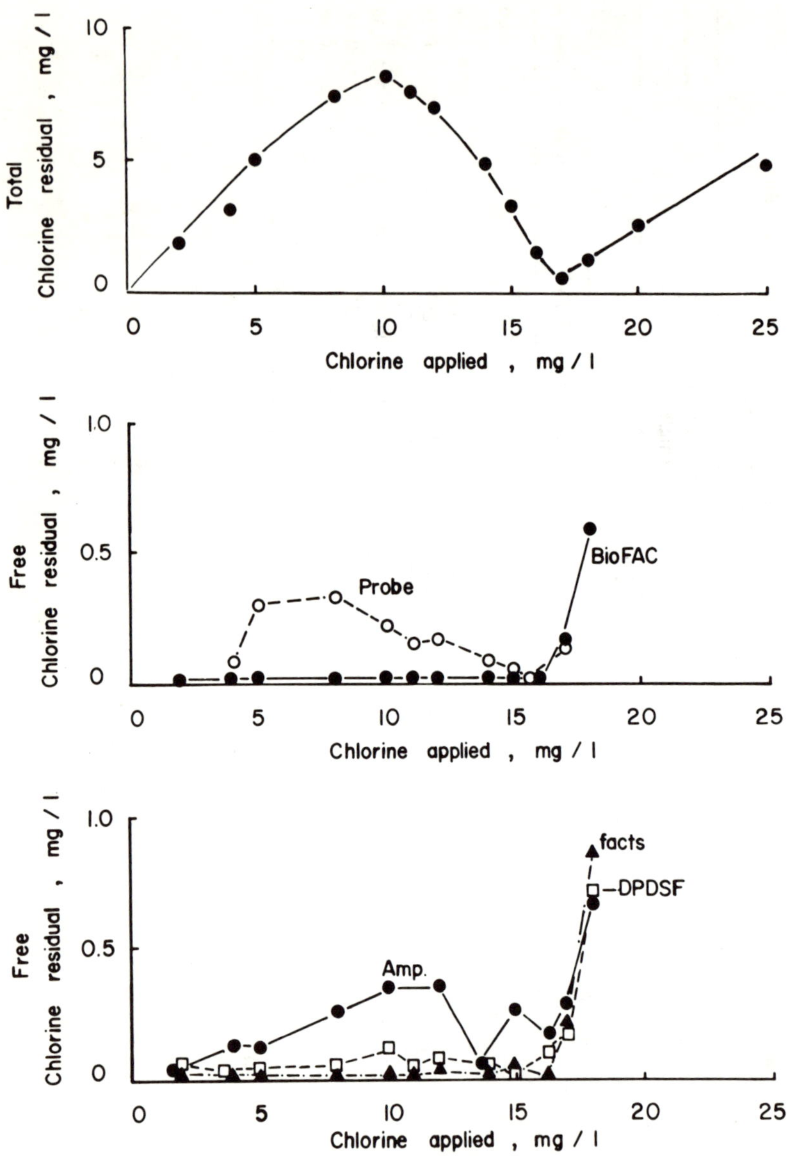

Figure 7. Breakpoint curve for a 2.0-mg/l NH_3-N solution (top); free chlorine measured by the electrode and Biofac procedures at points along the breakpoint curve (middle); and free chlorine measured by the FACTS, amperometric titration and DPD-Steadifac (tablet) procedures at points along the breakpoint curve (bottom).

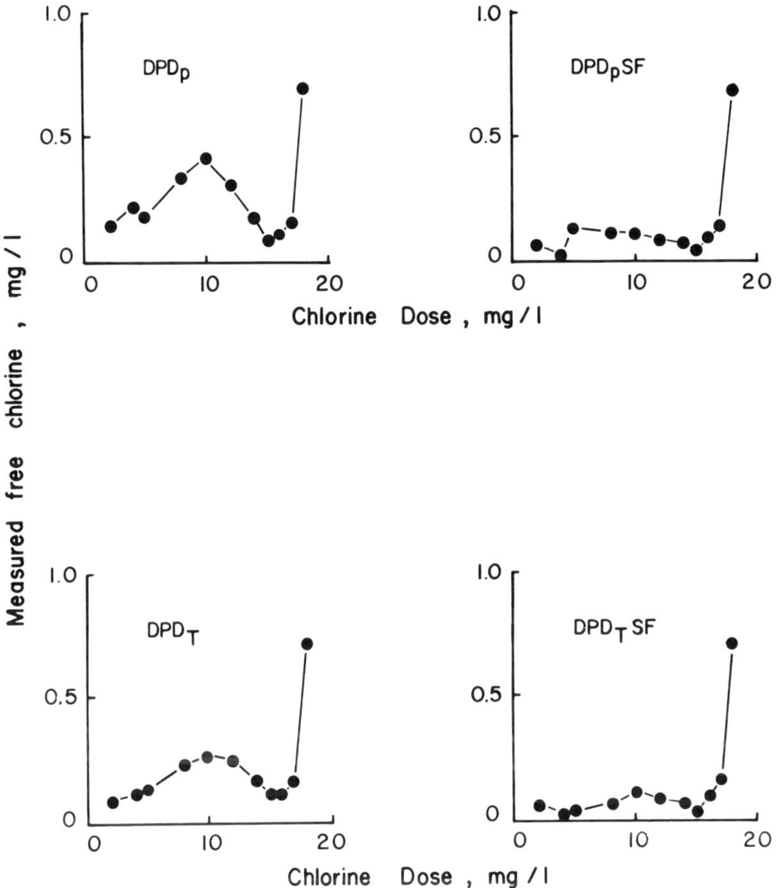

Figure 8. Free chlorine measured by the DPD procedures at points along the breakpoint curve of a 2.0-mg/l NH_3-N solution.

DISCUSSION

Biofac

One of the primary goals of this study was to develop a biological test for the quantitative determination of free chlorine. The results indicate that a system using the rate of inactivation of the bacterial virus, f2, as an indicator of free chlorine concentration yields quantitative calibration for pH 7.0. The

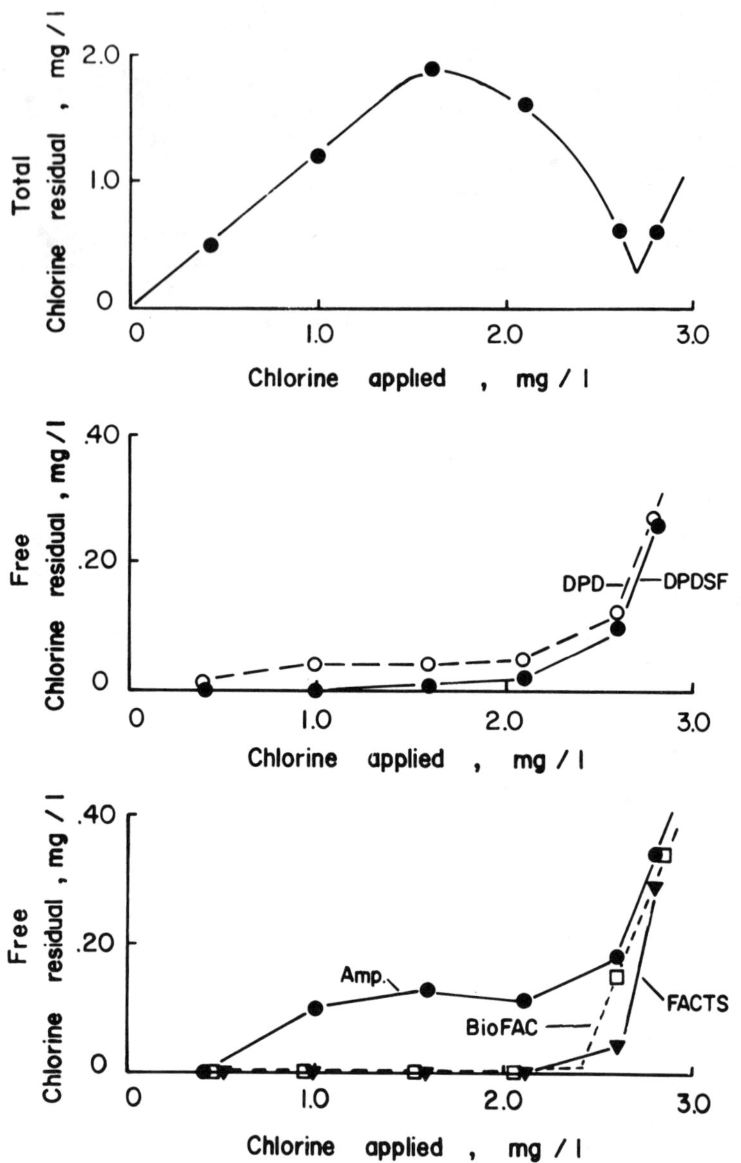

Figure 9. Breakpoint curve for a 0.30-mg/1 NH_3-N solution (top): free chlorine measured by the DPD and DPD-Steadifac (tablet) procedures at points along the breakpoint curve (middle); and free chlorine measured by the amperometric titration, FACTS and Biofac procedures at points along the breakpoint curve (bottom).

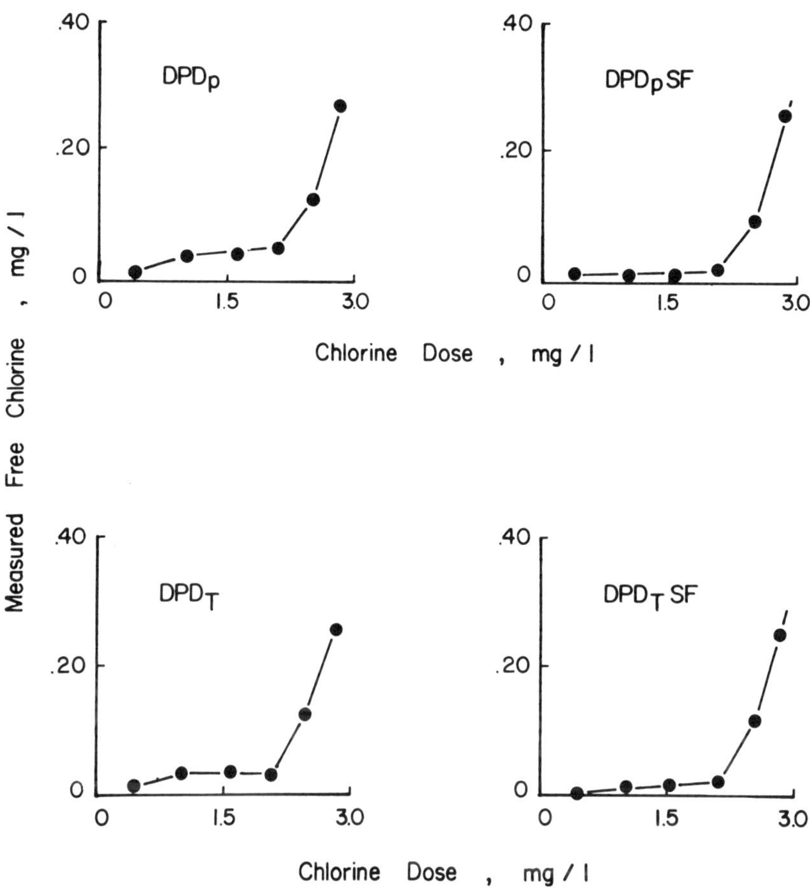

Figure 10. Comparison of the four DPD procedures for the experiment shown in Figure 9.

inactivation rate increases linearly with free chlorine concentration. A biological test system of this type provides the ideal referee technique for evaluation of test specificity, since there is no interference from monochloramine and dichloramine and the results obtained from the chlorine residual tests are intended to reflect the biocidal activity of the solution. An extension of this biological system to the measurement of other actively germicidal compounds (O_3, ClO_2, UV) should be possible. The procedure is suggested for evaluation of other tests only, not for routine use or monitoring, since there is a one-day delay in obtaining results.

Chloramines

The preparation of chemically defined chloramine solutions was essential for evaluation of the tests for chlorine. Previous work with monochloramine [16,22] has shown that solutions of these chloramines can be prepared with free chlorine absent. Monochloramine and dichloramine were sufficiently stable at the pH values and time periods used so that the studies could be performed without appreciable loss of either compound. The absence of free chlorine in the preparations was confirmed by the lack of inactivation of f2 by these solutions. Much less work has been done with trichloramine, primarily because of the lack of stability of the compound and because trichloramine is not thought to be as prevalent as monochloramine and dichloramine under conditions normally encountered in the field. A 30-fold molar excess of ammonia was used in the trichloramine preparations in this study to suppress free chlorine. Since the reaction between free chlorine and ammonia was rapid at higher values [32,33], any free chlorine formed by the decomposition of NCl_3 would be consumed. Chemical tests for determining the presence of free chlorine in trichloramine solutions are not adequate, since the tests respond to NCl_3 in much the same manner as to free chlorine. Spectrophotometric methods are not sensitive enough to detect small quantities of free chlorine in trichloramine solutions. The biological procedure also gives equivocal results. Trichloramine was found to be a better biocide than monochloramine or dichloramine, but not as potent as free chlorine. This intermediate result may reflect the true biocidal efficacy of NCl_3, or it may be due to the presence of trace quantities of free chlorine in the NCl_3 preparation. Biofac gave free chlorine levels of 8-18% of the NCl_3 level at pH 7.0.

DPD and DPD-Steadifac

The specificity of the DPD test was found to vary with the reagent (tablet or powder) used and with the species of chloramine. Monochloramine produced false positives with DPD powder at lower levels than with DPD tablets. At equivalent monochloramine concentrations, the powder reagent consistently gave higher false positive readings than the tablet reagent. DPD reagents are now commercially available in liquid form. The liquid reagent should be evaluated, since the results show that the form of the reagent may influence the specificity. The thioacetamide modification to the DPD procedure was found to be effective in reducing false positive readings. Palin [13] states that the thioacetamide "provides immediate dechlorination without having any effect on a previously developed DPD color from free chlorine." This implies that the time of addition of the thioacetamide is important. However, the results of this study indicate that thioacetamide selectively decolors the colored product resulting from the DPD-chloramine reaction, and that the time of addition is not critical. The color produced by the DPD-free chlorine reaction was not affected by thioacetamide.

Dichloramine was not found to be a significant interference in the DPD tests since the levels required to produce a false positive were greater than those which would be encountered in normal practice. The breakpoint experiments show the behavior of the test in the presence of mixtures of chloramine species. For the 2.0-mg/l NH_3-N breakpoint studies where the maximum total residual was approximately 8.0 mg/l, DPDT-Steadifac accurately measured the concentration of the active species. The other DPD procedures were not as specific and showed false positive free chlorine measurements.

All of the DPD procedures gave false positives with trichloramine. However, the significance of these false positives is debatable, since NCl_3 appears to be a better biocide than NH_2Cl and $NHCl_2$ and since NCl_3 is rarely found in the absence of free chlorine. Thus, a false positive with NCl_3 may lead to an overestimation of the free chlorine residual, if both are present. A procedure is available for the estimation of NCl_3 by DPD, but it is doubtful if it is routinely used by field personnel.

FACTS

The FACTS procedure was found to be specific for free chlorine at all levels of monochloramine (up to 22 mg/l) and dichloramine (up to 19 mg/l) tested. Additionally, FACTS accurately paralleled the Biofac procedure in measurements along the 2.0-mg/l NH_3-N breakpoint curve. False positive measurements were obtained with NCl_3. The significance of this interference is debatable, as indicated above. The greatest problem with the FACTS procedure was with the reagent itself. Some preparations of the FACTS indicator were relatively insensitive to free chlorine. The problem appears to be associated with the quality of the propanol used in the reagent. Different lots of propanol yielded reagents with different sensitivity. Although this does not pose a great problem in the laboratory, since reagents can be calibrated, a standard indicator is necessary where a color comparator is to be used.

Chorine Membrane Electrode

The electrode was not included in the experiments with the individual chloramine species but was included in the breakpoint experiments with 2.0-mg/l NH_3-N. False positive indications of free chlorine were obtained below the breakpoint. The response was 2-5% of the total chlorine level.

Johnson et al. [14] reported an interference of 3.0% and 1.3% from monochloramine and dichloramine, respectively. Under the pH conditions of the breakpoint studies, monochloramine and dichloramine were the predominant combined chlorine species present before the breakpoint. The electrode responded as expected from the results given [14]. The instrument used was a prototype model and the manufacturer indicated that the specificity can be improved by adjustment of the applied voltage.

Amperometric Titration

Amperometric titration has long been used in the United States as a referee technique for evaluation of other methods of chlorine residual measurement. The results obtained in this study show that the method is not absolutely specific for free chlorine. False positive measurements by the amperometric procedure were similar in magnitude to those obtained with the membrane electrode and the DPD procedures without thioacetamide. Although *Standard Methods* [8] states that the amperometric titration method is a "standard of comparison for the determination of free or combined chlorine," it is also noted that "monochloramine can intrude into the free chlorine fraction." Nicolson [34] reported 5-7% loss of free chlorine due to the high-speed stirrers employed on commercially available amperometric titrators, and expressed some doubt as to the completion of the reaction between free chlorine and phenylarsine oxide (PAO). At the PAO endpoint, positive colorimetric reactions were consistently observed.

FACTS was the most specific for free chlorine. However, DPD-Steadifac with tablet reagents approached FACTS in specificity.

The data reported are shown as significant false positives, with significance determined from the 95% confidence bands around the standard curves. In actual field practice, any color is probably interpreted by field personnel as positive for free chlorine. In cases where free chlorine levels are low (<0.2 mg/l), a measurement of total chlorine should be taken. If the total chlorine residual is high, the validity of the free chlorine measurement may be in question.

The breakpoint studies with 0.3-mg/l ammonia nitrogen solutions show that the colorimetric tests reflect the biocidal activity of the solution in the absence of high levels of chloramines. The levels of combined chlorine found in this experiment reflect those likely to be encountered in water treatment practices. Under unusual conditions of high chloramine concentration, the free chlorine residual measurements by any method must be carefully interpreted with consideration given to the type and level of interfering compounds.

CONCLUSIONS

All of the methods tested for measurement of free chlorine residuals (DPD, FACTS, amperometric titration and membrane electrode) yield false positive determinations with one or more of the inorganic chloramines. A fundamental understanding of chlorine chemistry and the particular test procedure employed is an absolute necessity.

The Steadifac modification of the DPD procedure reduces the frequency and magnitude of false positives obtained. Thioacetamide appears to selectively decolor the product of the DPD-monochloramine reaction.

The biological calibration procedure, Biofac, clearly differentiates and quantifies free chlorine, and may have possible extension for other disinfectants.

The FACTS procedure was the most specific for free chlorine. Variability of reagent preparations of FACTS limits the quantitative use of this test, in its present form, to the laboratory.

Specificity of the DPD reagent was found to vary with the form of the reagent, with the tablet reagent consistently yielding fewer false positives than the powder reagent.

The membrane electrode compares favorably with the other procedures for the measurement of free chlorine in the absence of combined chlorine, but was not as specific as DPD and FACTS in the presence of combined chlorine.

RECOMMENDATIONS

A total chlorine residual measurement should be taken along with a free chlorine measurement to aid in the interpretation of the free residual measurement.

Effort should be directed towards further development of the FACTS procedure to make the reagent more consistent.

The Steadifac modification of the DPD procedure appears to improve the specificity of the free chlorine measurement. The method should be evaluated by other laboratories before widespread application for field use.

The commercially available DPD reagents and preparations should be further evaluated for free chlorine specificity. Significant differences were observed for tablet and powder reagent in this study.

REFERENCES

1. Butterfield, C. T., E. Wattie, S. Megregian and C. W. Chambers. "Influence of pH and Temperature on the Survival of Coliform and Enteric Pathogens When Exposed to Free Chlorine," *Pub. Health Rep.* 58:1837-1866 (1943).
2. Butterfield, C. T., and E. Wattie. "Influence of pH and Temperature on the Survival of Coliform and Enteric Pathogens When Exposed to Chloramine," *Pub. Health Rep.* 61:157-192 (1946).
3. Kelly, S., and W. W. Sanderson. "The Effect of Chlorine in Water on Enteric Viruses," *Am. J. Pub. Health* 48:1323-1334 (1958).
4. Kelly, S., and W. W. Sanderson. "The Effect of Chlorine in Water on Enteric Viruses. II. The Effect of Combined Chlorine on Poliomyelitis and Cosxackie Viruses," *Am. J. Pub. Health* 50:14-20 (1960).
5. Kruse, C. W., Y. C. Hsu, A. C. Griffiths and R. Stronger. "Halogen Action on Bacteria, Viruses and Protozoa," in *Proceedings of the National Speciality Conference on Disinfection* (New York: American Society of Civil Engineers, 1970), pp. 113-136.
6. Olivieri, V. P., T. K. Donovan and K. Kawata. "Inactivation of Virus in Sewage," in *Proceedings of the National Speciality Conference on Disinfection* (New York: American Society of Civil Engineers, 1970), pp. 365-384.
7. "National Interim Primary Drinking Water Regulations," U.S. EPA (1975).
8. *Standard Methods for the Examination of Water and Wastewater*, 14th ed. (New York: American Public Health Association, 1975).
9. Cooper, W. J., E. P. Meier, J. W. Highfill and C. A. Sorber. "The Evaluation of Existing Field Test Kits for Determining Free Chlorine Residuals in Aqueous Solutions, Final Report," U.S. Army Medical Bioengineering Research and Development Laboratory Technical Report 7402 (1974).
10. Sorber, C., W. Cooper and E. Meier. "Selection of a Field Method for Free Available Chlorine," in *Disinfection: Water and Wastewater*, J. D. Johnson, Ed. (Ann Arbor, MI: Ann Arbor Science Publishers, Inc., 1975), pp. 91-112.
11. Meier, E. P., W. J. Cooper and J. W. Highfill. "Evaluation of the Specificity of the DPD-Glycine and FACTS Test Procedures for Determining Free Available Chlorine," paper presented before the Division of Environmental Chemistry, American Chemical Society, Anaheim, CA, March 13-17, 1978.
12. Strupler, N. "A Study of Interferences in the Measurement of Free and Combined Chlorine in Water by the DPD and Syringaldazine Methods," in *Proceedings of the American Water Works Association Water Technology Conference*, (Denver, CO: American Water Works Association, 1978), pp. 2A-6, 1-13.
13. Palin, A. T. "A New DPD-Steadifac Method for the Specific Determination of Free Available Chlorine in the Presence of High Monochloramine," *J. Inst. Water Eng. Sci.* 32:327 (1978).
14. Johnson, J. D., J. W. Edwards and F. Keesler. "Chlorine Residual Measurement Cell: The HOCl Membrane Electrode," *J. Am. Water Works Assoc.* 70:341-348 (1978).

15. Savage, T. G., and L. P. Stratton. "Bacteriological Evaluation of Two Test Methods for Chlorine in Swimming Pools," *Appl. Microbiol.* 22(5):809-811 (1971).
16. Snead, M. C. "Inactivation of f2 Bacterial Virus by Monochloramine," ScM Thesis, The Johns Hopkins School of Hygiene and Public Health, Baltimore, MD (1976).
17. Richfield, D. T. "Inactivation of f2 Bacterial Virus with Dichloramine," ScM Thesis, The Johns Hopkins School of Hygiene and Public Health, Baltimore, MD (1978).
18. Olivieri, V. P. "The Mode of Action of Chlorine on f2 Bacterial Virus," ScD Thesis, The Johns Hopkins School of Hygiene and Public Health, Baltimore, MD (1974).
19. Soper, F. G. "Action of Hydrogen Chloride on a Dry Solution of Chloramine," *J. Chem. Soc.* 125:768 (1924).
20. Olivieri, V. P. "Chlorine Dioxide and Protein Synthesis," MS Thesis, University of West Virginia, Morgantown, WV (1968).
21. Granstrom, M. L. "The Disproportionation of Monochloramine," PhD Thesis, Harvard University, Cambridge, MA (1954).
22. Johnson, J. D., and R. Overby. "Stabilized Neutral Orthotolidine, SNORT, Colorimetric Method for Chlorine," *Anal. Chem.* 41:1744 (1969).
23. Chapin, R. M. "Dichloramine," *J. Am. Chem. Soc.* 51:2112-2117 (1929).
24. Saguinsin, J. L. S., and J. C. Morris. "The Chemistry of Aqueous Nitrogen Trichloride," in *Disinfection: Water and Wastewater*, J. D. Johnson, Ed. (Ann Arbor, MI: Ann Arbor Science Publishers, Inc., 1975), pp. 277-299.
25. Galal-Gorchev, H., and J. C. Morris. "Formation and Stability of Bromanide, Bromimide and Nitrogen Tribromide in Aqueous Solution," *Inorg. Chem.* 4:899-905 (1965).
26. Loeb, T., and N. D. Zinder. "A Bacteriophage Containing RNA," *Proc. Nat. Acad. Sci., U.S.* 47:282 (1961).
27. Dennis, W. H., Jr. "The Mode of Action of Chlorine on f2 Bacterial Virus During Disinfection," ScD Thesis, The Johns Hopkins School of Hygiene and Public Health, Baltimore, MD (1977).
28. Adams, M. H. *Bacteriophages* (New York: Interscience, 1959).
29. Fair, G. M., and J. C. Geyer. *Water Supply and Wastewater Disposal* (New York: John Wiley & Sons, Inc., 1954).
30. Snedecor, G. W., and W. G. Cochran. *Statistical Methods*, 6th ed. (Ames, IA: The Iowa State University Press, 1967).
31. "Instruction Manual for the Chlorine Analyzer," Orion Research Inc., Cambridge, MA (1979).
32. Morris, J. C. "The Chemistry of Aqueous Chlorine in Relation to Water Chlorination," in *Water Chlorination: Environmental Impact and Health Effects, Vol. 1*, R. L. Jolley, Ed. (Ann Arbor, MI: Ann Arbor Science Publishers, Inc., 1978), pp. 21-35.
33. Weil, I., and J. C. Morris. "Kinetic Studies on the Chloramines. I. The Rates of Formation of Monochloramine, N-Chlormethylamine and N-Chlordimethylamine," *J. Am. Chem. Soc.* 71:1664-1671 (1949).
34. Nicholson, N. J. "An Evaluation of the Methods for Determining Residual Chlorine in Water," *Analyst* 90:187 (1965).

CHAPTER 20

CHLORINE DIOXIDE CHEMISTRY:
GENERATION AND RESIDUAL ANALYSIS

E. Marco Aieta and Paul V. Roberts

Department of Civil Engineering
Stanford University
Stanford, California

Chlorine dioxide is a strong oxidant which has been given increasing attention as an alternative to chlorine for the disinfection of water and wastewater. Chlorine dioxide shows promise of good disinfection performance, without the disadvantage of forming large quantities of undesirable halogenated by-products. A major disadvantage of chlorine dioxide is its cost. In water treatment practice, chlorine dioxide is generated at the point of use by reaction of its precursor, sodium chlorite, with another reactant, either a mineral acid or chlorine. Because of the high unit cost of the sodium chlorite reactant, it is essential that the yield of chlorine dioxide based on feed chlorite be maximized.

OBJECTIVES

The primary objectives of this work were to: (1) measure the yield and stoichiometry of generation of chlorine dioxide under conditions representative of those used in water treatment; (2) compare the yield of chlorine dioxide when alternative methods of generation are used; and (3) develop an analytical approach for measuring the concentrations of chlorine dioxide and other chlorine species that is suitable for the above tasks.

The secondary objectives were to: (1) evaluate analytical procedures for the determination of chlorine dioxide residuals in water and wastewater: and (2) determine the applicability of these procedures to disinfection process control in water and wastewater for which chlorine dioxide was the sole disinfecting agent.

MATERIALS AND METHODS

Chlorine and Chlorine Dioxide Stock Solution

Pure chlorine dioxide solutions for use in method development were generated daily according to the method described in *Standard Methods* [1], Section 411A (Figure 1). The concentrations of the sulfuric acid and the

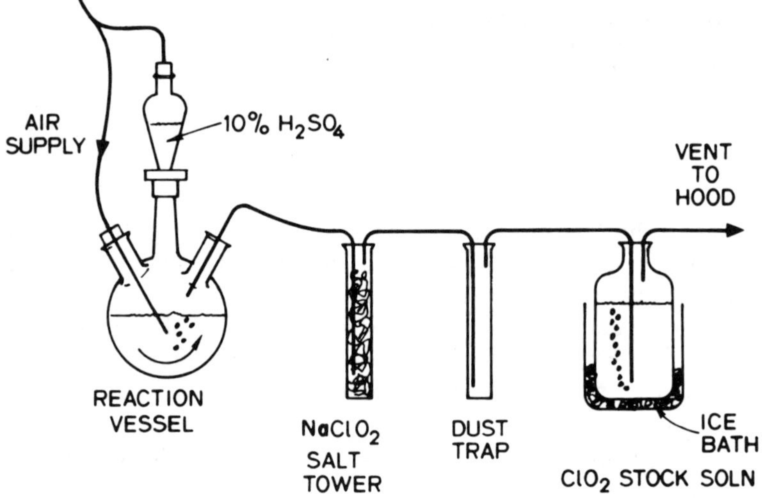

Figure 1. Chlorine dioxide generator using acid activation of a sodium chlorite solution to prepare an approximately 1000-mg/l ClO_2 stock solution.

sodium chlorite were doubled, compared to the recommendations in *Standard Methods*, to obtain higher chlorine dioxide concentrations in the stock solution. The concentration so obtained was 1000-1500 mg/l as ClO_2. Spectrophotometric analysis (Bausch and Lomb Spectronic 70) at 360 nm (the optimum for chlorine dioxide) was used to verify that the chlorine species in solution was indeed chlorine dioxide. Pure chlorine stock solutions were generated daily by passing 99.99% pure chlorine gas through chilled, deionized water. The concentrations so obtained were 3000-4000 mg/l chlorine.

Both chlorine and chlorine dioxide stock solutions were standardized immediately prior to experiments by the iodometric method described in *Standard Methods* [1], Sections 409A and 411A. There is controversy in the literature about the pH required to completely reduce chlorine dioxide to chloride, as is assumed in the iodometric standardization technique [2,3,4]. The standardization reaction is performed at pH 1.8; in the presence of excess iodide, chlorine dioxide is assumed to be reduced to chloride as it oxidizes iodide to iodine. The number of equivalents of sodium thiosulfate required to reduce the iodine can then be used to calculate the concentration of chlorine dioxide in the stock solution. To ensure that the chlorine dioxide was completely reduced to chloride and hence that the correct concentration of chlorine dioxide was determined in the stock solutions, the iodometric method was compared to the amperometric method. The amperometric analysis is analogous to the iodometric method with these exceptions [1]:

1. In the amperometric analysis the endpoint is detected by observing the change in current flow in the sample while adding a strong reducing agent as titrant.
2. In the amperometric analysis, phenylarsine oxide (PAO) is the titrant instead of sodium thiosulfate as used in the iodometric method.
3. In the amperometric analysis the determination is performed at pH 7. At this pH only one electron transfer is possible for chlorine dioxide, as shown below:

$$ClO_2 + e^- \rightarrow ClO_2^- \tag{1}$$

$$ClO_2 + 5e^- \rightarrow Cl^- + 2O^{2-} \tag{2}$$

Reaction 1 occurs at pH 7, Reaction 2 at pH < 2.

The analyses were performed in distilled water. The amperometric titrator was a prototype model (Fischer and Porter, similar in operation to the F & P Model 17T1010 amperometric titrator). The results of this comparison are given in Table I; the data were subjected to a significance test using analysis of variance, as shown in Table II.

The difference between the values obtained by the two methods is not significant [(P(F) > 1.12) = 0.32] according to the F-test. It can be inferred from these data that the iodometric method accurately measures the concentration of chlorine dioxide in stock solutions.

Chlorite, Chlorate and Chloride Stock Solutions

Chlorite, chlorate and chloride stock solutions were prepared from the sodium salts of the respective anions. Reagent-grade chemicals were used; the

Table I. Comparison of Amperometric and Iodometric Methods
for Determining Concentration of ClO_2

Trial	Iodometric (mg/l)	Amperometric (mg/l)
1	563.7	547.2
2	563.7	554.8
3	566.4	571.9
4	561.0	560.5
5	560.3	558.6
Mean	563.0	558.6
Std. Dev.	2.18	8.06
s^2	4.77	64.98

Table II. Comparison of Amperometric and Iodometric Methods
by Analysis of Variance

Source of Variance	Degrees of Freedom	Sum of Squares	Mean Square	F-Ratio
Between Methods	1	48.8	48.8	1.12
Error	8	348.7	43.6	
Total	9	397.5		

nominal concentrations were based on mass of reagent and supplier's composition analysis. These solutions were made up immediately prior to the experiment in which they were used.

MEASUREMENT OF CHLORINE DIOXIDE
GENERATION PRODUCT COMPOSITION

Chlorine dioxide for use in wastewater disinfection is generated by either a chlorine-sodium chlorite reaction or by a mineral acid-sodium chlorite reaction. Impurities of the reactants and incomplete reaction necessitate a methodology to identify all pertinent chlorine species. The relevant chlorine species in the final product are: chlorine (reported as Cl_2), chlorine dioxide (ClO_2), chlorite (ClO_2^-), chlorate (ClO_3^-) and chloride (Cl^-).

Analysis of Chlorine Species

The analytical methodology developed for this study to identify reactor product composition is essentially a sequential process in which successive measurements include one more chlorine species than the previous measurement. The sequential procedure consists of the following steps:

1. spectrophotometric measurement of chlorine dioxide;
2. measurement of the sum of chlorine dioxide and chlorine by amperometric titration at pH 7;
3. measurement of the sum of chlorine dioxide, chlorine and chlorite by iodometric titration at pH 2;
4. reduction of chlorine dioxide, chlorine, chlorite and chlorate to chloride by reaction with iodide in concentrated acid, followed by iodometric titration at pH 1 to pH 2; and
5. measurement of chloride using the mercuric nitrate method.

A more detailed description of the steps follows:

1. Measurement of chlorine dioxide: Chlorine dioxide absorbs ultraviolet (UV) radiation in the 240- to 440-nm wavelength. A wavelength of 360 nm was chosen for chlorine dioxide determinations because this corresponds to the peak of the absorbance curve; also, there is no interference from other chlorine species at this wavelength.

    A calibration curve was prepared by appropriate dilutions of a standardized stock solution to yield concentrations in the 5- to 60-mg/l range. The response of ClO_2 is linear in this concentration range.

    A sample of reaction product is diluted so that the concentration of ClO_2 to be analyzed is in the linear range (5-60 mg/l). The instrument is set to zero absorbance with a blank of the dilution water and the sample is then read. The concentration of ClO_2 in the diluted solution is calculated using the calibration equation and the appropriate dilution factor is applied. This is "Reading A."

2. Measurement of chlorine dioxide and chlorine: Chlorine dioxide and chlorine are measured by forward amperometric titration at pH 7. An appropriate sample size is chosen so that no more than 5 ml of PAO are used in the titration. To approximately 200 ml of distilled water, add 2 ml pH 7 buffer, the sample, and 2 ml of 5% KI solution, in that order. Perform the titration with 0.00564 N PAO and record the resulting volume (ml) of PAO solution as "Reading B."

3. Measurement of chlorine dioxide, chlorine and chlorite: Chlorine dioxide, chlorine and chlorite are measured by iodometric titration at pH 2. To 50 ml of distilled water, add 5 ml of concentrated acetic acid, 5 ml of sample and approximately 1 g of KI. Allow to react in the dark for 5 min. Titrate to starch endpoint with 0.1 N (standardized) sodium thiosulfate. This is "Reading C."

4.　　Measurement of chlorine dioxide, chlorine, chlorite and chlorate: To a 50-ml Erlenmeyer flask, add 5 ml concentrated hydrochloric acid and 0.5 g KI. To this mixture, add 5 ml of sample, seal and allow to react in the dark in a closed flask for 10 min. A blank must also be carried through the procedure as KI is oxidized by oxygen in this low-pH environment (pH < 0.1). After 10 min, dilute sample to ~50 ml and titrate with 0.1 N sodium thiosulfate. This reading minus the blank value is "Reading D."

5.　　Measurement of chloride: This procedure is essentially the mercuric nitrate method as given in *Standard Methods* [1] Section 408B, with the following exception: when measuring the chloride concentration in a reaction mixture containing high levels of chlorine dioxide, the sample must be diluted to yield a final chlorine dioxide concentration of about 10 mg/l. To 100 ml of this solution, 4.0 ml of indicator-acidifier reagent must be added instead of 1.0 ml as given in *Standard Methods* [1]. Titrate as usual. This reading minus the blank value is "Reading E." The relative standard deviation is reported to be 3.3% [1].

All calculations, except the chlorine dioxide-spectrophotometric measurement, are based on the equivalents of reducing titrant required to react with the equivalents of oxidants present. The equivalent weight of a compound is defined as that weight of the compound which contains one gram-mol of available electrons or

$$\text{Equivalent weight} = \text{molecular weight/number of electrons transferred}$$

For chlorine dioxide, the equivalent weight is pH-dependent; at pH 7 chlorine dioxide is reduced to chlorite, while at pH less than 2 chlorine dioxide reduces to chloride. Chlorite and chlorate also exhibit an equivalent weight pH-dependency. The equivalent weights used in the calculations are shown in Table III.

Under typical conditions of chlorine dioxide generation, pH of the product is in the range of pH 2 (acid activation) to 5 (chlorine-chlorite reaction). Moreover, any chlorine residual measured in the generator product stream is virtually certain to be in the free form, either as Cl_2 or HOCl. The speciation is governed by chlorine hydrolysis. In this work it is assumed that Cl_2 and HOCl are present in proportions corresponding to hydrolysis equilibrium.

Species Statistics

Since each determination is replicated, the variance of the analytical determinations can be calculated. However, because the determination of chlorine, chlorite and chlorate depend on previous measurement, the variance of a particular determination must reflect the variances of previous measurements. The values reported are the mean of replicates for a particular species.

Table III. Equivalent Weights for Calculating Concentrations on a Mass Basis

pH	Species	Molecular Weight (mg/mol)	Electrons Transferred	Equivalent Weight (mg/eq)
7	ClO_2	67,450	1	67,450
2, 0.1	ClO_2	67,450	5	13,490
7, 2, 0.1	Cl_2	70,900	2	35,450
2, 0.1	ClO_2^-	67,450	4	16,863
0.1	ClO_3^-	83,450	6	13,908

Species Calculations:

Chlorine dioxide: $mg/l\ ClO_2 = 58.26\ (absorbance) - 0.255 = A$

where 58.26 and 0.255 are instrument-specific calibration constants

$$\text{Chlorine:}\quad mg/l\ Cl_2 = \frac{35,450\ mg}{eq}\left(\frac{B(ml)\times normality(eq/l)}{sample(ml)} - \frac{mg/l\ ClO_2}{67,450\ mg/eq}\right)$$

$$\text{Chlorite:}\quad mg/l\ ClO_2^- = \frac{16,863\ mg}{eq}\left(\frac{C(ml)\times normality(eq/l)}{sample(ml)} - \frac{mg/l\ Cl_2}{35,450\ mg/eq}\right.$$

$$\left. - \frac{mg/l\ ClO_2}{13,490\ mg/eq}\right)$$

$$\text{Chlorate:}\quad mg/l\ ClO_3^- = \frac{13,908\ mg}{eq}\left(\frac{D(ml)\times normality(eq/l)}{sample(ml)} - \frac{mg/l\ ClO_2^-}{16,863\ mg/eq}\right.$$

$$\left. - \frac{mg/l\ Cl_2}{35,450\ mg/eq} - \frac{mg/l\ ClO_2}{13,490\ mg/eq}\right)$$

$$\text{Chloride:}\quad mg/l\ Cl^- = \frac{35,450\ mg/eq\times E(ml)\times normality(eq/l)}{sample(ml)}$$

where A, B, C, D, E denote the respective Readings A, B, C, D, E, as defined in the text.

This mean (e.g., the mean of chlorine dioxide measurements), is used in the calculations of the individual replicate values of the successive species (e.g., chlorine). The variance of the chlorine measurements is then the sum of the variance of the mean of the chlorine dioxide value plus the variance of the chlorine measurements. The square root of this combinate variance then gives the standard deviation of the measurements. Of course, for chlorine dioxide and chloride, the variance and standard deviations are calculated conventionally, since no prior measurements are involved. Formulas for calculating the variances of chlorine, chlorite and chlorate are given below [5].

Variance of chlorine measurements:

$$S^2_{Cl_2} = \frac{S^2_{ClO_2}}{n} + s^2_{Cl_2} \tag{3}$$

Variance of chlorite measurements:

$$S^2_{ClO_2^-} = \frac{S^2_{ClO_2}}{n} + \frac{S^2_{Cl_2}}{n} + s^2_{ClO_2^-} \tag{4}$$

Variance of chlorate measurements:

$$S^2_{ClO_3^-} = \frac{S^2_{ClO_2}}{n} + \frac{S^2_{Cl_2}}{n} + \frac{S^2_{ClO_2^-}}{n} + s^2_{ClO_3^-} \tag{5}$$

where S_i^2 = variance of species i

s_i^2 = variance of species i before consideration of successive species determination

n = number of replicates (all n need not be equal)

Test of Analytical Procedures

An analysis of pure solutions in deionized water of chlorine dioxide, chlorine, chlorite and chlorate is shown in Table IV. The values reported are the means of five replicate analyses plus and minus the standard deviation of the measurements. The variances were calculated as above.

YIELD AND STOICHIOMETRY IN GENERATION OF CHLORINE DIOXIDE

Chlorine dioxide generation by acid activation was characterized in both a laboratory reactor and a full-scale operating reactor. Chlorine dioxide generation by the chlorine-chlorite process was also studied under full-scale operating conditions.

For the laboratory generation study, characterization of chlorine species was completed within two hours of sampling the reactor product. For both full-scale operations, sample transportation necessitated a lapse of approxi-

Table IV. Measurement of Chlorine Species in Deionized Water[a]

| | Concentration (mg/l) | | | |
	Chlorine Dioxide	Chlorine	Chlorite	Chlorate
Added	1524	0	0	0
Found	1538 ± .25	2.5 ± 5.2	ND[b]	42 ± 26.9
Added	0	1041	0	0
Found	ND	1048 ± 6.3	ND	ND
Added	0	0	897	0
Found	ND	ND	897 ± 14.9	ND
Added	0	0	0	791
Found	ND	ND	ND	802 ± 3.1

[a] Values reported are mean of five replicates ± standard deviation of measurements, in mg/l.

[b] ND = none detected; detection limits are:

$$\text{Chlorine dioxide} = 2 \text{ mg/l} = 3.0 \times 10^{-5} \text{ mol/l}$$
$$\text{Chlorine (as } Cl_2) = 2 \text{ mg/l} = 5.6 \times 10^{-5} \text{ mol/l}$$
$$\text{Chlorite} = 20 \text{ mg/l} = 3.0 \times 10^{-4} \text{ mol/l}$$
$$\text{Chlorate} = 15 \text{ mg/l} = 1.8 \times 10^{-4} \text{ mol/l}$$

mately four hours before analysis was begun; chemical analysis was complete within six hours of sampling the reactor product. The concentration of chlorine dioxide only was measured onsite for both full-scale reactors; all chlorine species were analyzed after return to the laboratory. Comparing on-site measurements of chlorine dioxide with those obtained in the laboratory some four hours later indicated that for the acid activation generation scheme, the chlorine dioxide concentration of the reactor product was stable. For the chlorine-chlorite process, however, the chlorine dioxide concentration measured in the laboratory was significantly lower than that measured in the reactor product immediately after generation. The decrease in chlorine dioxide is believed to be due to reaction of excess unreacted chlorine with the generated chlorine dioxide [6].

Laboratory generation of chlorine dioxide differs significantly from continuous generation, as practiced at water treatment plants, in reactant purity and concentration and in reaction time. Concentration and purity of reactants may influence the reaction stoichiometry as well as the rate, because of the possibility of competitive reactions. Hence, reaction stoichiometries and rates must be interpreted with caution in attempting to predict the performance of full-scale generators from laboratory data.

In all cases discussed below, the reaction stoichiometries are based on a mass balance of chlorine atoms around the reactor of interest. Initial concentrations were determined by measuring all species in the process water and the concentrated reactant feed solutions. These concentrations are expressed as concentrations per liter of final reactor flow. Final concentrations were determined by analyzing the reactor product directly.

Acid Activation of Sodium Chlorite

When chlorine dioxide is generated from sodium chlorite by acid activation, chlorine dioxide is not the only end product. The final composition is dependent on several variables, including sodium chlorite concentration, purity of sodium chlorite used, acid concentration and pH of reaction mixture, reaction time, and temperature of reaction.

Several stoichiometries have been reported for acid activation of a chlorite solution [6]. The two stoichiometries that are most widely accepted are:

$$4ClO_2^- + 2H^+ = 2ClO_2 + Cl^- + ClO_3^- + H_2O \qquad (6)$$

and

$$5ClO_2^- + 4H^+ = 4ClO_2 + Cl^- + 2H_2O \qquad (7)$$

Reaction 6 is catalyzed by the chloride ion, which is also a product of the reaction. It has been reported that the reaction rate is accelerated by the product, chloride ion, and also that the stoichiometry is altered to that of the second reaction [6].

Reaction Stoichiometry in Laboratory Generation

The generation scheme used in determining reaction stoichiometry for laboratory generation of chlorine dioxide is that shown in Figure 1, with the exception that the chlorine dioxide gas was not purged from the reaction vessel and the reaction vessel was sealed and covered to prevent photodecomposition and loss of volatile constituents. The reaction was initiated by adding 4 ml of concentrated (36 N) sulfuric acid to 1 liter of reactant mixtures as shown in Table V. Under these conditions, the reaction was complete within three hours; no change in composition was observed subsequently.

The initial and final compositions of the laboratory reaction mixture are shown in Table V. The recovered mass of chlorine represents 104% of the initial mass of chlorine; considering experimental error, this value is not significantly different from the expected value of 100%.

Table V. Composition of Reactants and Products in Laboratory Generation of Chlorine Dioxide by Sulfuric Acid Activation of Sodium Chlorite

Species	Chlorine Atoms (mol/l)
Reactants	
ClO_2	0.0
ClO_2^-	0.083
ClO_3^-	0.0004
Cl_2	0.0
Cl^-	0.0177
Total	0.1011
Final Composition (pH 1.75)	
ClO_2	0.038
ClO_2^-	0.0037
ClO_3^-	0.0206
Cl_2[a]	0.0004
$HOCl$[a]	0.0005
Cl^-	0.0424
Total	0.1056

[a]Calculated, assuming equilibrium is attained in chlorine hydrolysis, $K_H = 3 \times 10^{-4}$.

If the initial composition of the reaction mixture is equated to the final composition, the following approximate balance for chlorine results:

$$0.083 \; ClO_2^- + 0.0004 \; ClO_3^- + 0.0177 \; Cl^- = 0.0037 \; ClO_2^- + 0.0206 \; ClO_3^- +$$
$$+ \; 0.0424 \; Cl^- + 0.0380 \; ClO_2 + 0.0002 \; Cl_2 + 0.0005 \; HOCl \qquad (8)$$

Collecting terms and relating all species to the conversion of four moles of chlorite, analogous to Reaction 6, gives the following equation, which is in approximate balance only for chlorine:

$$4 \; ClO_2^- = 1.92 \; ClO_2 + 1.02 \; ClO_3^- + 1.25 \; Cl^- + 0.04 \; Cl_2 \qquad (9)$$

This stoichiometry for chlorine species is essentially that of Reaction 6.

Reaction Stoichiometry in Full-Scale, Continuous Generation

A reaction mixture from a continuous, full-size generator (Rio Linda Chemical Company) was analyzed. This reactor utilized the sulfuric acid activation of sodium chlorite. The reactor consists of a section of poly-vinyl chloride (PVC) pipe, 10 cm in diameter and 66 cm long, packed with turbulence-inducing packing (Figure 2). Its nominal capacity is approximately 2 kg/hr chlorine dioxide. The residence time in the reactor is approximately 20 min. Process water flow was 0.33 liter/sec, the chlorite feed rate was 3.73 ml/sec of 1.92 M chlorite (ClO_2^-) solution, the acid feed rate was 1.42 ml/sec of commercial grade (40% by weight) sulfuric acid. Accordingly the mole ratio $[H_2SO_4]:[ClO_2^-]$ under these feed conditions was 1.06; H_2SO_4 was substantially in excess of the stoichiometric requirement according to Reactions 6 and 7. Product flow, defined as the sum of process water, chlorite feed and acid feed was 0.335 liter/sec. The presence of chlorate and chloride in the reactor feed is due to impurities in the chlorite feed solution.

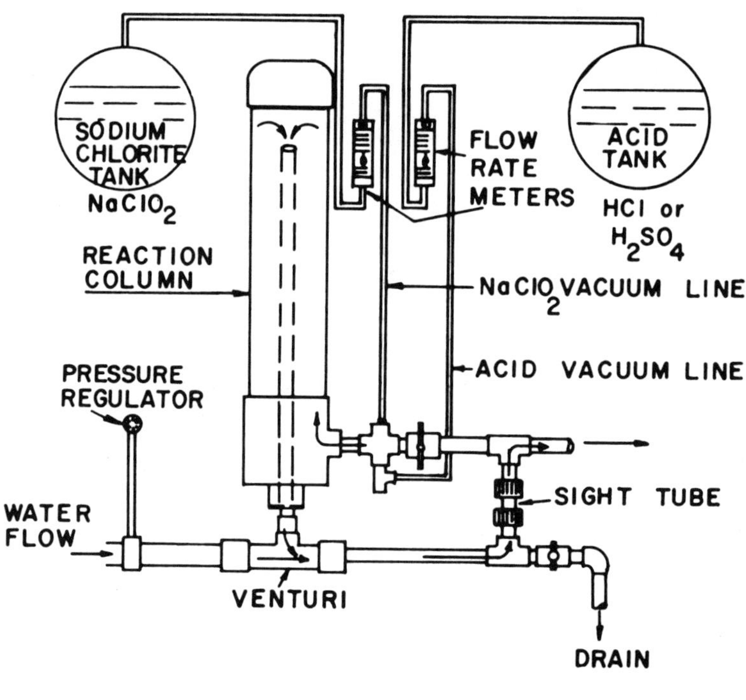

Figure 2. Chlorine dioxide generator flow diagram. Acid-chlorite process.

The initial and final composition of this reaction mixture is shown in Table VI. The recovered mass of chlorine atoms represents 104% of the initial mass; this is not significantly different from 100%. From Table VI, the stoichiometry based on the reaction of four moles of ClO_2^- is:

$$4\ ClO_2^- = 2.63\ ClO_2 + 0.53\ ClO_3^- + 1.01\ Cl^- + 0.05\ Cl_2 \qquad (10)$$

which can be compared to the result from the laboratory reaction (Equation 9), or, on the basis of five moles of ClO_2^- reacting:

$$5\ ClO_2^- = 3.29\ ClO_2 + 0.66\ ClO_3^- + 1.26\ Cl^- + 0.06\ HOCl \qquad (11)$$

Table VI. Composition of Reactants and Products from Chlorine Dioxide Generation in a Continuous, Full-Sized Acid-Chlorite Reactor Using H_2SO_4

Species	Chlorine Atoms (mol/l)
Reactants	
ClO_2	0.0
ClO_2^-	0.0214
ClO_3^-	0.0003
Cl_2	0.0
Cl^-	0.0021
Total	0.0238
Final Composition (pH 2.1)	
ClO_2	0.0109
ClO_2^-	0.0048
ClO_3^-	0.0025
$Cl_2{}^a$	<0.0001
$HOCl^a$	0.0002
Cl^-	0.0063
Total	0.0247

[a]Calculated, assuming equilibrium is attained in chlorine hydrolysis, $K_H = 3 \times 10^{-4}$.

Comparing Equations 10 and 11 with Equations 6 and 7, it appears that the reactor stoichiometry falls between the stoichiometries corresponding to Reaction 6 and 7. This may indicate that the overall reaction in a sulfuric acid-chlorite generator is a combination of these reactions (Equations 6 and 7). In addition, from Table VI, it can be seen that only 78% of the added chlorite is converted and of the 78%, 66% is converted to chlorine dioxide. This gives an overall yield of 51%: 51% of added chlorite is converted to chlorine dioxide, 10% is converted to chlorate, 22% is unreacted chlorite, and the remainder forms chloride and a trace of chlorine. Under the most favorable conditions, the maximum yield of chlorine dioxide found from the sulfuric acid activation of sodium chlorite was 50-55%, based on chlorite feed.

Since it had been reported [6] that the chloride ion catalyzes the reaction and alters the stoichiometry to the more favorable Reaction 7, the continuous reactor under study was converted from sulfuric acid activation to hydrochloric acid activation. To evaluate this generation process, a mass balance on chlorine atoms was again performed around the continuous reactor. In the HCl activation experiment, process water flow was 0.32 liter/sec, chlorite feed was 1.87 ml/sec of 1.92 M chlorite (ClO_2^-), acid feed was 0.79 ml/sec of commercial 20°Bé muriatic acid, which contains 32 wt% HCl. Accordingly the mole ratio $[HCl]:[ClO_2^-]$ was approximately 1.40, which corresponds to a 40% excess of HCl above the stoichiometric requirements according to Reactions 6 and 7. Product flow was 0.323 liter/sec. The pH of the resulting reaction mixture was approximately 2.2.

The results of the balance for acid activation with HCl are reported in Table VII. The recovered mass of chlorine represents 99.7% of the initial mass. When the stoichiometry of this mixture is determined and related to an equation in which five moles of chlorite react, the following approximate stoichiometry for acid activation with HCl is found:

$$5\ ClO_2^- = 4.02\ ClO_2 + 0.14\ ClO_3^- + 0.79\ Cl^- \tag{12}$$

Stoichiometry 12 very closely approximates that of Reaction 7, the stoichiometry proposed [6] as optimum for acid activation of sodium chlorite.

If HCl is used for activation, 97% of the chlorite is converted (Table VII), compared to only 78% conversion of chlorite in activation by sulfuric acid. Of the feed chlorite, 78.2% is converted to chlorine dioxide. The average yield of chlorine dioxide based on feed sodium chlorite observed in three field experiments was 78.5% (Table VIII). These measurements of reactor yields were made in the field immediately after sampling, while those made in the laboratory for the mass balance were made after a transportation and storage time on the order of six hours. With this consideration in mind it is reasonable to expect consistent reactor yields of 75-78% conversion of sodium chlorite to chlorine dioxide by acid activation with hydrochloric acid. The complete stoichiometry expected from hydrochloric acid activation of sodium chlorite is as follows:

$$5\ NaClO_2 + 4\ HCl = 4\ ClO_2 + 5\ NaCl + 2\ H_2O \tag{13}$$

Table VII. Composition of Reactants and Products from Chlorine Dioxide Generation in a Continuous, Full-Sized Acid-Chlorite Reactor Usinc HCl

Species	Chlorine Atoms (mol/l)
Reactants	
ClO_2	0.0
ClO_2^-	0.0110
ClO_3^-	0.0000
Cl_2	0.0000
Cl^-	0.0260
Total	0.0370
Final Composition (pH 2.2)	
ClO_2	0.0086
ClO_2^-	0.0003
ClO_3^-	0.0003
Cl_2 and HOCl	$<$0.0001
Cl^-	0.0277
Total	0.0369

Table VIII. Yield of Chlorine Dioxide from Acid Activation in Field Experiments Using HCl

Date	Chlorite in Feed Stream (mol/l ClO_2^-)	Chlorine Dioxide in Product Stream (mol/l ClO_2)	% Yield Based on Feed Chlorite
1-16-79	0.0313	0.0233	74.5
2-05-79	0.0307	0.0234	76.3
2-13-79	0.0216	0.0183	84.7

Chlorine-Chlorite Generation of Chlorine Dioxide

An alternative generation method for chlorine dioxide utilizes the reaction of sodium chlorite with chlorine. The simplified stoichiometry for this reaction is:

$$2\ ClO_2^- + Cl_2 = 2\ ClO_2 + 2\ Cl^- \tag{14}$$

From Equation 14, theoretically 100% of the sodium chlorite can be reacted to chlorine dioxide. An excess of chlorine above the stoichiometric requirement as predicted by equation 14 has been reported to increase yields of chlorine dioxide substantially and ensure complete conversion of sodium chlorite. Generally, twice the stoichiometric amount of chlorine is used [7]. It seems unreasonable to expect excess chlorine to increase chlorine dioxide yield, because it is known that chlorine and chlorine dioxide react rapidly in neutral solution [6]. Hypochlorite ion reacts with chlorine dioxide to produce chlorate ion according to the reaction:

$$2\ ClO_2 + OCl^- + H_2O = 2\ ClO_3^- + Cl^- + 2\ H^+ \tag{15}$$

Some researchers [6] have reported that more chlorine dioxide disappears in the presence of chlorine than predicted by Equation 15. This indicates that there may be other reactions that also consume chlorine dioxide over time. In highly acidic solution, however, the reaction (Equation 15) is extremely slow due to the low concentration of hypochlorite ion. While Reaction 15 is not instantaneous, even at more neutral pH, it may be sufficiently rapid to affect analyses for chlorine species several hours after generation.

The reactor utilized for the study of the chlorine-chlorite generation process was a commercially abailable unit (Rio Linda Chemical Company, Rio Linda, CA). The reactor is shown in Figure 3; it consists of a packed section of PVC pipe, 10 cm in diameter and 66 cm long. Its normal capacity is 2 kg/hr chlorine dioxide. Chlorine gas is fed directly to the concentrated chlorite solution as shown. After a short reaction time (several seconds), the reaction mixture is diluted by the process water and the reaction proceeds to completion in the larger reaction chamber which has a residence time of 2 min. The initial reaction period allows contact of the concentrated chlorite feed solution with the most reactive form of chlorine, diatomic chlorine gas. Process water flow was 0.32 liter/sec; chlorite feed was 1.87 ml/sec of 1.92 M chlorite (ClO_2^-); chlorine feed was 0.131 g chlorine gas/sec. Product flow was 0.322 liter/sec.

In this study, chlorine dioxide analyses immediately after generation of chlorine dioxide by the chlorine-chlorite process (within five minutes of sampling) indicated a 93-98% yield of chlorine dioxide based on feed chlorite with a chlorine feed of only 4% in excess of the stoichiometric amount. Analyses four to six hours later, after transportation to the laboratory, showed a chlorine dioxide yeild of only 73% based on chlorite feed. Reactions of the type shown in Equation 15 are presumably responsible for the apparent loss of chlorine dioxide. Since chlorine dioxide is generated immediately before use, these reactions should not significantly affect the concentration of chlorine dioxide at the point of application.

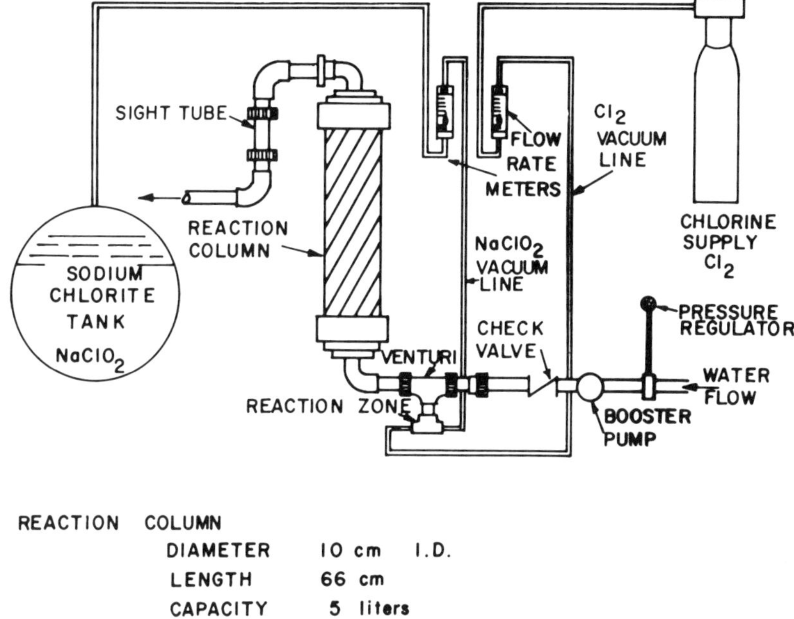

Figure 3. Chlorine dioxide generator flow diagram. Chlorine-chlorite process.

The results of the mass balance four hours after generation are shown in Table IX. The recovered mass of chlorine represents 96% of the initial chlorine, again not significantly different from 100% recovery. A reaction stoichiometry derived from this mass balance would not be indicative of reactor performance due to the reactions taking place in the time interval between generation and analyses for the mass balance. Two measurements of chlorine dioxide concentration were made in the field during continuous operation under the initial conditions as shown in Table IX. The chlorine dioxide concentrations were 724 mg/l (0.0107 mol/l) and 692 mg/l (0.0103 mol/l) corresponding to 98 and 93% conversion of feed chlorite to chlorine dioxide, respectively.

The practice of utilizing excess chlorine in the chlorine-chlorite generation of chlorine dioxide is unnecessary and deleterious. This practice results in an apparent increase in the chlorine dioxide yield of a sample analyzed at some later time, when in fact the excess chlorine serves only to retard Reaction 15 by increasing the acidity of the sample. More importantly, the primary rationale for utilizing chlorine dioxide as a water and wastewater disinfectant

**Table IX. Composition of Reactants and Products in Continuous Full Scale Generation
of Chlorine Dioxide by Chlorine Activation of Sodium Chlorite**

Species	Chlorine Atoms (mol/l)
Reactants	
ClO_2	0.0
ClO_2^-	0.0110
ClO_3^-	0.0
Cl_2	0.0114
Cl^-	0.0005
Total	0.0229
Final Composition (pH 5.35)	
ClO_2	0.0081
ClO_2^-	0.0002
ClO_3^-	0.0014
Cl_2	0.0002
Cl^-	0.0121
Total	0.0220

is to provide satisfactory disinfection without forming large quantities of
halogenated by-products. When chlorine dioxide is generated with chlorine in
excess, the problem still remains, i.e., the halogenated organic by-products
may be formed by the excess chlorine.

MEASUREMENT OF CHLORINE DIOXIDE RESIDUALS

There are several analytical methodologies for determining chlorine
residuals in water and waste water (*Standard Methods* [1] Section 409).
Amperometric titration has been the standard of comparison for the determination of free and combined chlorine in natural and treated waters and
iodometric back titration with amperometric endpoint detection has been
used for determination in waste waters. Most utilities use a continuous
amperometric titrator to control the disinfection process by monitoring
chlorine residuals. The amperometric method has been shown above to be
applicable to the determination of chlorine dioxide residuals in water.

Amperometric Determination of Chlorine Dioxide Residuals in Waste Water

Amperometric titration measures iodine (I_2) released into solution when iodide is oxidized by any powerful oxidant such as chlorine dioxide or chlorine. If this released I_2 comes into contact with organic constituents in the sample (as in the case of waste water), the I_2 will react with the organics. To prevent this, a back titration procedure is recommended, in which the liberated I_2 is immediately reacted with PAO, which has been added to the sample prior to the addition of iodide (as KI). The PAO is added in excess and the unreacted PAO is back-titrated with an I_2 solution of known concentration. The iodine titrant is standardized against the PAO primary standard. The concentration of chlorine dioxide (or chlorine) is then determined by difference.

The back titration procedure for determination of chlorine dioxide residuals is as follows [1]:

1. Add enough 0.00564 N PAO solution to a 250-ml beaker to be in excess of the expected chlorine dioxide residual.
2. Add 2 ml of pH 7 buffer.
3. Add 1 ml of 5% wt/vol potassium iodide solution.
4. Carefully pour a known volume of sample into the beaker, 150-175 ml. (Volume can be determined after analysis by subtracting volume of added reagents from total volume.)
5. Turn the titrator to the "Total" position and titrate with standardized iodine solution to an amperometric endpoint. The endpoint is denoted by a sudden needle deflection.

Residual is calculated by the following formulas:

$$\text{mg/l } ClO_2 = [\text{PAO} - (I_2 - \text{blank}) \times \text{CF}] \ \frac{0.00564 \ N \times 67450 \text{ mg } ClO_2/\text{equiv}}{\text{vol sample (ml)}}$$

where   PAO = volume of 0.00564 N PAO
         I_2 = volume I_2 titrant used (ml)
    blank = volume of I_2 found for blank (see below)
       CF = correction factor for I_2 titrant (see below)

The correction factor is found as follows:

1. Prepare a 250-ml beaker with 2 ml of the PAO solution, pH 7 buffer and potassium iodide solution.
2. Add 150 ml of deionized water.
3. Titrate as for a normal sample.

$$CF = \frac{vol\ PAO}{volume\ I_2\ titrant}$$

The blank correction value is found as follows:

1. Prepare a 250-ml beaker by adding 2 ml of PAO solution, pH 7 buffer and potassium iodide solution.
2. Add 150-175 ml of undisinfected sample water.
3. Titrate as for normal sample.

$$Blank = I_2\ volume - I_2\ volume\ from\ CF\ determination$$

The forward titration procedure for determination of chlorine dioxide residuals is as follows:

1. Prepare a 250-ml beaker with 2 ml pH 7 buffer and 2 ml potassium iodide solution.
2. Add 150-175 ml of sample.
3. Turn the titrator to the "Total" position and titrate with PAO solution to amperometric endpoint.

To evaluate the need for this back titration procedure, chlorine dioxide residual measurements were made with Palo Alto secondary effluent, comparing the amperometric method with and without back titration. The results of this experiment are reported in Table X. The experiment involved dosing 2 liters of Palo Alto secondary effluent to 4.50 mg/l with pure chlorine dioxide. The sample was then contacted for 2 minutes. After the contact period, sufficient KI and pH 7 buffer were added to the sample to react with all the chlorine dioxide present. In the case of I_2 back titration, sufficient PAO was added to swamp the released I_2. Eight replicates were run at approximately one-minute intervals. The experiment was repeated for filtered Palo Alto secondary effluent. The values reported have been corrected for blank titrations. In the case of the forward titration, no correction was necessary, but for the I_2 back titration, more I_2 solution was required in the blank than was indicated by the amount of PAO added. (The I_2 solution was standardized immediately prior to use.) Although not shown in Table X, the blank titrations (done in triplicate) showed the same variation as the forward or backward titration. The standard deviations shown in Table X are not statistically different from one another according to the F-test.

An analysis of variance (ANOVA) comparing forward titration with I_2 back titration showed that there is a significant difference between the values measured by the two titrations (Table XI). Based on the ANOVA results, amperometric back titration is required for the measurement of chlorine dioxide residuals in waste water.

Table X. Amperometric Titration Evaluation

| | Concentration[a] $(mg/l\ ClO_2)$ | | | |
| | Secondary Filtered | | Secondary Unfiltered | |
Sample	Forward Titration	Back Titration	Forward Titration	Back Titration
1	1.90	2.12	1.62	1.88
2	1.86	2.31	1.62	1.97
3	1.90	2.13	1.52	2.07
4	1.84	2.22	1.52	1.88
5	1.81	2.22	1.52	1.88
6	1.73	2.12	1.39	1.88
7	1.67	2.12	1.39	1.88
8	1.91	2.12	1.27	1.79
Mean	1.80	2.17	1.49	1.90
Std. Dev.	±0.083	±0.067	±0.109	±0.077
s^2	0.0069	0.0045	0.0119	0.0060

[a] Results are given as $mg/l\ ClO_2$ at 2-minute contact time; dose = 4.50 $mg/l\ ClO_2$. Titrations were performed at pH 7.

Table XI. Analyses of Variance for Amperometric Titration

Source of Variation	Degrees of Freedom	Sum of Squares	Mean Square	F-Ratio
ANOVA for Filtered Secondary Effluent Between forward and back titrations	1	0.54023	0.54023	82.79[a]
Error	14	0.09135	0.006525	
Total	15	0.63158		
ANOVA for Unfiltered Secondary Effluent Between forward and back titrations	1	0.67651	0.67651	66.35[a]
Error	14	0.14274	0.01020	
Total	15	0.81924		

[a] $P < 0.001$ that F is exceeded.

Chlorite Interference

To investigate the possible interference of chlorite in the determination of chlorine dioxide by the amperometric method, a solution of sodium chlorite was added to a sample containing chlorine dioxide in distilled water. Since the sodium chlorite used was only $\sim$95% pure, glycine (aminoacetic acid) was added to the sample to react any chlorine or bromine compounds present which would give an apparent positive result from the chlorite [3]. The results of this experiment (Table XII) showed no significant interference in the determination of chlorine dioxide caused by the presence of chlorite.

**Table XII. The Effect of Chlorite on the Determination of
Chlorine Dioxide by Amperometric Titration**

Sample	ClO_2	ClO_2 + 4.33 mg/l ClO_2^- + Glycine[a]
1	5.47	5.55
2	5.55	5.51
3	5.72	5.55
4	5.61	5.61
5	5.59	5.61
Mean	5.59	5.57

[a]Two ml of 10% w/v solution of glycine added to 200 ml of sample.

Amperometric Titration for Chlorine Dioxide
Disinfection Process Control

Commercially available amperometric titrators can be adapted to control the disinfection process when chlorine dioxide is the disinfectant of choice.

For control of the disinfection process with chlorine as the disinfectant in water or wastewater treatment where there is a requirement for a "free" chlorine residual, the concentration of free chlorine is determined directly in the amperometric cell with no addition of reagents. The control unit is standardized by calibration against a manual unit in which the free chlorine concentration is determined at pH 7 (no addition of potassium iodide). Where control is based on "combined chlorine," the sample stream is buffered at pH 4 to ensure quantitative response from mono- and dichloramine. The measurement of combined chlorine couples the reduction of chlorine to chloride with the oxidation of iodide to iodine. The introduction of iodide into the sample stream is achieved by adding potassium iodide to the pH 4 buffer solution. The control unit is standardized by calibration against a manual unit in which the combined residual is determined by the back titration procedure at pH 4.

Chlorine dioxide residuals can be monitored with the identical amperometric titration instruments by buffering the sample at pH 7 and adding sufficient potassium iodide to the buffer solution to couple the reduction of chlorine dioxide to chlorite with the oxidation of iodide to iodine in the sample stream. For water treatment, the control unit can be calibrated against a manual unit in which the chlorine dioxide residual is determined at pH 7 by the forward titration as described above. For wastewater disinfection process control, the control unit should be calibrated against a manual unit in which the chlorine dioxide residual is determined at pH 7 by the back titration method described above.

SUMMARY AND CONCLUSIONS

A sequential approach based in part on amperometric and iodometric techniques was developed to measure the concentrations of chlorine dioxide and related species that are of interest in evaluating the yield and selectivity of chlorine dioxide generation. The procedure consists of the following steps:

1. measurement of chlorine dioxide by UV absorption spectrophotometry at a wavelength of 360 nm;
2. measurement of the sum of chlorine dioxide and chlorine by amperometric titration at pH 7;
3. measurement of the sum of chlorine dioxide, chlorine and chlorite by iodometric titration at pH 2;
4. measurement of the sum of chlorine dioxide, chlorine, chlorite and chlorate by carrying out step 3, after allowing prior reaction with concentrated HCl; and
5. measurement of chloride by an adaptation of the mercuric nitrate method

Hence, the concentrations of chlorine dioxide and chloride were measured directly, whereas the concentrations of other chlorine species were estimated by difference calculations.

The stoichiometry and yield of chlorine dioxide generation by acid activation of sodium chlorite were studied using the analytical techniques described above. Acid activation was studied in both bench-scale experiments and by quantifying the performance of a full-scale, continuous reactor; both H_2SO_4 and HCl were used for activation. The mass balances for chlorine atoms—comparing reactor input and output—agreed within 3-5%. The molar yield of chlorine dioxide from chlorite was in the range of 50% when H_2SO_4 was used for activation: the yield was approximately 75-80% when HCl was used as the activating acid. The latter yield agrees closely with the predicted maximum yield (80%) for acid activation of sodium chlorite.

When chlorine dioxide was generated by the reaction of sodium chlorite with chlorine (4% excess), the yield of chlorine dioxide based on chlorite reactant was found to vary from 93 to 98% based on chlorite feed.

The practice of using excess chlorine in the chlorine-chlorite generation appears not to increase chlorine dioxide yield but rather only serves to retard subsequent reactions that consume the chlorine dioxide product. Indeed, the use of excess chlorine may be disadvantageous in another respect; the unreacted chlorine has the potential to react with organic constituents present in the water being disinfected, thereby forming undesirable halogenated organic by-products.

The conventional amperometric method was found acceptable for measuring chlorine dioxide residuals either manually or automatically for disinfection process control.

ACKNOWLEDGMENTS

The authors thank the staff of the Dublin-San Ramon Utility District and the Rio Linda Chemical Company for their cooperation in the field portion of this work. Dr. Gary Stevenson of the Fischer & Porter Co. kindly provided an amperometric titrator for use on this project. This research was supported by the Municipal Environmental Research Laboratory, U.S. Environmental Protection Agency, under Grant No. R-805426.

REFERENCES

1. *Standard Methods for the Examination of Water and Wastewater*, 14th ed. (Washington, DC: American Public Health Association, 1976).
2. Miltner, R. J. "Measurement of Chlorine Dioxide and Related Products," in Proceedings AWWA Water Quality Technology Conf., Dec. 6-7, 1976, San Diego, California. Paper No. 2A-5; AWWA, Denver, Colorado, 1977.
3. Granstrom, M., and G. Lee. "Generation and Use of Chlorine Dioxide in Water Treatment," *J. Am. Water Works Assoc.* 50:1453 (1958).
4. Ingols, R., and G. Ridenour. "Chemical Properties of Chlorine Dioxide in Water Treatment," *J. Am. Water Works Assoc.* 40:1207 (1948).
5. Sokal, R. R., and F. J. Rohlf. *Biometry* (San Francisco: W. H. Freeman and Company, 1969).
6. Gordon, G., R. G. Kieffer and D. H. Rosenblatt. "The Chemistry of Chlorine Dioxide," in *Progress in Inorganic Chemistry, Vol. 15*, S. J. Lippard, Ed. (New York: Wiley-Interscience, 1972).
7. White, G. *Handbook of Chlorination* (New York: Van Nostrand Reinhold, 1972).

CHAPTER 21

DETERMINATION OF CHLORINE DIOXIDE

Roberts G. Stevenson, Jr.,*
Leo L. Dailey and Brian J. Ratigan

Environmental Instrumentation Development
Fischer & Porter Company
Warminster, Pennsylvania

After the 1974 publication of the report by Harris [1] on the occurrence and relationship to cancer of chloroorganic compounds in drinking water derived from the Mississippi River in New Orleans, a great deal of research has been undertaken to study the role of treatment with chlorine in the production of these compounds [2-7]. Other research has been undertaken to find alternatives to the standard treatment of water with chlorine to minimize the formation of halomethanes and other compounds. One alternative being investigated is treatment with chlorine dioxide (ClO_2), which has been used in the past to solve special problems in water treatment [8-10]. Chlorine dioxide as a disinfectant for waste water has been evaluated by several workers [11-13]. Chlorine dioxide has several advantages: it does not seem to react with dissolved organics to produce trihalomethanes (THM) nor does it react with ammonia [8,10,14,15]. It does react with organic precursors which would otherwise lead to the formation of THM in treatment

---

*Present address: Leeds & Northup, North Wales, PA.

with chlorine. Further, ClO_2 maintains a longer lasting residual in water than does ozone. Finally there is a growing trend away from chlorine gas treatment in several European nations because of the hazards associated with its handling, and ClO_2 is being used to replace it. In view of the possible future widespread use of ClO_2 in greater amounts, it seems apparent that equipment and methodology for continuous, on-line measurement of ClO_2 residuals may soon be required for effective and economical control of its application to water. Present methods of analysis in common use tend to be batch processes, which are not suitable for continuous control. The object of this research is to investigate several approaches to automated or continuous analysis of ClO_2 in aqueous solutions.

Chlorine dioxide is manufactured at the site of its use by reactions involving sodium chlorite, chlorate, chlorine gas or hypochlorite, and sulfuric or hydrochloric acid. Consequently ClO_3^-, ClO_2^-, OCl^- or $HOCl$ will frequently be found occurring with it in solutions as by-products or unreacted starting materials [16]. These materials are oxidizing agents which react in ways similar to ClO_2. Indeed, previous methods of analysis often considered ClO_2 as only another type of chlorine residual, since they failed to discriminate between it and concomitant interferences. Other interferences are materials such as dichromate, permanganate, ozone and nitrate. It was felt that new procedures could be developed or modifications would be made that would allow a more selective determination of ClO_2 in the presence of these species. The effect of other variables such as pH, suspended solids and background color must be minimized or considered as well.

Feuss [17] and Myhrstad and Samdal [18] presented the analytical problems associated with ClO_2. The chemistry of ClO_2 and ClO_2^- is described in good detail by Gordon et al. [19] and Masschelein [20].

SPECTROPHOTOMETRY

Direct spectrophotometry is attractive for the analysis of ClO_2 solutions in that it is simple to use with flow-through systems, capitalizing on the inherent color of ClO_2. Chlorine dioxide posesses a greenish yellow color sufficiently intense to cause an observable tint in deep water even at the low concentration of 0.2–0.3 mg/1 [21]. Chlorine dioxide has an absorption maximum at about 360 nm (Figure 1) with a molar absorptivity of about 1100 liter/mol-cm-1 [19–21]. It has been shown that this value is reasonably independent of conditions such as temperature, salinity, ionic strength and acidity [19]. Unfortunately, it is somewhat low, and trace analysis by direct absorption spectrophotometry may be difficult.

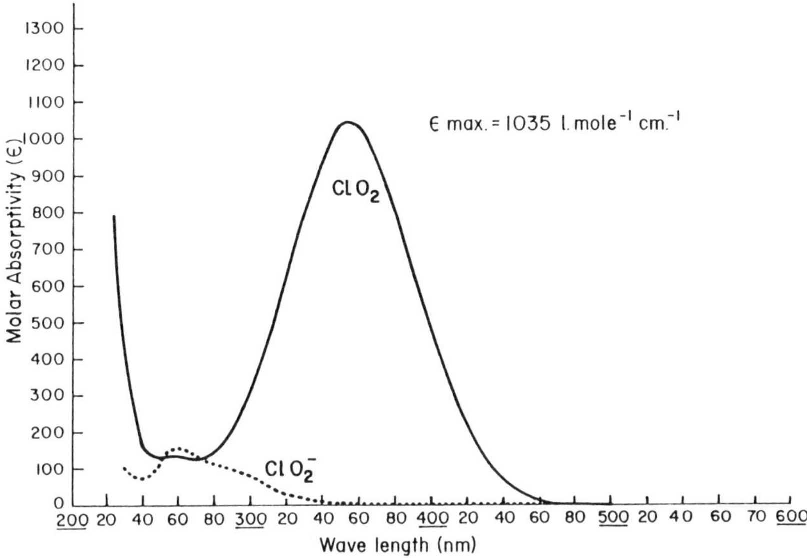

Figure 1. UV-Vis absorption spectra of ClO_2 and ClO_2^-.

The limits to direct determination are set by the available cell length and the theoretical error inherent in photometric analysis. The relative error of spectrophotometric procedures increases drastically when the absorbance is outside the limits of 0.1–1.0 units [24]. In a 1-cm cell, this limits the concentration range to 6.0–60.0 mg/l, working at a wavelength of 360 nm. Figure 2 displays the linearity achievable in this concentration region. Other ranges are possible if one is willing to accept errors larger than 1–2%, or is able to use shorter or longer cells. Many laboratory instruments can handle 5- or 10-cm cells. Cells down to 1 mm are available, but these tend to become air-bound if used as flow cells. Industrial photometers are available which can use very long cells (up to 1 meter) to determine ClO_2 in very dilute solutions. For high concentrations, one may operate at wavelengths which are off the 360-nm peak.

Equipment can be simple or complex, and a variety of configurations is possible. Direct spectrophotometry is capable of reasonable accuracy if care is taken to maintain the signal within proper photometric limits. The technique is reasonably selective since, as shown in Figure 1, ClO_2^- does not absorb appreciably at 360 nm.

Direct spectrophotometry will experience interference from other colored species which absorb appreciably at 360 nm. This might be minimized by

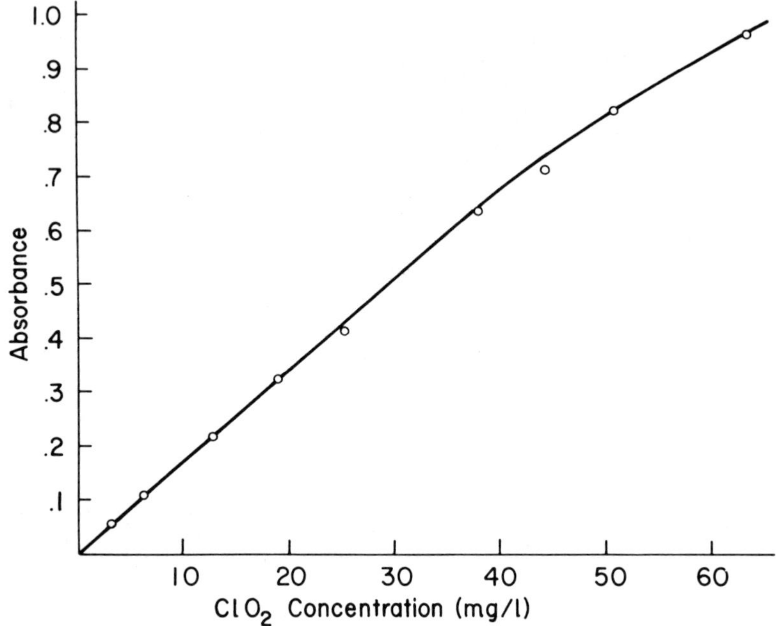

Figure 2. Absorbance of ClO_2 as a function of concentration (1-cm cell).

choosing another analytical wavelength, but sensitivity will be sacrificed. Turbidity-causing suspended solids will also interfere if not removed by filtration or otherwise accounted for. Furthermore, dirty optical surfaces will be interpreted as increased ClO_2 concentration.

Figure 3 represents the experimental setup to measure ClO_2 in the development of ClO_2 generators for use in water treatment. The concentrated effluent from the generators was diluted with tap water to bring the ClO_2 concentration into the 6- to 60-mg/l range. Most of the diluted solution was then passed to the drain, but a small portion was sent to a Perkin-Elmer/ Coleman Model 55 laboratory spectrophotometer equipped with a 1-cm flow-cell. Generators were developed using both the chlorine-chlorite and acid-chlorite methods. Yields of 95% and higher were obtained using stoichiometric ratios and based on $NaClO_2$ as the starting material.

The yields were checked by the following procedure:

1. With the spectrophotometer set at 250 nm, (the absorption maximum for ClO_2^-) feed water without any reactants ($NaClO_2$, Cl_2 or acid) was passed through the generator and a 10-cm measurement cell. When the spectrophotometric reading reached a steady, low

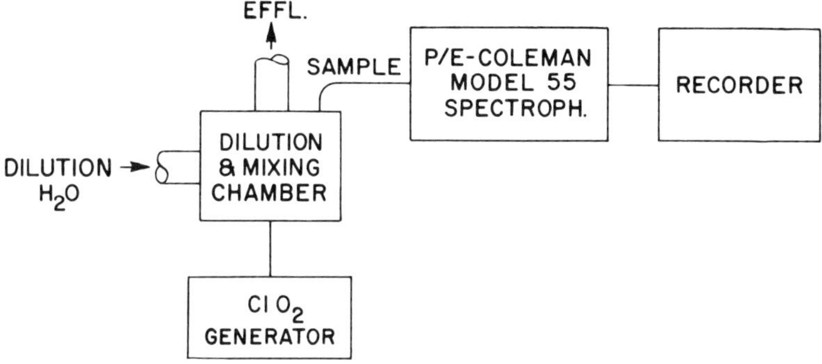

Figure 3. Laboratory measurement of ClO_2 using spectrophotometry.

value, indicating that all materials from previous experiments had been washed out, the instrument was adjusted to read 0.0 A (100% T).

2. The $NaClO_2$ feed solution pumps were then turned on and adjusted to the desired flowrate. The absorption at 250 nm was taken as the measure of ClO_2^- concentration.

3. The flow to the absorption cell was temporarily turned off while the rest of the system remained operating. The 10-cm cell was then replaced by a 1-cm cell, and the wavelength control set to 360 nm, the absorbance peak for ClO_2. The sample stream was then turned on and the critical reactant solution feed pump started, to begin ClO_2 production. The absorbance at 360 nm is taken as a measure of ClO_2 concentration.

4. The yields were calculated by the relationships:

$$Y = \frac{A_{360}}{A_{250}} \cdot \frac{K_{ClO_2^-}}{K_{ClO_2}}$$

where   A = absorbance read at 360 or 250 nm
           K = the slope of the spectrophotometric calibration plots for ClO_2 and
               ClO_2^- at the appropriate wavelength

To obtain the calibration plots for ClO_2 and ClO_2^-, stock solutions of each species were first prepared. Reagent grade $NaClO_2$ was weighed and dissolved to prepare a 1000-mg ClO_2^-/l solution. The exact concentration was determined by iodometric titration at pH 1.5. Appropriate quantities were added via a burette to 100-ml volumetric flasks and the absorbance of each solution was then read at 250 nm in a 10-cm cell. The ClO_2 stock solution was made using the *Standard Methods* ClO_2 generation technique [25]. Appropriate dilutions were made by adding portions of the stock solution to 100-ml volumetric flasks. After reading the absorption at 360 nm in a 1-cm cell, the dilute solution was immediately titrated iodometrically at pH 7.00 to obtain its true ClO_2 concentration.

It was discovered during the generation experiments that the flowrate of the ClO_2 solution to the absorption cell affected the output signal, which increased as the sample flow increased. This curious phenomenon can be explained by noting the fact that the tubing delivering the sample was stained yellow by ClO_2, indicating that ClO_2 can permeate Tygon. Apparently, at slow flow, enough ClO_2 passed through the walls of the tubing before the sample stream reached the spectrophotometer to significantly affect the results. Increasing the flow minimized this error.

SPECTROPHOMETRY WITH CHROMAGENIC REAGENTS

To extend the sensitivity of spectrophotometry to lower levels it is possible to add chromagenic reagents whose absorptivities after reaction with oxidizing agents may be quite large. Aston [25] developed the technique recommended by *Standard Methods* involving orthotolidine [26]. The method uses oxalic acid and arsenite to suppress interferences and to slow color development. It is quite sensitive to pH and color development time. Masschelein [27] used the decolorizing effect of ClO_2 on acid chrome violet K (ACVK) in which free chlorine presents minimal interference, presumably because of the high pH (8.1–8.4), where most HOCl has been converted to OCl^-. Although O_3 presents an intereference, ClO_2^- and ClO_3^- do not [20].

Knechtel, et al. applied ACVK to the determination of ClO_2 in sewage effluents [28]. Interfering solid material was centrifuged. The method was tested using electron spin resonance spectrometry as a reference method, taking advantage of the unpaired electron in the molecular structure of ClO_2.

Post and Moore determined ClO_2 in surface waters using 1-amino-8-naphthol-3,6-disulfonic acid, called "H acid," which develops an absorption peak at 525 nm in the presence of ClO_2 and HOCl [29]. HOCl can be masked using malonic acid. Because of the low pH (4.2) of the method, ClO_2^- interferes.

Palin (30, 31) developed a series of methods for active chlorine compounds using DPD as a colorimetric endpoint indicator and as a spectrophotometric reagent. The procedure for ClO_2, HOCl, OCL^-, chloramines and ClO_2^- is similar to the iodometric titration of Haller and Listek [32]. Palin uses malonic acid or glycine to destroy free chlorine before performing the ClO_2 titration with phenylarsine oxide (PAO). The method has been modified by the addition of thioacetamide to control ClO_2^- "breakthrough" at pH 7 [33]. The disadvantage of DPD in continuous monitoring is the requirement to control the pH and mask interferences by the addition of a host of reagents, making it a more complex method to automate and maintain.

The problem with most of the above methods is the lack of selectivity for ClO_2 in the presence of other oxidizing agents. Lott et al. developed chlorophenol red as a selective reagent for CLO_2 [34]. Hypochlorous acid (5 mg/l), OCl^- (10 mg/l), ClO_2^- (50 mg/l), Fe^{3+} (10 mg/l), CrO_4^{2-} (50 mg/l) and MnO_4^- (10 mg/l) do not interfere. The compound may be used either as a titrant or as a spectrophotometric reagent. Like ACVK, it is a "bleaching" method.

While capable of very good sensitivity and accuracy in the determination of ClO_2, chromagenic reagents possess several shortcomings for continuous monitoring. They require extra solutions which must be prepared and added to the sample stream. The sample also must frequently be "conditioned" by the addition of other reagents to control the pH and mask interferences. Time and temperature, as parameters in the color developing reaction, must also be controlled. Along with direct spectrophotometry, these methods may be sensitive to background color and turbidity in the sample, and may suffer as optical windows become dirty or cloudy.

TITRIMETRIC METHODS

The most common titrimetric methods for ClO_2 make use of the iodometric reactions shown below.

$$pH\ 7:\quad 2ClO_2 + 2I^- \longrightarrow 2ClO_2^- + I_2$$

$$pH\ 1.5:\quad 2ClO_2 + 8H^+ + 10I^- \longrightarrow 2Cl^- + 4H_2O + 5I_2$$

In the first reaction ClO_2 exhibits an equivalent weight equal to its molecular weight of 67.45 g/mol. In the second reaction, there are five equivalents per mole, or an equivalent weight of 13.49 g/equivalent. Chlorite reacts only slowly at pH 7; hence, it interferes only slightly. Nevertheless, even at pH 7, if present in large excess, it will react, and the determination of ClO_2 may seem to require further addition of titrant, even though the ClO_2 endpoint has been reached. Chlorite reacts with I^- at acid pH values, quickly giving its complete iodometric equivalent at pH 1.5, having an equivalent weight of 16.86 g/equivalent.

$$ClO_2^- + 4H^+ + 4I^- \longrightarrow Cl^- + 2H_2O + 2I_2$$

Other species capable of producing I_2 will also be titrated. These may be viewed as interferences, but it is possible to capitalize on the two reactions involving ClO_2 to arrive at the concentration of various concomitant species in a sample, such as ClO_2, ClO_2^-, $HOCl$, ClO_3^- and O_3. This is the basis of the method first described by Haller and Listek [32] and adapted for use in *Standard Methods* [26]. It consits of four distinct titrations with PAO. The endpoint is detected amperometrically. Four samples are drawn and treated. One sample is raised to pH 12.00, at which point ClO_2 dispro-portionates to ClO_2^- and ClO_3^-. After 10 minutes or longer, the pH is reduced to 7.0 and the sample is titrated directly to determine the free chlorine (HOCl) concentration. The addition of KI to a second sample, similarly treated, determines $HOCl$ and NH_2Cl together. Adding KI to a third sample at pH 7.0, without previously raising the pH to 12.0, allows one to titrate the sum of $HOCl$, NH_2Cl and ClO_2. A fourth titration, after adding KI to a sample at pH 2.0, gives the total iodine equivalent of the sample, including the contribution by chlorite.

There are some problems with this technique, however. It does not seem that pH 12.0 is sufficiently high to speedily decompose all the ClO_2 in the first two titrations. To investigate this possibility, the following experiments were done. Sodium hydroxide (50% solution) was added to two 100-ml samples of approximately 40 mg ClO_2/l. One was adjusted to pH 12.0, the other to pH 12.5. The decomposition of ClO_2 in the solutions was observed spectrophotometrically at 360 nm in a 1-cm cell. After more than an hour at ambient temperature the pH-12 sample still contained 25% of its original ClO_2. In the pH-12.5 sample, on the other hand, only 0.1% of the ClO_2 remained undecomposed at the end of 30 minutes. The residual ClO_2 in the former sample will interfere with the determination of NH_2Cl in the second titration, since it will give its usual iodine equivalent at pH 7. Also, in the free chlorine titration, even though the reaction between ClO_2 and PAO is so kinetically slow as to be insignificant, the amperometric signal used to detect the endpoint will be elevated, and may be off scale.

A second problem appears in the determination of total ClO_2 and ClO_2^-, the fourth titration at pH 2.0. There is reason to believe that pH 2.0 is not sufficiently low to bring about the complete conversion of ClO_2 and ClO_2^- to I_2. Miltner suggests going to pH 1.5 [14].

For concentrated solutions of ClO_2, visual endpoint detection using starch indicator is usually sufficient. Titration at the mg/l level can be made quite accurately using microburettes and amperometric endpoint detection. Titration procedures for a variety of mixtures involving ClO_2 and ClO_2^- can be found in Masschelein [20]. In waste water, compounds which present an iodine demand will cause serious errors. A back-titration procedure developed by Aieta et al. was effective in determining ClO_2 in such media [35].

Other titration procedures have been developed using colorimetric reagents such as DPD [30,31,33] and chlorophenol red [34]. DPD has not been as successfully applied to the measurement of ClO_2 in waste water [36], as mentioned previously. Chlorophenol red seems a likely candidate as a direct titrant for ClO_2 which is relatively free from other interferences but it has not yet been tested in waste water. Since ClO_2 can be determined amperometrically, it may be possible to improve on the accuracy of microtitration using CPR and an amperometric titrator such as the Fischer and Porter 17T2000 Titrator which incorporates a platinum/copper electrode pair biased at a potential where ClO_2 responds. Experimental work on this has not yet begun.

An important requirement in any analytical procedure is the care necessary to ensure that ClO_2 is not lost from the solution during sample manipulation. For example, it is poor procedure to allow the sample to fall through air or to splash vigorously. Sample delivery to the bottom of a narrow mouthed bottle, allowing it to overflow, is recommended, if convenient. When pipetting samples, the tip of the pipette should be submerged during delivery, and if possible, KI should be present to convert ClO_2 to I_2 immediately, if this is desirable.

The following procedure is used to determine ClO_2, ClO_2^- and HOCl in concentrated streams from ClO_2 generators operating in the acid-chlorite generation scheme [37]. To each of six BOD bottles is added 80 ml of distilled H_2O. To the first two bottles, labeled A, 20-ml portions of 1 M phosphate buffer at pH 7.7 are added and approximately 1 g KI is dissolved. The next two, labeled C, receive 1 g KI and 5 ml 85% H_3PO_4. The remaining bottles, D, receive 2 ml of 50% NaOH. An appropriate quantity of sample is pipetted to each bottle, and the bottles are stoppered. Solutions A and C are stored in the dark for 10 min while solution D may be kept in indirect light for 30 minutes until its color fades.

After 10 minutes, bottles A are titrated with 0.05 N thiosulfate to the starch endpoint. Then 5 ml of concentrated H_2SO_4 are added to each and the solutions are again stored in the dark for 10 minutes before titrating with 0.05 N thiosulfate. The results of the second titration are called B.

Solutions C are likewise titrated after 10 minutes with 0.05 N thiosulfate. The results of A and B should add to give C ± 0.1 ml.

After complete decoloration, 1 g of KI is dissolved in solutions D, followed by the addition of 5 ml 85% H_3PO_4. After 10 minutes in the dark, they are titrated with 0.05 N thiosulfate. The concentrations of each species are calculated below:

$$mg\ ClO_2/l = 112.5\ (C\text{-}D)$$
$$mg\ Cl_2/l = 178\ (A - 1/3\ (C\text{-}D))$$
$$mg\ NaClO_2/l = 113\ (B - 4/3\ (C\text{-}D))$$

The chemical reactions on which the method is based are given as follows:
(A) pH 7.7: ClO_2^- does not react.

$$Cl_2 + 2I^- \longrightarrow 2Cl^- + I_2$$
$$2ClO_2 + 2I^- \longrightarrow 2ClO_2^- + I_2$$

(B) pH about 1.5: Cl_2 and ClO_2 already reacted.

$$ClO_2^- + 4I^- + 4H^+ \longrightarrow Cl^- + 2H_2O + 2I^-$$

(Part of the ClO_2^- was generated in part A.)
   (C) pH about 1.5:   All of the species Cl_2, ClO_2 and ClO_2^- react as above, except ClO_2.

$$2ClO_2 + 10I^- + 8H^+ \ - \longrightarrow 2Cl^- + 4H_2O + 5I_2^-$$

(D) First part, at pH 12.5:

$$2ClO_2 + 20H^- \longrightarrow ClO_2^- + ClO_3^- + H_2O$$

Acidification to pH 1.5:

$$Cl_2 + 2I^- \longrightarrow Cl^- + I_2$$
$$ClO_2^- + 4H^+ + 4I^- \longrightarrow Cl^- + 2H_2O + 2I_2$$

ClO_3^- does not react.
   This method has been evaluated. While it is generally accurate, it may indicate as much as 4% of the total I_2 equivalent to be from Cl_2 in a Cl_2–free ClO_2 solution. Another error may arise in titration B if the starch in the solution following titration A presents a significant I_2 demand. To prevent this, the second titration should be carried out as soon as the 10-min wait has elapsed.

ELECTROCHEMICAL METHODS

Apart from amperometric endpoint detection, electrochemical means are seldom used for chlorine dioxide measurements, although they have great potential application. Hartley and Adams studied the polarographic characteristics of chlorite, which was shown to give a well defined, diffusion-limited wave at pH lower than 4.5 [38]. If the pH is lower than 4.2, the decomposition of $HClO_2$ proceeds too rapidly for accurate determination. Schwarzer and Landsberg investigated the voltametric behavior of ClO_2^- using a rotating graphite disk electrode in media of various pH values [39]. In the pH range between 5 and 9, well defined anodic waves were obtained which permitted the determination of ClO_2^-. Smooth platinum was found to be an unsatisfactory electrode because of the presence of Pt oxides in the electrode surface and a low O_2 overvoltage. On the other hand, Raspi and Pergola successfully applied a platinized Pt microelectrode with periodic renewal of the diffusion layer to the determination of ClO_2^- and ClO_2 [40]. At pH 7, a deaerated solution of ClO_2^- gave two distinct voltametric waves, an anodic one at $E_{1/2}$ = +0.705 volt and a cathodic wave at $E_{1/2}$ = +0.080 volt. The height of the anodic wave, representing the one electron oxidation of ClO_2^- to ClO_2 was one-fourth the height of the cathodic reduction wave. Chlorine dioxide alone in solution yielded two cathodic waves, one at $E_{1/2}$ = +0.705, representing the reduction of ClO_2 to ClO_2^-, the other at $E_{1/2}$ = 0.080. The charge transfer process between ClO_2 and ClO_2^-, which occurs irreversibly on smooth Pt, was shown to be reversible on platinized Pt.

Cauquis and Limosin determined ClO_2, ClO_2^-, $HOCl$, OCL^-, ClO_3^- and Cl^- by a five-step procedure [41]. Cl^- was determined by an ion-selective electrode; ClO_2^- and ClO_3^- by ion exchange separation and spectrophotometric measurement; ClO_2^- was determined by voltametry with a glassy carbon electrode; ClO_2, $HOCl$ and OCl^- were determined together, also with a glassy carbon electrode; and OCl^- and $HOCl$ were measured after stripping ClO_2 from the sample with N_2 gas.

Recently, Benga and Gordon determined ClO_2^- using differential pulse polarography [42]. Their calculated detection limit was 0.008 mg/l.

A German patent describes an apparatus for the measurement of ClO_2 and OCl^- by measurement of the redox potential [43]. A more specific method has been developed by Krunchak et al. for the measurement of ClO_2 and ClO_2^- in bleach solution using glass semiconducting electrodes which are selective with respect to the ClO_2/ClO_2^- couple [44].

CONTINUOUS ELECTROCHEMICAL ANALYZERS

In addition to the obvious parameters of accuracy and precision, an analytical procedure should possess other qualities such as simplicity, mechancial and electronic reliability, and economy if it is to be used for continuous or automated analysis and industrial control. Electrochemical instrumentation possesses many of these traits, and, as seen in the foregoing section, can be used to determine ClO_2 and $ClO_2{}^-$.

Two flow-through amperometric cells were studied as possible ClO_2 analyzers. In the first (Figure 4), two copper electrodes in direct contact with the sample stream were studied with ClO_2 solution only briefly. The device is presently used to monitor free chlorine in tap water and swimming pools. It is also used to monitor O_3 in clean water. It possesses several advantages: it is simple, has no moving parts and requires no reagents. The current sensitivity of the cell is about 20 $\mu A/(mg\ ClO_2/l)$, working with an impressed potential of 0.15 V. The cell monitored ClO_2 in a clean sample at concentrations from 0.6 to 1.10 mg/l for two weeks.

Since it is used to monitor chlorine and ozone in water, these substances are obvious interferents. Chloramine presents a minimal response and hexavalent chrome does not interfere up to 50 mg/l, the upper limit tested. It can only be used in relatively clean water and requires a filter to remove particles over 88 μ. Nevertheless, its simplicity makes it attractive and useful when these requirements are met.

A chlorine residual analyzer, also used for ozone, is shown in Figure 5 [46,47]. It incorporates a rotating gold measuring electrode and a copper counter electrode, and proved useful in measuring ClO_2, both directly and as the HOBr analog. Voltamograms from this cell at pH 4 and 7 are shown in Figures 6 and 7. These data were obtained using a Princeton Applied Research Model 174A Polarographic Analyzer. Examination of the ClO_2 curve at pH 7 shows a well defined wave with a half-wave potential of +0.495 V relative to the saturated calomel electrode. The plateau is relatively broad and flat. The foot of the dissolved oxygen wave begins to intrude at about +0.15 V. The current sensitivity of the cell is about 100 $\mu A/(mg\ ClO_2/l)$.

When 1 ml of 20% KBr is added to 200 ml of a ClO_2 solution, HOBr is produced, which gives a well defined wave with a half-wave potential of 0.485 V. The current sensitivity is decreased somewhat, being about 90 $\mu A/(mg\ ClO_2/l)$. Notice that the dissolved O_2 wave is shifted to a more negative potential and does not appear until about -0.05 V.

The curves remain essentially unchanged from pH 8 to 5, but below 5 they are less ideal, although ClO_2 is still directly measureable. The plateau regions in the curves are flat and broad, which allows freedom in choosing an impressed potential for the operation of the cell, should that be necessary

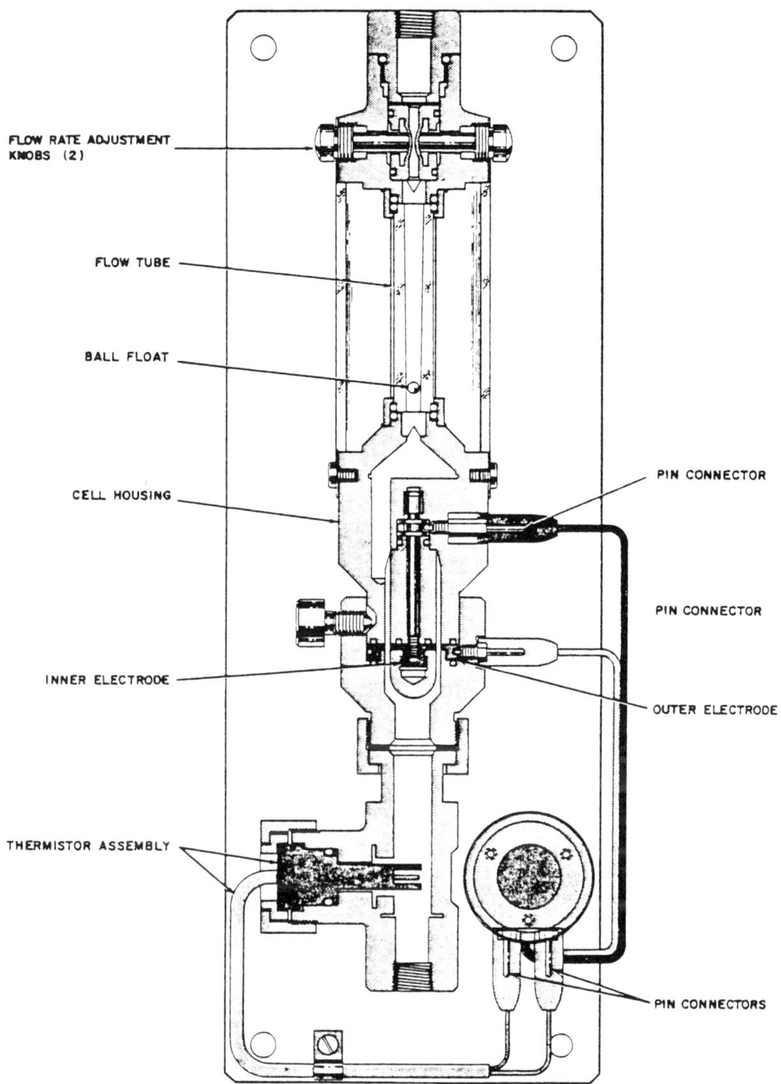

Figure 4. Flow-through amperometric cell.

to minimize interferences from substances such as dissolved O_2 or chloramines. However, glavanic operation is possible, and is probably to be preferred for simplicity. Chlorine, other halogens and ozone are serious interferences. The effect of chlorine can probably be minimized by operating in the im-

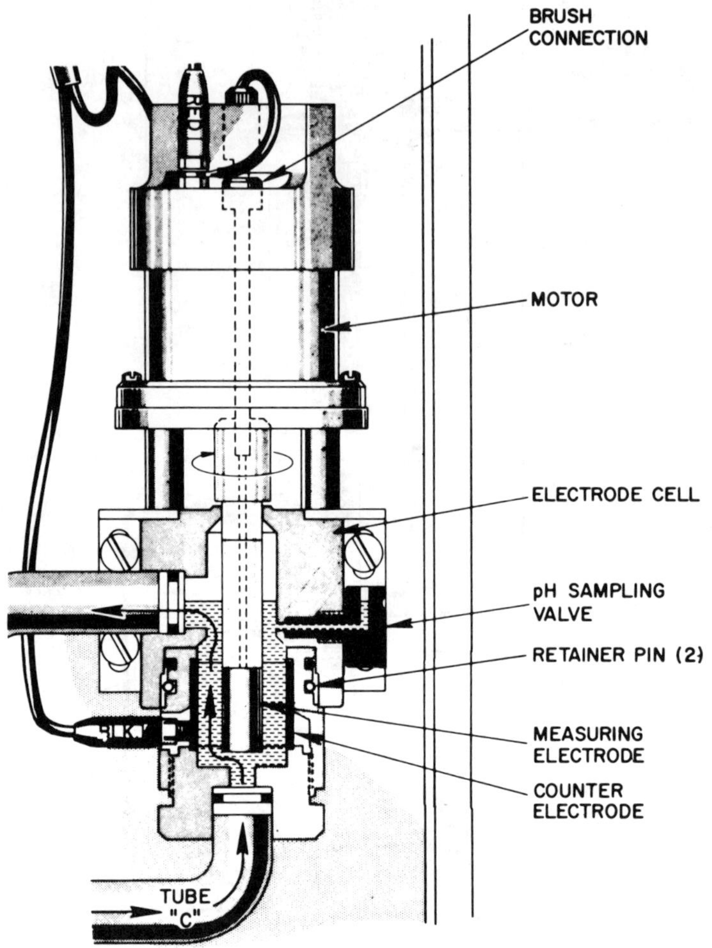

Figure 5. Chlorine residual analyzer.

pressed potential mode without addition of KBr. It might also be masked by the addition of NH_3, malonic acid or glycine, as is done in certain spectrophotometric methods. Chlorite does not give a response.

The cell operated for about two weeks on a stream of tap water to which was added ClO_2 at 1–2 mg/l. Its upper limit seems to be 150 mg/l. Further study is needed to determine its long-term reliability, especially in industrial applications.

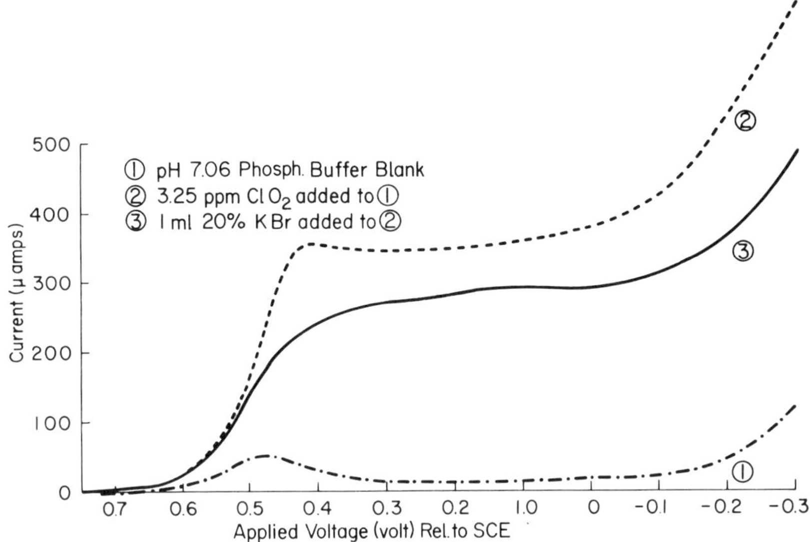

Figure 6. Voltammogram with Au working electrode, Cu counter electrode, SCE ref. Scan rate = 5 mV/sec, pH = 7.06. Phosphate buffer.

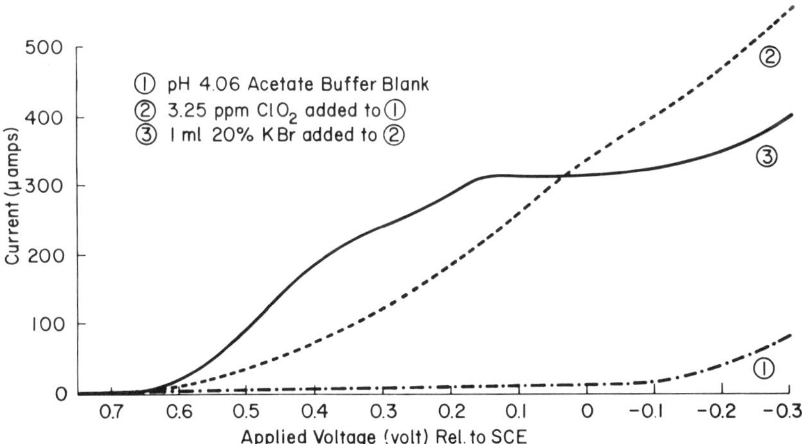

Figure 7. Voltammogram with Au working electrode, Cu counter electrode, SCE ref. Scan rate = 5 mV/sec, pH = 4.06. Acetate buffer.

Chlorine dioxide for these investigations was manufactured using a Fischer and Porter/Italy ClO_2 generator. The generator receives solutions of 32% HCl and 25% $NaClO_2$. These are diluted in a packed continuous-flow reactor with tap water to concentrations of 2.5% HCl and 1.7% $NaClO_2$ before reaction. Depending on the withdrawal rate from the generator, the reaction time varies from 5 to 50 minutes. Reaction yields of 94% were obtained. The generator uses the reaction below which is expected to give a chlorine free product [9,21,48]:

$$5\ NaClO_2 + 4\ HCl \longrightarrow 4\ ClO_2 + 5\ NaCl + 2\ H_2O$$

There is some controversy over that fact [49], but it was observed in this study that free chlorine was either not present, or impossible to detect using Palin's DPD titration or spectrophotometric procedures [31], and *Standard Methods* amperometric titration procedure [26]. The product stream was diluted with tap water to about 150 mg/l and sent to the analyzer which used a Au/Cu cell. Some of the diluted solution was bottled and saved for several months for use in other work. Further dilution to about 1 mg/l was done for the tests on the Cu/Cu electrode cell.

MEMBRANE AMPEROMETRIC PROBE

The amperometric analyzers described in the preceding are generally satisfactory but possess certain limitations which might be overcome using a membrane amperometric probe. For example, a probe would have no moving parts, and require no reagent feed since the reagents would be contained inside the cell and consumed only slowly. It could be placed directly in the process stream as an in-situ sensor. The membrane would protect the electrodes from the environment, making them less sensitive to poisoning effects. Finally, the membrane would permit such a device to be more selective since only those substances which can permeate it would be detected.

Membrane sensors for the measurement of oxygen are well known [50-53]. Poole and Morrow [54] describe a cell consisting of a gold measuring electrode, a copper counter electrode and 4 M KOH electrolyte. The oxygen-permeable membrane is made of FEP Teflon®. Johnson et al. developed a membrane sensor for HOCl determinations [55]. Recently, similar devices for ozone have been described [56,57].

A membrane amperometric cell for ClO_2 is depicted in Figure 8. It consists of a gold cathode held in tight contact with a gas-permeable polymer film, in this case, Celgard 2400 microporous polypropylene. The remote anode is pure silver. The electrolyte is a solution of 0.1 M KBr and 0.1 M

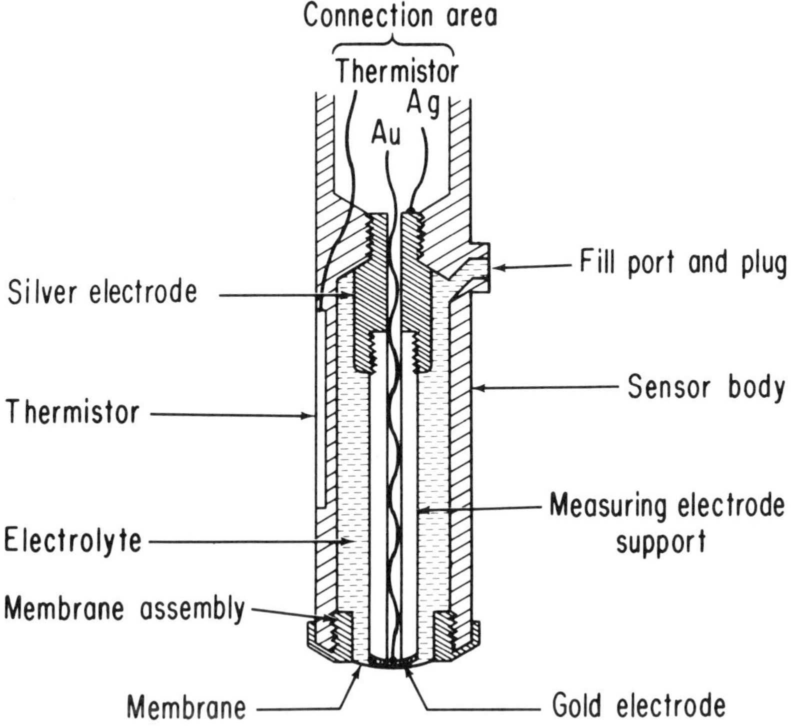

Figure 8. Membrane amperometric cell for ClO_2.

EDTA in distilled water, adjusted to pH 7 with reagent grade NaOH. The EDTA is provided to complex silver as it is oxidized, thus keeping the anode surface clean.

Figure 9 displays the current voltage curves obtained from the membrane cell using a P.A.R. Model 174A polarograph, with a scan rate of 5 mV/sec. As can be seen, a well-defined wave is obtained in the presence of ClO_2, having an $E_{1/2}$ = +0.53 V, and a plateau from about +0.45 to +0.1 V.

In operation, a potential is impressed on the cell so that the cathode potential is fixed at a value on the plateau. More positive potentials will permit the cell to be relatively free of interferences from HOCl and chloramine. Chlorine dioxide in the sample medium diffuses through the gas permeable membrane into the thin layer of electrolyte between the cathode and the membrane. At this point, two possible reactions may occur. The ClO_2 may be reduced to ClO_2^- directly at the electrode surface, or

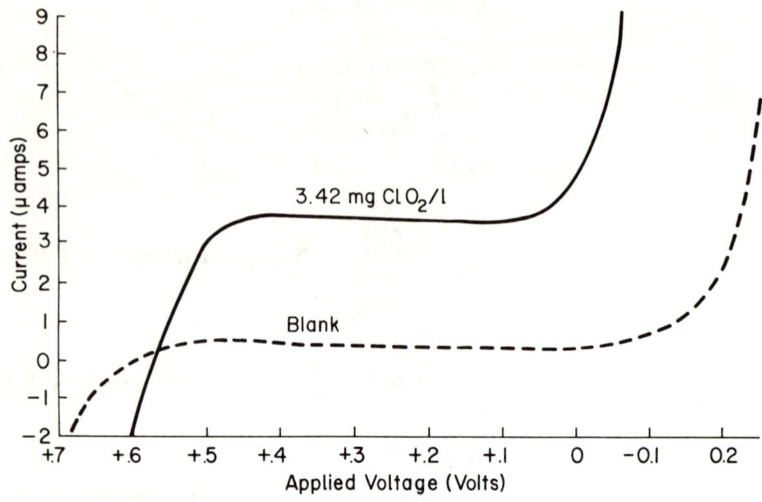

Figure 9. Voltammogram of amperometric ClO$_2$ probe.

it may react with bromide in the electrolyte to produce Br_2, which is then reduced at the electrode. Which mechanism is correct is not certain, but the first seems favored in that similar waves are observed when the electrolyte is replaced with a bromide-free solution of KC1.

The reduction of either ClO_2 or Br_2 will produce a current whose magnitude is proportional to the rate of diffusion of ClO_2 through the membrane, which in turn depends on the concentration of ClO_2 in the sample medium [50, 57]:

$$i_{ss} = nFA \frac{Pm}{b} C$$

where

i_{ss} = steady state current
n = no. of electrons/mole ClO_2 reduced
F = the Faraday
A = cathode area
Pm = membrane permeability
b = membrane thicknesss
C = ClO_2 concentration

The response of the device is shown in Figure 10, which depicts the output current at +0.4 V applied potential as a function of increasing ClO_2 concentration. The data were obtained by placing the cell in 1500 ml of

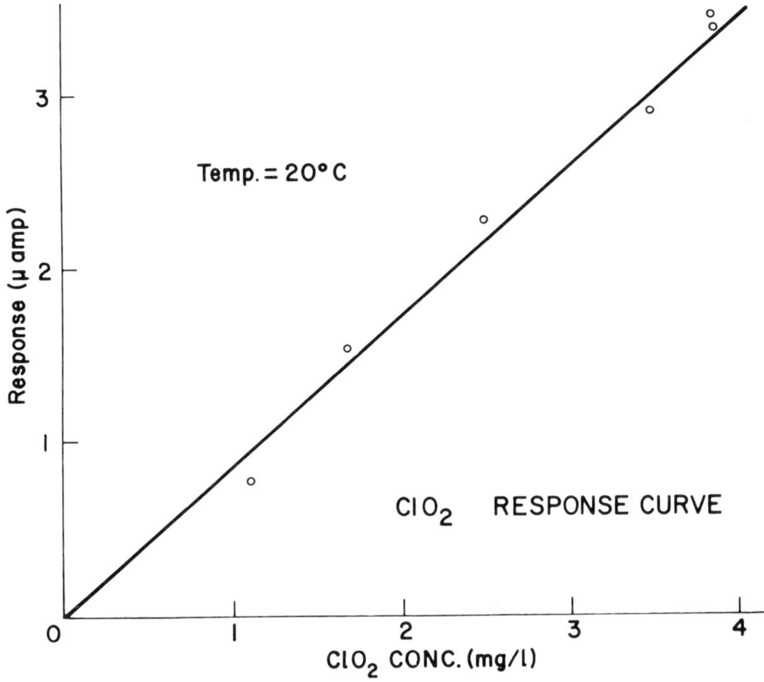

Figure 10. Response of the cell shown in Figure 8.

distilled H_2O and adding increments of a concentrated ClO_2 stock solution. The concentration of the stock solution was determined by amperometric titration immediately prior to the experiment. The response of the device to HOCl and NH_4Cl is about 10% of its ClO_2 response. These data were obtained in a similar fashion, adding appropriate increments of stock solution to 1500 ml of distilled H_2O. The sample was buffered at pH 7, so that the concentration of HOCl in equilibrium with OCl^- could be found by calculation knowing the temperature and equilibrium constants [58]. The chloramine stock solution was prepared by mixing stoichiometric amounts of HOCl and NH_3, buffered at pH 7.8, to make a liter of solution that was about 1000 mg/1. This was allowed to react overnight and was stored in the refrigerator.

The effect of temperature on the cell response can be seen in Figure 11. Temperature affects variables such as diffusion rate through the membrane, and must be compensated for in industrial applications where temperatures may not be controlled. To accomplish this, it is possible to place a tempera-

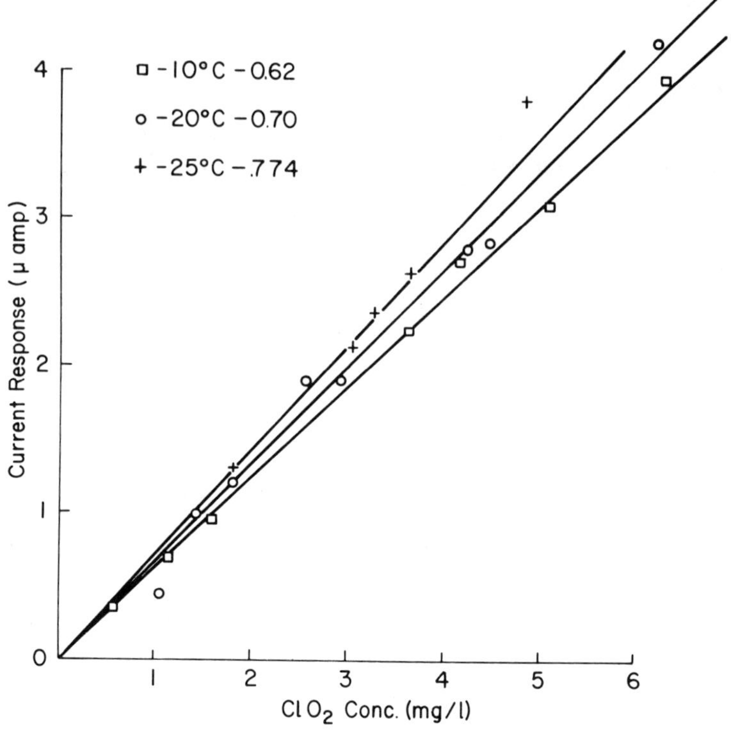

Figure 11. ClO_2 response curves.

ture-sensing thermistor in the probe. Electronically, the thermistor is in the feedback loop of an operational amplifier circuit which amplifies the output current from the cell. Changes in the resistance result in changes in the gain of the circuit which compensates for the temperature effect.

Because of the limitation of the membrane to pass only volatile, un-ionized molecules, the membrane cell is free from the interference of species such as OCl^-, ClO_2^-, ClO_3^-, $KMnO_4$, K_2CrO_4 and H_2O_2. It will experience interference from Br_2 should that species be present. Likewise, in acid media Cl_2 will present a serious interference. Surprisingly, O_3 was not detected in the experiment described here.

Field-testing of the device has been done by Fischer and Porter/Italy where it has been used to control the ClO_2 residual in a swimming pool and water at the Fischer and Porter plant. The response of the instrument drifts at first, becoming less sensitive. After a week of continuous operation, however, it produces reliable results with only occasional calibration read-justments.

REFERENCES

1. Harris, R. H., and T. Page. "The Implications of Cancer Causing Substances in Missassppi River Water," Environmental Defense Fund Report, Washington, DC (November 1974).
2. Morris, J. C. "The Chemistry of Aqueous Chlorine in Relation to Water Chlorination," in *Water Chlorination: Environmental Impact and Health Effects, Vol. 1,* R. L. Jolley, Ed. (Ann Arbor, MI: Ann Arbor Science Publishers, Inc., 1978), pp. 21-36.
3. Stevens, A. A., C. J. Slocum, D. R. Seeger and G. G. Robeck. "Chlorination of Organics in Drinking Water," in *Water Chlorination: Environmental Impact and Health Effects, Vol. 1,* R. L. Jolley, Ed. (Ann Arbor, MI: Ann Arbor Science Publishers, Inc., 1978), pp. 77-104.
4. Carlson, R. M., and R. Caple. "Organochemical Implications of Water Chlorination," in *Water Chlorination: Environmental Impact and Health Effects, Vol. 1,* R. L. Jolley, Ed. (Ann Arbor, MI: Ann Arbor Science Publishers, Inc., 1978), pp. 65-76.
5. Rook, J. J. "Chlorination Reactions of Fulvic Acids," *Environ. Sci. Technol.* 11:478 (1977).
6. Oliver, B. G., and J. H. Carey. "Photochemical Production of Chlorinated Organics in Aqueous Solutions Containing Chlorine," *Environ. Sci. Technol.* 11:893 (1977).
7. "Abstracts of Technical Papers," Conference on Water Chlorination: Environmental Impact and Health Effects, Oak Ridge National Laboratory, Oak Ridge, TN (October 1977).
8. White, C. *Handbook of Chlorination* (New York: Van Nostrand Reinhold, 1972), Chapter 11.
9. Symons, J. M. et al. "Ozone, Chlorine Dioxide and Chloramines as Alternatives to Chlorine for Disinfection of Drinking Water," paper presented at the Second Conference on Water Chlorination: Environmental Impact and Health Effects, Gatlinburg, TN, October 31-November 4, 1977.
10. Blanck, C. A. "Trihalomethane Reduction in Operating Water Treatment Plants," *J. Am. Water Works Assoc.* 71:525 (1979).
11. Berg, J. D., E. M. Aieta, P. V. Roberts and R. C. Cooper. "Effectiveness of Chlorine Dioxide as a Waste Water Disinfectant," in *Progress in Wastewater Disinfection Technology*, U.S. EPA Report 600/9-79-018 (1979) p. 61.
12. Gan, H. B., C. L. Lin and A. D. Venosa. "Dechlorination of Waste Water: State-of-the-Art Field Survey and Pilot Plant Studies," in *Progress in Wastewater Disinfection Technology*, U.S. EPA Report 600/9-79-018 (1979) p. 36.
13. Longley, K., B. Moore and C. Sorber. "Relative Waste Water Disinfection Efficiencies of Chlorine and Chlorine Dioxide in a Gravity Flow Contactor," paper presented at the 52nd Annual Conference of the Water Pollution Control Federation, Houston, TX, October 7-12, 1979.
14. Miltner, R. J. "The Effect of Chlorine Dioxide on Trihalomethanes in Drinking Water," MS Thesis, University of Cincinnati (1976).

15. Miltner, R. J. "The Measurement of Chlorine Dioxide and Related Products," U.S. EPA, Cincinnati, OH (January 1977).

16. "Chlorine Oxygen Acids and Salts: Chlorine Dioxide," in *Encyclopedia of Chemical Technology, Vol. V,* R. E. Kirk and D. F. Othmar, Eds. (New York: Wiley-Interscience, 1964), pp. 35-50.

17. Feuss, J. V. "Problems in Determination of Chlorine Dioxide Residuals," *J. Am. Water Works Assoc.* 56:607 (1964).

18. Myhrstad, J. A., and J. E.Samdal. "Behaviour and Determination of Chlorine Dioxide," *J. Am. Water Works Assoc.* 61:205 (1969).

19. Gordon, G., R. G. Kieffer and D. H. Rosenblatt. "The Chemistry of Chlorine Dioxide," in *Progress in Inorganic Chemistry, Vol. 15,* S. J. Lippard, Ed. (New York: John Wiley & Sons, Inc., 1972), pp. 202-286.

20. Masschelein, W. J. *Chlorine Dioxide: Chemistry and Environmental Impact of Oxychlorine Compounds* (Ann Arbor, MI: Ann Arbor Science Publishers, Inc., 1979).

21. Malpas, J. F. "Disinfection of Water Using Chlorine Dioxide," *Water Treat. Exam.* 22:209 (1973).

22. Hong, C. C., and W. H. Rapson. "Analyses of Chlorine Dioxide, Chlorous Acid, Chlorite, Chlorate, and Chloride in Composite Mixtures," *Can. J. Chem.* 46:2061 (1968).

23. Granstrom, M. L., and G. F. Lee. "Generation and Use of Chlorine Dioxide in Water Treatment," *J. Am. Water Works Assoc.* 50:1453 (1958).

24. Willard, H. H., L. L. Merritt and J. A. Dean. *Instrumental Methods of Analysis,* 5th ed. (New York: Van Nostrand Co., 1974), pp. 92-94.

25. Aston, R. N. "Developments in the Chlorine Dioxide Process," *J. Am. Water Works Assoc.* 42:151 (1950).

26. *Standard Methods for the Examination of Water and Wastewater,* 14th ed. (Washington, DC: American Public Health Association, 1976), pp. 349-358.

27. Masschelein, W. "Spectrophotometric Determination of Chlorine Dioxide with Acid Chrome Violet K," *Anal. Chem.* 38:1839 (1966).

28. Knechtel, J. R., E. G. Janzen and E. R. Davis. "Determination of Chlorine Dioxide in Sewage Effluents," *Anal. Chem.* 50:202 (1978).

29. Post, M. A., and W. A. Moore. "The Determination of Chlorine Dioxide in Treated Surface Waters," *Anal. Chem.* 31:1872 (1959).

30. Palin, A. T. "Methods for the Determination, in Water, of Free and Combined Available Chlorine, Chlorine Dioxide and Chlorite, Bromine, Iodine, and Ozone Using Diethyl-*p*-phenylene Diamine (DPD)," *Inst. Water Eng. J.* 21:537 (1967).

31. Palin, A. T. "Analytical Control of Water Disinfection with Special Reference to Differential DPD Methods for Chlorine, Chlorine Dioxide, Bromine, Iodine, and Ozone," *Inst. Water Eng. J.* 28:139 (1974).

32. Haller, J. F., and S. S. Listek. "Determination of Chlorine Dioxide and Other Active Chlorine Compounds in Water," *Anal. Chem.* 20:639 (1948).

33. Palin, A. T., and K. G. Darrall. "A Modified DPD Titrimetric Procedure for Determination of Chlorine Dioxide and Chlorite in Water," *Inst. Water Eng. Scientists J.* 33:467 (1979).

34. Lott, P. F., G. L. Wheeler and F. W. Yau. "A Rapid Microdetermination of Chlorine Dioxide in the Presence of Active Chlorine Compounds," *Microchem. J.* 23:160 (1978).

35. Aieta, E. M., B. Chow and P. V. Roberts. "Chlorine Dioxide: Analytical and Pilot Plant Evaluation," in *Progress in Wastewater Disinfection Technology*, U.S. EPA Report 600/9-79-018 (1979).

36. Berg, J., and P. V. Roberts, Department of Environmental Engineering, Stanford University. Personal communication (February 1978).

37. Luise, A., Fischer and Porter/Italy, Milan. Memorandum (April 1979).

38. Hartley, A. M., and A. C. Adams. "Polarographic Behaviour of Chlorite," *J. Electroanal. Chem.* 6:460 (1963).

39. Schwarzer, O., and R. Landsberg. "Polarographische Charakterisierung von Chlorite und Chlordioxide an der rotierenden schieben Elektrode," *J. Electroanal. Interfacial Chem.* 14:339 (1967).

40. Raspi, G., and F. Pergola. "Voltammetric Behaviour of Chlorites and Chlorine Dioxide on a Platinized Platinum Microelectrode with Periodical Renewal of the Diffusion Layer and Its Analytical Applications," *J. Electroanal. Interfacial Chem.* 20:419 (1969).

41. Cauquis, G., and D. Limosin. "Determination of Chloride and Oxychloride in Aqueous Media," *Analusis* 5:70 (1977).

42. Benga, J., and G. Gordon. "Determination of Chlorite Ion by Differential Pulse Polarography," paper presented at the 179th American Chemical Society National Meeting, Houston, TX, March 24-28, 1980.

43. "Apparatus for Determination of Chemicals in Test Streams," German Patent 2,416,716.

44. Krunchak, V. G. et al. "Potentiometric Method for Determining Chlorine Dioxide in Industrial Solutions," *Izv. Vyssh. Uchebn. Zuved, Lesn. Zh.* (1979) p. 95; *Chem Abst.* 91:194895t (1979).

45. "Instruction Bulletin for Series 17K1000 Chlortrol[TM] Free Residual Analyzer," Instruction Bulletin 17K1000 (Rev. 2), Fischer & Porter Co., Warminster, PA.

46. Morrow, J. J., and R. N. Roop. "Advances in Chlorine Residual Analysis," *J. Am. Water Works Assoc.* 67:184 (1975).

47. Dailey, L. L., and J. J. Morrow. "On Stream Analysis of Ozone Residual," First International Symposium on Ozone for Water and Waste Water Treatment, International Ozone Institute (December 1973) p. 69.

48. Staheli, V. T. "Die praktische Anwendung von Chlordioxyd fur Trinkwasser-Enkeimung," *Monatsbull. Schweiz. Ver. Gas Wasserfach.* 42:244 (1962).

49. Dowling, L. T. "Chlorine Dioxide in Potable Water Treatment," *Water Treat. Exam.* 23(2):190 (1974).

50. Mancy, K. H., and W. C. Westgarth. "A Galvanic Cell Oxygen Analyzer," *J. Water Poll. Control Fed.* 34:1037 (1962).

51. Keidel, F. A. "Coulometric Analyzer for Trace Quantities of Oxygen," *Ind. Eng. Chem.* 52:490 (1960).

52. Clark, L. C., Jr. et al. "Continuous Recording of Blood Oxygen Tension by Polarography," *J. Appl. Physiol.* 6:189 (1953).

53. Okun, D. A. et al. "Oxygen Detector," U.S. Patent 3,227,643 (January 1966).

54. Poole, R., and J. Morrow. "Improved Galvanic Dissolved Oxygen Sensor for Activated Sludge," *J. Water Poll. Control Fed.* 49:422 (1977).

55. Johnson, J. D., J. W. Edwards and F. K. Kesslar. "Chlorine Residual Measurement Cell: The HOCl Membrane Electrode," *J. Am. Water Works Assoc.* 70:341 (1978).

56. Stanley, J. H., and J. D. Johnson. "Amperometric Membrane Electrode for Measurement of Ozone in Water," *Anal. Chem.* 51:2144 (1979).

57. Smart, R. B., R. Dormand-Herrera and K. H. Mancy. "In Situ Voltammetric Membrane Ozone Electrode," *Anal. Chem.* 51:2315 (1979).

58. Morris, J. C. "The Acidic Ionization Constant of HOCl from 5 to 30°C," *J. Phys. Chem.* 70:3799 (1966).

CHAPTER 22

REDUCTION OF TRIHALOMETHANE PRODUCTION WITH OPTIMAL DISINFECTION THROUGH ALTERNATIVE DISINFECTION SYSTEMS

Robert F. Williams, Barbara E. Moore and
Karl E. Longley

> The Center for Applied Research & Technology
> The University of Texas at San Antonio
> San Antonio, Texas

Charles A. Sorber

> College of Engineering
> The University of Texas at Austin
> Austin, Texas

The use of chlorine as a disinfectant in potable water has been based on its long and successful history of making water sufficiently free of pathogenic organisms, thereby minimizing the probability of waterborne disease among the consuming public. Because of its effectiveness, relative ease of application and low cost, chlorination has become synonymous with disinfection in the United States. Indeed, the widespread use of chlorine has been one of the most important public health control measures of this century.

Recently, increased attention has been focused on organic substances in potable water supplies. Since the discovery of the production of trihalomethanes (THM) and other organochlorine compounds during the chlorination of natural unpolluted waters (1-5), efforts have been made to identify the precursor compounds and control the amounts of THM and organohalides produced (6-11). The exact nature of the precursor materials is not clear; however, humic and fulvic acids and algae have been implicated as contributors to THM formation. Chloroform is the major THM found in finished waters, but dibromochloromethane, bromodichloromethane and bromoform are found in varying quantities.

The primary reason for public concern about the formation of THM in drinking water is that these compounds may be carcinogenic to humans or hazardous to health in other ways. Page and Harris (12) have supported the hypothesis that there is a link between carcinogens in drinking water and cancer mortality. Tardiff (13) has argued that the basic criteria for establishing chloroform as a carcinogen have been satisfied. These criteria include statistically significant increases in neoplasia in treated animals over the spontaneous incidence in control animals, a dose-related increase in the incidence of cancer, time-to-tumor incidence data demonstrating a decrease in latency with increasing dose and reproducibility not only within the same strain and species but also within other species. At present there is no direct evidence of chloroform carcinogenicity in humans. However, on the basis of rodent data, the likely target organs for chloroform in man would be the liver and kidney. Tardiff [13] has estimated that if drinking water doses are 0.01 mg/kg/day, chloroform in water may be responsible for as much as 1.6% of the current yearly human liver cancer incidence in the United States and for as much as 1.44% of the yearly kidney cancer.

A maximum contaminant level of 100 μg/l total THM has been set by the U.S. Environmental Protection Agency (EPA) for systems serving populations of greater than 10,000 persons [14]. However, it has been pointed out that health risk assessments should be based on the total number of carbon-chlorine bonds rather than on the extent of chloroform or haloform formation only [6]. This recommendation was made because most halogenated compounds (volatile and nonvolatile) that are formed by chlorination are likely to show possible carcinogenic risks.

The ubiquitous nature of THM necessitates the development of techniques and methods to reduce THM and other organochlorine concentrations in finished waters. Nevertheless, it may be necessary to tolerate threshold levels of chloroorganic contaminants if alternative modes of disinfection are unacceptable. Techniques likely to reduce the concentration of chloroform and other THM are also likely to reduce the concentrations of other chlorinated organic compounds. Approaches to control organochlorine production include: utilization of alternative disinfectants, control of the THM formation potential and reduction of the THM concentration after formation.

The objective of this investigation was to develop techniques to reduce or eliminate THM and other organochlorine compounds in finished water without compromising the microbiological quality of the water. To accomplish this goal a rapid mixing system employing a plug flow reactor was examined with several disinfectants. The THM produced and disinfection efficiency of alternative disinfectant schemes (chlorine dioxide, chloramines and chlorine followed by ammonia) were compared to chlorine. The prototype disinfection system utilized nondisinfected water from a 75-gal/min side stream from a water treatment plant in Boerne, TX. To challenge the disinfection system, two seed organisms, *Escherichia coli* lys 147 and *E. coli* C, were employed in addition to the indigenous organisms found in the untreated water.

EXPERIMENTAL PROCEDURES

Water Treatment Plant

The site for this study was the new water treatment plant (WTP) of the City of Boerne, Texas. The WTP is located at a newly constructed reservoir and went on-stream in January 1979 when the water level in the reservoir was sufficient to permit withdrawal through the collection works. The reservoir is fed by runoff from the Cibolo Creek watershed, 12,500 ac characterized by a low population density with vegetation typically being grasses, oak and cedar. The principal use of the watershed is livestock grazing, although limited development within the watershed, particularly near the reservoir, is underway. Generally, the development is confined to single-unit dwellings utilizing septic tanks for domestic wastewater disposal. Limited recreational use is available at the reservoir. When full, the reservoir has an area of 200 ac and a maximum operating pool of 4000 ac-ft. Maximal depth of the reservoir is 42 ft.

The treatment train (Figure 1) of the 1.6-Mgal/day WTP consists of prechlorination, alum coagulation and sedimentation through an upflow clarifier, pressure sand filtration and postchlorination. Gas-fed chlorinators are used for chlorination. Facilities are available for fluoridation. Disposal of sludge, filtered backwash water and other plant liquid waste streams is to two nearby evaporation ponds.

As shown by the sampling matrix in Table I, the WTP was evaluated monthly during the study period to develop baseline disinfection and THM formation data relating to "standard" practice.

Prototype System

Filtered, nondisinfected water was used for this study. Normal prechlorination of the main water stream has not been consistently practiced, but whenever it was employed, prechlorination was discontinued in sufficient time prior to the initiation of a data gathering run to preclude prechlorination as a source of in-plant THM formation.

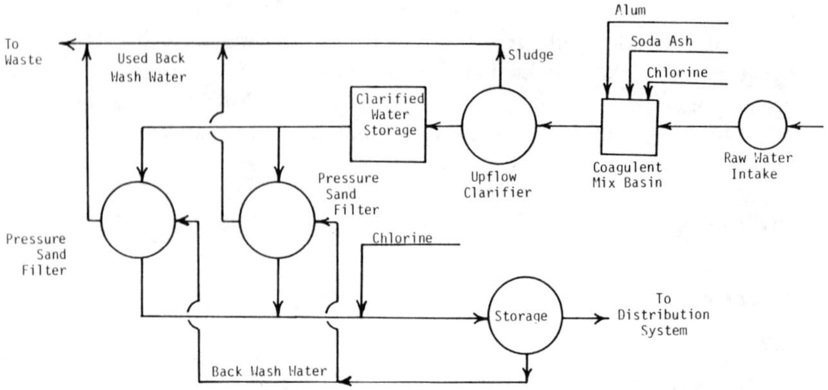

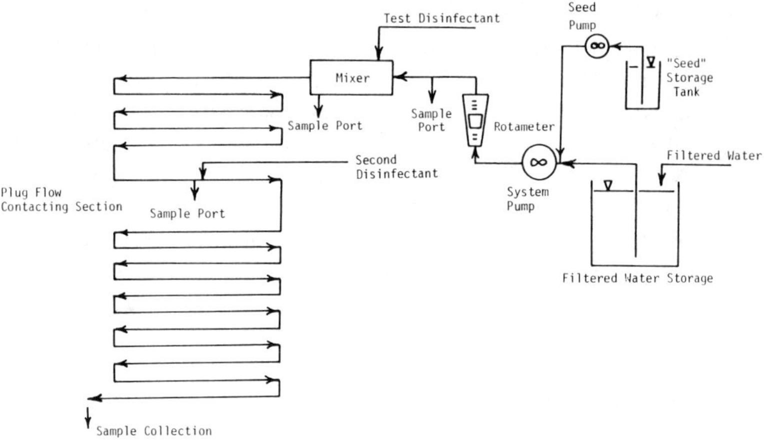

Figure 1. Schematic of Boerne water treatment plant.

Figure 2. Schematic of prototype disinfection system.

Figure 2 is a schematic of the prototype disinfection system. The filtered, nondisinfected water was fed to a 500-gal storage tank and pumped to the prototype system at a flowrate of 75 gal/min. Provisions were available for "seeding" the water stream prior to disinfection. Determination of seeding requirements for the prototype system was based on the levels of indigenous organisms found in the test water. Seeding experiments were performed to evaluate the prototype system, since the natural levels were insufficient to ensure statistical significance in the evaluation of disinfection effectiveness.

Table I. Sampling Matrix for Boerne Water Treatment Facility[a]

Sampling Point	Routine Parameters Monitored
Reservoir	Chlorine Residual
Pretreatment	Standard plate count
Predisinfection	Total coliform
Postdisinfection	Total trihalomethane (InstTHM)
Clear Well	Total trihalomethane (TermTHM)
Distribution System	Total trihalomethane formation potential (TTHMP)
	Chlorophyll-a
	NH_3 and TKN
	Temperature
	Alkalinity
	pH
	Turbidity and color
	Total organic carbon (TOC)

[a]Disinfectant dose was applied as system operated, sampling frequency was once per month.

Aqueous chlorine, chlorine dioxide and ammonia were peripherally injected into one of two plug flow mixers which had internal throat diameters of 0.8 in. and throat lengths of 3 ft. Downstream of each mixer the water stream passed through a 9-in. energy recovery section into a 4.0-in.-i.d. polyvinyl chloride (PVC) pipe (schedule 80) having a total length of 122.6 ft. For a flow rate of 75 gal/min. the Reynolds numbers of the 0.8 in. and 4.0-in. sections are 370,000 and 74,200, respectively. Studies involving the addition of chlorine followed by ammonia or ammonia followed by chlorine were performed by addition of the second chemical into the second mixer located 28.8 ft downstream from the first mixer. This permitted a 15-sec mean contact with free chlorine or ammonia prior to initiation of chloramine formation. Total mean contact time following the addition of the disinfectant to the discharge of the water stream was approximately 60 sec. Addition of a seed organism was accomplished at the suction side of the system pump (organisms were not introduced into the holding tank).

Sampling

Samples having contact times of approximately 1 sec were collected at the end of the mixer throat, and samples with mean contact times of 60 sec were collected at the discharge of the pipe contactor. Samples with desired contact times greater than 60 sec were collected in glass or plastic containers, as appropriate, at the discharge point. After the predetermined contact time had elapsed, a suitable reducing agent was added to quench the disinfecting agent. Additionally, a sampling port immediately prior to the point of ammonia (or chlorine) addition allowed sample collection for evaluation of disinfectant residual and THM formation immediately prior to either chlorine or ammonia addition for the chloramination studies. Overall control samples were obtained from a sample port immediately upstream from the first mixer.

The sampling matrix and the chemical and physical parameters monitored for the prototype disinfection system are shown in Table II.

Bacteriological Analyses

Collection of the bacteriological samples was accomplished in 500-ml polypropylene bottles that were cleaned, rinsed with distilled water and autoclaved to ensure sterility. Water samples containing residual chlorine were collected in sterile 500-ml bottles containing sodium thiosulfate (0.01%). Care was taken to avoid contamination at sample collection by proper

Table II. Sampling Matrix for Prototype Disinfection System Runs.

Disinfectant Dose	Disinfectant	Sampling Time	Routine Parameters Monitored
0.5 mg/l	Chlorine	0	Disinfectant residual
1.5 mg/l	Chlorine plus ammonia	1 sec	Standard plate count
5.0 mg/l	Ammonia plus chlorine	15 sec	Total coliform
	Chlorine dioxide	31 sec	Total trihalomethane (InstTHM)
		1 min	Total trihalomethane (TermTHM)
		15 min	Total trihalomethane formation potential (TTHMP)
		60 min	Chlorophyll-a
			NH_3-nitrogen and TKN
			Temperature
			Alkalinity
			pH
			Turbidity and color
			Total organic carbon
			Chlorate and chlorite[a]

[a]Runs which include chlorine dioxide as a disinfectant only.

handling of the bottle and cap. After labeling, the samples were placed in wet ice at 4°C until analysis.

Standard Plate Count

The appropriate serial dilutions of the sample were prepared and plated in triplicate. Inoculation into plate count agar was conducted at a temperature of 35 ± 0.5°C for 48 hours [15]. Plates were placed in the incubator without crowding. Only plates showing 30–300 colonies were considered in determining the standard plate count. The results as reported are the average of all plates falling within these limits. If colonies per plate in the highest dilution exceeded 300, results were recorded as greater than 300 times the appropriate dilution factor. Likewise, if no chosen dilution had colonies, the results were recorded as less than 1 colony per lowest dilution. Counts are reported as colony forming units per milliliter (CFU/ml).

Total Coliform

Total coliform densities were determined by the membrane filter technique [15]. Triplicate volumes for undiluted or diluted samples were handled using the enrichment technique. Plates showing 20–80 typical coliform colonies were used in the calculation of total coliforms/100 ml.

Seed Organisms

The coliform bacteria chosen as a seed organism for the prototype disinfection system is unique in that it carries a lambda (λ) phage integrated into its chromosome. In the case of *E. coli* lys 147, this prophage state is maintained by a heat-sensitive repressor. At 30°C this bacterium grows normally. However, if the organism is exposed to a higher temperature for a relatively short period of time (e.g. 42°C for 3-4 hr), the repressor protein is inactivated and the phage is excised from the bacterial chromosome. From this point a lytic phage infection is established, and ultimately the *E. coli* cell is lysed with concomitant release of infectious particles.

These attributes allow the use of a coliform bacteria in the disinfection tests which were measured as a viral plaque forming unit (PFU). At the same time, indigenous standard plate count were enumerated on the sample by "self-destructing" the seed organism.

In practice, an appropriate volume of sample was plated using a classical soft agar overlay technique [16] with *E. coli* 2eO1c (λ sensitive) as the indicator organism. Concurrently, each sample was assayed using the protocol for standard plate count [15]. All plates were incubated at 42°C for 3–4 hours followed by a shift down to 35°C. After approximately 18 hours, PFU reflecting the initial presence of *E. coli* lys 147 were counted. The level of indigenous bacteria observed as viable colony forming units (CFU) was recorded at 48 hours as the standard plate count.

Disinfectant Systems

Sodium thiosulfate was employed to quench the disinfectants at the desired times for all bacteriological and THM analyses. The quantity of reducing agent depended on the disinfectant dose, but was always in excess.

Disinfectants were introduced into the mixers through a vacuum ejector, and chlorine and ammonia were fed as gases. Chlorine dioxide was fed as an aqueous solution prepared by reacting sodium chlorite with hydrochloric acid in a contact chamber filled with glass beads (five-minute detention time).

Field analyses for residual chlorine, total chlorine, chloramines and chlorine dioxide were performed by either amperometric titration (Fisher and Porter Model 17T1010) or colorimetric determination with N,N-diethyl-*p*-phenylenediamine sulfate (DPD). Ammonia determinations in the field were done colorimetrically with Nessler's reagent.

Chemical Analyses

Temperature and pH were measured at the WTP sampling sites in the field. For prototype runs these parameters were measured at the field location as well as after transport back to the analytical laboratory. All samples for chemical analyses were collected in 1-liter plastic bottles and stored at 4°C for transport to the laboratory.

Alkalinity, turbidity (DRT-100, H. F. Instruments) and total organic carbon (Beckman Model 915A carbon analyzer, Model 865 infrared analyzer) were measured as described in *Standard Methods* [15]. Preservation of the TOC samples was required if analyses could not be performed immediately. TOC samples were preserved with 0.1 ml concentrated hydrochloric acid per 5 ml of sample and stored at 4°C.

Ammonia-nitrogen (NH_3-N) was determined by utilizing the automated alkaline phenolate procedure [17] modified for low-level concentrations for the Technicon Auto Analyzer® II system. Total Kjeldahl nitrogen (TKN) was determined by digestion (Technicon Block Digestor, DB-20), dilution to volume and analysis on the Technicon system [17]. Preservation, if necessary, was accomplished for both NH_3-N and TKN samples by addition of 0.03 ml of a 0.15 M solution of mercuric chloride per 30 ml of sample and storage at 4°C.

Chlorophyll-a concentrations were determined by the fluorometric assay (Farrand A-4 filter fluorometer, 5-60 excitation and 2-28 emission filters) method as described in *Standard Methods* [15].

Trihalomethane Analyses

Samples were collected in 40-ml vials with (instant THM) or without (terminal THM) sodium thiosulfate and sealed with Teflon®-faced septa. Instant THM samples were stored at 4°C and terminal THM samples were stored at room temperature in the dark until analysis or until the designated times for quenching by thiosulfate.

THM were measured by gas chromotography using the gas sparging technique developed by Bellar and Lichtenberg [18]. Samples (5 ml) were sparged for 20 minutes at 25°C with a 30-ml/min flow of grade six helium carrier gas and collected on a Tenax trap (30.5 cm long; 0.318 cm o.d.; 0.293 cm i.d.). Thermal desorbtion at 210°C with a 20-ml/min helium flow onto a 50°C analytical Tenax column (60/80 mesh; 2.4 cm long, 0.31 cm o.d. 0.216 cm i.d.) was then accomplished. The analytical column was held isothermally for 6 min and temperature-programmed to 250°C at 10°C/min. Retention times of 14.40, 16.38, 18.24 and 20.02 min for chloroform, dichlorobromomethane, chlorodibromomethane and bromoform, respectively, were exhibited for this analytical column. All analyses were performed on a Varian 3700 gas chromatograph equipped with a flame ionization detector and a CDS 111C reported integrator.

RESULTS

Water Treatment Plant

Baseline data collected during the normal operation of the WTP have shown little variability in chemical or bacteriological parameters during the past six months. Over the past year operation conditions have varied somewhat due primarily to problems associated with bringing the plant on-line and gaining experience with the most effective operation.

Table III illustrates the ranges of bacteriological and THM data observed since the sampling program began in early 1979. All of the high coliforms and standard plate count (SPC) values were observed during the first six months of operation. Since September 1979 there has been <0.33 CFU/ml total coliforms (TC) and <0.2 CFU/ml SPC at the postchlorination sampling site and in the simulated (24-hr hold) distribution samples. The reservoir exhibited a characteristic change in TC from a level of 150 CFU/100 ml in August 1979 (9000 CFU/ml, SPC) to 9 CFU/100 ml in March 1980 (120 CFU/ml, SPC) due to the colder weather.

The observed range of THM data (Table III) follows the same general trends with higher levels observed during the first six months of operation. For the past four months the chlorine dose (average 2.37 mg/1) and residuals (average 1.14 mg/l) have been relatively constant. Furthermore, TOC levels at the chlorination site have not changed greatly (2-5mg/l). Consequently, the terminal THM levels observed during this period have been constant (85.1 ± 5.3 μg/l).

Other chemical parameters have shown relatively little change during the year's operation. The greatest variation in the data was observed during the first six months of operation. Table IV shows the ranges of chemical parameters observed. The consistency of the reservoir water and the plant operation has resulted in a stable system. The total rainfall into the watershed was 12.3 inches from July 1979 to February 1980 (average 1.53 inches per month). The reservoir reached a maximum depth of 40 feet during July and has been declining steadily since. The relatively small amount of runoff to the reservoir may partially explain the stability observed, since organic addition to the system has been low.

Table III. Ranges of Bacteriological and Trihalomethane Data at Sampling Sites in the Boerne WTP for February 1979 through March 1980

Sampling Site	Total Coliform (CFU/100 ml)	Standard Plate Count (CFU/ml)	TTHM (μg/1)
Reservoir	<0.33–150	51–9000	<0.1
Pretreatment	0.33–38	80–380	<0.1
Postcoagulation	0.33–52	0.6–430	<0.1
Postfiltration	5–60	10–1700	<0.1–3
Postchlorination	<0.33–4	<0.2–43	58–137

Table IV. Ranges of Inorganic and Physical Parameters for Normal Operation of the Boerne WTP from March 1979 to March 1980

Sampling Point	Turbidity (NTU)	pH	TOC (mg/1)	Temp (°C)	TKN (mg/1)	NH_3 (mg/1)	Total Alkalinity (mg/1 $CaCO_3$)	Specific Conductance (µmhos/cm)	Chlorophyll-a (µg/l)
Reservoir	0.5–5.2	8.1–8.6	2–12	11–24	<0.1–0.5	<0.01–0.02	142–180	340–380	3.7–31.8
Pretreatment	0.3–1.2	7.5–8.1	2–6	11–24	<0.1–0.6	<0.01–0.03	142–187	343–400	2.3–8.6
Postcoagulation	0.26–5.2	7.4–8.0	2–6	11–24	<0.1–0.6	<0.01–0.02	140–180	345–420	<0.15–7.7
Postfiltration	0.14–1.3	7.5–8.1	2–5	11–24	<0.1–0.6	<0.01–0.02	139–180	346–410	<0.15
Postchlorination	0.05–1.3	7.5–8.0	2–5	11–25	<0.1–0.4	<0.1	135–182	346–411	<0.15
Simulated Distribution[a]	0.03–1.3	7.6–8.0	2–5	11–25	<0.1–0.5	<0.1	135–182	346–413	<0.15

[a]Held for 24 hours before analysis.

Preliminary Disinfection Studies

Bench-scale experiments were conducted to compare the disinfection response of *E. coli* lys 147 to a nonlysogenized coliform bacteria, *E. coli* C. One liter of water obtained at the Boerne WTP was placed in a two-liter trypsinizing flask containing a magnetic stirring bar. While rapidly mixing, the test organism was introduced; followed in 10 minutes by addition of a preselected amount of chlorine. During the disinfection period, samples were removed with a Cornwall® pipetting syringe directly into chilled brain heart infusion broth to quench all residual chlorine.

Representative results for the two coliform bacteria are shown in Figure 3. Both organisms showed a typical, biphasic response to chlorination. Notably, *E. coli* lys 147 was more sensitive to inactivation by chlorine as demonstrated by a nearly 2 $\log_{10}$ greater reduction at a contact time of 30 minutes. Nonetheless, this organism was chosen for use in further prototype testing because of the ability to coincidentally monitor indigenous standard plate count (a parameter receiving increasing support as an inplant measure of disinfection adequacy).

Prototype System

Disinfection runs with the disinfectant systems shown in Table V have been performed. All of the disinfection experiments shown are with a medium-free organism (collected from culture, washed and resuspended in saline) added to the system. The *E. coli* lys 147 seed was approximately 2×10^9 PFU/ml and the *E. coli* C was approximately 3×10^9 CFU/ml. organisms observed with the *E. coli* lys 147. Bench studies have shown *E. coli* to be more resistant to chlorine. Consequently, it was thought that a dual seed would provide more interpretable disinfection data.

The chemical parameters measured (except THM) do not show any changes or trends other than what has been observed for the normal WTP operation (Table IV). However, increased ammonia levels are observed for the runs with added ammonia. These observations have been true for all disinfection runs performed; the chemical data have remained invariant regardless of disinfectant, dose or biological seed.

Bacteriology

Table VI shows the results of several disinfection runs with four disinfectant systems at the 0.5-mg/l dose. Table VII shows the results of two disinfectant systems at the 1.5-mg/l dose. All systems function effectively

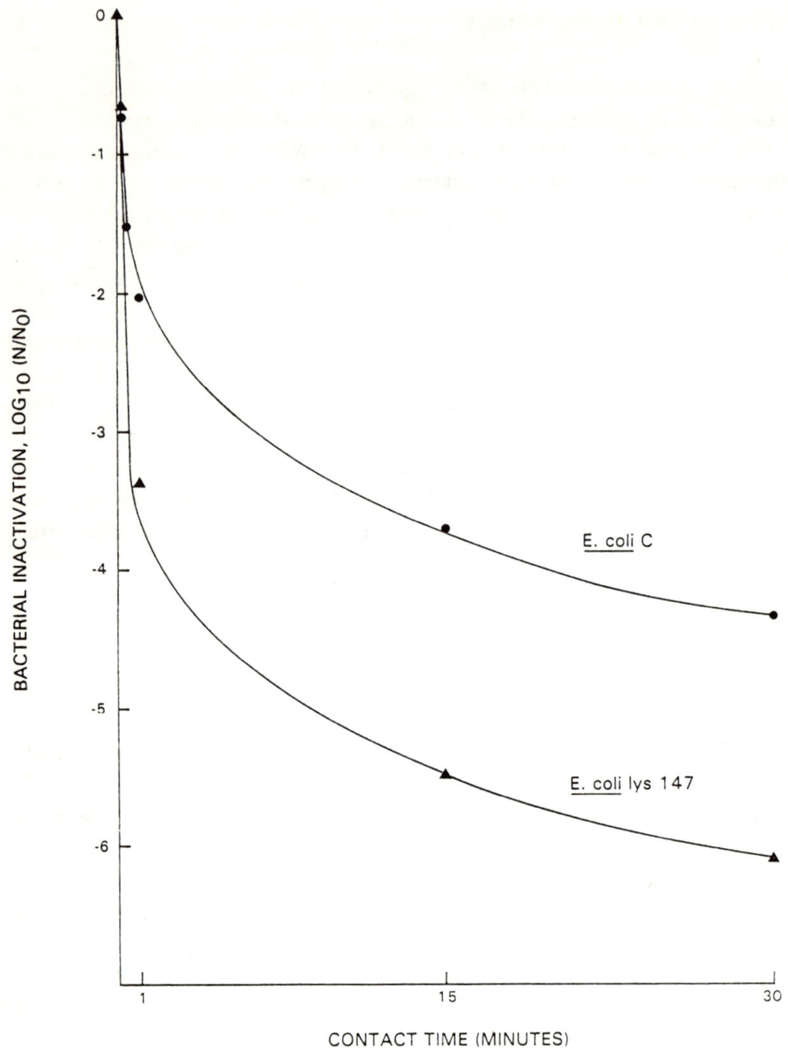

Figure 3. Comparative disinfection of *E. coli* lys 147 and *E. coli* C at a chlorine dose of 1.2 mg/l.

if a contact time of 15–60 minutes is allowed. Chlorine at both doses effectively reduces the seed organisms (as well as the indigenous organisms) within 15 sec (>6 log$_{10}$ inactivation).

Table V. Disinfection Runs with the Prototype System (Seeded)

Disinfectant	Disinfectant Dose (mg/1)	Seed Organism
Chlorine Followed	0.5	*E. coli* lys 147
by Ammonia	1.5	
	5.0	
Chlorine	0.2	*E. coli* lys 147
	0.5	plus *E. coli* C
	1.5	for doses of 0.2
	5.0	and 5.0 mg/1
Ammonia Followed	0.5	*E. coli* lys 147
by Chlorine	1.5	plus *E. coli* C
	5.0	
Chlorine Dioxide	0.2	*E. coli* lys 147
	0.5	plus *E. coli* C

Table VIII compares the effectiveness of chlorine at the highest dose employed (5mg/1) to chlorine (15-sec contact before ammonia added) followed by ammonia and ammonia followed by chlorine. The doses of chlorine and ammonia were equal. Effective disinfection occurs with all three systems within 30 seconds of contact with chlorine or chloramines at this dose.

To define better the disinfection process, lower doses (0.2 mg/1) of chlorine and chlorine dioxide were employed. Even at this low dose both chlorine and chlorine dioxide exhibit excellent inactivation within 15 sec for the *E. coli* lys 147 organism (Table IX). The chlorine dioxide exhibits a faster inactivation than chlorine as evidenced by the results for total plate count.

Trihalomethane Production

Table X demonstrates that THM production can be lowered significantly by lowering the chlorine dose. Furthermore, if ammonia is added, additional reduction in THM levels are observed at all doses. When chloramines are present (ammonia followed by chlorine), essentially no THM are produced. These results are expected with chlorine dioxide also; however, consistent but low levels of THM have been observed.

Table VI. Prototype Disinfection with Several Disinfectants at a Dose of 0.5 mg/l

Contact Time	Disinfectant System							
	Chlorine		Chlorine Followed by Ammonia[a]		Chlorine Dioxide		Ammonia Followed by Chlorine[b]	
	TPC^c	λ^d	TPC^c	λ^d	TC^e	λ^d	TC^e	λ^d
O	4.3	<0.33	81	<0.33	<0.33	<0.33	<0.33	<0.33
O^f	22	8.9×10^4	170	1700	6.2×10^5	1.2×10^5	3.2×10^5	1.1×10^5
1 sec	80	5.2×10^3	600	300	4.3×10^5	6.9×10^3	3.7×10^5	5.4×10^4
15 sec	46	<0.33	95	<0.33	<0.33	<0.33	3.9×10^5	8.7×10^4
16 sec	5.3	<0.33	130	<0.33	<0.33	<0.33	2.4×10^5	9.6×10^4
31 sec	8.0	<0.33	66	<0.33	1.0	<0.33	1.8×10^5	2.5×10^4
56 sec	<0.33	<0.33	g	g	<0.33	<0.33	1.1×10^5	1.5×10^4
72 sec	1.7	<0.33	83	<0.33	<0.33	<0.33	1.2×10^5	4.5×10^3
15 min	0.67	<0.33	61	<0.33	<0.33	<0.33	3.8×10^4	<0.33
60 min	3.7	<0.33	67	<0.33	<0.33	<0.33	2.0	<0.33
12 hr	2.3	<0.33	2	<0.33	<0.33	<0.33	g	g
24 hr	2.7	<0.33	0.33	<0.33	<0.33	<0.33	<0.33	<0.33
36 hr	g	g	g	g	<0.33	<0.33	g	g
48 hr	8.3	<0.33	<0.33	<0.33	<0.33	<0.33	<0.33	<0.33

[a] Ammonia added at 16 sec time point (dose equal to chlorine).
[b] Chlorine added at 16 sec time point (dose equal to ammonia).
[c] Total plate count (indigenous organisms, CFU/ml).
[d] *E. coli* lys 147 (PFU/ml).
[e] *E. coli* C added (CFU/ml).
[f] After addition of organisms.
[g] No data collected.

Table VII. Prototype Disinfection with Chlorine and Ammonia followed by Chlorine at a Dose of 1.5 mg/1

| Contact Time | Chlorine | | | | Ammonia followed by Chlorine[a] | | | |
| | E. coli lys 147 | | Total Plate Count | | E. coli lys 147 | | E. coli C | |
	(PFU/ml)	(N/N_O)	(CFU/ml)	(N/N_O)	(PFU/ml)	(N/N_O)	(CFU/ml)	(N/N_O)
O	<0.33		2.7		<0.33		<0.33	
O[b]	1.5×10^5	1.0	20	1.0	8.2×10^4	1.0	3.2×10^5	1.0
1 sec	<0.33	$<2.2 \times 10^{-6}$	2.7	0.14	9.7×10^4	1.2	3.0×10^5	0.94
15 sec	<0.33	$<2.2 \times 10^{-6}$	4.0	0.20	7.0×10^4	0.85	2.9×10^5	0.91
16 sec	<0.33	$<2.2 \times 10^{-6}$	<0.33	$<1.7 \times 10^{-2}$	2.4×10^4	0.29	1.7×10^5	0.53
31 sec	<0.33	$<2.2 \times 10^{-6}$	2.3	0.12	c	c	1.7×10^5	0.53
56 sec	<0.33	$<2.2 \times 10^{-6}$	0.33	1.7×10^{-2}	0.33	4×10^{-6}	1.3×10^5	0.41
72 sec	<0.33	$<2.2 \times 10^{-6}$	0.33	1.7×10^{-2}	<0.33	$<4 \times 10^{-6}$	1.1×10^5	0.34
15 min	<0.33	$<2.2 \times 10^{-6}$	<0.33	$<1.7 \times 10^{-2}$	<0.33	$<4 \times 10^{-6}$	<0.33	$<1 \times 10^{-6}$
60 min	<0.33	$<2.2 \times 10^{-6}$	0.33	1.7×10^{-2}	<0.33	$<4 \times 10^{-6}$	<0.33	$<1 \times 10^{-6}$
12 hr	<0.33	$<2.2 \times 10^{-6}$	<0.33	$<1.7 \times 10^{-2}$	<0.33	$<4 \times 10^{-6}$	<0.33	$<1 \times 10^{-6}$
24 hr	<0.33	$<2.2 \times 10^{-6}$	<0.33	$<1.7 \times 10^{-2}$	<0.33	$<4 \times 10^{-6}$	<0.33	$<1 \times 10^{-6}$
48 hr	<0.33	$<2.2 \times 10^{-6}$	<0.33	$<1.7 \times 10^{-2}$	<0.33	$<4 \times 10^{-6}$	<0.33	$<1 \times 10^{-6}$

[a] Chlorine added at 16 sec time point (dose equal to ammonia).
[b] After addition of organisms.
[c] No data collected.

Table VIII. Prototype Disinfection of *E. coli* lys 147 and Indigenous Organisms with a 5.0 mg/l Disinfectant Dose

| Contact Time | Disinfectant System | | | | | |
| | Chlorine | | Chlorine followed by Ammonia[a] | | Ammonia followed by Chlorine[b] | |
	TPC[c]	λ^d	TPC[c]	λ^d	TPC[c]	λ^d
0	37	<0.3	18	<0.3	16	<0.3
0[e]	72	1.1×10^5	170	5.8×10^4	180	6.0×10^4
1 sec	20	1.0	2	<0.3	210	8.1×10^4
15 sec	10	<0.3	2	<0.3	210	5.0×10^4
31 sec	f	f	2	<0.3	4	<0.3
56 sec	4	<0.3	3	<0.3	1	<0.3
15 min	f	f	0.3	<0.3	0.3	<0.3
60 min	f	f	0.7	<0.3	2	<0.3

[a] Ammonia added at 16 sec time point (dose equal to chlorine).
[b] Chlorine added at 16 sec time point (dose equal to chlorine).
[c] Total plate count (indigenous organisms CFU/ml).
[d] *E. coli* lys 147 (PFU/ml).
[e] After addition of organisms.
[f] No data collected.

DISCUSSION

The normal operation of the WTP plant exhibited chemical, bacteriological and THM values that were essentially identical to those parameters measured during the operation of the prototype system. Consequently, the information gained from the prototype should be applicable to plant operation. For example, a typical TTHM of the WTP was 107 μg/l and a TTHM from the prototype (with chlorine disinfection) at approximately the same dose and residual was 118 μg/l. The similarity between systems indicates that the rapid mixing techniques and the addition of a biological seed did not appreciably alter the chemical parameters characteristic of the water.

Most investigative work concerning rapid mixing and water streams has been directed towards either the role of rapid mixing associated with the addition of a coagulant to optimize floc formation, or the role of rapid mixing·associated with the addition of chlorine to enhance the biocidal action of chlorine in a wastewater stream. Longley [19], working with chlorination of a wastewater stream, achieved excellent bactericidal efficiencies using a plug flow reactor of similar design to the study reactor.

Table IX. Prototype Disinfection of *E. coli* lys 147 and Indigenous Organisms with a 0.2 mg/1 Disinfectant Dose

Contact Time	Disinfectant System							
	Chlorine				Chlorine Dioxide			
	E. coli lys 147		Total Plate Count		*E. coli* lys 147		Total Plate Count	
	(PFU/ml)	(N/N_O)	(CFU/ml)	(N/N_O)	(PFU/ml)	(N/N_O)	(CFU/ml)	(N/N_O)
0	<0.3		55		<0.3		41	
0[a]	4.7×10^4	1.0	95	1.0	8.9×10^4	1.0	81	1.0
1 sec	4.9×10^4	1.0	95	1.0	2.1×10^4	0.24	78	0.96
15 sec	1	2.1×10^{-5}	74	0.78	1	1.1×10^{-5}	1	1.2×10^{-2}
31 sec	<0.3	$<6.4 \times 10^{-6}$	74	0.78	b	b	b	b
56 sec	<0.3	$<6.4 \times 10^{-6}$	46	0.48	<0.3	$<3.4 \times 10^{-6}$	1	1.2×10^{-2}
15 min	<0.3	$<6.4 \times 10^{-6}$	2	2.1×10^{-2}	<0.3	$<3.4 \times 10^{-6}$	0.3	3.7×10^{-3}
60 min	<0.3	$<6.4 \times 10^{-6}$	2	2.1×10^{-2}	$<0 3$	$<3.4 \times 10^{-6}$	1	1.2×10^{-2}

[a] After addition of organism.
[b] No data collected.

Table X. Terminal THM Production from Several Disinfectant Systems in the Presence of Seed Organisms

Disinfectant Dose (mg/l)	Disinfectant System	THM Production (μg/l)				
		$CHCl_3$	$CHCl_2Br$	$CHClBr_2$	$CHBr_3$	TTHM
0.0	None	<0.1	<0.1	<0.1	<0.1	<0.1
0.2	Chlorine	1.06	1.80	1.43	<0.1	4.29
	Chlorine dioxide	0.14	0.28	<0.1	<0.1	0.42
0.5	Chlorine	1.87	2.17	1.18	1.07	6.29
	Chlorine + ammonia[a]	0.99	0.87	0.61	<0.1	2.47
	Ammonia + chlorine[a]	0.13	<0.1	<0.1	<0.1	0.13
	Chlorine dioxide	0.17	2.51[b]	0.04	<0.1	2.72[b]
1.5	Chlorine	52.0	34.4	24.5	7.0	118.9
	Chlorine + ammonia[a]	2.10	2.33	1.63	1.31	7.37
	Ammonia + chlorine[a]	0.19	0.10	0.14	<0.1	0.34
5.0	Chlorine	95.5	51.7	24.2	8.0	179.3
	Chlorine + ammonia[a]	3.06	4.27	1.88	1.03	10.24
	Ammonia + chlorine[a]	1.60	1.27	0.55	0.89	4.31

[a] Chlorine and ammonia at the same doses.
[b] Possible contamination of chlorine dioxide with chlorine.

He noted, however, that the bactericidal efficiencies achieved in the conventional chlorine contact chamber, only after the long contact time (approximately 12–15 minutes), approached those achieved with a plug flow reactor.

As demonstrated in Tables VI to VIII, very effective disinfection was achieved at low doses of disinfectant due to the efficiency of the tubular plug flow reactor. Stenquist and Kaufman [20] have cited two reasons for the superiority of a tubular over a backmixed reactor:

1. Backmixing apparently results in a chlorine residual with poorer bactericidal effectiveness because of reaction of the chlorine entering the reactor with residuals previously formed.
2. Short-circuiting, or the presence in the effluent from the reactor of fluid particles with a very short chlorine contact time relative to the mean contact time, can greatly affect overall performance.

The results of this study support these observations for bacterial inactivation and indicate that a tubular plug flow reactor can be used to advantage.

The seed organisims utilized in this study were not sufficiently resistant to any disinfection system (with the exception of the chloramines) to provide disinfection curves similar to that generated in static laboratory testing (Figure 3). Clearly, the rapid hydraulic mixing was responsible for the more rapid inactivation. Control experiments without disinfection demonstrated that there was no loss of viable organisms due to any mechanical effects of the prototype system.

Chlorine, chlorine dioxide and chlorine followed by ammonia showed rapid inactivation of viable organisms within the first few minutes of contact (Tables VI to VIII). The greater the dose, the more rapid the inactivation. A dose of 1.5 mg/l, chlorine was capable of achieving total inactivation in approximately 16 sec (Table VII). Even the chloramine system (ammonia followed by chlorine) showed essentially total inactivation within 15 min for a 1.5-mg/l dose and within 1 min for a 5-mg/l dose (Tables VII and VIII). A dose of 0.5 mg/l chloramine required a longer contact time (60 min) for inactivation (Table VI). Disinfection efficiency was excellent for all systems.

Of greater importance was the substantial reduction in THM production observed with the prototype system. As Table X demonstrates, lower disinfectant doses produced lower TTHM levels for chlorine. However, there was less of a difference between TTHM formed in the chlorine followed by ammonia system (2.47–10.24 μg/l) than was observed with the chlorine system (6.29–179.3 μg/l) for the same doses. The 15 sec of contact with free chlorine before ammonia addition produced higher levels of TTHM when compared to the addition of ammonia followed by chlorine (chloramines) at the same dose but also showed better disinfection. Thus, the ammonia addition was able to prevent additional THM from being produced.

The two best systems for minimizing THM production were chlorine dioxide and ammonia plus chlorine. The low level THM observed for chlorine dioxide may have been due to contaminants in the system. Significant reductions of THM were observed with chloramine and chlorine dioxide as the disinfectants. A 275-fold reduction in THM production was observed at a 1.5-mg/l dose of chloramine when compared to the same dose of chlorine. Disinfection efficiencies (Table VII) are not identical; however, essentially total inactivation occurred within a few minutes with the chloramine system (compared to one minute for the chlorine system).

The change of disinfectant application to a rapid mixing system and the selection of an alternative disinfectant system provide reasonable and significant treatment alternatives to chlorination without sacrificing the microbiological quality of the water. Clearly these results are applicable to moderately clean water; however, wastewater applications may be feasible.

CONCLUSIONS

Disinfection effectiveness was significantly improved by rapid hydraulic mixing. Optimal disinfection can be achieved with low to moderate doses of chlorine dioxide, chloramines and chlorine followed by ammonia with the test system studied. Chlorine dioxide was observed to be a slightly better disinfectant than was chlorine under the test conditions. Significant reduction in THM formation was achieved with chlorine dioxide, chloramines and chlorine followed by ammonia at all doses employed.

The best choice for optimal disinfection and THM minimization was a 1.5-mg/l ammonia dose (15 sec contact time) followed by a 1.5-mg/l chlorine dose. Essentially, total disinfection was achieved within 15 min contact, and the reduction of THM was 275-fold compared to a 1.5 mg/l chlorine dose alone.

ACKNOWLEDGMENTS

This study was funded by grant Number R-80604-01-0 from the U.S. Environmental Protection Agency, Drinking Water Research Division. The authors wish to thank the project liaison, Dr. Gary S. Logsdon, for his assistance. In addition, special appreciation is extended to Ms. Lou Jean Floyd, Mr. Donald Dudley, Mr. Charles L. Lockett, Ms. Kathryn F. Briley, Ms. Margaret S. Hulsey, Mr. Michael Ibarra and Ms. Sara Valenzuela from the staff of the Center for Applied Research and Technology, The University of Texas at San Antonio for their technical assistance in chemical and bacterial analyses.

Appreciation is also extended to Mr. Edgar Schwarz, Jr. and Mr. John B. Moring, Jr. from the City of Boerne, Texas for their cooperation.

REFERENCES

1. Kleopfer, R. D., and B. J. Fairless. "Characterization of Organic Components in a Municipal Water Supply," *Environ. Sci. Technol.* 6(12): 1036 (1972).
2. Bellar, T. A., J. J. Lichtenberg and R. C. Kroner. "The Occurrence of Organohalides in Chlorinated Drinking Waters," *J. Am. Water Works Assoc.* 66(12):703 (1974).
3. Symons, A. A., T. A. Bellar, J. K. Carswell, J. Dellarco, C. J. Slocum, B. L. Smith and A. A. Stevens. "National Organics Reconnaissance Survey for Halogenated Organics," *J. Am. Water Works Assoc.* 67 (1):634 (1975).
4. Coleman, W. E., R. D. Lingg, R. G. Melton and F. C. Kopfler. "The Occurrence of Volatile Organics in Five Drinking Water Supplies Using Gas Chromatography/Mass Spectrometry," in *Identification and Analysis of Organic Pollutants in Water*, L. H. Keith, Ed. (Ann Arbor, MI: Ann Arbor Science Publishers, Inc., 1977).
5. Dowty, B. J., D. R. Carlisle and J. L. Laseter. "New Orleans Drinking Water Sources Tested by Gas Chromatography/Mass Spectrometry," *Environ. Sci. Technol.* 9(8):762 (1975).
6. Morris, J. C., and B. Baum. "Precursors and Mechanisms of Haloform Formation in the Chlorination of Water Supplies," in *Water Chlorination: Environmental Impact and Health Effects, Vol. 2*, R. L. Jolley, H. Gorchev and D. H. Hamilton, Jr., Eds. (Ann Arbor, MI: Ann Arbor Science Publishers, Inc., 1978).
7. Stevens, A. A., C. J. Slocum, D. R. Seeger and G. G. Robeck. "Chlorination of Organics in Drinking Water," *J. Am. Water Works Assoc.* 68(11):615 (1976).
8. Hoehn, R. C., C. W. Randall, F. A. Bell, Jr. and P. T. B Shaffer. "Trihalomethanes and Viruses in a Water Supply," *J. Environ. Eng. Div., ASCE* 103(EE5):803 (1977).
9. Rook, J. J. "Haloforms in Drinking Water," *J. Am. Water Works Assoc.* 68(3):168 (1976).
10. Hoehn, R. C., R. P. Goode, C. W. Randall and P. T. B. Shaffer. "Chlorination and Water Treatment for Minimizing Trihalomethanes in Drinking Water," in *Water Chlorination: Environmental Impact and Health Effects, Vol. 2*, R. L. Jolley, H. Gorchev and D. H. Hamilton, Eds. (Ann Arbor, MI: Ann Arbor Science Publishers, Inc., 1978).
11. Rook, J. J. "Chlorination of Fulvic Acids in Natural Waters," *Environ. Sci. Technol.* 11(5):478 (1977).
12. Page, T., and R. H. Harris. "Drinking Water and Cancer Mortality in Louisiana," Science 193:55 (1976).
13. Tardiff, R. G. "Health Effects of Organics: Risk and Hazard Assessment of Ingested Chloroform," *J. Am. Water Works Assoc.* 69(12): 158 (1977).

14. *Federal Register* 44(231):68624-68707 (1979).
15. *Standard Methods for the Examination of Water and Wastewater,* 14th ed. (Washington, DC: American Public Health Association, 1975).
16. Adams, M. H. *Bacteriophages* (New York: Interscience Publishers, 1959).
17. "Methods for Chemical Analysis of Water and Wastes," U.S. EPA, Environmental Monitoring and Support Laboratory, Environmental Research Center, Cincinnati, OH (1976).
18. Bellar, T. A., and J. J. Lichtenberg. "Determining Volatile Organics at Microgram-per-Litre Levels by Gas Chromatography," *J. Am. Water Works Assoc.* 66(12):703 (1974).
19. Longley, K. E. "Mixing and Chemical Disinfection," PhD Thesis, The Johns Hopkins University, Baltimore, MD (1974).
20. Stenquist, R. J., and W. V. Kaufman. "Initial Mixing in Coagulation Processes," U.S. EPA Report 72-053, University of California, Berkeley, CA (1972).

CHAPTER 23

INSOLUBLE POLYMERIC CONTACT DISINFECTANTS: AN ALTERNATIVE APPROACH TO WATER DISINFECTION

Gilbert E. Janauer

> State University of New York at Binghamton
> Binghamton, New York

Charles P. Gerba

> Baylor College of Medicine
> Texas Medical Center
> Houston, Texas

William C. Ghiorse

> Cornell University
> Ithaca, New York

Michael Costello and Eva-Maria Heurich

> State University of New York at Binghamton
> Binghamton, New York

There is no question that the disinfection of water for human consumption is a necessity in any modern society. In the United States and other advanced nations the microbiological safety of potable water is assured by specific legislation, mandatory testing and monitoring, and water disinfection on a large scale.

With the exception of ultraviolet (UV) light irradiation, all water disinfection processes now in use—or contemplated for possible future use—rely on the addition of soluble reactive chemicals. Germicidal action is, in most cases, associated with oxidation potential. For instance, hypochlorous acid is a much stronger disinfectant than hypochlorite ion, which is stronger than monochloramine.

The most widely used disinfectant is chlorine in one chemical form or another. It has been proven effective, economic and safe over many years of use in the United States. Only recently have doubts been raised concerning a potential threat to human health due to the presence of haloorganic compounds in drinking water [1,2]. This risk is perceived as small by many in comparison to the benefits derived [3,4] and can, probably, be reduced significantly by incorporating changes in conventional water treatment trains while retaining essential advantages of chlorination, particularly the residual disinfectant capacity required in large distribution systems. Effective and economically feasible alternatives to conventional chlorination in large treatment systems are not likely to become available very soon, although considerable research efforts are being made. However, passive and active water reuse will of necessity increase [5], and this may prompt more work on alternative ways to ensure microbiological and chemical water safety and quality.

It must be recognized that there exists a very wide spectrum of disinfection needs. For instance, water disinfection in a spacecraft—a closed system, small and designed for total reuse—obviously requires an approach entirely different from a typical multimillion gallon-per-day drinking water plant serving a large city using a large river as the supply. Between those two extremes fall many so-called "small water treatment systems" [6], more than 200,000 in the United States alone, of which many have a poor record of compliance with mandated regulations [7]. In certain states, small freshwater systems do not use any disinfection, because residents object to the taste of disinfectants in their drinking water. An increase in the occurrence of waterborne disease has been noted recently in this country [8,9].

In addition to the need for potable water disinfection for human consumption there exist many other important applications, e.g., in health-related fields and in diverse industries that use water for a multitude of purposes. Not only pathogens, but a variety of nuisance microorganisms must be controlled under widely different conditions. In a number of circumstances, chlorination or other classical disinfection processes are difficult to use, for example, because of potential metal corrosion or other practical problems.

What are possible alternatives to employing dissolved, reactive chemicals? Excluding heat, UV and ionizing radiation—all energy-intensive, physical processes—two principally different approached remain. Both would offer advantages if they could be made effective, practical and economical.

FILTRATION/REMOVAL/SEPARATION OF MICROORGANISMS

One way of eliminating pathogenic or noxious microorganisms is, obviously, to effect complete physical removal/separation. This objective can be achieved, in principle, either by a mechanical-hydrodynamic method or process (such as membrane filtration) or by quantitative sorption on some surface-active material packed in a bed, filter or column. A spectrum of possible in-between or combination process devices may rely in part on a size/filtration effect, on differential migration and on strong surface adsorption. Antimicrobial properites of such an adsorptive material would be incidental, if any, so that the microorganisms would only be concentrated/separated, but not inactivated. While membrane filtration is very widely used in diverse applications (including the preconcentration of viruses) it is not an economical process for drinking water disinfection. Filtration using large beds as carried out in water treatment plants is very important for finished water quality. However, it will not remove all microorganisms, although filtration in conjunction with other conventional treatment steps very substantially improves microbiological quality together with other quality parameters [10]. Of a number of filtration/adsorption/removal systems proposed or used the only ones to be discussed here involve synthetic ion exchange resins, because they are also the basis of the insoluble polymeric contact disinfectants (IPCD) to be described later.

The first reference available goes back to 1960 when Gillissen [11] studied the potential of strong base anion exchange resins and strong acid cation exchange resins for the removal of *Escherichia coli* from aqueous suspensions. The anion exchanger removed considerable amounts of bacteria, but even high columns always passed some viable cells. Since cation exchangers were not effective it was concluded that the electrostatic charge of the exchanger was important. Interestingly, Gillissen found that resin columns performed better after more and more volume during any run—the first few fractions would never be free of bacteria. Furthermore, column height rather than total amount of resin was important for removal efficiency. Regeneration with NaOH was easily achieved and seemed to improve performance, but regeneration with NaOCl—while possible in principle—resulted in loss of resin performance. The latter regenerant was meant to make the approach suitable for pathogen removal, but proved to be impractical with results worse the higher the concentration of NaOCl. Finally, the presence of NaCl decreased the removal efficiency of the resin. Similar results were later obtained by Daniels and Kempe [12] with six different bacteria including gram-positive and gram-negative ones. Anion exchangers, as well as—to some extent—cation exchangers, sorbed bacteria, but a strong effect of pH (greater or lesser isoelectric point) and of salt concentration was noted. Binding of surface-negative bacteria to the positively charged quaternary ammonium groups on strong base exchange resins was postulated. Essentially the same principle combined with very large pore size is the basis of the new commercial "Ambergard XE 352 Filter" offered by Rohm and Haas [13,14] as a filtering resin for bacteria. Since pore diameters of the macroreticular strong base resin particles are on the order of 7μ, bacteria are admitted and strongly held by electrostatic forces. Electrolytes and other impurities are said to have only a minor effect at reasonable concentrations. The Ambergard filters use fairly high packed columns, and are quite efficient even though complete removal is not claimed by the manufacturer. One major proposed application is the use of Ambergard filters prior to postfiltration by submicron membrane filter. This is expected to extend the life of the membrane filter significantly. Further advantages include diminished effects of any present pyrogenics (removed by Ambergard), hardness of water and presence of other contaminants.

In summary, while complete disinfection by sorptive anion exchange filters seems not likely, bacteria removal is quite efficient, and the approach can be complementary to other methodologies.

INSOLUBLE CONTACT DISINFECTANTS: GREAT EXPECTATIONS

The second alternative to dissolved, reactive disinfectants is insoluble contact disinfectants (ICD), materials that can inactivate or kill target microorganisms by mere contact without adding any reactive agent to the bulk phase being disinfected and (hopefully), without themselves becoming irreversibly inactivated for further use. The concept of killing germs by contact with an insoluble antimicrobial is not entirely new [15], but attempts at realizing it have resulted in only partial success to this date.

During the last few years research has been conducted in this laboratory with the objective of producing insoluble polymeric contact disinfectants (IPCD), i.e., ICD based on inert polymeric materials which allow the incorporation or permanent attachment of germicidal moieties. Before presenting some representative IPCD results it may be useful to summarize what ought to be the essential attributes of an ideal ICD, and then to discuss briefly some work done by others in the same general direction.

Here is what could be considered an ideal ICD.

1. The ICD composition would, by definition, be completely water-insoluble (or immiscible, if a liquid), even in the presence of any natural impurities (such as dissolved electrolytes, etc.) commonly found in water supplies. A corollary of insolubility/nonreactivity would be that harmful

products of chemical reactions (such as chlorinated hydrocarbons) would not be formed, and no odor or taste would be introduced. On the other hand, there would not be provided any residual disinfectant to protect the water against reinfection during subsequent distribution. Therefore, major applications would be for point-of-use and water recycling systems, unless some soluble disinfectant were added after the contact disinfection step.

2. "Instant kill" of target microorganisms would be desirable. In any event, a relatively short contact time at temperatures encountered in the natural environment must suffice to achieve total disinfection in order for an ICD to be of practical value.

3. The perfect ICD would be a highly potent broad-spectrum germicide not only one hundred precent effective against all pathogens of major concern—bacteria, viruses, protoza and fungi—but also against any potential nuisance organisms, including saprophytic microorganisms, which are a bane of various conventional water treatment systems. Species selectivity would not usually be needed in potable water disinfection.

4. The biocidal functions/sites of the ICD must be easily accessible, even attractive, to all target microorganisms. Any repulsion or hindrance preventing contact would, obviously, defeat the purpose. It will be seen later that attractive forces may not be a trivial, but an essential, consideration among properties of ICD.

5. The antimicrobial functions of the ICD should retain their germicidal activity for a long time and no permanent "poisoning" or deactivation of the bioactive surface or sites should occur during the disinfection process. Ideally, one might hope that irreparable damage to cells (leading, ultimately, to lysis) would be inflicted on collision (see point 2 above) followed by immediate detachment (from sites) of damaged cells and/or debris. Similar to a catalyst in a true catalytic reaction, the bioactive functions would reemerge and continue to inactivate new target cells. Should this hope prove unrealistic, the ICD will eventually become "exhausted" during disinfection, probably, by "fouling" with dead cells and fragments. In that case, one would hope at least for a high disinfection capcity and regenerability of the active sites by a simple procedure.

6. Good resistance against "fouling" by diverse inorganic and organic matter present in raw water would be most desirable. The disinfectant sites should not bind any chemical species irreversibly nor be prone to be "covered" by (adsorption of) common impurities such as humic and fulvic acids.

It seems obvious that materials combining all—or at least several—of the above properties would offer distinct advantages for small, point-of-use potable water systems and might provide superior solutions to a host of other disinfection problems as well [16]. Recent dramatic successes with immobilized enzyme fermentation and, more recently—perhaps more to the point—results obtained with immobilized bacteria [17] have demonstrated the viability of bioactive surfaces. There is good reason to believe that effective contact biocidal surfaces can be prepared equally as well.

DEMAND-RELEASE DISINFECTANTS

While ICD may constitute the ultimate achievement, solving the practical problem of creating truly insoluble/nonreactive disinfectants has remained elusive. Sometimes, investigators were led to believe that they had accomplished this feat only to have later to retract. However, demand-release disinfectants (DRD) are the next best thing to ICD, plus they provide a disinfectant residual.

A fair number of attempts have been made at trying to insolubilize/immobilize antimicrobial materials. Perhaps, the most widely known systems have been based on inert matrices (e.g., charcoal) impregnated with silver. The following brief discussion will not include merely inhibitory or bacteriostatic systems but only some major ones among those for which *biocidal* action has been established.

Recent examples are macroreticular resin compositions impregnated with a silver salt of very low solubility, or with in situ reduced metallic silver, as described by Costin [18]. The starting materials for these "microbiocidal" compositions were commercial macroreticular anion exchange resins with strong base functionalities. There is an advantage in using ion exchange resins having high surface area and very large pores to facilitate interaction of microbes with silver (or another sparingly soluble germicide) deposited on the polymer surface: intimate contact with the antimicrobial agent is assured, and the microorganisms are retained/removed and inactivated. According to Costin [18], the leakage of toxic silver from his preparations is slow and can be further reduced by adding 200 ppm of sodium chloride to the influents before passing through the columns. Impressive reductions in viable cell counts (*E. coli* and *Salmonella faecalis*) were reported.

An entirely different disinfectant system was earlier investigated by Lambert and Fina [19]. The idea was to use strongly held germicidal counterions (rather than physically or chemically deposited antimicrobials) on conventional (gel type) strong base anion exchange resins. The fact that polyiodide ions are very selectively exchanged onto polystyrene resins vs almost any other anionic species is well known to all ion exchange workers and may be attributed to charge transfer type interactions and/or the very high polarizability of the polyhalide species. In any event, Dowex 1 type resins charged with triiodide or other polyhalide counterions were found to be highly effective in disinfecting bacterial suspensions passed through columns packed with these materials. A number of patents were granted for variations on that theme [20-23]. Hatch and Lambert [24] discussed operational and mechanistic aspects. Various applications for polyhalide resin compositions have been suggested. Thus, a strong base anion exchange resin triiodide formulation is being used in flow-through cups for travelers according to Lambert and Fina [25], and the polybromide form of conventional anion exchange resins is being used for potable water disinfection on sea-going vessels [26,27], a very important application. Initial hopes that the triiodide systems may constitute (defined above as) ICD were not fulfilled [28]. Indeed, strong base anion exchange resins loaded with tightly held polyhalide counterions are typical "demand-release" disinfectants. Among the compositions described by Costin [18] was also a binary combination of a silver impregnated macroreticular resin combined with a polyiodide resin, which showed significantly reduced release of *both* germicides into treated water while exhibiting impressive bactericidal action. Although some iodine disinfectant residual is desirable in principle, applications to public water systems have so far not been realized, probably, because of potential dietary implications. The same holds true for silver-containing compositions. However, as with polyiodide-loaded anion exchange resins, a "scavenger" may be employed to significantly reduce the release of potentially harmful compounds (50 ppb of Ag must not be exceeded in effluents).

Recently, the usefulness of silver composition has, again, been questioned, even when restricted to the objective of combating saprophytic microorganisms on activated charcoal beds. It has been stated that there may be no significant reduction in nonpathogenic bacteria in charcoal filter effluents in the presence of safe concentrations of silver, and that the saprophytic microorganisms are of little importance from the point of view of human health [29]. However, many silver-containing "filter" devices have been registered, and some European countries are advocating the use of silver in ion exchange water treatment systems [30].

In the same class of DRD mention should be made of the proposed use of conventional strong acid cation exchange resins containing quaternary ammonium ions as their counterions [31]. In that case the intrinsic antimicrobial functions would be expected to disinfect the water, as the "quats" would be displaced slowly from the resins by exchange against other positively charged counterions. From a number of other DRD type compositions two more shall be mentioned here to further illustrate the concept. One is the use of a functionally modified polystyrene divinyl benzene, which can slowly release active chlorine so as to achieve better than 99% effective bacteria kill after 2–4 min of bed contact with 900X its volume of the bacteria suspended in river water [32]. In that case the sulfonic acid groups of a conventional strong acid exchanger were converted to sulfonamide groups which were then chlorinated by means of hypochlorite with both mono- and dichlorosulfonamide groups formed. From 0.9 to 6.9 mmol active chlorine per gram dry resin were incorporated and the dichlorosulfonamide form worked best. It was possible to regenerate the resin with some loss in disinfectant action resulting.

An interesting array of polymers was synthesized and tested against various microorganisms by Donaruma and Vogl [33]. Their most recently studied materials were polythiosemicarbazides which bind Cu(II) very strongly [34] and release it over an extended period of time. Such compounds have been proposed as promising agents for *Schistosomiasis* prophylaxis [35] because they can control the disease-carrying snail population in natural waters. The compounds also exhibited some antibacterial properties.

SURFACE-BONDED ANTIMICROBIALS

There has been over the last decade an enormous surge of work in the area of surface-bonded functionalities. Especially, chelating groups and other strongly metal binding groups have been surface bonded. Most important, perhaps, have been applications in chromatography [36]. Obviously, there is no reason why surface-attachment should be limited to chemically reactive groups. The adding of bactericidal or bacteriostatic substances to a number of different surfaces has been successfully carried out and reported by Isquith and co-workers [37,38]. Of course, these preparations are ICD in the true sense, if they are indeed stable and insoluble. The Dow Corning workers were able to show that 3-(trimethyoxysilyl)-propyldimethyl octadecyl ammonium chloride (Si-QAC) exhibits antimicrobial activity as such *and* retains its antimicrobial properties once it is chemically bonded to almost any suface. Test organisms *E. coli* and *S. faecalis* suspensions were efficiently disinfected by glass surfaces and cotton wool cloth and cellulose acetate sheets treated with Si-QAC [37]. The use of 14 C-labeled Si-QAC permitted a stringent check on whether antimicrobial moieties were leached from the treated (and rinsed) surfaces during the tests. The conclusion that the antimicrobial activity resided at the surfaces proper was further substantiated by bioassays. A preliminary survey of surfaces that may be treated with Si-QAC and their antimicrobial spectrum was given [37]. The work was then extended to testing algicidal activity of Si-QAC bonded substrates with very encouraging results [38], and applications, e.g., to aquaria and to large industrial water systems, were suggested. Si-QAC is now an EPA-registered pesticide under the name Dow Corning® 5700 Antimicrobial Agent and has found application as an inhibitor of micro-organisms in textiles (e.g., socks). While there may remain certain questions with respect to the bonding aspects, strength and durability of adhesion to hydrophilic surfaces by means of silane additives, these problems are amenable to practical solutions putting into practice a new working theory of adhesion at the polymer-mineral interface [39].

A very important, fundamental problem arises from the fact that for all classical disinfectants their germicidal activity is defined in terms of rate of kill per concentration of compound *in solution*. Thus, there is really no way of comparing biocidal efficacy (1) among different ICD; and (2) between any ICD and a classical disinfectant—what would be the "phenol coefficient" for a Si-QAC treated glass surface? This problem was addressed and solved by Isquith and McCollum [40] by introducing a surface kinetic test method which allows determination of the rate of kill by any antimicrobial solid. The new method constitutes a major contribution to this new field in that the measurement of kill rate is based on the number of surviving cells at specified times of exposure to various amounts of total treated *surface area* of test substrate. This way, a direct comparison for all kinds of antimicrobial surfaces becomes possible in terms of what might be called a "specific kill rate" per unit area. The senior author of this chapter believes that this may become the standard for all new ICD. In the following section it will be seen that the authors had to face the same problem, but the concept of "surface concentration" was introduced to use at least a semiquantitive physical parameter in order to compare the results to those obtained with the soluble analogs of IPCD used. It is interesting to note that Isquith and collaborators chose a quaternary ammonium group for their first bonded antimicrobial surfaces as did the authors, coming from the direction of the ion-exchange and solution chemistry fields and not too knowledgeable about microbiological aspects at the outset. The rationale for choosing "quats" with resins will be given below.

INSOLUBLE POLYMERIC CONTACT DISINFECTANTS

Insoluble polymeric contact disinfectants are, like surface-bonded antimicrobials, a special case under the main category of ICD outlined above. The idea of creating IPCD resins was conceived at SUNY-Binghamton when two faculty members and students discussed this possibility, which turned into a research project later funded under the NSF-SOS program. Preliminary work was performed by a student team [41] and then was followed by systematic model studies carried out by Janauer and co-workers [42-44]. Among promising candidate germicides, the group of quaternary ammonium compounds (quats) was selected first because of its wide spectrum anti-microbial activity [45], low toxicity for mammals and the relative ease in synthesizing a variety of insoluble, polymeric quat resins—one of the investigators was an ion exchange chemist.

It was found that "Resin 12" (Figure 1), a weakly cross-linked chloro-methylated polystyrene functionalized with N,N-dimethyldodecylamine, was highly effective in disinfecting vegetative cell suspensions of *Bacillus subtilis* during short contact in passage through very small resin beds. Additional studies with *B. subtilis* and *E. coli*, in batch, showed that dis-infection is achieved during very short contact times of Resin 12 with resting bacteria cultures even in the presence of nutrient broth. Prolonged equili-bration over extended time periods (up to 48 hours) failed to result in any growth on culture plates with samples from supernatant suspensions of either *B. subtilis* or *E. coli*. Column runs with *E. coli* were equally successful except that flowrates had to be lowered from 5-10 ml/min to about 2 ml/min with the standard resin mini beds. Synthesis conditions were found to be more crucial: some miniscule "leakage" of live cells was found for *E. coli*

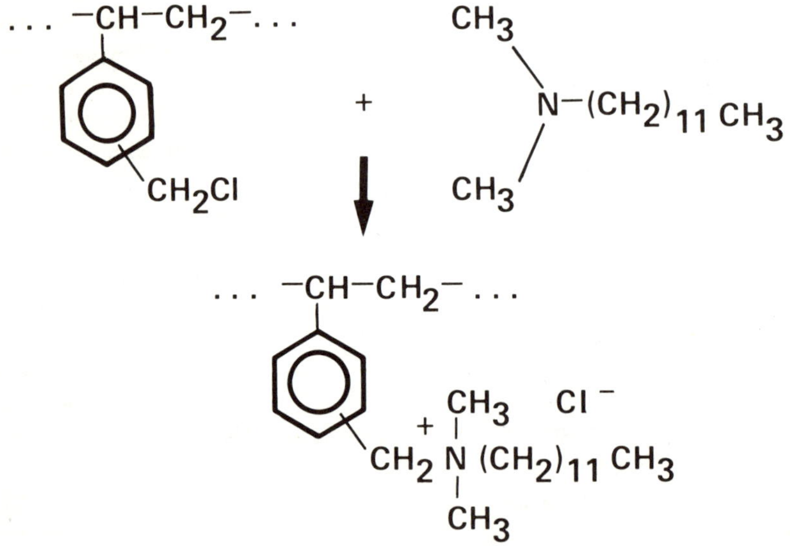

Figure 1. Resin 12.

with a different batch of Resin 12 in one standard column operation. This was attributed to a slightly different synthesis procedure–a parametric study revealed strong dependence of resin efficacy on synthesis conditions.

A set of typical data obtained wtih *B. subtilis* in column operations using standard minibeds (1 ml wet volume) is shown in Table I. Flowrates were 2-10 ml/min. As mentioned above, experiments with resting cultures of *E. coli* and *B. subtilis* [44] also yielded very encouraging results (Figures 2 and 3), suggesting the possibility of applying IPCD for disinfection in water storage tanks and closed-loop cycling systems of all kinds. It was shown later that *E. coli* bacteriophage MS-2 (Table II) and poliovirus type 1 (Table III) are very effectively removed from aqueous suspensions and tap water [46]. There was strong indication that the viruses and phages are being not only removed but also inactivated. Recent preliminary data with simian rotavirus SA-11 are highly encouraging, because even higher degrees of inactivation of eluted virus were found after relatively short exposure times. A first study of algicidal properties of Resin 12 and other IPCD gave impressive results. Finally, it is expected that the earlier demonstrated activity of soluble quats against the cysts of *Entamoeba histolytica* [47,48], will be reproduced and enhanced with the analog IPCD, judging from the dramatic disinfectant action against bacteria and viruses. The testing of Resin 12 and other IPCD against protozoa such as *Giardia lamblia* will be worthwhile.

Of course, the primary objective is to develop an effective, economical method which affords point-of-use potable water disinfection by removing/ destroying/inactivating all deleterious microorganisms without addition of soluble reactants and with traces in effluents (if any) below acceptable limits. Analyses are being carried out by gas chromatography/mass spectroscopy (GC/MS) in order to ascertain the purity of the treated water [49],

Table I. Disinfectant Capacity of Resin 12*

Temp ($^\circ$C)	Resin Batch	Amount Cells Appli.	Disinfectant Cap.	% Viable
24°C	1A	9.4 x 10^8/2300 ml	8.2 x 10^8	1
	1B			
	2A	9.6 x 10^8/1800 ml	Total	0
35°C				
	2B			
	2A	9.4 x 10^8/2300 ml	Total	1
24°C				
	2B			
24°C	3A	2.4 x 10^9/1600 ml	Total	0

* Using standard 1-ml resin bed, 10 ml/min flow.

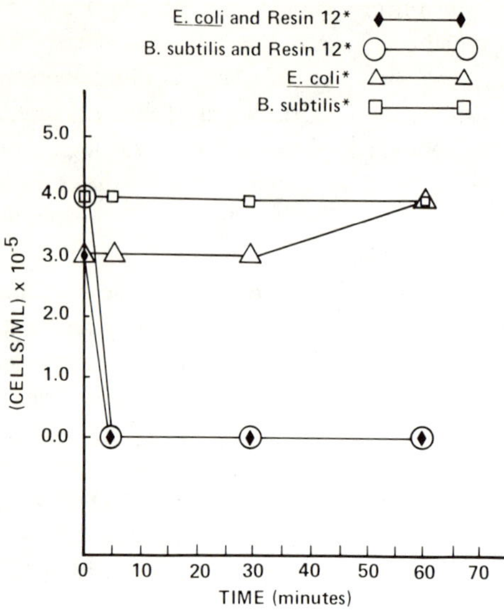

Figure 2. Disinfection of resting cultures by Resin 12.

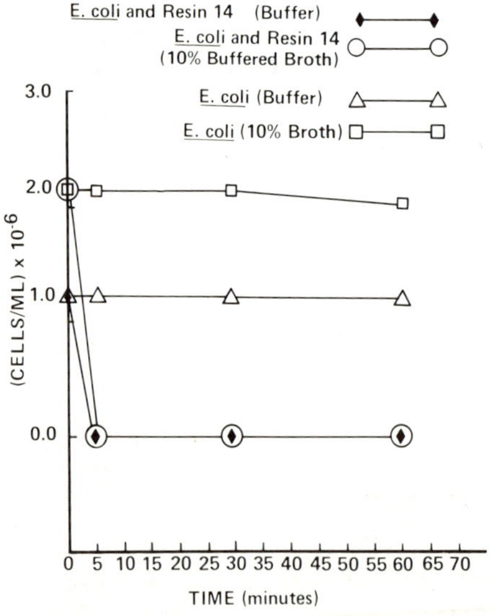

Figure 3. Disinfection of resting cultures by Resin 14.

Table II. Removal of *E. coli* Bacteriophage MS-2 by Resin 12 at pH 7.2 or 9.0 at 25°C

Total Volume Passed Through Column (ml)	PFU/ml		% Removal	
	pH 7.2[a]	pH 9.0[b]	pH 7.2[a]	pH 9.0[b]
100	1.5×10^3	8.0×10^4	99.6	85.0
200	2.4×10^3	1.5×10^4	99.4	97.2
300	2.1×10^3	1.8×10^3	99.5	99.3
400	1.7×10^3	1.2×10^4	99.6	97.7
500	4.2×10^d		99.0	

[a] 500 ml Tris buffer at pH 7.2 containing 4.2×10^5 PFU/ml passed through 2-cm beds of Resin 12 in 0.7- x 13-cm glass columns at flowrates of 10–12 ml/min. No significant inactivation occurred in the reservoirs during the course of the experiments.

[b] 400 ml glycine buffer at pH 9.0 containing 5.2×10^5 PFU/ml passed through 2-cm beds of resin 12 in 0.7- x 13-cm glass columns at flowrates of 10–12 ml/min. No significant inactivation occurred in the reservoirs during the course of the experiments.

Table III. Removal of Poliovirus Type 1 (LSc) from Dechlorinated Houston Tap Water by Resin 12-5[a]

Total Volume Passed Through Column (ml)	Viral PFU/ml	% Removal of Virus
Reservoir	2.2×10^4	
10	6.5×10^3	70
20	6.0×10^3	73
30	2.7×10^3	88
50	3.0×10^2	98.6
70	2.5×10^2	98.9
100	$< 1.0 \times 10^2$	99.5

[a] Virus was suspended in tap water and passed through a 2-cm length of resin in a 0.7- x 13-cm glass column at a flowrate of 5 ml/min. A total of 2.2×10^6 PFU of virus was thus passed through the resin column. After passage of the 100-ml volume, 5 ml of a 10% solution of bovine serum albumin at pH 10.0 was passed through the column to elute viruses adsorbed to the resin. This eluate was found to contain 1.2×10^6 PFU of virus, indicating that during the brief course of the experiment (20 min) perhaps 50% of the virus on the resin was inactivated.

although the polymer is a polystyrene type anion exchange resin very similar in nature to resins used safely for potable water treatment all over the world.

All results obtained to this date have shown that Resin 12 (and other IPCD) exhibit qualitatively similar, but disproportionally stronger, antimicrobial activity when compared to aqueous solutions of the soluble quat analogs N,N-dimethylalkylbenzyl chlorides. In addition, some special effects (e.g., high surface charge density) on protective hydration sheaths of hydrophilic viruses may be responsible for the observed disinfection of organisms not affected by soluble quats.

As with surface-bonded antimicrobials, the exact meaning of disinfectant "concentration" is somewhat problematic with IPCD based on gel-type polymers, but some useful conventions have been introduced. First, it must be kept in mind that only those biocidal groups which are at the outer surface of the polymer beads are accessible to organisms of bacterial size— the pore diameter in Resin 12 is estimated at 10-20Å. It follows that (1) the degree of functionalization (substitution) with the quat group, and (2) the bead diameter (mesh size) for the given amount of resin in a standard bed will together determine the total number of active sites a suspension of microorganisms will be exposed to during passage through the bed. A given flowrate translates into a certain pro forma "contact time" (6 sec with 10 ml/min used in most experiments to date), assuming uninterrupted contact between each volume element of suspension and the resin surface during passage (which is a somewhat unrealistic assumption). In addition to a "contact time" one needs to know, however, a "concentration" (and not the amount) of active sites in the columns. If the standard beds are packed with spherical beads of identical size distribution in all cases, and the degree of functionalization (e.g., with quats) of the polymer has been determined, a pro forma "surface molarity" (SM) can indeed be calculated. This SM can serve the useful purpose of allowing consistent and meaningful comparison among all monofunctional, gel-type disinfectant resins, and relative to the antimicrobial activity of solutions containing known concentrations of soluble disinfectants, such as iodine. The concept of SM also helps to give a physical picture of the nature of the resin/solution interface. For Resin 12, SM caluclates—from nominal resin wet molarity based on capacity/volume—to values of around 1 M in quat. This compares to customary quat solution concentrations of five to a few hundred ppm at the most— a factor of more than 10^4. No wonder that viruses are inactivated which are completely unaffected by any reasonable concentrations of the dissolved analogs.

It must be mentioned that the SM approximation while reasonable for gel-type spherical-shaped resins will be much cruder for nonspherically shaped particles of IPCD having macroporous parent matrices (now being explored), and for multifunctional IPCD. Still, at least some rational basis for comparison can be established in the context of a standard protocol for quantitative testing of a number of these IPCD. The foregoing considerations are of the essence from the point of view of developing optimum materials for practical disinfection in small water treatment systems. A standard test procedure similar to that developed by Isquith and McCollum [40] must be used.

In addition to determining the disinfectant capabilities of IPCD resins, the mechanism(s) of their activity is also of fundamental interest to microbiologists. It seems most likely that quat type resins either affect permeability by disrupting cell membranes or can interfere with cell metabolism by inhibiting enzyme reactions or both. Examination of attached bacteria on quat resins using scanning electron microscopy (Figures 4 to 6) combined with measurements of cellular components (DNA, RNA, protein) in effluents can show the effects of the resins on cell integrity. Inactive cells will be tested for various enzyme activities known to be involved in metabolic reactions. Such investigations can be expected to lead to a better understanding of the fundamental mechanism(s) of disinfection by quat type resins and may aid in the design of superior IPCD.

Concerning synthetic aspects, work by the group at Binghamton has led to a reasonably standardized synthetic procedure in which chloromethylated microporous polystyrene is reacted with a tertiary amine containing a twelve-carbon side chain to produce Resin 12, which has shown excellent disinfectant properties (see above). Resin 12 has only 50–60% of all its chloromethyl groups aminated, but the outer surface of the resin beads is believed to be fully functionalized. The flow characteristics of Resin 12 (in columns) are acceptable to good, but may be further improved for practical applications. The disinfectant action, although excellent, can probably be surpassed by other IPCD being synthesized.

A systematic study is aimed at preparing the most effective disinfectant resin in the benzyldimethylmonoalkyl series by attaching amines of various chain lengths (i.e., 10, 12, 14, 16 carbons) and varying the degree of functionalization. These syntheses can be conveniently monitored in terms of total nitrogen and ionic vs total halide in order to determine the degree of functionalization in the outer and inner regions of the resin beads. Monitoring of effectiveness against the indicator bacterium *E. coli* and selected pathogens serves to correlate structure with activity. A commercial mixture of long chain tertiary amines will also be used in the synthesis of IPCD and tested since this "mixed" resin would be economically desirable and might result in even stronger disinfectant properties. Too much functionalization of the resin interior will probably have to be avoided in all cases of gel-type IPCD since a highly functionalized resin may show slower flowrate characteristics

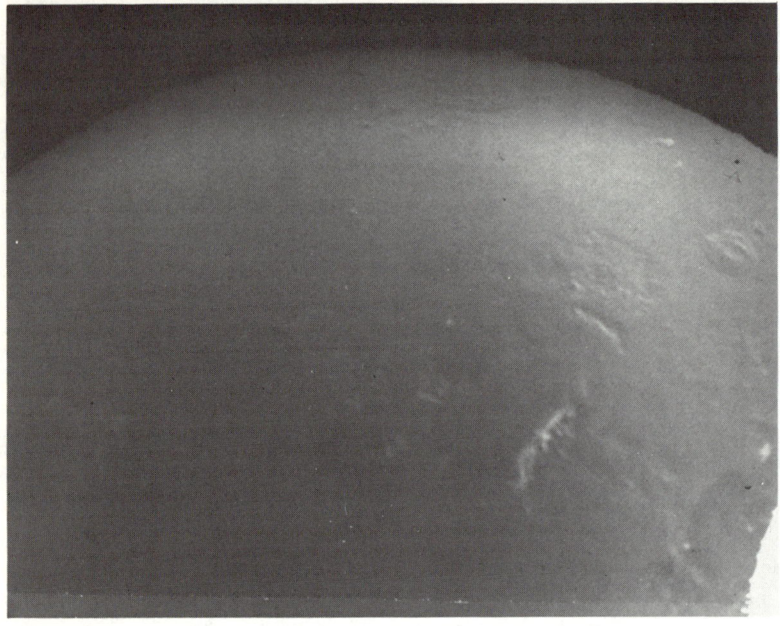

Figure 4. Scanning electron micrograph of Resin 12 bead before exposure to *E. coli* suspension (2071X magnification).

according to some preliminary results obtained. This will probably not be a problem with the macroporous preparations used here.

The IPCD initially investigated are all quaternary ammonium types. Some IPCD are being considered which contain not only different quaternary ammonium functions, but also other groups which may enhance biocidal action of the quats to give synergistic effects. In general, synthesis conditions are quite crucial.

As was pointed out above, the high antimicrobial efficacy of Resin 12 and other long-chain alkyl quat resin types may be, to a large degree, a direct consequence of the extremely high surface concentrations of the quat moieties at the resin/liquid interface (only short contact times needed for disinfection). Furthermore, it may be suggested that two major forces are instrumental in that they (1) "drive" microorganisms from the bulk liquid phase toward the bead surface; and (2) keep them near the surface long enough for the quats (or other antimicrobial functions when using different IPCD) to act on the cells.

The results of ion exchange selectivity studies by Boyd and Larsen [50], Janauer [51] and Janauer and Turner [52] demonstrated the importance of "hydrophobic interactions" in ion exchange equilibrium involving organic species in aqueous solutions. Entropy is gained when large, poorly hydrated ions such as $[CH_3(CH_2)_n]_4N^+$ or $R(CH_2)_nX^-$ (R = alkyl or aryl; $X^- = SO_3^-$; OSO_3^- or CO_2^-) are "squeezed out" from the structured water phase and exchanged against smaller, more hydrated species onto/into a (much less ordered) polymer phase. The same effect—only much stronger—may be expected when larger entities (viruses, bacteria, etc.) suspended in an aqueous medium are offered access to a comparatively less structured environment. Thus, the microorganisms are "driven" to the "quat surface" of Resin 12 (or any hydrophobic surface) by a general, mainly entropic effect. Since most microorganisms carry at least *some* negative surface charge in the intermediate pH range, they will also be attracted by coulombic forces if the ICPD is an anion exchanger (as all quat resins are) due to only partially screened positive charges on the polymer which are "seen" by the microorganisms as they approach. It will be remembered that Gillissen found bacteria to have a much smaller tendency to adsorb on conventional cation exchanger columns than on columns packed with anion exchange resins [11]. If the microorganisms can thus be forced to remain close enough to such an IPCD surface for a sufficient length of time, we may expect specific interactions with antimicrobial moieties.

With regard to practical applications some parameters, such as temperature, type of counterions and water hardness, are being studied now and very promising results are forthcoming. However, much remains to be done. For instance, a reasonable regeneration procedure is needed—the quick reactivation of Resin 12 with HCl-EtOH [42] is not practical for real-world applications—before IPCD are ready to be field-tested.

In any event, it is felt that IPCD incorporating antimicrobial quat moieties may become an important, practical alternative to classical water disinfectants, especially for small point-of-use systems. There exist other excellent candidate antimicrobials that may also retain their intrinsic biocidal properties when incorporated in IPCD with enhancement factors expected to be similar to those described. There is a host of potential applications besides potable water disinfection as was pointed out earlier [43].

ICD are a new concept, but there is now sufficient experimental evidence to warrant greater effort. Hopefully, the research community will accept the challenge and realize the great potential which we hope to have substantiated.

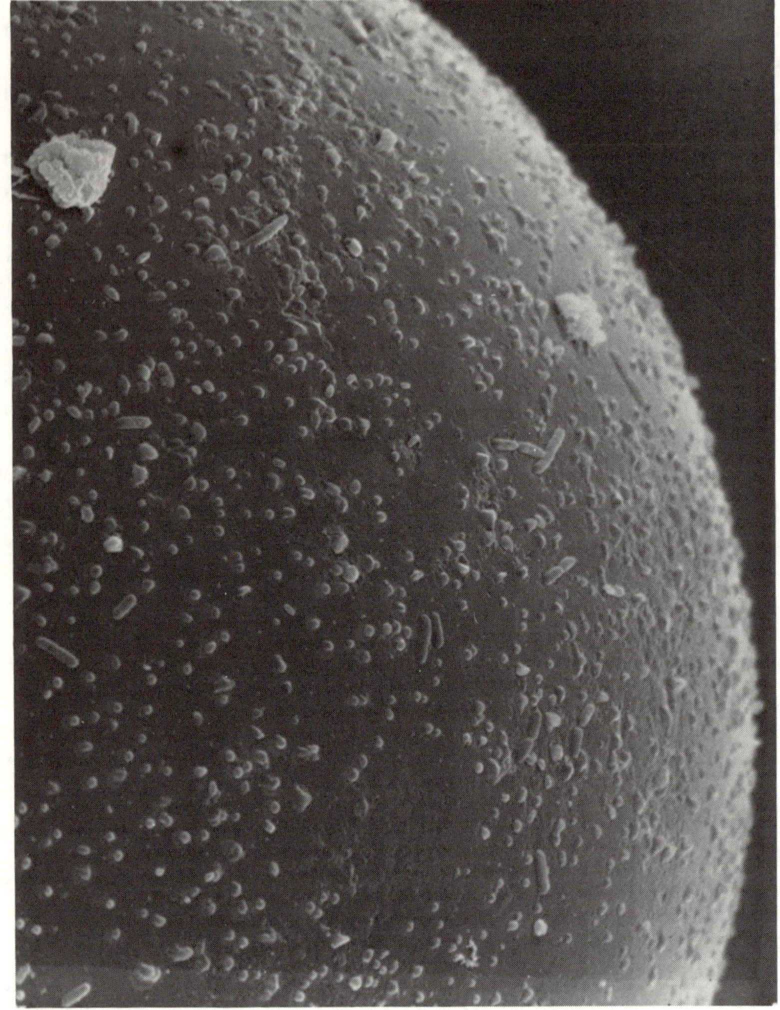

Figure 5. Scanning electron micrograph of Resin 12 bead after exposure to *E. coli* suspension (3227X magnification).

ACKNOWLEDGMENTS

The first author wishes to thank Ms. Ilona Figura-Walfish for her enthusiasm and dedication throughout the earlier stages of the experimental work. Important literature references for this chapter were provided by Dr. Hatch, Dr. Abrams and Dr. Isquith.

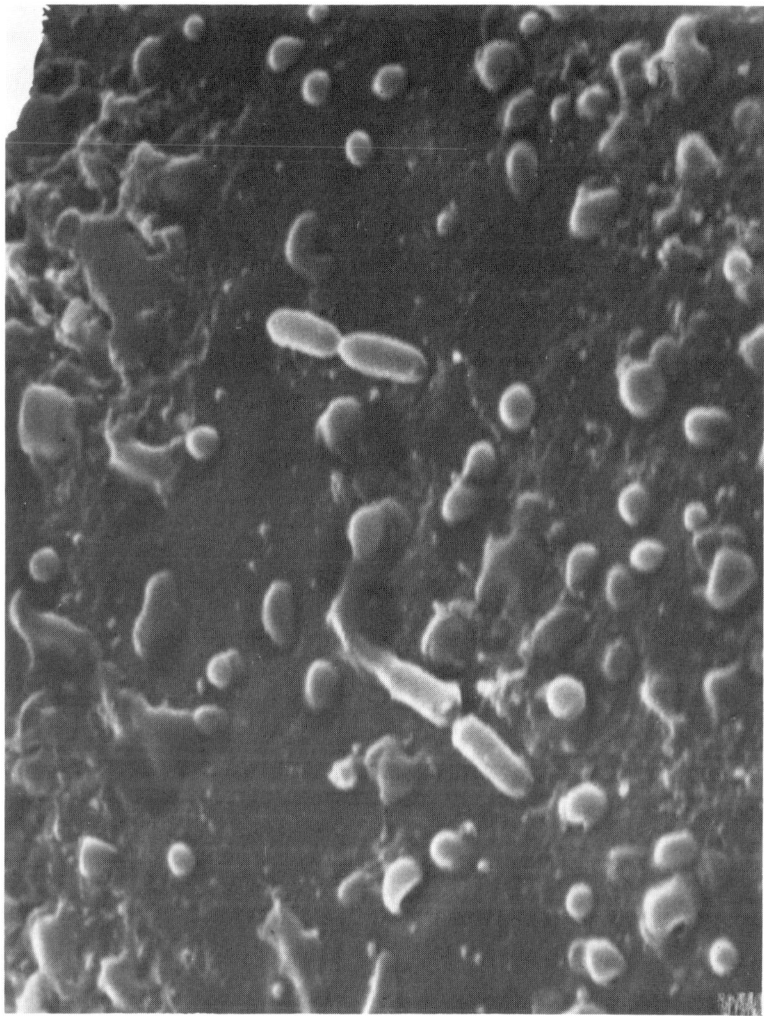

Figure 6. Scanning electron micrograph showing *E. coli* cells and other debris, possibly remnants of disrupted cells on the surface of a Resin 12 bead (11,600X magnification).

REFERENCES

1. Bellar, T. A., J. J. Lichtenstein and R. Kron. *J. Am. Water Works Assoc.* 66:703 (1974).
2. "Preliminary Assessment of Suspected Carcinogens in Drinking Water," U.S. EPA Interim Report to Congress (1975), Appendix VII.
3. *Water Quality and Health Significance of Bacterial Indicators of Pollution*, W. O. Pipes, Ed. (Philadelphia, PA: Drexel University, 1978).
4. Hoff, J. C. Personal communication (1979).
5. "Research Needs for the Potable Reuse of Municipal Wastewater," U.S. EPA Report EPA-600/9-75-007 (1975).
6. "State of the Art of Small Water Treatment System," U.S. EPA Office of Water Supply, Washington, DC (1977).
7. "National Interim Primary Drinking Water Regulations," U.S. EPA Report EPA-570/9-76-003 (1976).
8. Craun, G. F., L. J. McCabe and J. M. Hughes. *J. Am. Water Works Assoc.* 68:420 (1976).
9. Craun, G. F. *J. Water Poll. Control. Fed.* 49:1268 (1977).
10. Robeck, G., N. A. Clarke and K. A. Dostal. *J. Am. Water Works Assoc.* 54:1275 (1962).
11. Gillissen, G. "Entfernung von Microorganismen mittels Ionenaustauscher," *Gesundheitsingenieur* 81(7):207-210 (1960).
12. Daniels. S. L., and L. L. Kempe. "The Separation of Bacteria by Adsorption onto Ion-Exchange Resins," *Chem. Eng. Prog. Symp. Ser.* 62(69):142-148 (1966).
13. Costin, C. R., and F. L. Slejko. "Removal of Biological Matter from Aqueous Streams via Ion Exchange Filtration," *Proc. Parenteral Drug Assoc.* (Spring 1979).
14. "Ambergard XE 352," Rohm & Haas Technical Bulletin IE-246 (June 1978).
15. Klarmann, E. G., E. S. Wright and V. A. Shternov. "Prolongation of the Antibacterial Potential of Disinfected Surfaces," *Appl. Microbiol.* 1:19-23 (1953).
16. Janauer, G. E., I. H. Walfish, M. Ganz, M. Lalia and S. Marcus. "Quaternary Ammonium Resins as Water Disinfectants," paper presented at the Water Reuse Symposium, American Water Works Association Research Foundation, Washington, DC (1979).
17. *Immobilized Microbial Cells*, K. Venkatsubramanian, Ed. (Washington, DC: American Chemical Society, 1979).
18. Costin, C. R. "Microbiocidal Macroreticular Ion Exchange Resins, Their Method of Preparation and Use," U.S. Patent No. 4,076,622 (1978).
19. Lambert, J. L., and L. R. Fina. U.S. Patent No. 3,817,860 (1974).
20. Lambert, J. L., and L. R. Fina. U.S. Patent No. 3,923,665 (1975).
21. Mills, J. F. U.S. Patent No. 3,462,363 (1975).
22. Mills, J. F. U.S. Patent No. 3,462,363 (1969).
23. Hatch. G. L. "Mixed-Form Polyhalide Resins for Disinfecting Water," U.S. Patent No. 4,187,183 (1980).

24. Hatch, G. L., and J. L. Lambert. "Some Properties of the Quaternary Ammonium Anion-Exchange Resin-Triiodide Disinfectant for Water," *Ind. Eng. Chem. Prod. Res. Devel.* 19:259-263 (1980).

25. Lambert, J. L., and G. T. Fina. "Preparation and Properties of Tri-iodide-, Pentaiodide-, and Heptaiodide-Quaternary Ammonium Strong Base Anion-Exchange Resin Disinfectants," *Ind. Eng. Chem. Prod. Res. Devel.* 19:256-258 (1980).

26. "Everpure Bromination Systems," Everpure Technical Bulletin No. 618 (1976).

27. Goodenough, R. D., J. F. Mills and J. Place. "Anion Exchange Resin (Polybromide From) as a Source of Active Bromine for Water Disfection," *Environ. Sci. Technol.* 3(9):854-856 (1976).

28. Hatch, G. L. Personal communication.

29. "Dynamics and Significance of Bacteria in Everpure Precoat Filters," Everpure Engineering Bulletin (1980).

30. Abrams. I. Personal communication.

31. Nishino, A. U.S. Patent No. 3,872,013 (1975).

32. Emerson, D. W., D. T. Shea and E. M. Sorensen. "Functionally Modified Polystyrene Divinyl Benzene Preparation, Characterization and Bacterial Action," *Ind. Eng. Chem. Prod. Res. Devel.* 17(3):269-273 (1978).

33. Donaruma, L. G., and O. Vogl, Eds. *Polymeric Drugs* (New York, Academic Press, Inc., 1978), p. 349.

34. Donaruma, L. G. Personal communication.

35. Cheng, T. C. *Molluscicides in Schistosomiasis Control* (New York: Academic Press, Inc., 1974).

36. Karasek, F. W. "Chemically Bonded Chromatographic Columns," *Res./Devel.* (1974) pp. 34-37.

37. Isquith, A. J., E. A. Abbott and P. A. Walters. "Surface-Bonded Antimicrobial Activity of an Organosilicon Quaternary Ammonium Chloride," *Appl. Microbiol.* 24(6):859-863 (1972).

38. Walters, P. A., E. A. Abbott and A. J. Isquith. "Algicidal Activity of a Surface-Bonded Organosilicon Quaternary Ammonium Chloride," *Appl. Microbiol.* 25(2):253-256 (1973).

39. Plueddemann, E. P. "Reactive Silanes as Adhesion Promoters to Hydrophilic Surfaces," Dow Corning Corp., Midland, MI (1972), pp. 381-400.

40. Isquith, A. J., and C. J. McCollum. "Surface Kinetic Test Method for Determining Rate of Kill by an Antimicrobial Solid," *Appl. Environ. Microbiol.* 35(5):700-704 (1978).

41. Nudel, R. "Final Summary Report of Student Summer Project, NSF-SOS-547A," (1977).

42. Walfish, I. H., and G. E. Janauer. Unpublished results (1978, 1979).

43. Janauer, G. E., I. H. Walfish, M. Ganz, M. Lalia and S. Marcus. "Quaternary Ammonium Resins as Water Disinfectants," in *Proceedings of the Water Reuse Symposium, Vol. 2* (Washington, DC: American Water Works Association, 1979), pp. 1427-1434.

44. Walfish, I. H., and G. E. Janauer. "A New Approach to Water Disinfection. I. N,N-Dimethyl-n-alkylbenzyl Polystyrene Type Anion Exchange Resins as Contact Bactericides," *Water, Air Soil Poll.* 12:477 (1979).

45. Petrocci, A. N. In: *Disinfection, Sterilization and Preservation,* S. S. Block, Ed. (Philadelphia, PA: Lea and Febiger, 1977), pp. 325-347.

46. Gerba, C. P., and G. E. Janauer. Unpublished results (1979,1980).

47. Kessel, J. F., and F. J. Moore. *Am. J. Trop. Med. Hyg.* 26(3):345 (1946).

48. Fair, G. M., S. L. Chang, M. P. Taylor and M. A. Wineman. *Am. J. Public Health* 35(2):228 (1945).

49. Costello, M. J., and G. E. Janauer. Unpublished results (1979).

50. Boyd, G. E., and Q. V. Larsen. "The Binding of Quaternary Ammonium Ions by Polystryrenesulfonic Acid Type Cation Exchangers," *J. Am. Chem. Soc.* 89(24):6038 (1967).

51. Janauer, G. E. "Anionenaustausch—Selektivitätssequenzen in Homologen Reihen Hydrophober Ionen," *Mh. Chem.* 103(4):605 (1972).

52. Janauer, G. E., and I. M. Turner. "The Selectivity of a Polystyrenebenzyltrimethylammonium-Type Anion Exchange Resin for Alkanesulfonates," *J. Phys. Chem.* 73(7):2194 (1969).

SECTION 5
SWIMMING POOLS

CHAPTER 24

POLLUTION IN SWIMMING POOL WATER

J. Alan Beech*

Florida International University
Miami, Florida 33199

Swimming pools hold a number of records as far as water reuse is concerned. They represent the earliest successful reuse of water on a large scale and involve the largest number of water reuse installations of one kind. There are more than 1.8 million pools in the United States alone [1]. Because their water is continuously circulated and continually chlorinated, swimming pools probably hold bodies of water that have been repeatedly reused more than in any other application.

In regions where the water freezes in winter, open-air pools are usually at least partially emptied at the end of the swimming season. Southern California, Hawaii and south Florida have year-around pool use. These states contain 30% of all pools in the United States. At least another 10% of the U.S. pools are located in areas with mild winters, and do not need to be emptied annually [1]. The water in some year-round pools may be retained for several seasons. In these, it is only completely changed after gross contamination or when maintenance, such as painting, is needed. In some parts of the country where groundwater levels are high, pools are not emptied because of the danger of cracking or displacement.

Water may be lost from pools by filter backwashing, evaporation, splashout and occasionally by overflow after heavy rains. Solutes may or may not be

*Present address: Gulliver Preparatory School, Miami, FL.

lost, depending on the mode of water loss and the volatility of the solute. Despite replacement of these losses with makeup water, it is generally accepted in the swimming pool industry that the amount of dissolved material gradually increases in pool water. This is particularly the case when cartridge filters are used, because backwashing is unnecessary, and less makeup water is needed.

Open-air swimming pools may be contaminated from the air or from swimmers. Airborne pollutants are mainly water-insoluble. They include plant detritus such as leaves, pollen, twigs and grass clippings. Some of the organic matter comes from insects and some from organics on dust particles. Swimmer-borne pollutants are urine, sweat, hair, feces, suntan preparations and other cosmetics. Due to the large volume of urine that is likely to pollute a pool with a high bather load, it may even be relevant to consider unusual urinary components such as drugs and their metabolites excreted in urine.

Systematic examination of the chemistry and toxicology of pollutants in swimming pool water is a new field of environmental studies. In this chapter, common pool pollutants will be examined, together with the possible influence of pool water pH values on the concentrations of trihalomethanes, soluble heavy metals, nitrates and chlorates in the water.

METHODS

The water samples, unless otherwise mentioned, were obtained from open-air commercial swimming pools in Dade County, FL. All the pools were originally filled with municipal drinking water.

Samples for trihalomethane (THM) analysis were collected according to EPA recommendations [2] in cleaned 250-ml glass vials filled to the brim to exclude air and immediately sealed with Teflon®-faced silicone seals. Samples were kept at 4°C and analyzed within EPA-recommended holding times using the purge and trap method of Bellar and Lichtenberg [3]. Analyses were performed with a Tracor 560 gas chromatography unit fitted with a Tenax column, an integrator and a Tracor 700 Hall electrolytic conductivity detector. A standard solution of THM was run daily for comparison purposes.

Samples for inorganic anion analysis were collected in clean glass containers. Samples for metals analysis were collected in glass containers previously rinsed with pure nitric acid. They were brought to pH 2 with dilute nitric acid, stored at 4°C and analyzed within 48 hours of collection. Ion chromatographic analyses were made with a Dionex model 10 unit, using $0.003\,M$ NaHCO$_3$, $0.0024\,M$ Na$_2$CO$_3$ as eluent. Due to its high concentration in swimming pool water, chloride was measured by the argentometric method of *Standard Methods* [4]. Metal analyses were made with a Perkin-Elmer Model 503 atomic absorption unit and an HGA220 graphite furnace where applicable. The procedures for metal leaching experiments are included in the text.

Nitrate was initially measured by ion chromatography and confirmed by a standard method [4] (Cd/Hg reduction followed by diazotization and coupling). Although standard nitrate solutions and some pool water samples showed excellent agreement between the two methods, other samples gave much greater apparent nitrate peaks by ion chromatography. The anomalous peaks were found to be caused by chlorate interference.

Chlorate was initially confirmed by the iodometric distillation method of Williams and Meeker [5] as modified by Jacobs [6]. Serial dilutions were made of nitrate, chlorate and mixed nitrate and chlorate solutions. Analysis of these by ion chromatography showed that the peak height of the mixture was an additive function of peak heights of the individual ions. All pools were then resampled. Nitrate was measured by the standard method [4], and chlorate was measured by ion chromatography, deducting the nitrate contribution to the mixed peak. This method could not be used for samples from seawater pools due to interference from chloride and bromide.

TRIHALOMETHANES

Pool filters are seldom cleaned more than once weekly, even in the best maintained pools. In general, while the water is circulating, it flows continuously past organic detritus of plant or animal origin trapped on the prefilter or filter surface. Humic acids and other soluble organic substances are speculated to leach gradually from the trapped detritus and become available for reaction in the main body of the pool water.

Swimming pool water maintained at recommended pH values is a good solublizing medium for organic acids of low solubility. It is generally slightly alkaline in reaction. White [7] recommends a pH between 7.5 and 8.0. In the pool handbook published by the Center for Disease Control [8] the range of pH 7.2-8.0 is suggested. Alkalinity buffering of 80-120 mg/liter as calcium carbonate is also practiced to guard against pH fluctuations. The pH range for pool water is close to 7.4, the average pH value for human tears, and is believed to cause minimal eye irritation to swimmers. Corrosion of metal is negligible at this pH and algae growth is slower than at more acid pH values.

The chemical constitution of dead plant matter is not completely known. The alkali-soluble fraction was defined by Oden [9] as fulvic acids, humic acids and hymatomelanic acids. These were first associated by Rook [10] with formation of THM in drinking water. The chemistry of THM formation has been described by Morris [11]. He suggests that hypochlorite reacts with exposed acetyl groups in humic acids. Successive replacement of hydrogen by chlorine atoms occurs and chloroform is formed by hydrolysis. When bromide ions are present in the water, hypobromous acid is formed, and the haloform reaction produces partially or completely brominated THM.

The rate of THM formation is pH-dependent. Stevens et al. [12] showed that chloroform was formed faster and in higher yield at higher pH values. Babcock and Singer [13] compared the amount of chloroform obtained from fulvic acids and humic acids and showed that the latter fraction gave higher yields. Beech et al. [14] reported that water in public and commercial open-air saline pools contained up to 1287 μg/l of THM, mainly bromoform (average 657 μg/l) and freshwater pools contained up to 430 μg/l of THM, mainly chloroform (average 125 μg/l). Some pool water showed peaks due to other volatile halogenated compounds, but these were not fully characterized. An example of a chromatogram with unidentified peaks is shown in Figure 1.

Swimming pool water is usually chlorinated and pH-balanced daily during the swimming season. In addition, the filters are generally cleaned on a bi-weekly basis. For this reason, water samples taken from a pool at the beginning of the swimming season and several months later were not expected to contain similar amounts of THM. The analytical results from 14 pools sampled at intervals of several months are shown in Table I. Except for pool No. 30, which was emptied and repainted between samples, all the analyses from a particular pool were of the same order of magnitude.

Dilling et al. [15] showed that chloroform in dilute aqueous solution evaporates rapidly from an open surface. Chloroform at 25°C has a vapor pressure of 200 mm Hg. In a U.S. Environmental Protection Agency (EPA) publication [16], it was estimated that above water at 25°C containing 100 μg/l of chloroform, the concentration of chloroform in the air was 1.93×10^{-2} μg/l. However, this figure was based on steady-state closed system conditions, which do not apply in open-air swimming pools. Pools are subject to environmental factors such as sun, wind and swimmer loading that reduce the concentration of THM in the air above the water. This in turn increases their rate of loss from the water. (Samples considered here were taken on weekdays when the pools were generally not in use.)

At constant temperature the concentration C of THM found in pool water is:

$$C = \frac{F - L}{V}$$

where   F = amount formed
               L = amount lost
               V = the pool volume

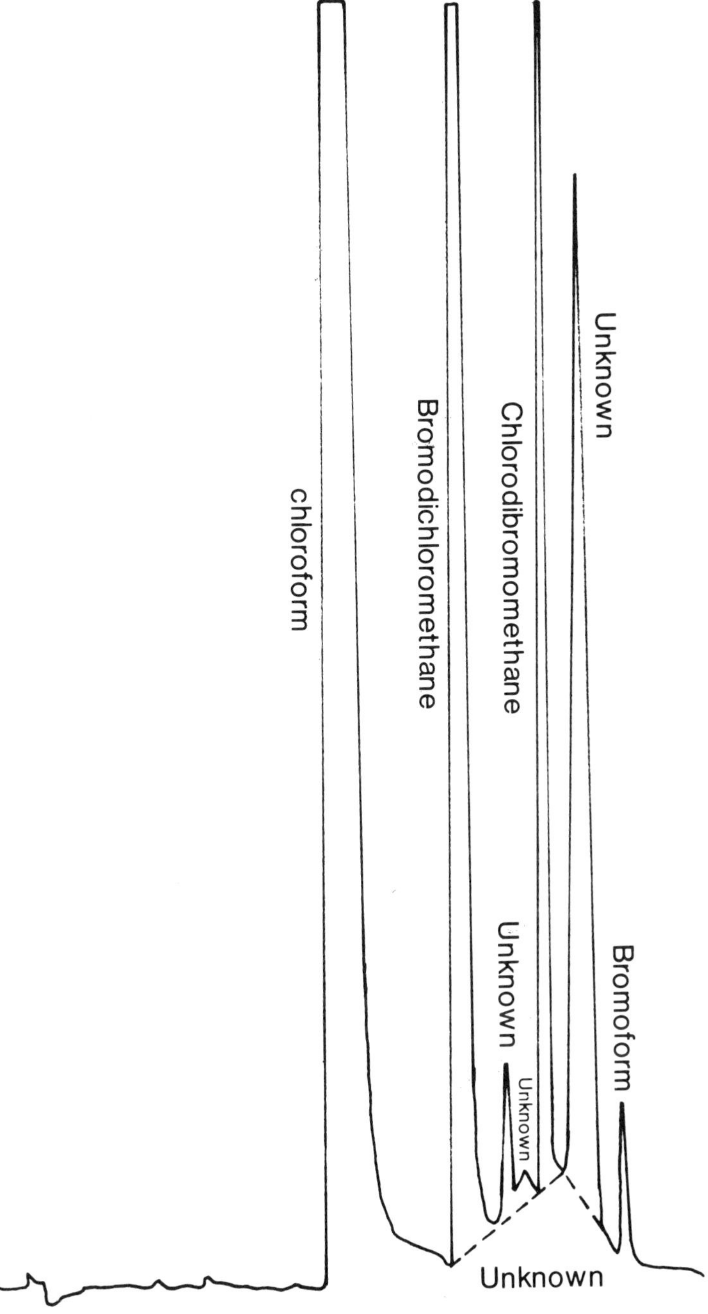

Figure 1. Chromatogram of swimming pool water with unidentified volatile halogenated substances.

Table I. THM Analyses of Swimming Pool Water Samples Collected from the Same Pools on Different Days in 1978

Pool No.	Date of Sample	Chloroform (μg/l)	Bromodichloro- methane (μg/l)	Chlorodibromo- methane (μg/l)	Bromoform (μg/l)
3	1/20	38.5	4.9	0.6	
	1/20	56.3	6.6	0.8	
	4/10	40.9	3.1	0.3	
	4/10	77.1	2.3		
8	4/6	64.9	7.5	0.9	
	4/6	62.6	7.8	0.9	
	9/28	69.8	6.4	1.7	
13	4/12	188.0	51.9	12.3	1.1
	4/12	115.5	54.6	13.0	2.4
	9/21	191.7	48.8	11.5	0.8
15	4/20	$<$1			
	4/20	$<$1			
	7/21	13.0			
16	4/18	10.4			
	4/18	6.7			
	7/21	7.0			
30	4/24	46.3	2.9		
	4/24	46.1	2.5		
	7/7	363.0	4.6	$<$0.5	
34	5/9	106.2	10.3	0.4	
	9/20	92.2	16.7	2.5	$<$0.5
35	5/9	13.4	0.9		
	9/1	$<$1	33.9		
37	5/9	11.9	2.2	0.3	
	7/8	32.3	6.8	0.4	
	9/20	32.9	7.1	1.6	
38	5/9	185.3	34.0	3.5	0.4
	9/25	225.9	26.9	12.1	0.3
39	6/5	66.6	12.2	1.1	
	9/28	112.7	17.4	2.0	
40	5/15	19.4	2.7	5.3	
	9/21	8.1	3.1	1.5	
41	5/15	20.8	5.3	1.6	
	9/20	53.2	18.2	4.5	
42	5/15	131.0	12.3	3.5	
	5/15	99.1	11.9	5.3	2.2
	9/21	96.8	4.0		

For a particular pool, the volume V will be constant. From Table I, values of C were generally constant. Therefore, F - L = VC = constant, i.e., the amount of THM formed equals the amount lost plus a pool constant.

Raw drinking water contains a definite concentration of organic precursors of THM (its THM formation potential). Swimming pool water, on the other hand, may continue to solublize more organic precursors as detritus continues to be trapped by the filters and water continues to circulate past them. Because of this and because the same water is rechlorinated daily, it is believed that constantly replenished sources of THM precursors are available for reaction in swimming pool water.

If THM precursors are present in excess, other parameters must determine the amount of THM formed in pool water. Based on the findings of Stevens et al. [12], it is suggested that these are the temperature and pH of the water. This is a speculative conclusion because pH readings were not taken for the samples analyzed in Table I. It will be reexamined in the section of this chapter on chlorates.

HEAVY METALS

Beech [17] has examined heavy metals in powdery sediments collected from the bottoms of commercial swimming pools alongside highways when the water was not circulating. The sediments contained up to 15,900 mg/kg of lead (average 3407 mg/kg) and up to 218 mg/kg of cadmium (average 35 mg/kg). Other metals analyzed were copper, zinc, manganese, iron, chromium and nickel.

The heavy metals in these sediments may constitute a toxicity hazard in two ways. Particles smaller than 10 μ in diameter will pass through some swimming pool filters, and particles smaller than 1 μ in diameter will pass through nearly all pool filters. Fine particles are likely to stay suspended in turbulent water for a long time and may be ingested with water by swimmers.

The composition of sediments in swimming pools alongside highways was shown by Beech [17] to be comparable with highway dustfall found by other authors. Shaheen [18] and Biggins and Harrison [19] have shown that the finest particles of highway dustfall are richest in lead. Scanning electron photomicrographs were made of a number of pool sediments. Three of these from pool No. 86 show typical features found in sediments from pools alongside busy highways (Figures 2 and 3).

At the lowest magnification, aggregates of fine crystalline particles can be seen. At higher magnifications, crystal shapes can be seen and measurements made. Many of the crystals are less than 10 μ in diameter and some are in the 1- to 2-μ range. The finest particles are likely to remain suspended in circulating water and to pass through most pool filters.

Figure 2. Sediment from pool No. 86 (150X).

Figure 3. Sediment from pool No. 86 (left) 1500X, (right) 7500X.

The other possibly toxic hazard from sediments rich in heavy metals is that heavy metals might dissolve and be swallowed in solution. Solubilization is more likely to occur with fine particles, especially if the reaction of the pool water changes from alkaline to acid. (Heavy metals may also be solublized by deliberate addition of a sequestering agent, a number of which are sold for pool use. When sequestering agents are used the toxic potential of the solublized heavy metal adduct is likely to greatly exceed that of the reactants.)

Several factors may reduce the pH of swimming pool water. When gaseous chlorine is used as disinfectant, the pH will fall rapidly if it is not constantly adjusted. This is because chlorine undergoes immediate hydrolysis in water to form hydrochloric and hypochlorous acids:

$$Cl_2 + H_2O \rightarrow H^+ + Cl^- + HOCl$$

The pH of pool water may also be reduced by human sweat (which has an average pH range of 4.0-6.8 [20] raw makeup water, acid rain, plant detritus and overzealous use of hydrochloric acid. (The acid is used to reduce the pH of pools that become too alkaline and to remove carbonate scale from tiles).

Beech [17] and Black et al. [21] both reported approximately 10% of Florida pools tested to have pH values less than 7.0. The author has tested commercial pool water with pH 6.3 and has been assured by colleagues that they have tested pool water with even lower pH values. For this reason, a study was designed to measure the leachability of pool sediments as the water became more acid. All metal analyses were by atomic absorption spectrophotometry.

Approximately 4 g of freshly sampled sediment associated with two liters of water from the same pool was used. The sediment was suspended by shaking and the liquid was rapidly poured off in 50-ml aliquots into 150-ml Erlenmeyer flasks.

Hydrochloric acid (0.1 N) was added to successive flasks in the following amounts: 0.0, 0.05, 0.01, 0.2, 0.4, 0.8, 1.6 and 3.2 ml. Sodium hydroxide (0.1 N) was added to other flasks in the following amounts: 0.05, 0.1, 0.2 and 0.4 ml. Magnetic stirrers were placed in each flask and were left to stir continuously, with daily pH checks of the supernatant liquid. When the pH of the supernatant liquid remained constant ($\pm$ 0.1 pH unit) for 48 hours, equilibrium leaching of the sediment was presumed to have occurred. The supernatant liquids were then removed by filtration through a 0.45 μ filter. The filtrate was brought to pH 2 by dropwise addition of concentrated nitric acid and analyzed for metals by atomic absorption.

The above experiment was repeated for slurries of sediment and water from six pools. Exact amounts of sediment were not used, as saturation of the supernatant liquid with respect to the metallic ions was expected to occur.

Cadmium and iron showed similar equilibrium curves for all six sediments. Examples are shown in Figures 4 and 5. The solubility of cadmium increased rapidly as the leaching solution dropped below pH 7. The iron results were the most reproducible. With few exceptions, all the analytical values from the six experiments in this series fell on the same curve of pH vs iron concentration at equilibrium.

The amount of lead, copper and manganese varied greatly for different sediments in solution at the equilibrium pH. The lead concentration of pool No. 1 leachate at pH 5 was approximately six times the concentration of pool No. 3. For copper and manganese, the highest concentrations at pH 5 were approximately three times the lowest ones. These results are shown in Figures 6 to 8.

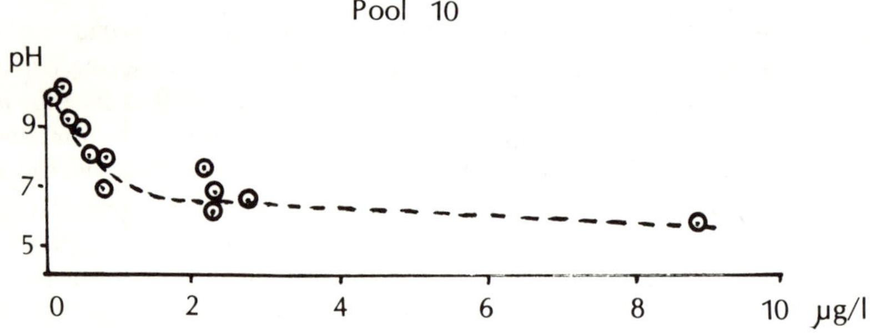

Figure 4. Typical equilibrium leaching curve for cadmium.

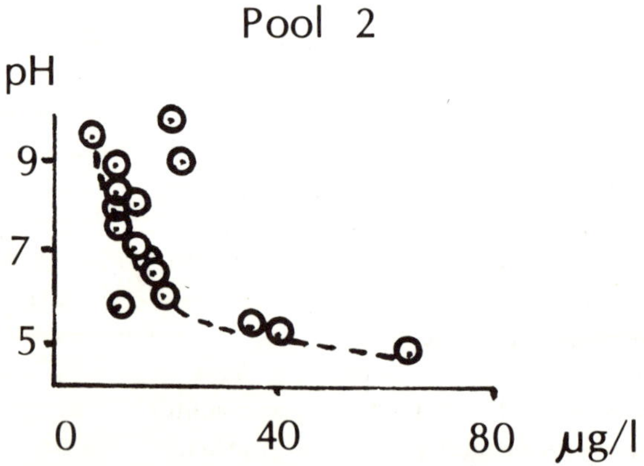

Figure 5. Typical equilibrium leaching curve for iron.

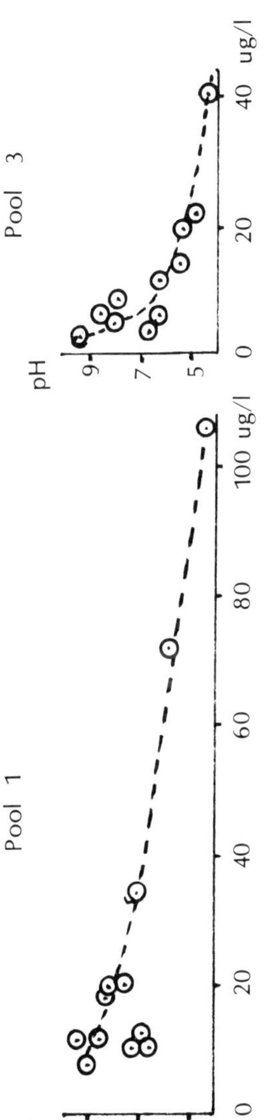

Figure 6. Two examples of different equilibrium leaching curves for lead.

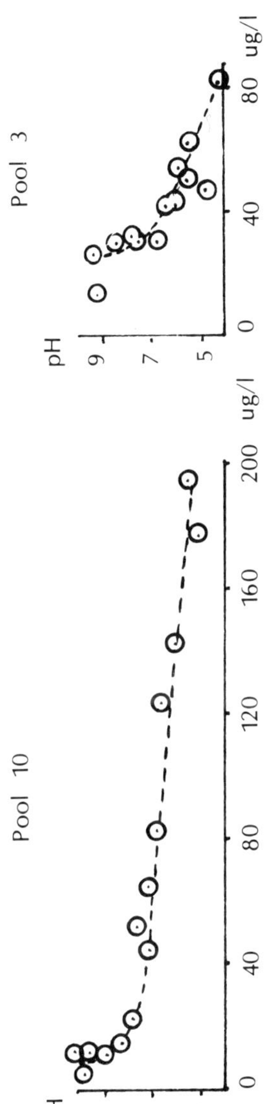

Figure 7. Two examples of different equilibrium leaching curves for copper.

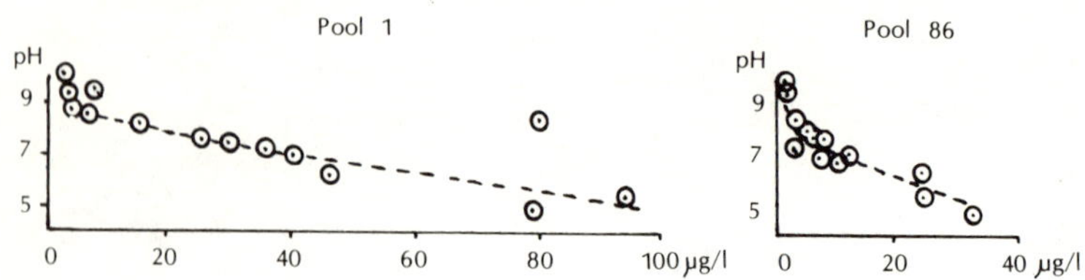

Figure 8. Two examples of equilibrium leaching curves for manganese.

The results show that under the conditions of the experiment, using swimming pool sediments and water from the corresponding pools, the shoulder of the equilibrium solubility curve for manganese occurred at about pH 8. Below this value, manganese solubility increased rapidly with falling pH. The corresponding shoulders for copper and cadmium were at about pH 7 and for iron were at about pH 6. The shoulder of the lead solubility curve appeared to occur below pH 7 but the solubility of the lead in sediment from pool No. 1 appeared to increase almost linearly as the equilibrium pH was reduced.

The differences in the pH of the shoulder of the equilibrium solubility curve for lead from different pools and in the amount of lead, copper and manganese dissolved from different sediments could be due to differences in the total carbonate alkalinity and salinity of the water or to differences in the chemical composition of the sediments.

Hem and Durum [22] showed that the solubility of lead in water at a given pH value decreased as the concentration of carbonate increased. Rohatgi and Chen [23] found that heavy metals were more readily released from sediment in saline than fresh water. Sibley and Morgan [24] reported that the highest concentrations of most dissolved metals and free metal ions occurred when the fresh:sea water ratio was between 10:1 and 1:1 (i.e., salinity of 1.5% and 0.15% as NaCl). The average salinity of pool water was 0.08% as NaCl, based on sodium analyses [17]. The initial pH, sodium, total dissolved solid concentrations of water from various pools used in these leaching experiments is listed in Table II.

Soil and dust in the Dade County area contains oolite, a deposit of marine origin rich in calcium carbonate, as its main component. The following experiment was performed to find the percentage of the sediments soluble in dilute hydrochloric acid.

Serial aliquots 0.05, 0.1, 0.2, 0.4, 0.8 and (where possible) 1.6 g of sediments No. 1, 3 and 86 were suspended in 50 ml of distilled water and brought to pH 5.0 by dropwise addition of 0.01 N hydrochloric acid. The flasks were then left to shake overnight. The next day, more 0.1 N hydrochloric acid was added to bring the liquid back to pH 5.0. Shaking and additions were con-

Table II. Composition of Pool Water Used in Leaching Experiments

	Pool No.				
	1	**3**	**10**	**86**	**110**
Initial pH	7.8	7.7	6.3	8.4	8.3
Total Alkalinity as $CaCO_3$ (mg/l)	105	109	21	107	79
Sodium (mg/l)	500	600	108	256	400
Total Dissolved Solids (mg/l)	885	1560	428	1050	1100

tinued until the overnight pH remained at 5.0 ± 0.2. The total amount of 0.1 N hydrochloric acid used was plotted against the amount of sediment. The results are shown in Figure 9.

The amount of hydrochloric acid used, calculated as carbonate, was 5% for sediment No. 1, 23% for sediment No. 4 and 35% for sediment No. 86. This result suggests that carbonate in the sediment may have reduced the solubility of lead in the sediment. If this is a correct assumption, lead may be more easily solublized from sediments in parts of the country that do not have high carbonate in the soil.

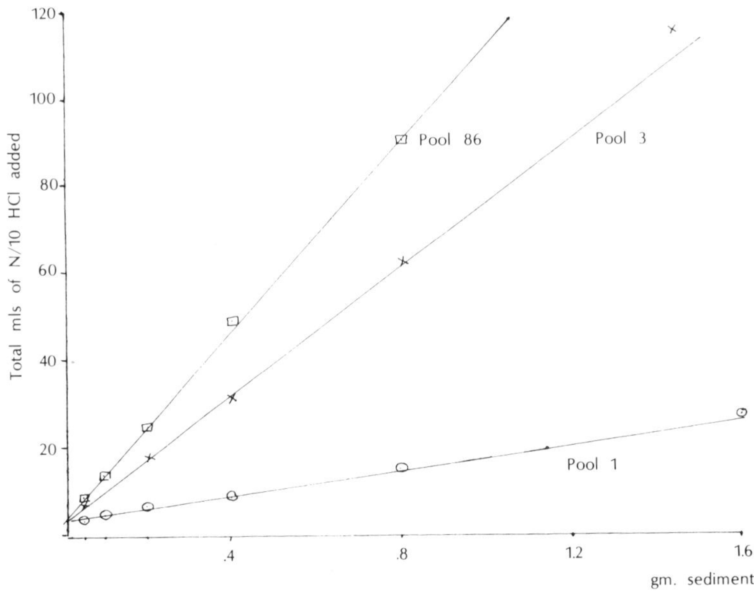

Figure 9. Hydrochloric acid needed for equilibrium leaching of pool sediments at pH 5 ± 0.3.

NITRATE

The principal source of nitrate in swimming pool water is probably chlorination of urea from the urine and sweat of swimmers. In this section, an estimate will be made of the average pollutant loading in pool water attributable to sweat and urine.

No reliable studies have been made on which to base the amount of urine voided in the water by swimmers. Warren and Ridgeway [25] estimated it to be 25-50 ml per swimmer. Inquiries by the author suggest that the figure is much higher for children under 10 years of age. For the purpose of this calculation, a figure of 50 ml per swimmer has been adopted.

Kuno [26] reported that an active swimmer in water at 24°C, when the air temperature was 38°C lost approximately one liter of sweat per hour. For the following calculation the sweat loss of an average swimmer is estimated to be 10% of this volume, i.e., 100 ml/hr. The average time spent in a pool is estimated to be two hours. This number is based on a report by Rylander et al. [27] that children use pools more often than adults and remain in them for a longer time. In an EPA memorandum [28] it was estimated that child swimmers aged 5-9 years spend 3 hours in pools at a time, teenagers spend 6 hours and adults 1 hour at a time. The figure of 2 hours has been adopted because the EPA figures are considered to be above average swimmer immersion times.

Based on average excretion of 50 ml urine and 200 ml sweat (100 ml for 2 hours) from each swimmer, a list of pollutant solutes from these sources has been derived in Table III. Because they are excretion products, the composition of urine and sweat will vary for different individuals and for different diets. Average amounts of solutes were calculated from ranges in Consolazio et al. [20].

Many municipal pools have loads of several thousand swimmers a week during summer vacations. In Britain, swimmer loads up to 4000 per day have been reported [7]. Water in pools with high swimmer loads or pools that are not emptied between seasons may achieve high solute concentrations from sweat and urine.

Pressley et al. [29] found that breakpoint chlorination of ammonia nitrogen in the pH range 6.5-7.5 oxidized 95-99% of the ammonia to nitrogen gas, but small amounts of nitrite and nitrate were also formed. These amounts increased rapidly when the pH of the water increased above pH 8.0.

Samples [30] studied the chlorination of urea in aqueous solution. He showed that stepwise chlorination occurred until nitrogen trichloride was formed. At pH 4 nitrogen trichloride was the final product, but at pH 8

Table III. Estimated Average Pollutant Loading From
Urine and Sweat for each Swimmer

	Urine Solutes (mg)	Sweat Solutes (mg)	Total (mg)
Sodium	150	280	430
Chloride	250	300	550
Phosphate	190	40	230
Sulfate	180	60	240
Potassium	90	60	150
Calcium + Magnesium	15	30	45
Ammonium	50	20	70
Urea	1200	140	1340
Creatinine	80	2	82
Amino Acids	50	60	110
Hippuric Acid	40		40
Uric Acid	20	2	22
Creatine	5		5

nitrate was the final product. Samples postulated the following sequence of reactions for the chlorination of urea:

$$\text{Urea} + \text{HOCl} \rightarrow \text{N-chlorourea} + H_2O$$

$$\text{N-chlorourea} + \text{HOCl} \rightarrow \text{N,N}'\text{-dichlorourea} + H_2O$$

$$\text{N,N}'\text{-dichlorourea} + \text{HOCl} \rightarrow \text{N,N,N}'\text{-trichlorourea} + H_2O$$

$$\text{N,N,N}'\text{-trichlorourea} + \text{HOCl} \rightarrow \text{N,N,N}'\text{N}'\text{-tetrachlorourea} + H_2O$$

$$\text{N,N,N}'\text{N}'\text{-tetrachlorourea} + \text{HOCl} \rightarrow H^+ + Cl^- + CO_2 + NCl_3 + (NCl)$$

$$(NCl) + (OH)^- \rightarrow NOH + Cl^-$$

$$2NOH \rightarrow H_2N_2O_2$$

$$H_2N_2O_2 \rightarrow N_2O + H_2O$$

$$NCl_3 + HOCl + 2H_2O \rightarrow NO_3^- + 4Cl^- + 5H^+$$

Samples [30] estimated that at slightly alkaline pH values, each mole of urea in solution could yield 0.5 mole of nitrate. Using that figure, breakdown of 1340 mg urea would yield 690 mg nitrate pollution per swimmer. For a 500,000-liter swimming pool that has 2000 swimmers per week over a 14-week season and loses 30% of its nitrate by splashout and backwashing, the concentration of nitrate at the end of the season is estimated to be:

$$\frac{690 \times 2000 \times 14 \times 7}{500,000 \times 10} = 27 \text{ mg NO}_3/\text{liter}$$

Beech et al. [14] found average concentrations of 4.5 mg NO_3 (0.9-18.7 mg/l) in samples of water from municipal swimming pools taken 1-6 weeks after the pools were filled at the beginning of the season. Nitrate averaged 16.5 mg/l in pools of condominiums and apartment houses, the highest concentration found was 54.9 mg/l.

CHLORATE

Sidgwick [31] states that hypochlorous acid can decompose in two ways:

$$2HOCl \rightarrow 2HCl + O_2$$

$$3HOCl \rightarrow 2HCl + HClO_3$$

Both reactions are accelerated by sunlight. The photodecomposition of hypochlorite to form chlorate was studied by Buxton and Subhani [32]. D'Ans and Freund [33] examined the kinetics of chlorate formation from hypochlorite. They concluded that it involved a rapid reaction between undissociated hypochlorous acid and hypochlorite ions to form an intermediate complex ion, which decomposed slowly to form chlorate, chloride and hydrogen ions:

$$2HClO + ClO^- \rightleftarrows [H_2Cl_3O_3]^-$$

$$[H_2Cl_3O_3]^- \rightleftarrows ClO_3^- + 2Cl^- + 2H^+$$

The amount of chlorate formed is a function of the square of the concentration of undissociated hypochlorous acid, which increases with decreasing pH values. It follows that chlorate formation in swimming pool water is favored when the water has an acid reaction.

Chlorate concentrations in swimming pool water were measured by Beech et al. [14]. They found concentrations from 0 to 124 mg/l with no apparent correlation between chlorate level and type of pool (hotel or motel, apartment house, condominium or municipal).

Chloride concentrations were also measured in pool water samples. The chloride and chlorate concentrations for 70 pools are compared in Figure 10.

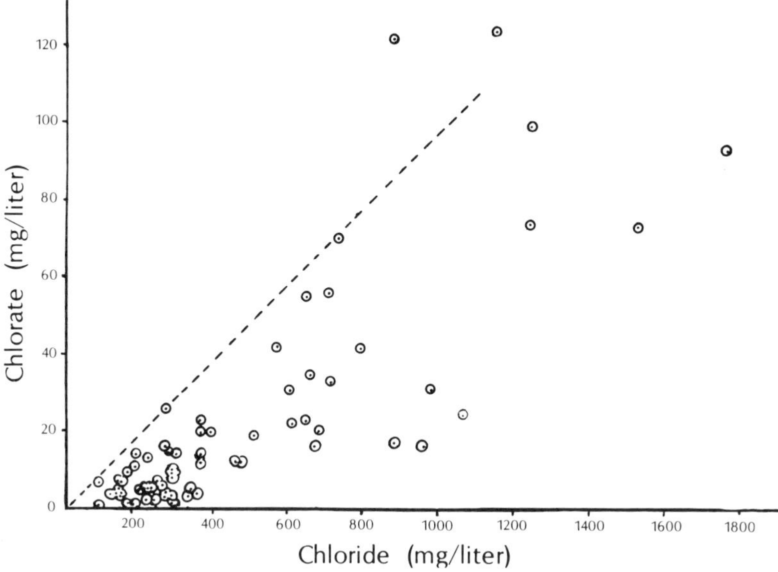

Figure 10. Chlorate and chloride values for swimming pool water.

Except for two pools that had very high chlorate values (122 and 124 mg/l), all the points on the graph fall below a line defined by Cl = 10.4 ClO_3 (mg/l) or Cl = 24.5 ClO_3 (mol). Because hydrochloric acid is commonly added to pool water to reduce a too-alkaline pH and remove carbonate stains, the ratios in Figure 10 are unlikely to have theoretical significance.

The simultaneous chlorate and THM concentrations of pool water are more interesting. These are compared in Figure 11 for 100 pools. The samples showing maximal values for THM concentration as a function of chlorate concentration fell on a smooth curve. The curve was analyzed using a non-linear model by a modified Gauss-Newton technique. It satisfied the equation of expotential decay

$$y = ae^{-bx^2}$$

where y = chlorate concentration, mg/l
     x = trihalomethane concentration, μg/l
     a = 124.55
     b = 0.0000152

The average deviation around the curve S = 6.74 mg/l as chlorate.

In the section on THM, the factors considered to contribute to loss of trihalomethanes from pool water were (1) water temperature (2) weather

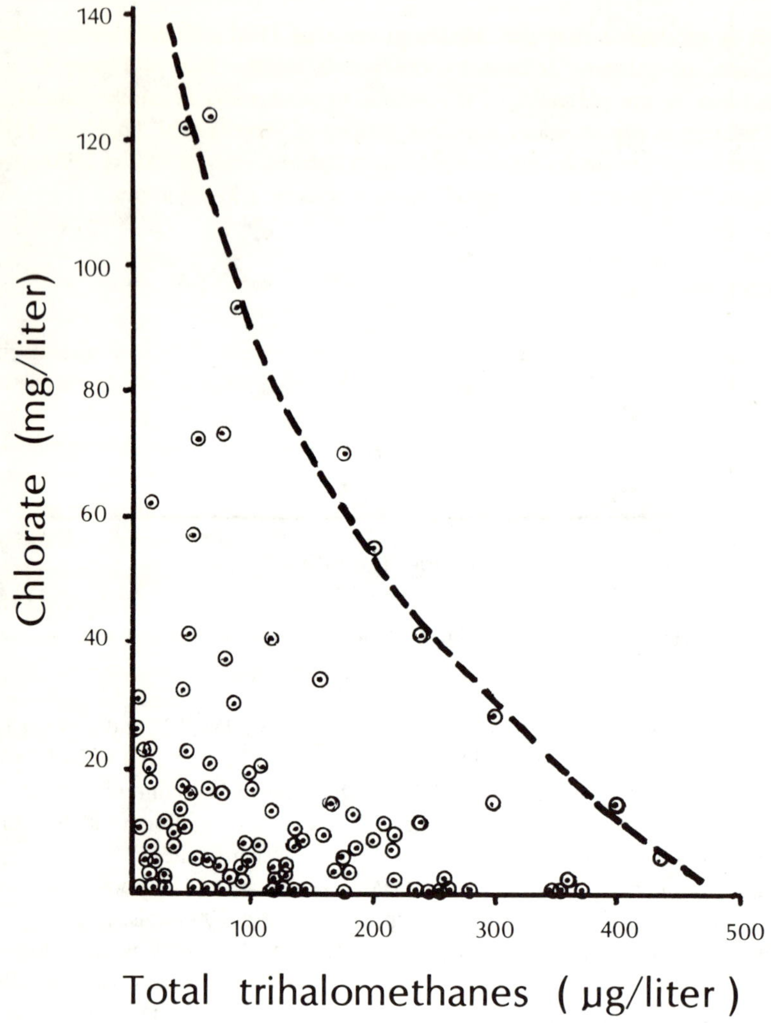

Figure 11. Chlorate and THM values for swimming pool water.

exposure and (3) swimmer load. The points on the line in Figure 11 are considered to represent pools for which the contribution of environmental factors and swimmer load to THM loss is minimal. Under these conditions, losses of THM would be mainly a function of their vapor pressure at the temperature of the water. Pools in Figure 11 that contained less THM for the same chlorate concentration are speculated to be ones where environmental factors and swimmer load added significantly to the rate of THM loss by evaporation or absorption.

It is speculated that the maximum value of THM and the corresponding amounts of chlorate defined by the points on the line of Figure 11 are functions of the pH value of the swimming pool water at equilibrium. This speculation is supported by the observations of Stevens et al. [12] that THM formation is favored by alkaline pH values and the observations of D'Ans and Freund [33] that chlorate formation is favored by acid pH values.

SIGNIFICANCE

Reported findings indicate that alkaline pool water favors formation of THM and nitrates, and acid pool water favors formation of chlorates and solublization of heavy metals. Only one registered disinfectant additive has been considered, namely, chlorine in aqueous solution (hypochlorous acid). However, in the United States there are 135 synthetic chemicals or mixtures registered by the EPA as disinfectants for use in swimming pool water or pool related surfaces. Some of these may interact to form substances with toxic potential. Ultraviolet radiation may catalyze breakdown reactions in outdoor pool water. Hypochlorous acid is a very reactive compound. At this time the EPA does not require studies of the toxicity of breakdown or chlorination products of registered disinfectants used in or around swimming pools.

An obvious similarity exists between the treatment of water for swimming and for drinking. In both cases the water is filtered and clarified, solute concentrations and pH are adjusted and the water is chlorinated (Figure 12).

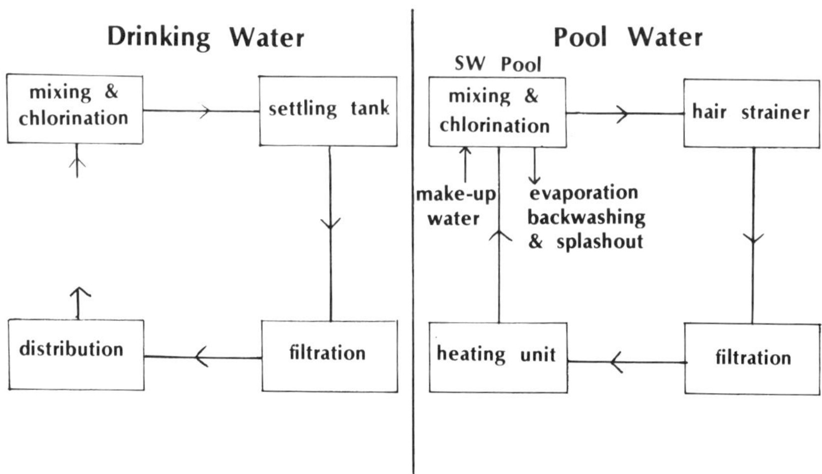

Figure 12. Comparison of drinking and swimming pool water treatments.

Although swimming pool water is not ingested in the same volume as drinking water, it has a much greater potential for percutaneous absorption of lipophilic substances and inhalation of volatiles evaporating from the water surface. Beech [34] calculated that a six-year-old boy swimming for three hours in a pool containing 500 μg/l of THM would attain a body burden of 2.8 mg, i.e., 14 times the recommended EPA daily limit for THM in drinking water.

Unlike drinking water, there are no federally regulated standards for pool water quality. Regulation of small residential pools is impractical. The Center for Disease Control (CDC) has an advisory role over water quality in public and commercial pools, with standards set by individual states. The CDC pool guide [8] states that "the water supply of a pool must be of quality that will permit adherence to the rigid bacterological standards . . . (set by the state) . . . for pool water." Also, "The chemical and physical properties of the water generally should be the same as drinking water."

There is a need for further research about chemical pollution in swimming pool water and regulation of the amount of pollutants in them. In France, the physicochemical purity of commercial swimming pool water has been controlled by legislation since 1978 [35]. Pool usage is probably much greater in the United States. According to the U.S. Water Resources Council [36], there were 0.51 billion swimming pool activity events in pools in the United States in 1975. By 1985, it is estimated that there will be more than 0.62 billion U.S. activity events each year in swimming pools.

REFERENCES

1. *Swimming Pool Industry Market Report* (Ft. Lauderdale, FL: Hoffman Publications, 1979).
2. "Analysis of Trihalomethanes in Finished Waters by the Purge and Trap Method," U.S. EPA, Cincinnati, OH (1977).
3. Bellar, T. A., and J. J. Lichtenberg. "Determining Volatile Organics at Microgram-per-liter Levels by Gas Chromatograph," *J. Am. Water Works Assoc.* 66:739-744 (1974).
4. *Standard Methods for the Examination of Water and Wastewater*, 14th ed. (Washington, DC: American Public Health Association, 1976).
5. Williams, D., and C. C. Meeker. "Determination of Chlorate in Caustic Soda," *Ind. Eng. Chem. Anal. Ed.* 17:535-538 (1945).
6. Jacobs, M. B. "The Analytical Toxicology of Industrial Inorganic Poisons," in *Chemical Analysis, Vol. 22* (New York: Interscience, 1967), pp. 646-648.
7. White, G. C. *Handbook of Chlorination* (New York: Van Nostrand Reinhold Co., 1972), p. 466.
8. "Swimming Pools—Safety and Disease Control through Proper Design and Operation," U.S. DHEW Publication No. (CDC) 76-8319 (1976).

9. Oden, S. "Humic Acids," *Kolloidchem. Beineftc.* 11:74 (1919).
10. Rook, J. "Formation and Occurence of Haloforms in Drinking Water," in Proc. 95th Ann. AWWA Conference (1975), Paper 32-4.
11. Morris, J. C. "The Chemistry of Aqueous Chlorine in Relation to Water Chlorination," in *Water Chlorination, Environmental Impact and Health Effects, Vol. 1*, R. J. Jolley, Ed. (Ann Arbor, MI: Ann Arbor Science Publishers, Inc., 1978), pp. 21-33.
12. Stevens, A. A., C. J. Slocum, D. R. Seeger and G. G. Robeck. "Chlorination of Organics in Drinking Water," in *Water Chlorination, Environmental Impact and Health Effects, Vol. 1*, R. J. Jolley, Ed. (Ann Arbor, MI: Ann Arbor Science Publishers, Inc., 1978), pp. 77-101.
13. Babcock, D., and P. Singer. "Chlorination and Coagulation of Humic and Fulvic Acids," in Proc. 97th Ann. AWWA Conf. (1977), paper 16-6.
14. Beech, J. A., R. Diaz, C. Ordaz and B. Palomeque. "Nitrates, Chlorates and Trihalomethanes in Swimming Pool Water," *Am. J. Public Health* 70:79-82 (1980).
15. Dilling, W. L., N. B. Terfertiller and G. J. Kallos. "Evaporation Rates and Reactivities of Methylene Chloride, Chloroform, 1, 1, 1-Trichloroethane, Trichloroethylene, Tetrachloroethylene and Other Chlorinated Compounds in Dilute Aqueous Solution," *Environ. Sci. Technol.* 9(9):833-838 (1975).
16. "Identification and Evaluation of Waterborne Routes of Exposure from Other Than Food and Drinking Water," U.S. EPA Report EPA-440/4-79-016 (1979).
17. Beech, J. A. "Are Metals from Highway Dustfall Hazardous in Swimming Pools?" *J. Environ. Health* 42:328-331 (1980).
18. Shaheen, D. G. "Contributions of Urban Roadway Usage to Water Pollution," U.S. EPA Report EPA-600/2-75-004 (1975).
19. Biggins, P. D., and R. M. Harrison. "Chemical Speciation of Lead Compounds in Street Dusts," *Environ. Sci. Technol.* 14(3):336-339 (1980).
20. Consolazio, C. F., R. E. Johnson and L. J. Pecora. *Physiological Measurements of Metabolic Functions in Man* (New York: McGraw-Hill, 1963).
21. Black, A. P., M. A. Klien, J. J. Smith, G. M. Dykes, Jr. and W. E. Harlan. "The Disinfection of Swimming Pools, II. A Field Study of the Disinfection of Public Swimming Pools," *Am. J. Public Health* 60(4):740-750 (1970).
22. Hem, J. D., and W. H. Durum. "Solubility and Occurrence of Lead in Surface Water," *J. Am. Water Works Assoc.* 65(8):562-568 (1973).
23. Rohatgi, N., and K. Y. Chen. "Transport of Trace Metals by Suspended Particulates on Mixing with Sea Water," *J. Water Poll. Control Fed.* 47(a):2298-2316 (1975).
24. Sibley, T. H., and J. J. Morgan. "Equilibrium Speciation of Trace Metals in Freshwater: Seawater Mixtures," Int. Conf. on Heavy Metals in the Environment, Toronto (1975), pp. 319-338.
25. Warren, I. C., and J. Ridgeway. "Swimming Pool Disinfection," Technical Report No. 90, Water Research Laboratory, Marlow, England (1978) p. 16.
26. Kuno, Y. *Human Perspiration* (Springfield, IL: C. C. Thomas, 1956).
27. Rylander, R., K. Victorin and S. Sorensen. "The Effect of Saline on the Eye Irritation Caused by Swimming Pool Water," *J. Hyg.* 71:587-592 (1973).

28. Zweig, G. "Cyanuric Acid, Oral Exposure to Swimmers," U.S. EPA Memorandum to F. T. Arnold (April 26, 1979).
29. Pressley, T. A., D. O. Bishop and S. G. Roan. "Ammonia-Nitrogen Removal by Breakpoint Chlorination," *Environ. Sci. Technol.* 6:662-628 (1972).
30. Samples, W. R. "A Study of the Chlorination of Urea," PhD Thesis, Harvard University, Cambridge, MA (1959).
31. Sidgwick, N. V. *The Chemical Elements and Their Compounds, Vol. 2* (Oxford: Oxford Press, 1950), p. 1216.
32. Buxton, G. V., and M. S. Subhani. "Radiation Chemistry and Photochemistry of Oxychlorine Ions, Part 2. Photodecomposition of Aqueous Solutions of Hypochlorite Ions," *J. Chem. Soc. Farad.* II:958-969 (1972).
33. D'Ans, J., and H. E. Freund. "Kinetische Untersuchungen. 1, Uber die Chloratbildung aus Hypochlorit," *Z. Electrochem.* 61:10-18 (1957).
34. Beech, J. A. "Estimated 'Worst Case' Trihalomethane Body Burden of a Child Using a Swimming Pool," *Medical Hypotheses*, 6:303-308 (1980).
35. *Degremont Water Treatment Handbook*, 5th (English) ed. (New York: John Wiley and Sons, 1979).
36. "Outdoor Recreation Requirements and Related Problem Issues, Appendix D. Nationwide Analysis Report," U.S. Water Resources Council, Washington, DC (1975).